Hydrogen-Transfer Reactions

Edited by
James T. Hynes,
Judith P. Klinman,
Hans-Heinrich Limbach,
Richard L. Schowen

1807–2007 Knowledge for Generations

Each generation has its unique needs and aspirations. When Charles Wiley first opened his small printing shop in lower Manhattan in 1807, it was a generation of boundless potential searching for an identity. And we were there, helping to define a new American literary tradition. Over half a century later, in the midst of the Second Industrial Revolution, it was a generation focused on building the future. Once again, we were there, supplying the critical scientific, technical, and engineering knowledge that helped frame the world. Throughout the 20th Century, and into the new millennium, nations began to reach out beyond their own borders and a new international community was born. Wiley was there, expanding its operations around the world to enable a global exchange of ideas, opinions, and know-how.

For 200 years, Wiley has been an integral part of each generation's journey, enabling the flow of information and understanding necessary to meet their needs and fulfill their aspirations. Today, bold new technologies are changing the way we live and learn. Wiley will be there, providing you the must-have knowledge you need to imagine new worlds, new possibilities, and new opportunities.

Generations come and go, but you can always count on Wiley to provide you the knowledge you need, when and where you need it!

William J. Pesce
President and Chief Executive Officer

Peter Booth Wiley
Chairman of the Board

Hydrogen-Transfer Reactions

Volume 2

Edited by
James T. Hynes, Judith P. Klinman,
Hans-Heinrich Limbach, Richard L. Schowen

WILEY-VCH Verlag GmbH & Co. KGaA

The Editors

Prof. James T. Hynes
Department of Chemistry and Biochemistry
University of Colorado
Boulder, CO 80309-0215
USA

Prof. Judith P. Klinman
Departments of Chemistry and
Molecular and Cell Biology
University of California
Berkeley, CA 94720-1460
USA

Département de Chimie
Ecole Normale Supérieure
24 rue Lhomond
75231 Paris
France

Prof. Hans-Heinrich Limbach
Institut für Chemie und Biochemie
Freie Universität Berlin
Takustrasse 3
14195 Berlin
Germany

Prof. Richard L. Schowen
Departments of Chemistry, Molecular
Biosciences, and Pharmaceutical Chemistry
University of Kansas
Lawrence, KS 66047
USA

Cover
The cover picture is derived artistically from
the potential-energy profile for the dynamic
equilibrium of water molecules in the hydration
layer of a protein (see A. Douhal's chapter in
volume 1) and the three-dimensional vibrational
wavefunctions for reactants, transition state,
and products in a hydride-transfer reaction
(see the chapter by S.J. Benkovic and S. Hammes-
Schiffer in volume 4).

Library of Congress Card No.: applied for

British Library Cataloguing-in-Publication Data
A catalogue record for this book is available
from the British Library.

**Bibliographic information published by
Die Deutsche Bibliothek**
Die Deutsche Nationalbibliothek lists this
publication in the Deutsche Nationalbibliografie;
detailed bibliographic data are available in the
Internet at <http://dnb.d-nb.de>.

© 2007 WILEY-VCH Verlag GmbH & Co. KGaA,
Weinheim

Typesetting   Kühn & Weyh, Freiburg;
Asco Typesetters, Hongkong
Printing   betz-druck GmbH, Darmstadt
Bookbinding   Litges & Dopf GmbH, Heppenheim
Cover Design   Adam-Design, Weinheim

Printed in the Federal Republic of Germany.
Printed on acid-free paper.

ISBN: 978-3-527-30777-7

Foreword
The Remarkable Phenomena of Hydrogen Transfer

Ahmed H. Zewail*
California Institute of Technology
Pasadena, CA 91125, USA

Life would not exist without the making and breaking of chemical bonds - chemical reactions. Among the most elementary and significant of all reactions is the transfer of a hydrogen atom or a hydrogen ion (proton). Besides being a fundamental process involving the smallest of all atoms, such reactions form the basis of general phenomena in physical, chemical, and biological changes. Thus, there is a wide-ranging scope of studies of hydrogen transfer reactions and their role in determining properties and behaviors across different areas of molecular sciences.

Remarkably, this transfer of a small particle appears deceptively simple, but is in fact complex in its nature. For the most part, the dynamics cannot be described by a classical picture and the process involves more than one nuclear motion. For example, the transfer may occur by tunneling through a reaction barrier and a quantum description is necessary; the hydrogen is not isolated as it is part of a chemical bond and in many cases the nature of the bond, "covalent" and/or "ionic" in Pauling's valence bond description, is difficult to characterize; and the description of atom movement, although involving the local hydrogen bond, must take into account the coupling to other coordinates. In the modern age of quantum chemistry, much has been done to characterize the rate of transfer in different systems and media, and the strength of the bond and degree of charge localization. The intermediate bonding strength, directionality, and specificity are unique features of this bond.

* The author is currently the Linus Pauling Chair Professor of
chemistry and physics and the Director of the Physical Biology
Center for Ultrafast Science & Technology and the National
Science Foundation Laboratory for Molecular Sciences at Caltech
in Pasadena, California, USA. He was awarded the 1999
Nobel Prize in Chemistry.
Email: zewail@caltech.edu
Fax: 626.792.8456

Hydrogen-Transfer Reactions. Edited by J. T. Hynes, J. P. Klinman, H. H. Limbach, and R. L. Schowen
Copyright © 2007 WILEY-VCH Verlag GmbH & Co. KGaA, Weinheim
ISBN: 978-3-527-30777-7

The supreme example for the unique role in specificity and rates comes from life's genetic information, where the hydrogen bond determines the complementarities of G with C and A with T and the rate of hydrogen transfer controls genetic mutations. Moreover, the not-too-weak, not-too-strong strength of the bond allows for special "mobility" and for the potent hydrophobic/hydrophilic interactions. Life's matrix, liquid water, is one such example. The making and breaking of the hydrogen bond occurs on the picosecond time scale and the process is essential to keeping functional the native structures of DNA and proteins, and their recognition of other molecules, such as drugs. At interfaces, water can form ordered structures and with its amphiphilic character, utilizing either hydrogen or oxygen for bonding, determines many properties at the nanometer scale.

Hydrogen transfer can also be part of biological catalysis. In enzyme reactions, a huge complex structure is involved in bringing this small particle of hydrogen into the right place at the right time so that the reaction can be catalytically enhanced, with rates orders of magnitude larger than those in solution. The molecular theatre for these reactions is that of a very complex energy landscape, but with guided bias for specificity and selectivity in function. Control of reactivity at the active site has now reached the frontier of research in "catalytic antibody", and one of the most significant achievements in chemical synthesis, using heterogeneous catalysis, has been the design of site-selective reaction control.

Both experiments and theory join in the studies of hydrogen transfer reactions. In general, the approach is of two categories. The first involves the study of prototypical but well-defined molecular systems, either under isolated (microscopic) conditions or in complexes or clusters (mesoscopic) with the solvent, in the gas phase or molecular beams. Such studies over the past three decades have provided unprecedented resolution of the elementary processes involved in isolated molecules and en route to the condensed phase. Examples include the discovery of a "magic solvent number" for acid-base reactions, the elucidation of motions involved in double proton transfer, and the dynamics of acid dissociation in finite-sized clusters. For these systems, theory is nearly quantitative, especially as more accurate electronic structure and molecular dynamics computations become available.

The other category of study focuses on the nature of the transfer in the condensed phase and in biological systems. Here, it is not perhaps beneficial to consider every atom of a many-body complex system. Instead, the objective is hopefully to project the key electronic and nuclear forces which are responsible for behavior. With this perspective, approximate, but predictive, theories have a much more valuable outreach in applications than those simulating or computing bonding and motion of all atoms. Computer simulations are important, but for such systems they should be a tool of guidance to formulate a predictive theory. Similarly for experiments, the most significant ones are those that dissect complexity and provide lucid pictures of the key and relevant processes.

Progress has been made in these areas of study, but challenges remain. For example, the problem of vibrational energy redistribution in large molecules, although critical to the description of rates, statistical or not, and to the separation of intra and intermolecular pathways, has not been solved analytically, even in an approximate but predictive formulation. Another problem of significance concerns the issue of the energy landscape of complex reactions, and the question is: what determines specificity and selectivity?

This series edited by prominent players in the field is a testimony to the advances and achievements made over the past several decades. The diversity of topics covered is impressive: from isolated molecular systems, to clusters and confined geometries, and to condensed media; from organics to inorganics; from zeolites to surfaces; and, for biological systems, from proteins (including enzymes) to assemblies exhibiting conduction and other phenomena. The fundamentals are addressed by the most advanced theories of transition state, tunneling, Kramers' friction, Marcus' electron transfer, Grote-Hynes reaction dynamics, and free energy landscapes. Equally covered are state-of-the-art techniques and tools introduced for studies in this field and including ultrafast methods of femtochemistry and femtobiology, Raman and infrared, isotope probes, magnetic resonance, and electronic structure and MD simulations.

These volumes are a valuable addition to a field that continues to impact diverse areas of molecular sciences. The field is rigorous and vigorous as it still challenges the minds of many with the fascination of how the physics of the smallest of all atoms plays in diverse applications, not only in chemistry, but also in life sciences. Our gratitude is to the Editors and Authors for this compilation of articles with new knowledge in a field still pregnant with challenges and opportunities.

Pasadena, California   *Ahmed Zewail*
August, 2006

Contents

Hydrogen-Transfer Reactions. Edited by J. T. Hynes, J. P. Klinman, H. H. Limbach, and R. L. Schowen
Copyright © 2007 WILEY-VCH Verlag GmbH & Co. KGaA, Weinheim
ISBN: 978-3-527-30777-7

Preface

As one of the simplest of chemical reactions, pervasive on this highly aqueous pla-
net populated by highly aqueous organisms, yet still imperfectly understood, the
transfer of hydrogen as a subject of scientific attention seems hardly to require
defense. This claim is supported by the readiness with which the editors of this
series of four volumes on *Hydrogen-transfer Reactions* accepted the suggestion that
they organize a group of their most active and talented colleagues to survey the
subject from viewpoints beginning in physics and extending into biology. Furthermore, forty-nine authors and groups of authors acceded, with alacrity and grace,
to the request to contribute and have then supplied the articles that make up these
volumes.

Our scheme of organization involved an initial division into physical and chemical aspects on the one hand, and biological aspects on the other hand (and one
might well have said biochemical and biological aspects). In current science, such
a division may provide an element of convenience but no-one would seriously
claim the segregation to be either easy or entirely meaningful. We have accordingly felt quite entitled to place a number of articles rather arbitrarily in one or the
other category. It is nevertheless our hope that readers may find the division adequate to help in the use of the volumes. It will be apparent that the division of
space between the two categories is unequal, the physical and chemical aspects
occupying considerably more pages than the biological aspects, but our judgment
is that this distribution of space is proper to the subjects treated. For example,
many of the treatments of fundamental principles and broadly applicable techniques were classified under physical and chemical aspects. But they have powerful implications for the understanding and use of the matters treated under biological aspects.

Within each of these two broad disciplinary categories, we have organized the
subject by beginning with the simple and proceeding toward the complex. Thus
the physical and chemical aspects appear as two volumes, volume1 on simple systems and volume 2 on complex systems. Similarly, the biological aspects appear
as volume 3 on simple systems and volume 4 on complex systems.

Volume 1 then begins with isolated molecules, complexes, and clusters, then
treats condensed-phase molecules, complexes, and crystals, and finally reaches

Hydrogen-Transfer Reactions. Edited by J. T. Hynes, J. P. Klinman, H. H. Limbach, and R. L. Schowen
Copyright © 2007 WILEY-VCH Verlag GmbH & Co. KGaA, Weinheim
ISBN: 978-3-527-30777-7

treatments of molecules in polar environments and in electronic excited states. Volume 2 reaches higher levels of complexity in protic systems with bimolecular reactions in solution, coupling of proton transfer to low-frequency motions and proton-coupled electron transfer, then organic and organometallic reactions, and hydrogen-transfer reactions in solids and on surfaces. Thereafter articles on quantum tunneling and appropriate theories of hydrogen transfer complete the treatment of physical and chemical aspects.

Volume 3 begins with simple model (i.e., non-enzymic) reactions for proton-transfer, both to and from carbon and among electronegative atoms, hydrogen-atom transfer, and hydride transfer, as well as the extension to small, synthetic peptides. It is completed by treatments of how enzymes activate C-H bonds, multiple hydrogen transfer reactions in enzymes, and theoretical models. Volume 4 moves then into enzymic reactions and a thorough consideration of quantum tunneling and protein dynamics, one of the most vigorous areas of study in biological hydrogen transfer, then considers several specific enzyme systems of high interest, and is completed by the treatment of proton conduction in biological systems.

While we do not claim any sort of comprehensive coverage of this large subject, we believe the reader will find a representative treatment, written by accomplished and respected experts, of most of the matters currently considered important for an understanding of hydrogen-transfer reactions. I am enormously grateful to James T. (Casey) Hynes and Hans-Heinrich Limbach, who saw to the high quality of the volumes on the physical and chemical aspects, and to Judith Klinman, who gave me a nearly free pass as her co-editor of the volumes on biological aspects. We are all grateful indeed to the authors who contributed their wisdom and eloquence to these volumes. It has been a very great pleasure to be assisted, encouraged, and supported at every turn by the outstanding staff of VCH-Wiley in Weinheim, particularly (in alphabetical order) Ms. Nele Denzau, Dr. Renate Dötzer, Dr. Tim Kersebohm, Dr. Elke Maase, Ms. Claudia Zschernitz, and – of course – Dr. Peter Gölitz.

Lawrence, Kansas, USA, September 2006 *Richard L. Schowen*

Preface to Volumes 1 and 2

These volumes together address the subject of the physical and chemical aspects of hydrogen transfer, volume 1 focusing on comparatively simple systems and volume 2 treating relatively more complex ones.

Volume 1 comprises three parts, commencing with Part I, dealing with hydrogen transfers of polyatomic molecules and complexes in relatively isolated conditions. In the first three contributions, the transfer is a coherent tunneling process rather than a rate process, characterized by "tunnel splittings" or delocalized hydrogen nuclei, for which electronic and vibrational spectroscopies are common and potent tools. The molecular systems discussed are malonaldehyde and tropolone (Redington, Ch. 1), carboxylic acid dimers (Havenith, Ch. 2) and strongly hydrogen-bonded systems such as $(H_2O...H...OH_2)^+$ (Asmis, Neumark and Bauman, Ch. 3). Kühn and Gonzales (Ch. 4) consider theoretically the more active role of infrared radiation in controlling hydrogen dissociation dynamics in e.g. OHF^-.

The five contributions of Part II focus on condensed matter. If the barriers are large, the hydrogen transfer becomes a rate process which may involve incoherent tunneling. Ceulemans (Ch. 5) examines proton abstraction by alkanes from strongly acidic alkane radical cations in inert matrices. Limbach (Ch. 5) follows the kinetics of single and multiple hydrogen and deuteron transfers in liquids and solids via NMR. Optical methods are applied by Douhal (Ch. 6) to systems embedded in a nanocavity, and embedded in liquids and polymer matrices by Waluk (Ch. 7), with a contrast to coherent hydrogen transfer in supersonic jets. Finally, Vener (Ch. 9) compares theory and experiment for anharmonic vibrations of strong hydrogen bonds in crystals.

Part III, comprising four chapters, commences the examination of hydrogen transfer – here proton transfer – in polar environments. The strong electrostatic proton-environment interaction guarantees incoherent rate phenomena. Kiefer and Hynes (Ch. 10) lay out the theoretical description for such reactions. The next three chapters exploit the greatly enhanced acidity of aromatic acids in the excited electronic state. Lochbrunner, Schriever and Riedle (Ch. 11) focus on the role of the motion of the groups between which the proton transfers, Pines and Pines (Ch. 12) thoroughly examine the insight to be gained from Förster cycle and free

Hydrogen-Transfer Reactions. Edited by J. T. Hynes, J. P. Klinman, H. H. Limbach, and R. L. Schowen
Copyright © 2007 WILEY-VCH Verlag GmbH & Co. KGaA, Weinheim
ISBN: 978-3-527-30777-7

energy analyses, while Tolbert and Solnstev (Ch. 13) pursue related themes for "super" photoacids in the concluding chapter of volume 1.

Volume 2 opens with Part IV dealing with hydrogen transfer in protic systems. Generally, a larger number of solvent molecules is involved, and hence multiple protons may be transferred. The first two chapters elucidate molecular details of proton transfer in solution via ultrafast infrared spectroscopy. Nibbering and Pines (Ch. 14) examine the transfer between acid-base pairs for the acid in the excited electronic state, while Elsaesser (Ch. 15) discusses coherent low frequency motions coupled to related proton transfers as well as in hydrogen-bonded complexes. The final two chapters in Part IV deal with proton transfer coupled to electron transfer, with Hammes-Schiffer (Ch. 16) expounding and illustrating the theory for these, while Hodgkiss, Rosenthal and Nocera (Ch. 17) discuss these reactions with a special emphasis on the connection to hydrogen atom transfer.

Part V, consisting of four chapters, opens with a discussion of the kinetics and mechanisms of proton abstraction from carbon in organic systems by Koch (Ch. 18) and then turns to a presentation by Williams (Ch. 19) on free energy relationships for proton transfer, as informed by various theoretical approaches. The final two chapters are devoted to hydrogen and dihydrogen mobility in the coordination sphere of transition metal complexes, where the transition from coherent to incoherent H-tunneling can be observed, with a review of the field given by Kubas in Ch. 20 and a discussion of insights from NMR studies presented by Buntkowsky and Limbach in Ch. 21.

In the first three of the five chapters of Part VI, hydrogen transfer is examined in assorted complex solids of importance in various applications: zeolites by Sauer in Ch. 22, fuel cells by Kreuer in Ch. 23 and ice bilayers by Aoki in Ch. 24. Attention is then turned to hydrogen transfer at metal surfaces in Ch. 25 by Christmann and in metals in Ch. 26 by Hempelmann and Skripov.

Volume 2 concludes in Part VII with contributions on the variational transition state theory approach to hydrogen transfer in various contexts (Truhlar and Garrett, Ch. 27), on experimental evidence of hydrogen atom tunneling in simple systems (Ingold, Ch. 28), and finally on a theoretical perspective for multiple hydrogen transfers (Smedarchina, Siebrand and Fernández-Ramos, Ch. 29).

JTH acknowledges the support of grant CHE-0417570 from the US National Science Foundation. HHL thanks the Deutsche Forschungsgemeinschaft, Bonn, and the Fonds der Chemischen Industrie, Frankfurt, for financial support.

Boulder and Paris, September 2006   *James T. Hynes*
Berlin, September 2006   *Hans-Heinrich Limbach*

List of Contributors to Volumes 1 and 2

Katsutoshi Aoki
Synchroton Radiation Research
Center
Kansai Research Establishment
Japan Atomic Energy Research
Institute
Kouto 1-1-1
Mikazuki-cho
Sayo-gun
Hyogo
Japan

Knut R. Asmis
Department of Molecular Physics
Fritz-Haber-Institut der Max-Planck-
Gesellschaft
Faradayweg 4–6
14195 Berlin
Germany

Joel M. Bowman
Department of Chemistry and
Cherry L. Emerson Center for
Scientific Computation
Emory University
Dickey Drive
Atlanta, GA 30322
USA

Gerd Buntkowsky
Department of Chemistry
FSU Jena
Helmholtzweg 4
07743 Jena
Germany

Jan Ceulemans
Department of Chemistry
K.U. Leuven
Celestijnenlaan 200-F
3001 Leuven
Belgium

Klaus Christmann
Institut für Chemie und Biochemie
Physikalische und Theoretische Chemie
Freie Universität Berlin
Takustrasse 3
14195 Berlin
Germany

Abderrazzak Douhal
Departamento de Química Físca
Sección de Químicas
Facultad de Ciencias del Medio
Ambiente
Universidad de Castillo-La Mancha
Avda. Carlos III
S.N. 45071 Toledo
Spain

Hydrogen-Transfer Reactions. Edited by J. T. Hynes, J. P. Klinman, H. H. Limbach, and R. L. Schowen
Copyright © 2007 WILEY-VCH Verlag GmbH & Co. KGaA, Weinheim
ISBN: 978-3-527-30777-7

Thomas Elsaesser
Max-Born-Institut für Nichtlineare
Optik
und Kurzzeitspektroskopie
Max-Born-Strasse 2A
12489 Berlin
Germany

Antonio Fernández-Ramos
Department of Physical Chemistry
Faculty of Chemistry
University of Santiago de Compostela
15706 Santiago de Compostela
Spain

Bruce C. Garrett
Chemical Sciences Division
Pacific Northwest National Laboratory
Richland, WA 99352
USA

Leticia González
Institut für Chemie und Biochemie
Freie Universität Berlin
Takustrasse 3
14195 Berlin
Germany

Sharon Hammes-Schiffer
Department of Chemistry
104 Chemistry Building
Pennsylvania State University
University Park, PA 16802-4615
USA

Rolf Hempelmann
Institute of Physical Chemistry
Saarland University
66123 Saarbrücken
Germany

Justin Hodgkiss
Department of Chemistry
Massachusetts Institute of Technology
77 Massachusetts Avenue
Cambridge, MA 02139-4307
USA

James T. Hynes
Department of Chemistry
and Biochemistry
Pacific Northwest National Laboratory
University of Colorado
Boulder, CO 80309-0215
USA
and
Ecole Normale Supérieure
CNRS UMR 8640 PASTEUR
Département de Chimie
24, rue Lhomond
75231 Paris
France

Keith U. Ingold
National Research Council
Ottawa, ON K1A 0R6
Canada

Philip M. Kiefer
Department of Chemistry and
Biochemistry
University of Colorado
Boulder, CO 80309-0215
USA
and
Ecole Normale Supérieure
CNRS UMR 8640 PASTEUR
Département de Chimie
24, rue Lhomond
75231 Paris
France

Heinz F. Koch
Department of Chemistry
Ithaca College
Ithaca, NY 14850
USA

Klaus-Dieter Kreuer
Max-Planck-Institut für
Festkörperforschung
Heisenbergstrasse 1
70569 Stuttgart
Germany

Gregory J. Kubas
Los Alamos National Laboratory
Chemistry Division
MS J514
Los Alamos, NM 87545
USA

Oliver Kühn
Institut für Chemie und Biochemie
Freie Universität Berlin
Takustrasse 3
14195 Berlin
Germany

Hans-Heinrich Limbach
Institut für Chemie und Biochemie
Freie Universität Berlin
Takustrasse 3
14195 Berlin
Germany

Stefan Lochbrunner
Department of Physics
Ludwig Maximillians University
Oettingenstrasse 67
80538 München
Germany

Daniel M. Neumark
Department of Chemistry
University of California
Berkeley, CA 94702
USA

Erik T. J. Nibbering
Max-Born-Institut für nichtlineare
Optik und Kurzzeitspektroskopie
Max-Born-Strasse 2A
12489 Berlin
Germany

Daniel G. Nocera
Department of Chemistry
Massachusetts Institute of Technology
77 Massachusetts
Cambridge, MA 02139-4307
USA

Dina Pines
Department of Chemistry
Ben-Gurion University of the Negev
P.O.B. 653
Beer Sheva 84105
Israel

Ehud Pines
Department of Chemistry
Ben-Gurion University of the Negev
P.O.B. 653
Beer Sheva 84105
Israel

Richard L. Redington
Department of Chemistry and
Biochemistry
Texas Tech University
Mail Stop 1061
Lubbock, TX 79409
USA

Eberhard Riedle
Department of Physics
Ludwig Maximilians University
Oettingenstrasse 67
80538 München
Germany

Joel Rosenthal
Department of Chemistry
Massachusetts Institute of Technology
77 Massachusetts Avenue
Cambridge, MA 02139-4307
USA

Joachim Sauer
Institut für Chemie
Humboldt-Universität zu Berlin
Unter den Linden 6
10099 Berlin
Germany

C. Schriever
Department of Physics
Ludwig Maximilians University
Oettingenstrasse 67
80538 München
Germany

Willem Siebrand
Steacie Institute for Molecular Sciences
National Research Council of Canada
Vorontsova pole 10
Ottawa, K1A 0R6
Canada

Alexander Skripov
Institute of Metal Physics
Urals Branch of the Academy of
Sciences
Ekaterinburg 620219
Russia

Zorka Smedarchina
Steacie Institute for Molecular Sciences
National Research Council of Canada
Ottawa, K1A 0R6
Canada

Kyril M. Solntsev
School of Chemistry and Biochemistry
Georgia Institute of Technology
Atlanta, GA 30332-0400
USA

Laren M. Tolbert
School of Chemistry and Biochemistry
Georgia Institute of Technology
Atlanta, GA 30332-0400
USA

Donald G. Truhlar
Department of Chemistry
University of Minnesota
Minneapolis, MN 55455-0431
USA

Mikhail V. Vener
Department of Quantum Chemistry
Mendeleev University of Chemical
Technology
Miusskaya Sq. 9
Moscow 125047
Russia

Jacek Waluk
Institute of Physical Chemistry
Polish Academy of Sciences
Kasprzaka 44/52
01-224 Warsaw
Poland

Ian H. Williams
Department of Chemistry
University of Bath
Bath BA2 7AY
UK

Part IV
Hydrogen Transfer in Protic Systems

This Section addresses, in two sets of two chapters, two recent developments in proton transfer: the ultrafast vibrational spectroscopic probing of the microscopic details of proton transfer in water and elsewhere, and the emergence of proton-coupled electron transfer reactions as a major reaction class.

Nibbering and Pines open in Ch. 14 with ultrafast acid-base reaction dynamics in aqueous solutions, using femtosecond (fs) resolved infrared (IR) spectroscopy to monitor an electronically excited aromatic photoacid, an acetate ion base, and the solvated proton. These optical pump-IR probe experiments thus involve "vibrational markers" to follow the dynamics. At high acetate levels, a bimodal proton transfer emerges between acid and base pairs in close proximity: a sub-150fs proton transfer between directly bonded pairs, and a sequential stepwise proton transfer on a picosecond (ps) time scale for acid-base pairs separated by a water bridge. These studies bear on the well-known Eigen mechanism for acid-base chemistry, as well as the Grotthus mechanism of proton transport, here involved in a chemical reaction context.

In Ch. 15, Elsaesser continues the theme of ultrafast vibrational spectroscopy as a powerful tool to elucidate hydrogen dynamics. The focus here is on, first, hydrogen bond dynamics in the ground electronic state of complexes, some of which exhibit proton transfer in the excited electronic state, and, second, the direct observation of low frequency vibrations coupled to intramolecular proton transfer in the excited electronic state. Inert rather than protic solvents are used to clearly expose the detailed key vibrational features, unobscured by the broad spectral features of protic solvents. Among the central observations (for a particular enol-keto tautomerization) are, first, that the ground state intramolecular hydrogen bond in this molecule is strongly modulated by a weakly damped low frequency mode, and, second, that the excited state ultrafast proton transfer requires motion among the low frequency vibrational modes, which in fact determines the time scale of the transfer.

Hammes-Schiffer expounds in Ch. 16 her group's theoretical formulation for proton-coupled electron transfer (PCET) mechanism and rates, pointing out the similarities with the separate special limits of electron transfer and (tunneling) proton transfer, and emphasizing the new features of PCET. The latter include the

Hydrogen-Transfer Reactions. Edited by J. T. Hynes, J. P. Klinman, H. H. Limbach, and R. L. Schowen
Copyright © 2007 WILEY-VCH Verlag GmbH & Co. KGaA, Weinheim
ISBN: 978-3-527-30777-7

simultaneous importance of two solvent (or environmental) coordinates associated with the electron and proton aspects of the rate process. The resulting rate expressions feature the simultaneous appearance of reorganization energy contributions and of low frequency vibrational contributions, the two naturally following from the involvement of both proton and electron transfer. The power of this approach is illustrated by a number of applications for rate constants and kinetic isotope effects, including an enzyme-catalyzed PCET.

Hodgkiss, Rosenthal and Nocera pursue the PCET theme in Ch. 17 in a detailed exposition, illustrated with examples from many areas of chemistry and biochemistry, many involving transition metals. The authors emphasize the connection with hydrogen atom transfer, in its "pure" limit a synchronous, strongly adiabatic coupled transfer of the electron and the proton, without the development of charge separation, i.e. the standard image of hydrogen atom transfer. Away from this limit, the various signatures of the assorted types of PCET are classified, discussed and illustrated, with a particular stress on the simultaneous consideration of mechanism and geometry.

14

Bimolecular Proton Transfer in Solution

Erik T. J. Nibbering and Ehud Pines

14.1
Intermolecular Proton Transfer in the Liquid Phase

Proton transfer between Brønsted acids and bases in aqueous solution is a key chemical process [1–4]. Proton transfer occurs between water molecules at close range e.g. in the autoionization of water [5], in the proton conductivity in water (von Grotthuss mechanism) [6–13], and in acid dissociation in water [14–18]. Neutralization reactions between acid and bases often occur in aqueous solutions and are mediated by water [19]. Water actively takes part in the channelling of protons through membrane proteins [20–22] or in the dynamics of photosensor proteins [23]. It is thus of great importance to investigate the dynamics of proton transfer and the role that the solvent water plays. The ultimate goal is to grasp the microscopic mechanisms determining the time scales and efficiencies of proton transfer.

Since the days of "flash spectroscopy" much effort has been dedicated to the elucidation of the dynamics of proton transfer induced by an optical trigger pulse, allowing a well-defined time zero in time-resolved studies [2, 24]. Here, a pump pulse tuned in the ultraviolet or visible (UV/vis) region of the electromagnetic spectrum promotes a molecule to an electronically excited state with a significantly altered electronic charge distribution, causing a prompt hydrogen or proton shift away from the donor side to accepting groups. An early limitation on deciphering the inherent ultrafast dynamics of proton transfer has been the time resolution of time-resolved spectroscopy. The development of picosecond UV/vis spectroscopy in the early 80s marks the beginning of the era of time-resolved proton transfer studies. In the following decade femtosecond spectroscopy became a common tool. In these studies the proton transfer dynamics have been followed typically by monitoring electronic transitions through emission (time correlated single photon counting and up-conversion techniques) or absorbance changes using pump–probe methods with UV/vis probe pulses.

Ultrafast UV/vis spectroscopy has until now been used extensively in the case of excited state intramolecular hydrogen or proton transfer (ESIHT/ESIPT) [25–27], where donor and acceptor groups, linked by the hydrogen or proton, are part of the same molecule [28, 29]. The dynamics and microscopic mechanisms of

Hydrogen-Transfer Reactions. Edited by J. T. Hynes, J. P. Klinman, H. H. Limbach, and R. L. Schowen
Copyright © 2007 WILEY-VCH Verlag GmbH & Co. KGaA, Weinheim
ISBN: 978-3-527-30777-7

ESIHT or ESIPT have been studied in fine detail, see also Chapters 9, 14 and 20, without interference from any additional processes where translation or reorientation of donor and acceptor groups are required to facilitate the hydrogen or proton transfer reaction.

This is unlike the situation of intermolecular proton transfer where mutual diffusion of the donor (acid) and acceptor (base) makes up part of the overall reaction dynamics. Since the seminal works of Weller and Eigen, intermolecular proton transfer in solution is understood to be assisted by mutual diffusion of the acid and base molecules towards forming an encounter pair [2–4], after which the proton is exchanged by an "on-contact" reaction rate k_r (Fig. 14.1) Only rough estimations of k_r have been made due to lack of direct observation methods. Often values in the range of $(10 \text{ ps})^{-1}$–$(0.1 \text{ ps})^{-1}$ have been suggested when acid and base are at a reaction contact radius a of 6 to 8 Å (see Fig. 14.1). Direct access to the actual proton transfer, however, has remained problematic due to the slower diffusion process, limiting the overall dynamics to the diffusion assisted reaction rate k_D. Only when the "on-contact" transfer rate becomes much slower does a transition from diffusion-limited to activation-controlled reaction dynamics occur. As a result, the nature of the contact reaction pair in aqueous solutions has remained an unsolved problem. This is also caused by the many different roles water plays in facilitating bimolecular proton transfer reactions in aqueous solutions. Proton transfer may occur when a proton donor (acid) and a proton acceptor (base) are directly connected by a hydrogen bond, forming the proton transfer reaction coordinate. Alternatively, proton transfer may be achieved through bridging water molecules, either in a concerted fashion or by a sequential, von Grotthuss-type of hopping mechanism.

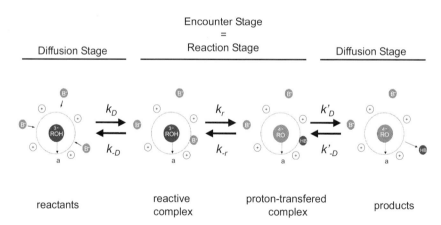

Figure 14.1 Eigen–Weller model for proton transfer reactions between acids and bases, that react when at contact distance a. (Adapted from Ref. [136].)

14.2
Photoacids as Ultrafast Optical Triggers for Proton Transfer

Photoacids, first investigated by Förster and Weller [2, 30–32], are a class of molecules that exhibit a strong change in acidity upon electronic excitation [33–36]. Examples of photoacid compounds are naphthols, pyrenols and aminopyrenes [34, 37]. Typically, the pK_a value of ground state photoacids and the pK_a^* value of electronically excited photoacids in their first singlet states differ by 5–10 pK_a units [38]. A stronger, "super", photoacidic behavior can be accomplished with electronegative sidegroups enhancing the photoacidity of the functional acidic group of the aromatic molecular system [39–41], see also Chapter 19. As a result photoacids are an ideal means for optically triggering proton transfer reactions [42], where the acids may be excited with optical light pulses as short as a few femtoseconds. Photoacids have been used in geminate recombination and acid–base neutralization studies [2–4] (see following sections), where elementary stages in bimolecular proton transfer can be investigated [37 , 43], and may be even used in pH-jump studies affecting biomolecular systems [44–46].

Upon electronic excitation a photoacid releases its proton to the solvent or a scavenging acceptor, while converting to its conjugate photobase. The origin of photoacidity has until now been a subject of intense debate. The nature of the photoacid electronically excited state charge distribution must be considered in conjunction with the proton accepting solvent, that typically is water. Correlations between rate, equilibrium constant and free energy of proton transfer can be made [47]. Solvatochromic shifts in absorbance and emission can be analysed with the Kamlet–Taft approach [48], providing insight into the dipole moment interactions and the hydrogen bond donating and accepting capabilities of the photoacid states [38, 49–53].

Traditionally the nature of the acidity of the photoacid S_1-state has been ascribed to intramolecular charge transfer (CT) from the nonbonding orbital of the hydroxyl oxygen to the aromatic ring π^*-orbital [33, 36, 54, 55]. The enhanced acidity in the S_1-state is ascribed to the Coulombic repulsion between the partial positive charge on the OH group and in particular on the H-atom as a result of partial CT to the aromatic backbone. In this picture the excited state proton transfer dynamics resemble the conventional dynamics of proton transfer in the ground state. It is described by an activated transition in a two-state reaction model (Fig. 14.2(a)), where an optically excited photoacid state converts into the excited conjugated photobase upon proton transfer.

A second model invokes the occurrence of and the internal conversion between nearby lying electronically excited levels of the photoacid (Fig. 14.2(b)). Typically two energetically nearby spectroscopically accessible states can be reached for aromatic molecules upon electronic excitation, with light polarized either along the through-bond axis (^1L_b-state) or along the through-atom axis (^1L_a-state). The ^1L_b state is the lowest singlet level in the gas phase, while the more polar ^1L_a state is thought to be the more stable singlet state in polar solvents. Thus, a singlet–singlet level crossing may occur in polar solvents like water when the vertical excita-

(a) Two-state reaction model

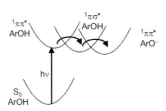

(b) Three-state LE-CT reaction model

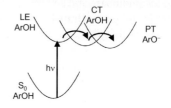

(c) Three-state hydrogen transfer model

(d) Three-state strong coupling reaction model

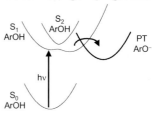

Figure 14.2 Different models to explain excited state photoacidity: the two state model (a), the three state model with nonadiabatic level crossing between LE (^1L_b) and CT (^1L_a) states, the excited state hydrogen transfer model (c), and the three state model with strongly coupled ^1L_b and ^1L_a states (d). (Adapted from Ref. [76].)

tion is to the ^1L_b state and the ensuing solvent relaxation process is capable of shifting the excited-state population of the photoacid to the ^1L_a state. For 1-naphthol the $^1L_b \rightarrow {}^1L_a$ transition has been considered to be the proton transfer rate determining step [56–58]. In contrast, in a recent combined experimental/theoretical study of pyranine (8-hydroxy-1,3,6-trisulfonate-pyrene; HPTS) [59–61] it has been concluded that the rate determining step in the excited state proton dissociation reaction is not the conversion from the optically accessible locally excited (LE) state (resembling the properties of the ^1L_b-state of naphthalene) to the electronically excited CT state (resembling the ^1L_a-state character of naphthalene), but the transition of the photoacid CT to photobase CT states.

A third model, where excited state hydrogen transfer (ESHT) rather than proton transfer is supposed to occur [62, 63] (Fig. 14.2(c)), has recently emerged from excited state dynamics studies of gas phase phenol clustered with ammonia or water molecules [64]. In this model a level crossing between the initially excited $^1\pi\pi^*$ state and a $^1\pi\sigma^*$-state leads to a concerted migration of an electron and a proton from the photoacid to the solvent, with a net transfer of a hydrogen atom as a result. This model has been invoked in an experiment on 7-hydroxyquinoline where donor and acceptor groups are connected through a wire of ammonia molecules [65–67]. A conical intersection of the $^1\pi\sigma^*$-state with the S_0 state leads to an efficient internal conversion pathway for phenol-ammonia clusters [64], and in 2-amino-pyridine clusters [68]. Net proton transfer on the other hand should involve at least one more step with an electron back-transfer to the photoacid, producing the photobase and solvated proton as separate species.

Typically the proton transfer reaction has been followed by probing electronic transitions of the photoacid S_1-state and of the photobase S_1-state, using UV/vis pump–probe or time-resolved fluorescence. Electronic transitions are strongly sensitive to solvent reorganization (solvation dynamics) [69–71]. For time-dependent changes of electronic bands it remains however problematic to disentangle the contributions of solvent reorganization from those of level crossings, including the proton transfer event. Vibrational transitions are typically less affected by solvent reorganization, with the hydroxyl stretching oscillator in a hydrogen bond as the exception to the rule [72–74]. As a result vibrational spectroscopy may enable a clear distinction between the time scales of electronic state transitions and of solvent reorganization at early pulse delays [75].

Recently, the excited state characteristics of HPTS have been probed with mid-infrared pulses providing insight into state-specific vibrational modes [76]. In Fig. 14.3 the absorbance changes in the fingerprint region of HPTS are shown to be solvent dependent. The fact that these vibrational band patterns appear within the time resolution of 150 fs, without any additional changes up to several tens of picoseconds, indicates that previous observations of a 2.5 ps time component in UV/vis pump–probe experiments [59, 60] previously assigned to a $^1L_b \rightarrow {}^1L_a$ level crossing in HPTS are more likely be due to solvation dynamics. In addition, the time-dependent frequency position and magnitude of the hydroxyl stretching band of HPTS indicate the significant impact of solvent reorganization on the solute–solvent interactions of the hydroxyl group of the photoacid compounds [53]. Comparison with results obtained on the methoxy derivative of HPTS reveals that these absorbance changes are strongly affected by a solvent dependent electronic

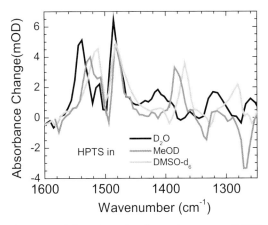

Figure 14.3 Solvent dependent fingerprint spectra of HPTS in the excited state as indicated by the positive absorbance changes after excitation at 400 nm. Negative signals indicate bleach contributions due to vibrational transitions in the electronic ground state. (Adapted from Ref. [76].)

state configuration. A strong solvent dependent coupling between the energetical-ly nearby lying ^1L_b and ^1L_a-states of HPTS (Fig. 14.2(d)), in similar fashion as in the description of gas phase 1-naphthol–ammonia clusters [56, 58], has been pos-tulated as the underlying reason for these observations. Future developments in comparison of experimental vibrational mode patterns with quantum chemical calculations may reveal the molecular origins of photoacidity.

14.3
Proton Recombination and Acid–Base Neutralization

Acids are in equilibrium with their conjugate bases in protic solvents, where the relative concentrations depend on the pK_a value. The observed dynamics of an electronically excited photoacid, typically interpreted as the proton transfer rate to the (protic) solvent [77, 78], is thus governed by the equilibration dynamics to the new configuration – as long as the photoacid and conjugate photobase remain in the electronically excited state – as dictated by the new excited state pK_a^* value. Depending on the pH of the solvent one can observe the reversible time-depen-dent geminate recombination of the photobase with the released proton [79–83], or even the reaction of the photobase with other protons present in solution.

Proton transfer dynamics of photoacids to the solvent have thus, being reversible in nature, been modelled using the Debye–von Smoluchowski equation for diffu-sion-assisted reaction dynamics in a large body of experimental work on HPTS [84–87] and naphthols [88–92], with additional studies on the temperature depen-dence [93–98], and the pressure dependence [99–101], as well as the effects of spe-cial media such as reverse micelles [102] or chiral environments [103]. Moreover, results modelled with the Debye–von Smoluchowski approach have also been reported for proton acceptors triggered by optical excitation (photobases) [104, 105], and for molecular compounds with both photoacid and photobase functional-ities, such as 10-hydroxycamptothecin [106] and coumarin 4 [107]. It can be expected that proton diffusion also plays a role in hydroxyquinoline compounds [108–112]. Finally, proton diffusion has been suggested in the long time dynamics of green fluorescent protein [113], where the chromophore functions as a photoa-cid [23, 114], with an initial proton release on a 3–20 ps time scale [115, 116].

The diffusive kinetics of geminate pairs have been predicted to show a $t^{-3/2}$ time-dependent decaying behavior [117–122]. Early experiments showed, in con-trast, a t^{-a} decay, with a being dependent on the proton concentration [123]. Experiments on longer time ranges with improved sensitivity are prerequisites for an accurate determination of the asymptotic behavior [124]. In fact, recent mea-surements on HPTS have demonstrated the validity of the theoretically predicted $t^{-3/2}$ decay law (see Fig. 14.4) [125]. For 5-(methanesulfonyl)-1-naphthol a kinetic transition from power law to exponential has been reported due to a short photo-base lifetime [126].

Neutralization of the photoacids as a result of direct proton transfer to (scaven-ging) bases has also been explored in time-resolved studies. Whereas initial work

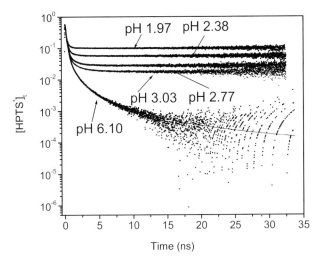

Figure 14.4 Semi-logarithmic plot of normalized fluorescence decay of excited HPTS. Points are experimental data (λ_{ex} = 375 nm, λ_{em} = 420 nm) in water acidified by HClO$_4$ after lifetime correction. The geminate recombination data (pH = 6) is fitted by a numerical solution of the Debye–von Smoluchowski equation convoluted with the instrument response function after lifetime correction. (Adapted from Ref. [125].)

used a Stern–Volmer quenching analysis [127, 128], it has been realized in a time-resolved fluorescence study that when working with high base concentrations, the apparent "on-contact" reaction rate can be obtained, despite the typically much slower diffusion rates [129]. The apparent "on-contact" reaction rates of naphthol and pyrenol derivatives with acetate or formate bases were found to range from (6 ps)$^{-1}$ for 5-cyano-1-naphthol and (500 ps)$^{-1}$ for 1-pyrenol. Debye–von Smoluchowski dynamics [130] have been included in UV/vis pump–probe neutralization studies of HPTS with acetate [131] and 2-naphthol-6-sulfonate and acetate [132, 133]. A more extensive listing of "on-contact" reaction rates has recently been published [134].

14.4
Reaction Dynamics Probing with Vibrational Marker Modes

The bimolecular reaction dynamics of geminate recombination or acid–base neutralization have until recently been studied with time-resolved techniques probing electronic transitions. Time-resolved fluorescence using time-correlated single photon counting detection is limited to a time resolution of a few picoseconds. UV/vis pump–probe experiments, in principle, may have a time resolution of a few tens of femtoseconds, but may be hampered by overlapping contributions of bleach (stimulated emission) and transient excited-state absorbance. A major disadvantage is that the reaction dynamics can only be inspected on the side of the photoacid/photobase with optical pulses. The experimental results thus only give information on the time scales when the proton dissociates from the photoacid. In particular, the arrival time when the proton arrives at a mineral base, such as acetate, cannot be determined in practice with optical pulses. Vibrational spectroscopy on the other hand can be used to follow the dynamics of both acid and base molecules [75]. Disadvantages include the lower cross sections for vibrational transitions and the frequent overlap with solvent bands. In particular water has a substantial steady-state absorbance throughout the mid-infrared, limiting the sample thicknesses to less than 100 μm.

Recently, the potential of ultrafast mid-infrared spectroscopy has been demonstrated for the study of these aqueous proton transfer reactions [135, 136]. Here the reaction dynamics have been followed by inspection of vibrational marker modes of the photoacid HPTS, the conjugate photobase HPTS⁻, and the conjugate acid of the base acetate ⁻OAc, i.e. acetic acid. When exciting HPTS dissolved at a fixed concentration of 20 mM in D_2O, with varying amounts of the base acetate added (ranging from 0.25–4 M), the deuteron transfer reaction can be followed in real time by inspection of the decay of the 1486 cm⁻¹ marker mode of the photoacid, the rise of signal at the 1503 cm⁻¹ transition of the conjugated photobase, and the rise of the C=O stretching band of acetic acid at 1720 cm⁻¹ (Fig. 14.5). At low base concentrations (0.25–0.5 M) deuteron transfer to the solvent, followed by deuteron pick-up by acetate, dominates the dynamics, as can be learned from a faster rise of the photobase signal than of the acetic acid band (Fig. 14.5). This feature cannot be observed in time-resolved fluorescence or UV/vis pump–probe measurements where it was

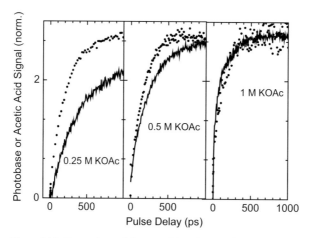

Figure 14.5 Comparison of the rise of the vibrational marker bands of the conjugate photobase of HPTS in the electronically excited state (dots) and of acetic acid (solid lines) at low base concentrations, indicating the initial deuteron release to the solvent and subsequent deuteron pick-up by the base.

indirectly studied by monitoring the proton-scavenging effect of a base on diminishing the geminate recombination reaction of the photoacid [129, 131].

At high base concentrations (> 1 M) practically identical rise times for photobase and acetic acid are observed, indicating a dominant direct deuteron transfer mechanism from the photoacid to the base. At these high concentrations the observed reaction dynamics are bimodal (Fig. 14.6). The two contributions to the signals can be ascribed to HPTS $\cdots^-$OAc complexes with a pre-formed hydrogen bond along the reaction coordinate, and initially uncomplexed HPTS that first has to form an encounter pair with $^-$OAc before a reaction can proceed. The preformed complex shows deuteron transfer faster than 150 fs. In contrast, for the fraction of initially uncomplexed HPTS, where the reaction coordinate is estab-

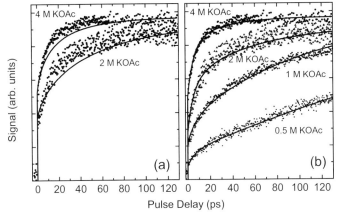

Figure 14.6 Rise of the carbonyl stretching marker mode of acetic acid for different concentrations of acetate (dots) in the reaction with HPTS. The solid lines denote calculated signals including contributions by "tight" complexes and diffusion assisted kinetics (a), or by "tight" complexes and "loose" complexes in addition to diffusion assisted kinetics (b). (Adapted from Ref. [136].)

lished by the diffusion of the reactants and by solvent fluctuations, a much slower bimolecular reaction rate on contact ($a = 6.3$ Å) of (12 ps)$^{-1}$ M^{-1} is found from data analysis using the theory of diffusion-controlled bimolecular reaction dynamics as given by von Smoluchowski with Szabo–Collins–Kimball (SCK) radiative boundary conditions [130], assuming fully screened potential at the high base concentrations used (Fig. 14.6(a)) [135]. A better fit is achieved when a static reaction component (with a time constant of 6 ps) is added to the SCK model (Fig. 14.6(b)), describing a fraction of reactive pairs already at close range (but not directly complexed) to each other at the time of initiation of the reaction, and thus not delayed by diffusion [136]. In addition, the correlation between the intrinsic (bimolecular) proton transfer constant of the SCK model (k_0) and the intrinsic unimolecular proton recombination constant of the Eigen–Weller model (k_r) was demonstrated for the first time to be in accordance with the analysis of Shoup and Szabo [137].

The fact that the deuteron transfer reaction in the pre-formed "tight" HPTS$\cdots^-$OAc complex is at least 2 orders of magnitude faster than the "loose" HPTS$\cdots(H_2O)_n\cdots^-$OAc encounter complex leads to the important finding that the deuteron transfer mechanism as initially suggested by Eigen and Weller has to be refined. One explanation follows the line of argument that the acid and base in the encounter complex can only react after substantial rearrangement of water molecules in the solvation shells before acid and base reach direct contact (Fig. 14.7), and the slower reaction rate of the encounter complex is due to this bottleneck solvent rearrangement dynamics. An alternative explanation lies in the possibility of a von Grotthuss-type of hopping of the proton from the acid to the base via solvation shell water molecules (Fig. 14.8), in which case acid and base never reach direct contact and the overall proton transfer reaction is considerably slower than the hopping time of the proton along a single hydrogen-bond.

Encounter Stage	Reaction Stage	Encounter Stage

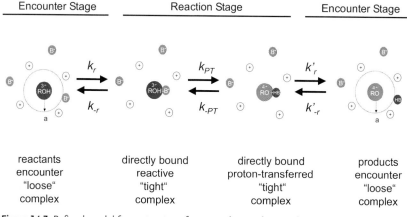

reactants	directly bound	directly bound	products
encounter	reactive	proton-transferred	encounter
"loose"	"tight"	"tight"	"loose"
complex	complex	complex	complex

Figure 14.7 Refined model for proton transfer between acids and bases with a three-stage mechanism consisting of diffusion, encounter and reaction stages. In this model the Eigen–Weller "on-contact" reaction rate k_r is to be understood as a solvent reorganization rate. The "loose" complexes rearrange then into "tight" complexes, that promptly react with the proton transfer rate k_{PT}. (Adapted from Ref. [136].)

Recent experimental results on the acid–base neutralization reaction between HPTS and the carboxylic bases mono-, di- and trichloracetate have revealed the underlying mechanisms of proton transfer of the "loose" complexes [138]. It turns out that a sequential, von Grotthuss-type of hopping occurs through a water molecule bridging the HPTS photoacid and the carboxylic base. Figure 14.9 shows the transient spectra obtained with a solution of 20 mM HPTS in D_2O with 1 M of monochloroacetate $^-$OAc-Cl added. At early pulse delays about 20% of HPTS has released its deuteron, as is indicated by the appearance of the HPTS$^-$ photobase marker band at 1435 cm^{-1} within the time resolution. A vibrational marker band at 1850 cm^{-1} indicates the transient existence of hydrated deuterons. Comparison with literature values for hydrated proton species with well-defined surroundings

Encounter Stage
=
Reaction Stage

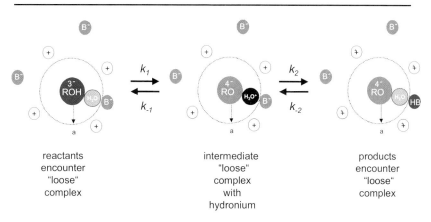

reactants
encounter
"loose"
complex

intermediate
"loose"
complex
with
hydronium

products
encounter
"loose"
complex

Figure 14.8 Proton transfer mechanism of the "loose" complexes $ROH \cdots (H_2O)_n \cdots B^-$ with a sequential, von Grotthuss-type, hopping of protons through water bridges. For HPTS and monochloroacetate the first transfer to the water bridge forming the hydronium ion H_3O^+ is ultrafast, and the second transfer to the base is slower. (Adapted from Ref. [136].)

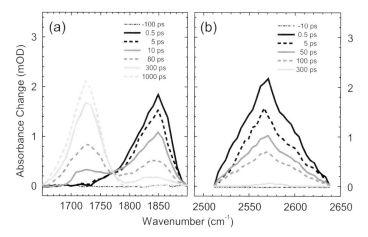

Figure 14.9 Transient spectra of the reaction of HPTS with 1 M $^-$OAc-Cl in D_2O, showing the response of the C=O stretching band of DOAc-Cl at 1720 cm^{-1} and the hydronium O–D stretching band at 1850 cm^{-1} (a). For comparison the transient response of the hydronium O–H stretching band at 2570 cm^{-1} measured with 1 M $^-$OAc-Cl in H_2O is shown in (b). (Adapted from Ref. [138].)

[139–144] strongly suggests that the hydrated deuterons exists as hydronium ions, D_3O^+. A similar transient band is detected at 2570 cm^{-1} for measurements performed in H_2O, in full accordance with an H/D isotope effect on the transition frequency of H_3O^+ vs. D_3O^+. Kinetic modelling shows that the initial transfer in the "loose" HPTS $\cdots$ (D_2O) $\cdots$ $^-$OAc-Cl complex occurs faster than the time resolution. The second transfer, generating HPTS$^-\cdots$(D_2O)$\cdots$DOAc-Cl, has a much slower rate of (25 ps)$^{-1}$.

The findings of the proton transfer reaction of HPTS with $^-$OAc-Cl are reminiscent of the results of theoretical studies for acid dissociation of hydrogen halides in water [14–17] Here the proton transfers in a hopping fashion from the first to the secondary solvation shells, away from the dissociating hydrogen halide, with the rate determining step being the cleavage of a hydrogen bond in the proton-accepting water molecule, which must change its coordination number from 4 to 3. This mechanism for proton transfer appears to be a general one: the proton transmission through water, named after von Grotthuss as a dedication [7] to his landmark electrolysis measurements of aqueous ion solutions [6], also occurs by sequential hopping to neighboring water molecules and involves similar rearrangements of the hydrogen-bonding network, although the full details of the Grotthuss mechanism in water are yet to be fully deciphered. Perhaps the most fundamental aspect of proton solvation in pure water have come from *ab initio* molecular dynamics simulations [10, 12, 13] that have demonstrated a continuous exchange between the Eigen cation $H_3O^+(H_2O)_3$ (i.e. $H_9O_4^+$) [3] and the Zundel cation $H_5O_2^+$ (i.e. $H_2O \cdots H^+ \cdots OH_2$) [145] in liquid water. In the acid–base reaction between HPTS and $^-$OAc-Cl the observed marker bands of D_3O^+ at 1850 cm^{-1} and of H_3O^+ at 2570 cm^{-1} indicate that in the encounter (reactive) complex between HPTS$^-$ and $^-$OAc-Cl (that is water depleted to facilitate the acid–base reaction) the Eigen solvation core, the H_3O^+ cation, plays a key role. In particular the frequency position of the hydrated proton band observed for the ionic complex HPTS$^-\cdots H_3O^+ \cdots ^-$OAc-Cl is very similar to that observed for the Eigen cation $H_3O^+(H_2O)_3$, as recently measured in dedicated experiments of hydrated proton clusters [144], and calculated for the Eigen cation in the proton wire of bacteriorhodopsin [143]. This finding strongly suggests that in our experiments the hydrated proton is the Eigen solvation core with a symmetric hydrogen bonding configuration, i.e. $(H_3O^+)L_3$ with L hydrogen accepting groups. Here the intermediate HPTS$^-\cdots D_3O^+/H_3O^+\cdots ^-$OAc-Cl complex appears to contain only the hydronium ion, hydrogen bonded in an almost symmetric three-fold way and thus resembling the hydronium core of the Eigen cation. A role of the Zundel cation $H_5O_2^+$ in the proton transfer dynamics is suggested less likely, since its vibrational transitions are located at other frequency positions [143, 144, 146].

The present results on the reaction dynamics between HPTS and the family of acetate bases demonstrate that a base-induced sequential proton transfer mechanism at close acid–base proximities is at the heart of aqueous acid–base proton transfer reactions. The experimentally found, relatively long lived, intermediate ionic complex consisting of photobase, hydrated proton and carboxylic base, indicates the special property of carboxylic bases, a finding with important implications not only for proton protein channels, where the dynamics of various intermediate stages are dictated by glutamic and aspartic amino acid functionalities [20–22] but also for wild type green fluorescent protein (GFP), in which the photoacid chromophore releases a proton through a water bridge to the carboxylic side group of Glu222 [23, 147].

Acknowledgment

We cordially acknowledge the important contributions by our present and former group members Matteo Rini, Omar F. Mohammed, Ben-Zion Magnes, Dina Pines. We also thank the German-Israeli Foundation for Scientific Research and Development, the Israel Science Foundation, and the Egyptian Government for financial support (Project GIF 722/01, Project ISF 562/04 for EP and a long term mission fellowship for OFM).

References

1 R. P. Bell, *The Proton in Chemistry*, Chapman and Hall, London, 2nd edn., **1973**.

2 A. Weller, *Prog. React. Kinet.* **1961**, *1*, 187–213.

3 M. Eigen, *Angew. Chem. Int. Ed.* **1964**, *3*, 1–19.

4 M. Eigen, W. Kruse, G. Maass, L. DeMaeyer, *Prog. React. Kinet.* **1964**, *2*, 285.

5 P. L. Geissler, C. Dellago, D. Chandler, J. Hutter, M. Parrinello, *Science* **2001**, *291*, 2121–2124.

6 C. J. T. de Grotthuss, *Ann. Chim.* **1806**, *LVIII*, 54–74.

7 H. Danneel, *Z. Elektrochem. Angew. Phys. Chem.* **1905**, *11*, 249–252.

8 N. Agmon, *Chem. Phys. Lett.* **1995**, *244*, 456–462.

9 J. T. Hynes, *Nature* **1999**, *397*, 565.

10 D. Marx, M. E. Tuckerman, J. Hutter, M. Parrinello, *Nature* **1999**, *397*, 601–604.

11 R. Vuilleumier, D. Borgis, *J. Chem. Phys.* **1999**, *111*, 4251–4266.

12 U. W. Schmitt, G. A. Voth, *J. Chem. Phys.* **1999**, *111*, 9361–9381.

13 H. Lapid, N. Agmon, M. K. Petersen, G. A. Voth, *J. Chem. Phys.* **2005**, *122*, 014506.

14 K. Ando, J. T. Hynes, *J. Phys. Chem. B* **1997**, *101*, 10464–10478.

15 K. Ando, J. T. Hynes, *Adv. Chem. Phys.* **1999**, *110*, 381–430.

16 K. Ando, J. T. Hynes, *J. Phys. Chem. A* **1999**, *103*, 10398–10408.

17 A. Al-Halabi, R. Bianco, J. T. Hynes, *J. Phys. Chem. A* **2002**, *106*, 7639–7645.

18 R. Bianco, J. T. Hynes, *Theor. Chem. Acc.* **2004**, *111*, 182–187.

19 F. Hibbert, *Adv. Phys. Org. Chem.* **1986**, *22*, 113–212.

20 H. Luecke, H. T. Richter, J. K. Lanyi, *Science* **1998**, *280*, 1934–1937.

21 W. Kühlbrandt, *Nature* **2000**, *406*, 569–570.

22 T. E. DeCoursey, *Physiol. Rev.* **2003**, *83*, 475–579.

23 M. Zimmer, *Chem. Rev.* **2002**, *102*, 759–781.

24 M. Eigen, in *Nobel Lectures, Chemistry 1963–1970*, Elsevier, Amsterdam, **1972**, pp. 170–203.

25 T. Elsaesser, in *Femtosecond Chemistry*, Vol. 2 , J. Manz, L. Wöste (Eds.), VCH, Weinheim, Germany, **1995**, p. 563.

26 A. Douhal, F. Lahmani, A. H. Zewail, *Chem. Phys.* **1996**, *207*, 477–498.

27 T. Elsaesser, in *Ultrafast Hydrogen Bonding Dynamics and Proton Transfer Processes in the Condensed Phase*, Vol. 23,

T. Elsaesser, H. J. Bakker (Eds.), Kluwer Academic Publishers, Dordrecht, **2002**, pp. 119–153.

28 W. Klöpffer, *Adv. Photochem.* **1977**, *10*, 311–358.

29 S. J. Formosinho, L. G. Arnaut, *J. Photochem. Photobiol. A* **1993**, *75*, 21–48.

30 T. Förster, *Die Naturwissenschaften* **1949**, *36*, 186–187.

31 T. Förster, *Z. Elektrochem* **1950**, *54*, 42–46.

32 T. Förster, *Z. Elektrochem* **1950**, *54*, 531–535.

33 E. Vander Donckt, *Prog. React. Kinet.* **1970**, *5*, 273–299.

34 J. F. Ireland, P. A. H. Wyatt, *Adv. Phys. Org. Chem.* **1976**, *12*, 131–221.

35 I. Y. Martynov, A. B. Demyashkevich, B. M. Uzhinov, M. G. Kuz'min, *Russ. Chem. Rev. [Usp. Khim.]* **1977**, *46*, 1–15 [13–31].

36 L. G. Arnaut, S. J. Formosinho, *J. Photochem. Photobiol. A* **1993**, *75*, 1–20.

37 E. Pines, D. Pines, in Ref. [27], pp. 155–184.

38 E. Pines, in *Chemistry of Phenols*, Z. Rappoport (Ed.), Wiley, New York, **2003**, pp. 491–529.

39 L. M. Tolbert, J. E. Haubrich, *J. Am. Chem. Soc.* **1990**, *112*, 8163–8165.

40 L. M. Tolbert, J. E. Haubrich, *J. Am. Chem. Soc.* **1994**, *116*, 10593–10600.

41 L. M. Tolbert, K. M. Solntsev, *Acc. Chem. Res.* **2002**, *35*, 19–27.

42 E. M. Kosower, D. Huppert, *Annu. Rev. Phys. Chem.* **1986**, *37*, 127–156.

43 N. Agmon, *J. Phys. Chem. A* **2005**, *109*, 13–35.

44 E. Pines, D. Huppert, *J. Phys. Chem.* **1983**, *87*, 4471–4478.

45 M. Gutman, *Methods Biochem. Anal.* **1984**, *30*, 1–103.

46 M. Gutman, E. Nachliel, *Annu. Rev. Phys. Chem.* **1997**, *48*, 329–356.

47 E. Pines, G. R. Fleming, *J. Phys. Chem.* **1991**, *95*, 10448–10457.

48 M. J. Kamlet, J. L. M. Abboud, M. H. Abraham, R. W. Taft, *J. Org. Chem.* **1983**, *48*, 2877–2887.

49 N. Barrash-Shiftan, B. B. Brauer, E. Pines, *J. Phys. Org. Chem.* **1998**, *11*, 743–750.

50 K. M. Solntsev, D. Huppert, L. M. Tolbert, N. Agmon, *J. Am. Chem. Soc.* **1998**, *120*, 7981–7982.

51 K. M. Solntsev, D. Huppert, N. Agmon, *J. Phys. Chem. A* **1998**, *102*, 9599–9606.

52 B.-Z. Magnes, D. Pines, N. Strashnikova, E. Pines, *Solid State Ionics* **2004**, *168*, 225–233.

53 E. Pines, D. Pines, Y.-Z. Ma, G. R. Fleming, *ChemPhysChem* **2004**, *5*, 1315–1327.

54 M. Barroso, L. G. Arnaut, S. J. Formosinho, *J. Photochem. Photobiol. A* **2003**, *160*, 227–227.

55 N. Agmon, W. Rettig, C. Groth, *J. Am. Chem. Soc.* **2002**, *124*, 1089–1096.

56 R. Knochenmuss, I. Fischer, D. Lührs, Q. Lin, *Isr. J. Chem.* **1999**, *39*, 221–230.

57 B.Z. Magnes, N. V. Strashnikova, E. Pines, *Isr. J. Chem.* **1999**, *39*, 361–373.

58 R. Knochenmuss, I. Fischer, *Int. J. Mass Spectrom.* **2002**, *220*, 343–357.

59 T. H. Tran-Thi, T. Gustavsson, C. Prayer, S. Pommeret, J. T. Hynes, *Chem. Phys. Lett.* **2000**, *329*, 421–430.

60 T.-H. Tran-Thi, C. Prayer, P. Millié, P. Uznanski, J. T. Hynes, *J. Phys. Chem. A* **2002**, *106*, 2244–2255.

61 J. T. Hynes, T. H. Tran-Thi, G. Granucci, *J. Photochem. Photobiol. A* **2002**, *154*, 3–11.

62 A. L. Sobolewski, W. Domcke, C. Dedonder-Lardeux, C. Jouvet, *Phys. Chem. Chem. Phys.* **2002**, *4*, 1093–1100.

63 W. Domcke, A. L. Sobolewski, *Science* **2003**, *302*, 1693–1694.

64 O. David, C. Dedonder-Lardeux, C. Jouvet, *Int. Rev. Phys. Chem.* **2002**, *21*, 499–523.

65 C. Tanner, C. Manca, S. Leutwyler, *Science* **2003**, *302*, 1736–1739.

66 C. Manca, C. Tanner, S. Coussan, A. Bach, S. Leutwyler, *J. Chem. Phys.* **2004**, *121*, 2578–2590.

67 C. Tanner, C. Manca, S. Leutwyler, *J. Chem. Phys.* **2005**, *122*.

68 T. Schultz, E. Samoylova, W. Radloff, I. V. Hertel, A. L. Sobolewski, W. Domcke, *Science* **2004**, *306*, 1765–1768.

69 G. R. Fleming, M. Cho, *Annu. Rev. Phys. Chem.* **1996**, *47*, 109–134.

70 W. P. de Boeij, M. S. Pshenichnikov, D. A. Wiersma, *Annu. Rev. Phys. Chem.* **1998**, *49*, 99–123.

71 M. Glasbeek, H. Zhang, *Chem. Rev.* **2004**, *104*, 1929–1954.

72 C. Chudoba, E. T. J. Nibbering, T. Elsaesser, *Phys. Rev. Lett.* **1998**, *81*, 3010–3013.

73 C. Chudoba, E. T. J. Nibbering, T. Elsaesser, *J. Phys. Chem. A* **1999**, *103*, 5625–5628.

74 E. T. J. Nibbering, C. Chudoba, T. Elsaesser, *Isr. J. Chem.* **1999**, *39*, 333–347.

75 E. T. J. Nibbering, H. Fidder, E. Pines, *Annu. Rev. Phys. Chem.* **2005**, *56*, 337–367.

76 O. F. Mohammed, J. Dreyer, B.-Z. Magnes, E. Pines, E. T. J. Nibbering, *ChemPhysChem* **2005**, *6*, 625–636.

77 S. P. Webb, L. A. Philips, S. W. Yeh, L. M. Tolbert, J. H. Clark, *J. Phys. Chem.* **1986**, *90*, 5154–5164.

78 R. Krishnan, J. Lee, G. W. Robinson, *J. Phys. Chem.* **1990**, *94*, 6365–6367.

79 E. Pines, D. Huppert, *Chem. Phys. Lett.* **1986**, *126*, 88–91.

80 E. Pines, D. Huppert, *J. Chem. Phys.* **1986**, *84*, 3576–3577.

81 E. Pines, D. Huppert, N. Agmon, *J. Chem. Phys.* **1988**, *88*, 5620–5630.

82 N. Agmon, E. Pines, D. Huppert, *J. Chem. Phys.* **1988**, *88*, 5631–5638.

83 N. Agmon, *J. Chem. Phys.* **1988**, *88*, 5639–5642.

84 D. Huppert, E. Pines, N. Agmon, *J. Opt. Soc. Am. B* **1990**, *7*, 1545–1550.

85 N. Agmon, D. Huppert, A. Masad, E. Pines, *J. Phys. Chem.* **1991**, *95*, 10407–10413.

86 A. Masad, D. Huppert, *Chem. Phys. Lett.* **1991**, *180*, 409–415.

87 S. Y. Goldberg, E. Pines, D. Huppert, *Chem. Phys. Lett.* **1992**, *192*, 77–81.

88 I. Carmeli, D. Huppert, L. M. Tolbert, J. E. Haubrich, *Chem. Phys. Lett.* **1996**, *260*, 109–114.

89 D. Huppert, L. M. Tolbert, S. Linares-Samaniego, *J. Phys. Chem. A* **1997**, *101*, 4602–4605.

90 K. M. Solntsev, D. Huppert, N. Agmon, *J. Phys. Chem. A* **1999**, *103*, 6984–6997.

91 K. M. Solntsev, D. Huppert, N. Agmon, L. M. Tolbert, *J. Phys. Chem. A* **2000**, *104*, 4658–4669.

92 C. Clower, K. M. Solntsev, J. Kowalik, L. M. Tolbert, D. Huppert, *J. Phys. Chem. A* **2002**, *106*, 3114–3122.

93 B. Cohen, D. Huppert, *J. Phys. Chem. A* **2000**, *104*, 2663–2667.

94 B. Cohen, P. Leiderman, D. Huppert, *J. Phys. Chem. A* **2002**, *106*, 11115–11122.

95 B. Cohen, P. Leiderman, D. Huppert, *J. Phys. Chem. A* **2003**, *107*, 1433–1440.

96 E. Pines, B.-Z. Magnes, T. Barak, *J. Phys. Chem. A* **2001**, *105*, 9674–9680.

97 B. Cohen, D. Huppert, *J. Phys. Chem. A* **2001**, *105*, 2980–2988.

98 B. Cohen, J. Segal, D. Huppert, *J. Phys. Chem. A* **2002**, *106*, 7462–7467.

99 N. Koifman, B. Cohen, D. Huppert, *J. Phys. Chem. A* **2002**, *106*, 4336–4344.

100 L. Genosar, P. Leiderman, N. Koifman, D. Huppert, *J. Phys. Chem. A* **2004**, *108*, 309–319.

101 L. Genosar, P. Leiderman, N. Koifman, D. Huppert, *J. Phys. Chem. A* **2004**, *108*, 1779–1789.

102 B. Cohen, D. Huppert, K. M. Solntsev, Y. Tsfadia, E. Nachliel, M. Gutman, *J. Am. Chem. Soc.* **2002**, *124*, 7539–7547.

103 K. M. Solntsev, L. M. Tolbert, B. Cohen, D. Huppert, Y. Hayashi, Y. Feldman, *J. Am. Chem. Soc.* **2002**, *124*, 9046–9047.

104 B. Cohen, D. Huppert, *J. Phys. Chem. A* **2002**, *106*, 1946–1955.

105 S. R. Keiding, D. Madsen, J. Larsen, S. K. Jensen, J. Thøgersen, *Chem. Phys. Lett.* **2004**, *390*, 94–97.

106 K. M. Solntsev, E. N. Sullivant, L. M. Tolbert, S. Ashkenazi, P. Leiderman, D. Huppert, *J. Am. Chem. Soc.* **2004**, *126*, 12701–12708.

107 B. Cohen, D. Huppert, *J. Phys. Chem. A* **2001**, *105*, 7157–7164.

108 E. Bardez, I. Devol, B. Larrey, B. Valeur, *J. Phys. Chem. B* **1997**, *101*, 7786–7793.

109 E. Bardez, *Isr. J. Chem.* **1999**, *39*, 319–332.

110 S. Kohtani, A. Tagami, R. Nakagaki, *Chem. Phys. Lett.* **2000**, *316*, 88–93.

111 O. Poizat, E. Bardez, G. Buntinx, V. Alain, *J. Phys. Chem. A* **2004**, *108*, 1873–1880.

112 T. G. Kim, M. R. Topp, *J. Phys. Chem. A* **2004**, *108*, 10060–10065.

113 P. Leiderman, M. Ben-Ziv, L. Genosar, D. Huppert, K. M. Solntsev, L. M. Tolbert, *J. Phys. Chem. B* **2004**, *108*, 8043–8053.

114 K. Brejc, T. K. Sixma, P. A. Kitts, S. R. Kain, R. Y. Tsien, M. Ormo, S. J. Remington, *Proc. Natl. Acad. Sci. USA* **1997**, *94*, 2306–2311.

115 M. Chattoraj, B. A. King, G. U. Bublitz, S. G. Boxer, *Proc. Natl. Acad. Sci. USA* **1996**, *93*, 8362–8367.

116 K. Winkler, J. R. Lindner, V. Subramaniam, T. M. Jovin, P. Vöhringer, *Phys. Chem. Chem. Phys.* **2002**, *4*, 1072–1081.

117 A. Szabo, R. Zwanzig, *J. Stat. Phys.* **1991**, *65*, 1057–1083.

118 A. Szabo, *J. Chem. Phys.* **1991**, *95*, 2481–2490.

119 W. Naumann, N. V. Shokhirev, A. Szabo, *Phys. Rev. Lett.* **1997**, *79*, 3074–3077.

120 I. V. Gopich, N. Agmon, *Phys. Rev. Lett.* **2000**, *84*, 2730–2733.

121 N. Agmon, I. V. Gopich, *J. Chem. Phys.* **2000**, *112*, 2863–2869.

122 I. V. Gopich, A. A. Ovchinnikov, A. Szabo, *Phys. Rev. Lett.* **2001**, *86*, 922–925.

123 D. Huppert, S. Y. Goldberg, A. Masad, N. Agmon, *Phys. Rev. Lett.* **1992**, *68*, 3932–3935.

124 K. M. Solntsev, D. Huppert, N. Agmon, *J. Phys. Chem. A* **2001**, *105*, 5868–5876.

125 D. Pines, E. Pines, *J. Chem. Phys.* **2001**, *115*, 951–953.

126 K. M. Solntsev, D. Huppert, N. Agmon, *Phys. Rev. Lett.* **2001**, *86*, 3427–3430.

127 M. Lawrence, C. J. Marzzacco, C. Morton, C. Schwab, A. M. Halpern, *J. Phys. Chem.* **1991**, *95*, 10294–10299.

128 L. M. Tolbert, S. M. Nesselroth, *J. Phys. Chem.* **1991**, *95*, 10331–10336.

129 E. Pines, B. Z. Magnes, M. J. Lang, G. R. Fleming, *Chem. Phys. Lett.* **1997**, *281*, 413–420.

130 A. Szabo, *J. Phys. Chem.* **1989**, *93*, 6929–6939.

131 L. Genosar, B. Cohen, D. Huppert, *J. Phys. Chem. A* **2000**, *104*, 6689–6698.

132 B. Cohen, D. Huppert, N. Agmon, *J. Am. Chem. Soc.* **2000**, *122*, 9838–9839.

133 B. Cohen, D. Huppert, N. Agmon, *J. Phys. Chem. A* **2001**, *105*, 7165–7173.

134 M. Barroso, L. G. Arnaut, S. J. Formosinho, *J. Photochem. Photobiol. A* **2002**, *154*, 13–21.

135 M. Rini, B.-Z. Magnes, E. Pines, E. T. J. Nibbering, *Science* **2003**, *301*, 349–352.

136 M. Rini, D. Pines, B.-Z. Magnes, E. Pines, E. T. J. Nibbering, *J. Chem. Phys.* **2004**, *121*, 9593–9610.

137 D. Shoup, A. Szabo, *Biophys. J.* **1982**, *40*, 33–39.

138 O. F. Mohammed, D. Pines, J. Dreyer, E. Pines, E. T. J. Nibbering, *Science* **2005**, *310*, 83–86.

139 B. S. Ault, G. C. Pimentel, *J. Phys. Chem.* **1973**, *77*, 57–61.

140 J. M. Williams, in *The Hydrogen Bond: Recent developments in Theory and Experiments, Vol. II. Structure and Spectroscopy*, P. Schuster, G. Zundel, C. Sandorfy, (Eds.), North Holland, Amsterdam, **1976**, pp. 655–682.

141 L. Delzeit, B. Rowland, J. P. Devlin, *J. Phys. Chem.* **1993**, *97*, 10312–10318.

142 E. S. Stoyanov, C. A. Reed, *J. Phys. Chem. A* **2004**, *108*, 907–913.

143 R. Rousseau, V. Kleinschmidt, U. W. Schmitt, D. Marx, *Angew. Chem. Int. Ed.* **2004**, *43*, 4804–4807.

144 J. M. Headrick, E. G. Diken, R. S. Walters, N. I. Hammer, R. A. Christie, J. Cui, E. M. Myshakin, M. A. Duncan, M. A. Johnson, K. D. Jordan, *Science* **2005**, *308*, 1765–1769.

145 G. Zundel, *Adv. Chem. Phys.* **2000**, *111*, 1–217.

146 K. R. Asmis, N. L. Pivonka, G. Santambrogio, M. Brummer, C. Kaposta, D. M. Neumark, L. Wöste, *Science* **2003**, *299*, 1375–1377.

147 D. Stoner-Ma, A. A. Jaye, P. Matousek, M. Towrie, S. R. Meech, P. J. Tonge, *J. Am. Chem. Soc.* **2005**, *127*, 2864–2865.

15
Coherent Low-frequency Motions in Condensed Phase Hydrogen Bonding and Transfer

Thomas Elsaesser

15.1
Introduction

Hydrogen bonding represents a local interaction which determines the fluctuating structure of liquids forming extended molecular networks, e.g., water and alcohols, as well as macromolecular structure of biological relevance. Hydrogen bonds also play a key role in hydrogen and proton transfer processes in both electronic ground and excited states [1, 2]. The structural dynamics of hydrogen bonds and proton transfer processes are determined by motions along nuclear coordinates which are characterized by vibrational periods in the femtosecond time domain. For instance, the period of an O–H stretching vibration is of the order of 10 fs whereas low-frequency modes of hydrogen bonds display periods of up to several hundred femtoseconds. In general, vibrational modes of hydrogen bonds show pronounced coupling to each other, resulting in a highly complex dynamics of structural changes.

The dynamics of hydrogen bonded systems cover a wide range in time, from about 50 fs up to tens of picoseconds [2]. Linear vibrational spectroscopy, a standard tool of hydrogen bond research, provides the steady-state, i.e., time-averaged infrared and Raman spectra, giving very limited insight into the processes underlying such dynamics. In most cases, there is no quantitative understanding of vibrational line shapes and the different broadening mechanisms, in spite of extensive theoretical work on molecular potential energy surfaces and vibrational couplings. Much more infomation is available from studies of the nonlinear vibrational response in which the macroscopic vibrational polarization and/or changes in vibrational absorption display a higher order dependence on the amplitude of the radiation fields interacting with the sample. Nonlinear vibrational spectroscopy in the femtosecond time domain allows one to observe ultrafast hydrogen bond dynamics in real-time and to separate different microscopic couplings in the nonlinear response [3, 4]. Quantum coherent vibrational dynamics of hydrogen bonds in liquids is a topic of substantial current interest [5] and both coherent nuclear motions, i.e., vibrational wavepackets, and processes of vibrational dephasing and relaxation have been studied recently by ultrafast pump–probe and

Hydrogen-Transfer Reactions. Edited by J. T. Hynes, J. P. Klinman, H. H. Limbach, and R. L. Schowen
Copyright © 2007 WILEY-VCH Verlag GmbH & Co. KGaA, Weinheim
ISBN: 978-3-527-30777-7

photon echo techniques. In photoinduced hydrogen transfer, measurements of transient vibrational spectra in electronically excited states provide direct information on the molecular structure of reaction products and on reaction pathways [6–8].

In this chapter, coherent low-frequency motions and their role in hydrogen bond dynamics and hydrogen transfer are discussed. In Section 15.2, the basic vibrational excitations and couplings in a hydrogen bond are introduced. Recent results on coherent low-frequency motions of intra- and intermolecular hydrogen bonds in the electronic ground state are presented in Section 15.3. The role of low-frequency motions in excited state intramolecular hydrogen transfer is addressed in Section 15.4, followed by some conclusions (Section 15.5).

15.2
Vibrational Excitations of Hydrogen Bonded Systems

The formation of hydrogen bonds results in pronounced changes of the vibrational spectra of the molecules involved [1, 9]. In an X–H $\cdots$ Y bonding geometry, the stretching mode of the X–H donor group displays the most prominent modifications, i.e., a red-shift and – in most cases – a substantial spectral broadening and reshaping. The red-shift reflects the reduced force constant of the oscillator and/or the enhanced anharmonicity of the vibrational potential along the X–H stretching coordinate, i.e., an enhanced diagonal anharmonicity. The red-shift has been used to characterize the strength of hydrogen bonds [10]. Spectral broadening can arise from a number of mechanisms, among them anharmonic coupling to low-frequency modes, Fermi resonances with overtone and combination tone levels of fingerprint modes, vibrational dephasing, and inhomogeneous broadening due to different hydrogen bonding geometries in the molecular ensemble [11–15].

In the weak attractive potential between hydrogen donor and acceptor groups, new modes occur which are connected with motions of the heavy atoms and affect the geometry of the hydrogen bond. The small force constants and the large reduced mass of such hydrogen bond modes result in low frequencies between 50 and 300 cm^{-1}, corresponding to vibrational periods of 110 to 660 fs. Such periods are much longer than that of the X–H stretching mode and, thus, there is a clear separation of the time scales of hydrogen bond and X–H stretching motions.

Third- and higher order terms of the vibrational potential containing mixed products of vibrational coordinates result in a coupling of different modes. Anharmonic coupling exists between the high-frequency X–H stretching mode and low-frequency hydrogen bond modes and has been considered a potential broadening mechanism of the X–H, e.g., O–H stretching band [9, 11, 12]. The separation of time scales of the low- and high-frequency modes allows for a theoretical description in which the different states of the O–H stretching oscillator define adiabatic potential energy surfaces for the low-frequency modes [Fig. 15.1 (a)], similar to the separation of electronic and nuclear degrees of freedom in the Born–Oppenheimer picture of vibronic transitions. Vibrational transitions from different levels of the low-frequency oscillator in the $v_{OH} = 0$ state to different low-frequency levels in the $v_{OH} = 1$ state with a shifted origin of the potential result in a progression of lines which is centered at the pure X–H stretching transition and display a mutual line separation by one quantum $\hbar\Omega$ of the low-frequency mode [Fig. 15.1 (b)]. The absorption strength is determined by the dipole moment of the $v_{OH} = 0 \rightarrow 1$ transition of the O–H stretching mode and the Franck–Condon factors between the optically coupled levels of the low-frequency mode. With increasing difference in quantum number of the low-frequency mode in the $v_{OH} = 0$ and 1 states, the Franck–Condon factors decrease and the progression lines become weaker for larger frequency separation from the progression center.

For each low-frequency mode coupling to an O–H stretching oscillator, an independent progression of lines occurs. Such mechanisms can result in a strong broadening and/or spectral substructure of the overall O–H stretching band, even for a small number of absorption lines with large Franck–Condon factors. Third-order coupling strengths of the O–H stretching mode and hydrogen bond modes in acetic acid dimers of up to 100 cm^{-1} have been calculated recently [16].

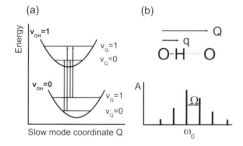

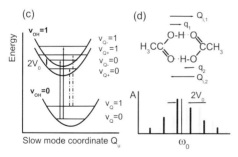

Figure 15.1 Anharmonic coupling of the O–H stretching mode q and a low-frequency hydrogen bond (O...O) mode Q. (a) Potential energy diagram for the low-frequency mode in a single hydrogen bond. The potential energy surfaces as defined by the stretching mode and the quantum levels of the low-frequency mode are plotted for the $v_{OH} = 0$ and 1 states as a function of the slow-mode coordinate Q. (b) Progression of vibrational lines centered at the pure O–H stretching transition ω_0 with a line separation Ω, the frequency of the mode Q. (c) Potential energy diagram for two excitonically coupled O–H stretching oscillators. The two $v_{OH} = 1$ potentials are separated by $2V_0$ (V_0: excitonic coupling). (d) Progressions of vibrational lines resulting from the coupling scheme in (c).

In systems with several identical O–H groups and/or hydrogen bonds, excitonic coupling occurs between resonant oscillators. Such interaction which can be mediated by through-bond or through-space, e.g., dipole–dipole interactions, results in vibrational excitations delocalized over the different oscillators. Cyclic carboxylic acid dimers represent important model systems with coupled oscillators which have been analyzed theoretically by Marechal and Witkowski [12]. In their approach, the $v_{OH} = 0$ states of the two O–H oscillators are considered degenerate and the coupling comes into play whenever one of the oscillators is excited. The excitonic coupling leads to a splitting of the $v_{OH} = 1$ states [Fig. 15.1 (c)]. Considering both anharmonic and excitonic coupling, the coupled system has been described by taking into account the C_2 symmetry of the cyclic dimer and introducing symmetrized vibrational coordinates $q_{g,u} = (1/\sqrt{2})(q_1 \pm q_2)$ and $Q_{i,g,u} = (1/\sqrt{2})(Q_{i,1} \pm Q_{i,2})$ ($i = 1,2,..$) for the stretching and the low-frequency modes i, respectively. Taking into account this symmetry and evaluating the dipole selection rules, one finds that transitions between $|v_u (q_u)>$ states are infrared active whereas transitions between $|v_g (q_g)>$ states contribute to the Raman band of the O–H stretching mode ($v_{u,g}$: vibrational quantum numbers) [12, 14, 15]. The $v_u = 1$ potential energy surface along the gerade $Q_{i,g}$ coordinate remains unaltered whereas the excitonic coupling V_0 leads to a splitting of the $v_u = 1$ potential energy surface along the ungerade $Q_{i,u}$ coordinate by $2V_0$. The resulting line shape consists of two different progressions between the $v_{i,Q} = 0$ level of the $Q_{i,u}$ mode in the $v_u = 0$ state and the $v_{i,Q}$ levels in the $v_u = 1$ state as well as between the $v_{i,Q} = 1$ level in the $v_u = 0$ state and the v_{i,Q^+} levels in the $v_u = 1$ state [Fig. 15.1 (d)]. Simultaneously, the number of quanta in the $Q_{i,g}$ mode can be changed when exciting the system to the $v_{OH} = 1$ state, introducing an additional degeneracy of the lines in the respective progression. An individual molecule displays only one progression, depending on whether the $v_{i,Q} = 0$ or $v_{i,Q} = 1$ level in the $v_u = 0$ state is populated. In an ensemble of molecules at finite vibrational temperature, both levels are populated and both series of lines contribute to the overall vibrational band.

The excitonic coupling strength of the O–H stretching oscillators in carboxylic acid dimers has remained uncertain. Early work [12] has suggested values of $V_0 = -85$ cm^{-1} whereas a later analysis assumed much smaller values [13]. For the C=O stretching oscillators of acetic acid dimers, a coupling strength of approximately 50 cm^{-1} has been reported [17].

Fermi resonances between the $v_{OH} = 1$ state of the stretching mode and overtones or combination bands of modes in the fingerprint range result in a splitting of the O–H stretching transition into different components with a separation determined by the respective coupling. For large couplings, Fermi resonances have a strong influence on the line shape of the O–H stretching band, leading to features like the so-called Evans window. Studies of the linear absorption band of the O–H stretching mode in carboxylic acid dimers have suggested Fermi resonances between the $v_{OH} = 1$ level and the $\delta_{OH} = 2$ bending level, as well as between $v_{OH} = 1$ and combination tones of δ_{OH} with v_{C-O} and $v_{C=O}$ stretching modes [13]. More recent theoretical work suggests an absolute value of the third order coupling for such modes of the order of 100 cm^{-1} and attributes the coarse shape of the O–H stretching band to Fermi resonances, without, however, allowing for a full quantitative understanding [18, 19]. Very recently, nonlinear two-dimensional (2D) infrared spectroscopy provided direct evidence for Fermi resonances through off-diagonal peaks in the 2D spectra [20]. A theoretical analysis of such experiments gave couplings of 40 to 150 cm^{-1} between the O–H stretching mode and combination and overtones of fingerprint vibrations. This has allowed for a quantitative modeling of the linear O–H stretching absorption spectrum [21].

In summary, the couplings discussed so far transform the hydrogen stretching oscillator into a vibrational multi-level system with a multitude of transition lines. The interaction with the fluctuating surrounding leads to an additional broadening of the individual lines by vibrational dephasing [22–25]. Nonlinear vibrational spectroscopy allows one to separate the different couplings in the nonlinear time-resolved response following femtosecond vibrational excitation. In particular, the coherent vibrational dynamics can be isolated from processes of population relaxation [26] and energy redistribution.

15.3
Low-frequency Wavepacket Dynamics of Hydrogen Bonds in the Electronic Ground State

The multi-level character of X–H stretching excitations in hydrogen bonds allows the preparation of quantum-coherent superpositions of states. Excitation of a set of transitions within the O–H stretching band by a broadband femtosecond pulse creates a wavepacket moving along the low-frequency vibrational coordinates contributing to this superposition. In many liquids, such motions are rapidly damped due to coupling with the fluctuating environment. In recent pump–probe experiments, however, underdamped, i.e., oscillatory motions along hydrogen bond modes have been observed for picoseconds after impulsive excitation by 100 fs pulses. In the following, we review such results.

15.3.1
Intramolecular Hydrogen Bonds

Low-frequency wavepacket dynamics were first observed in intramolecular hydrogen bonds with a well-defined geometry [27–31]. The enol tautomer of 2-(2′-hydroxyphenyl)benzothiazole (HBT) represents such a system: in the electronic ground state, an O–H$\cdots$N or – in the deuterated compound HBT-D (Fig. 15.2) – an O–D$\cdots$N bond is formed with a strongly red-shifted and broadened hydrogen/deuterium donor stretching band. In this enol ground state, the hydrogen bond dynamics have been studied in mid-infrared pump–probe experiments with a time resolution of 100 fs. The pump pulse created a vibrational excitation on the O–H or O–D stretching band and the resulting change of O–H/O–D stretching absorption was measured by weak probe pulses [31]. In Fig. 15.3, results of a pump–probe study of HBT-D dissolved in toluene are summarized. The spectrally

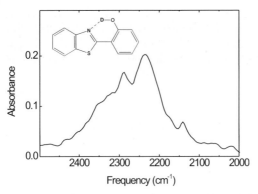

Figure 15.2 Molecular structure of HBT-D together with the linear O–D stretching band (concentration of HBI in toluene c = 0.15 M).

resolved change of vibrational absorption [Fig. 15.3 (a)] exhibits a decrease in absorption (bleaching) with a maximum at 2210 cm^{-1} and an enhanced absorption at higher frequencies. The bleaching is due to the depletion of the $v_{OD} = 0$ state and stimulated emission from the $v_{OD} = 1$ state, whereas the blue-shifted absorption originates from a hot ground state formed by femtosecond relaxation of the $v = 1$ state of the O–D stretching oscillator. In the hot ground state, the $v_{OD} = 1$ state has been depopulated and other anharmonically coupled vibrations have accepted the excess energy supplied by the pump pulse. On a time scale of several tens of picoseconds, the hot ground state cools by energy transfer to the surrounding solvent.

In Fig. 15.3 (b), the change in vibrational absorption is plotted as a function of pump–probe delay for two different frequency positions in the probe spectrum. The signals at negative delay times and around delay zero are dominated by the perturbed free induction decay of the vibrational polarization and the coherent pump–probe coupling. At positive delay times, i.e., for a sequential interaction of the molecules with pump and probe pulses, the transients exhibit strong oscillations with a period of 280 fs. Oscillations with this period occur throughout the O–D stretching band. The corresponding Fourier transform [Fig. 15.3 (c), lower trace] peaks at 120 cm^{-1}.

The electric field envelope of the femtosecond pump pulse which is short compared to the period of the oscillations in Fig. 15.3 (b) covers a frequency range much broader than the energy spacing of individual levels of the low-frequency mode. In other words, the pump spectrum overlaps with several lines of the vibrational progression depicted in Fig. 15.1 (b). As a result, impulsive dipole excitation from the $v_{OD} = 0$ to 1 state creates a nonstationary superposition of the wavefunctions of low-frequency levels in the $v_{OD} = 1$ state with a well-defined mutual phase. This quantum-coherent wavepacket oscillates in the $v_{OD} = 1$ state with the frequency Ω of the low-frequency mode and leads to a modulation of O–H stretching absorption which is measured by the probe pulses. In addition to the wavepacket in the $v_{OD} = 1$ state, impulsive Raman excitation within the spectral envelope of

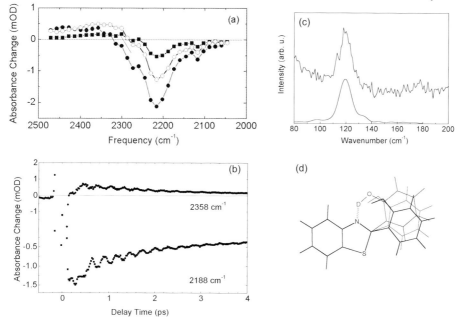

Figure 15.3 (a) Transient O–D absorption spectra of HBT-D for delay times of 0.3 ps (points), 1 ps (circles), and 4 ps (squares). The change of absorbance $\Delta A = -\log(T/T_0)$ is plotted as a function of probe frequency (T,T_0: transmission with and without excitation). (b) Time-resolved absorbance changes measured at probe frequencies of 2188 and 2358 cm^{-1}. Pronounced oscillatory signals are observed. (c) Fourier transform of the oscillatory absorbance changes of (b) (lower trace) and resonance Raman spectrum of HBT displaying a low-frequency band at 120 cm^{-1} (upper trace). (d) Microsopic elongations connected with a wavepacket motion along the 120 cm^{-1} in-plane mode.

the pump pulse creates a wavepacket in the $\nu_{OD} = 0$ state, also undergoing oscillatory motion and modulating the O–H stretching absorption.

The oscillatory absorbance change is observed over a period of 1 to 2 ps, pointing to a comparably slow vibrational dephasing, i.e., loss of mutual phase of the wavefunctions contributing to the underlying wavepacket. The wavepacket in the $\nu_{OD} = 1$ state is damped effectively by population relaxation on a time scale of several hundreds of femtoseconds and, consequently, makes a minor contribution to the long-lived oscillations. In contrast, the wavepacket in the $\nu_{OD} = 0$ state is exclusively damped by fluctuating forces exerted by the liquid environment and/or other intramolecular modes of HBT. Obviously, such damping is comparatively weak, resulting in an underdamped character of the low-frequency motions. The phase of the oscillatory pump–probe signal displays a change by approximately π at the maximum of the O–D stretching band, even for pump pulses centered in the wing of the linear absorption band [31]. This finding demonstrates a resonant enhancement of the Raman generation process by the O–D stretching transition moment, pointing again to a strong anharmonic coupling of the low- and high-frequency modes.

The frequency of the coherent motions agrees very well with the position of a low-frequency Raman band of HBT [Fig. 15.3 (c), upper trace]. Motion along this mode is connected with the microscopic elongations shown in Fig. 15.3 (d) which lead to a strong modulation of the geometry of the intramolecular hydrogen bond. Thus, our time-resolved data give a direct real-time image of hydrogen bond motions.

The impulsive excitation scheme of low-frequency wavepackets applied here is not mode-specific. In principle, all modes displaying a finite anharmonic coupling to the O–H/O–D stretching mode and a vibrational frequency which is smaller than or comparable to the pump bandwidth are excited. This subset of modes can include vibrations not affecting the hydrogen bond geometry directly. In HBT, low-frequency wavepacket motion is dominated by a single mode modulating the length and strength of the hydrogen bond. In the next section, acetic acid dimers displaying coherent motions along several low-frequency modes will be discussed.

15.3.2
Hydrogen Bonded Dimers

Cyclic dimers of carboxylic acids represent important model systems forming two coupled intermolecular hydrogen bonds [Fig. 15.1 (d), inset of Fig. 15.4]. The linear vibrational spectra of carboxylic acid dimers have been studied in detail, both in the gas and the liquid phase, and a substantial theoretical effort has been undertaken to understand the line shape of their O–H and/or O–D stretching bands. In contrast, there have been only a few experiments on the nonlinear vibrational

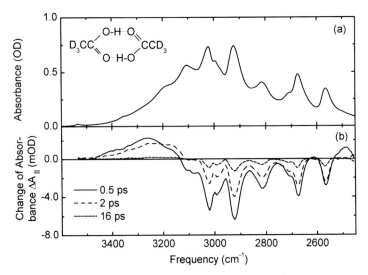

Figure 15.4 (a) Linear O–H stretching band of cyclic acetic acid dimers. (b) Transient vibrational absorption spectra measured for different pump–probe delays. The change of vibrational absorbance $\Delta A_{\parallel}$ for pump and probe pulses of parallel linear polarization is plotted as a function of the probe frequency.

response. The coupling of the two carbonyl oscillators in acetic acid dimers has been investigated by femtosecond pump–probe and photon-echo measurements [17] and vibrational relaxation following O–H stretching excitation has been addressed in picosecond pump–probe studies [32]. In the following, recent extensive pump–probe studies of cyclic acetic acid dimers in the femtosecond time domain are presented [16, 33, 34].

Dimer structures containing two O–H$\cdots$O (OH/OH dimer) or two O–D$\cdots$O (OD/OD dimer) hydrogen bonds were dissolved in CCl_4 at concentrations between 0.2 and 0.8 M. Two-color pump–probe experiments with independently tunable pump and probe pulses were performed with a 100 fs time resolution. Approximately 1% of the dimers present in the sample volume were excited by the 1 µJ pump pulse. After interaction with the sample, the probe pulses were spectrally dispersed to measure transient vibrational spectra with a spectral resolution of 6 cm^{-1}.

The steady state and the transient O–H stretching absorption spectra of OH/OH dimers are displayed in Fig. 15.4 (a) and (b), respectively. The transient spectra show a strong bleaching in the central part of the steady-state band and enhanced absorption on the red and blue wing. The bleaching which consists of a series of comparably narrow spectral dips, originates from the depopulation of the $v_{OH} = 0$ state and stimulated emission from the $v_{OH} = 1$ state. The enhanced absorption at small frequencies is due to the red-shifted $v_{OH} = 1 \rightarrow 2$ transition and decays by depopulation of the $v_{OH} = 1$ state with a lifetime of approximately 200 fs. The enhanced absorption on the blue side is caused by the vibrationally hot ground state formed by relaxation of the $v_{OH} = 1$ state, similar to the behavior discussed for HBT-D. This transient absorption decays by vibrational cooling on a 10 to 50 ps time scale. Transient spectra have also been measured for the OD/OD and the mixed OH/OD dimers – both on the O–H and O–D stretching bands – and display very similar behavior.

The time evolution of the nonlinear O–H stretching absorption shows pronounced oscillatory signals for all types of dimers studied. In Fig. 15.5, data for OD/OD dimers are presented which were recorded at 3 different spectral positions in the O–D stretching band. For positive delay times, one finds rate-like kinetics which is due to population and thermal relaxation of the excited dimers and, more importantly, superimposed by very strong oscillatory absorption changes. In contrast to the intramolecular hydrogen bonds discussed above, the time-dependent amplitude of the oscillations displays a slow modulation with an increase and a decrease on a time scale of several hundreds of femtoseconds. Such features of a beatnote demonstrate the presence of more than one oscillation frequency. In Fig. 15.6 (a), the Fourier transforms of the oscillatory signals are plotted for the 3 spectral positions. There are 3 prominent frequency components, a strong doublet with maxima at 145 and 170 cm^{-1} and a much weaker component around 50 cm^{-1}. Comparative pump–probe studies of OH/OH dimers reveal a similar doublet at 145 and 170 cm^{-1} with slightly changed relative intensities of the two components. The 50 cm^{-1} component is practically absent in the OH/OH case.

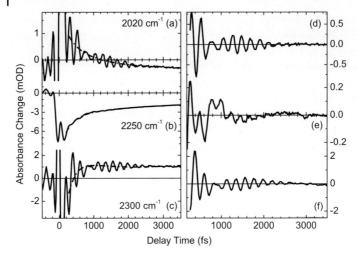

Figure 15.5 (a) – (c) Time-resolved change of O–D stretching absorbance as a function of the pump–probe delay for 3 different probe frequencies (solid lines). Around delay zero, coherent pump–probe coupling leads to a strong signal. The absorbance changes for positive delay times consist of rate-like components due to population relaxation of the O–D stretching oscillator and oscillatory contributions. Dash-dotted lines: Numerical fits of the rate-like signals. (d) – (f) Oscillatory signals after subtraction of the rate-like components. The oscillations are due to coherent wavepacket motions along several low-frequency modes.

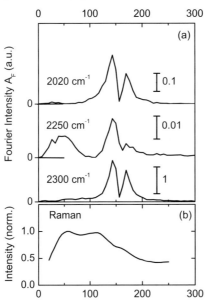

Figure 15.6 (a) Fourier spectra of the oscillatory absorbance changes of Fig. 15.5 (d) – (f). The spectra are scaled relative to each other and display 3 low-frequency modes. (b) Low-frequency spontaneous Raman spectrum of acetic acid (taken from Ref. [35]).

The two stretching oscillators in the OH/OH and OD/OD dimers should display an excitonic coupling resulting in a splitting of their $v = 1$ states, on top of the anharmonic coupling to low-frequency modes. In the linear absorption spectrum of the ensemble of dimers, this results in two separate low-frequency progressions originating from the $v_Q = 0$ and $v_Q = 1$ levels in the $v = 0$ state of the stretching vibrations [cf. Fig. 15.1 (d)]. In thermal equilibrium, a particular dimer populates only one of the v_Q levels at a certain instant in time and, thus, only one of the progressions can be excited. Consequently, a quantum coherent nonstationary superposition of the split $v_{OH} = 1$ states of the stretching mode cannot be excited in an individual dimer and quantum beats due to excitonic coupling are absent in the pump–probe signal. This behavior is also evident from the identical oscillatory response of OD/OD and OH/OD dimers, the latter displaying negligible excitonic coupling because of the large frequency mismatch between the O–H and the O–D stretching oscillator.

A contribution of quantum beats between states split by Fermi resonances can also be ruled out. There are different Fermi resonances within the O–H and O–D stretching bands [13, 21]. Depending on the spectral positions of pump and probe, this should lead to a variation of the oscillation frequencies, in particular when comparing O–H and O–D stretching excitations. Such behavior is absent in the experiment demonstrating identical oscillation frequencies for O–H and O–D stretching excitation which remain unchanged throughout the respective stretching band.

The oscillatory absorption changes are due to coherent wavepacket motions along several low-frequency modes which anharmonically couple to the stretching modes. Wavepackets in the $v = 0$ state of the O–H or O–D stretching oscillators which are generated through an impulsive resonantly enhanced Raman process, govern the oscillatory response whereas wavepackets in the $v = 1$ states are strongly damped by the fast depopulation processes. Low-frequency modes of acetic acid have been studied in a number of Raman experiments. The spectrum in Fig. 15.6 (b) was taken from Ref. [35] and displays three maxima around 50, 120 and 160 cm^{-1}. The number of subbands in such strongly broadened spectra and their assignment have remained controversial [36]. Recently, the character of the different low-frequency modes and their anharmonic coupling to the O–H stretching mode have been studied in normal mode calculations based on density functional theory [16]. In Fig. 15.7 (a), the calculated Raman transitions (solid bars) and the respective cubic force constants for coupling to the hydrogen stretching mode (hatched bars) are shown for the OH/OH and OD/OD dimers. There are four vibrations, the methyl torsion at 44 cm^{-1} [Fig. 15.7 (b)], the out-of-plane wagging mode at 118 cm^{-1}, the in-plane bending mode around 155 cm^{-1} [Fig. 15.7 (c)], and the dimer stretching mode at 174 cm^{-1} [Fig. 15.7 (d)]. In this group, the in-plane bending and the dimer stretching modes couple strongly to the hydrogen/deuterium stretching mode via a third-order term in the vibrational potential that dominates compared to higher order terms. The coupling of the methyl torsion is much weaker, that of the out-of-plane wagging mode even negligible. Such theoretical results are in good agreement with the experimental find-

(a)

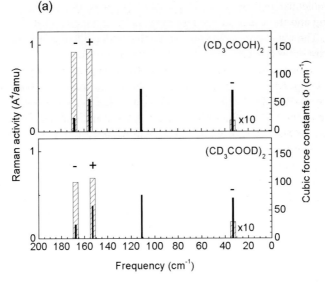

(b)

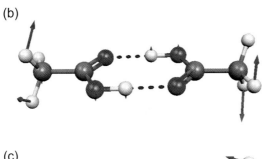

(c)

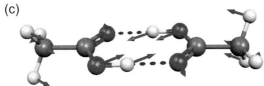

(d)

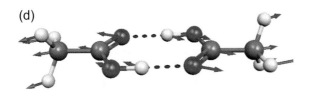

Figure 15.7 (a) Calculated low-frequency Raman spectra (solid bars, left ordinate) and cubic force constants Φ describing the coupling to the O–H or O–D stretching modes (hatched bars, right ordinate scale). Plus and minus signs indicate the sign of the force constants. (b) – (d) Microscopic elongations of the methyl torsion at 50 cm^{-1}, the dimer in-plane bending at 155 cm^{-1} and the dimer stretching at 170 cm^{-1}.

ings: the strong doublet in the Fourier spectra [Fig. 15.6 (a)] is assigned to the in-plane dimer bending and the dimer stretching, the weak band around 50 cm^{-1} to the methyl torsion. The out-of-plane wagging is not observed at all. It should be noted that the spectra derived from the oscillatory pump–probe signals, i.e., time domain data, allow a much better separation of the low-frequency modes coupling than the steady-state spontaneous Raman spectra. The calculated anharmonic couplings Φ of the O–H/O–D stretching vibrations and the 3 low-frequency modes observed are of the same order of magnitude as the couplings calculated for Fermi resonances between the $v_{OH} = 1$ state and combination and overtones of the O–H bending and other fingerprint modes [21].

In conclusion, the results presented here demonstrate how nonlinear pump–probe spectroscopy allows isolation of the anharmonic couplings of hydrogen bond modes and the O–H/O–D stretching mode. Such couplings underlie oscillatory wavepacket motions contributing to the pump–probe signals, whereas excitonic couplings and Fermi resonances play a minor role. The results for cyclic acetic acid dimers demonstrate coherent intermolecular motions for several picoseconds. This should allow the generation of tailored vibrational wavepackets by excitation with phase-shaped infrared pulses and may pave the way towards controlled infrared-induced hydrogen transfer in the electronic ground state.

15.4
Low-frequency Motions in Excited State Hydrogen Transfer

Transient vibrational spectra of electronically excited molecules give insight into local changes of molecular geometries due to photoinduced hydrogen transfer which occur in systems like HBT. An early picosecond infrared study of the photo-induced enol-keto transformation of HBT [Fig. 15.8 (a)] has revealed new vibrational bands at 1535 cm^{-1} and 2900 cm^{-1} [7]. The band at 1540 cm^{-1} was attributed to the stretching vibration of the carbonyl (C=O) group formed by hydrogen transfer and being part of a strong hydrogen bond with the newly formed N–H group of the keto tautomer. The comparatively low frequency of the new band is due to the fact that this mode involves, in addition to the carbonyl stretch, elongations of bonds in the phenyl ring. Correspondingly, the band around 2900 cm^{-1} was interpreted as an N–H stretching band. The new bands were formed within the time resolution of the experiment of 5 ps, pointing to a much faster hydrogen transfer process.

A similar study with substantially improved time resolution has been reported recently [8, 37, 38]. In such experiments, HBT was excited to the enol-S_1 state by a 40 fs pulse at 350 nm, i.e., nearly resonant to the S_0–S_1 transition. Transient vibrational spectra were measured with 100 fs mid-infrared probe pulses which were spectrally dispersed after interaction with the sample. Such spectra are displayed in Fig. 15.8 (b) for different delay times, together with the stationary vibrational spectrum in the enol ground state of HBT [Fig. 15.8 (c)]. In agreement with Ref. [7], the spectra display a prominent new band around 1530 cm^{-1} with a spectral

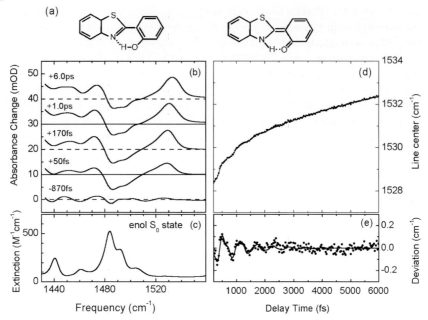

Figure 15.8 (a) Molecular structures of the enol (left) and keto (right) tautomer of HBT. (b) Transient vibrational spectra of HBT after femtosecond excitation of the enol tautomer at 335 nm. The change in vibrational absorbance in mOD is plotted as a function of probe frequency for different time delays after electronic excitation. The spectra show the build-up of the carbonyl stretching band of the keto tautomer at 1535 cm⁻¹. The absorbance changes at lower frequencies are due to skeletal modes. (c) Ground state vibrational spectrum of enol HBT. (d) Frequency position of the center of the carbonyl stretching band as a function of delay time. Data (points) and numerical rate equation fit of the blue shift with time constants of 0.5 and 5 ps (solid line). (e) Deviation of the center position from the rate-like blue-shift of (d) (points). The solid line represents a calculated oscillatory response with oscillation frequencies of 60 and 120 cm⁻¹ and a damping time of 1 ps.

width of about 15 cm⁻¹. The formation of this band occurs with a delay of 30 to 50 fs, representing a measure of the hydrogen transfer time. With increasing time delay, this band shows a continuous blue shift by about 5 cm⁻¹ which is superimposed by weak oscillations of the line center position [Fig. 15.8 (d) and (e)]. The oscillations consist of a prominent 120 cm⁻¹ frequency component and a weak 60 cm⁻¹ contribution.

The carbonyl stretching and other fingerprint bands of keto HBT show only minor changes of their spectral envelope and the spectrally integrated absorption as a function of time. In particular, contributions from the $v = 1 \rightarrow 2$ transition of the respective mode are absent. Thus, the carbonyl group of the keto tautomer is formed without excitation of its stretching motion and the fingerprint modes at frequencies between 1000 and 1500 cm⁻¹ remain in their $v = 0$ states. On the other hand, the energy difference between the enol S_1 and the keto S_1 state of approximately 3000 cm^{-1} is released as excess energy in the hydrogen transfer process. Such excess energy is mainly contained in low-frequency modes of the keto species, as has been discussed in detail in Ref. [8].

Resonance Raman studies of HBT with excitation in the range of the S_0–S_1 absorption band have demonstrated large Franck–Condon factors of the in-plane mode at 120 cm^{-1} as well as other low-frequency modes at 266, 293, 505, and 537 cm^{-1} [39]. Thus, the 120 cm^{-1} oscillator is elongated upon photoexcitation to the S_1 state and undergoes an underdamped oscillatory wavepacket motion. The latter becomes visible through oscillations of the frequency position of the keto carbonyl stretching vibration to which the 120 cm^{-1} mode couples anharmonically. As the carbonyl stretching mode is a clear signature of the keto reaction product formed in the excited state, a potential contribution of coherent motions in the enol ground state (created through a pump-induced Raman process) is not visible. It is important to note that the low-frequency oscillations persist much longer than the hydrogen transfer time of 30 to 50 fs, showing that the transfer reaction does not result in a damping of such in-plane motion.

In an indirect approach not providing structural information, femtosecond excited state hydrogen transfer has been studied via the transient electronic spectra of the initial and the product species [38, 40–43] . In most cases, the predominant ground state species was excited in the range of its S_0'–S_1' absorption band and the onset of stimulated emission or fluorescence [44] on the S_1–S_0 transition of the product species was monitored. Some of such studies have been performed with sub-30 fs pump and probe pulses, i.e., a somewhat higher time resolution than the infrared experiment discussed above. Oscillatory signals superimposed on the rise of product emission have been observed for the first time with the benzotriazole compound TINUVIN P [41]. In TINUVIN P, oscillation frequencies of 250 and 170 cm^{-1} were found, reflecting quantum-coherent wavepacket motions along two low-frequency modes with large Franck–Condon factors. Due to the coupling of those modes to the electronic transition of the keto-type product species, the product emission exhibits oscillations. The two modes observed strongly modulate the intramolecular hydrogen bonding geometry. In HBT, systematic pump–probe studies show a femtosecond rise time of keto emission depending on the spectral position within the emission band, i.e., ranging from about 60 fs at 530 nm to 170 fs around 650 nm [40, 42, 43]. Such kinetics are superimposed by oscillatory wavepacket motions with frequencies of 118, 254, 289 and 529 cm^{-1} [42, 43].

The hydrogen transfer occurring on a 50 fs time scale points to an essentially barrierless excited state potential along the reaction pathway. The transfer appears, however, much slower than the period of the O–H stretching vibration of approximately 10 fs. This fact demonstrates that hydrogen transfer does not involve a simple stretching motion towards the acceptor atom but requires motion along vibrational modes at low frequencies. Taking into account both femtosecond pump–probe data and the results of resonance Raman studies, the following qualitative picture of excited state hydrogen transfer emerges:

1. Excitation of the enol species to the S_1 state induces a redistribution of electronic charge. For a strong electronic coupling between the initially excited vibronic states and the keto excited state, this charge redistribution occurs on a time scale much faster than 20 fs and establishes an excited state potential energy surface with a minimum for the keto-type configuration of the molecule.

2. The initial dynamics of hydrogen transfer on this potential energy surface are determined by the propagation of the vibrational wavepacket created upon electronic excitation. This wavepacket is made up of Raman active modes with high Franck–Condon factors which are excited within the spectral width of the electric field envelope of the pump pulse. In particular, low-frequency modes including the 120 cm^{-1} in-plane vibration contribute. In contrast, the O–H stretching mode with a negligible Franck-Condon factor is not part of the wavepacket. The low-frequency wavepacket oscillations persist for 1–2 ps, demonstrating that the vibrational potential of such modes is not changed significantly upon hydrogen transfer.

3. The non-instantaneous (30–60 fs) rise of both the carbonyl stretching absorption and the keto emission shows that the excited state reaction pathway involves propagation along low-frequency modes. In the initial Franck–Condon window where the wavepacket is created by electronic excitation, a barrier exists along the hydrogen coordinate preventing a direct hydrogen transfer along this coordinate. With increasing time, the motion of the wavepacket along individual or a combination of low-frequency modes brings the system into a range of the excited state potential where a barrierless channel exists for motion along a high-frequency coordinate. In this range, the hydrogen is transferred from the enol to the keto configuration. The overall time for hydrogen transfer is set by a fraction of the period of a low-frequency mode. The fact that the 120 cm^{-1} mode with a period of 280 fs strongly modulates the hydrogen bond geometry and displays a pronounced anharmonic coupling to both the O–H stretching mode of enol-HBT (cf. Section 15.3.1) and the carbonyl stretching of keto-HBT suggests a prominent role of this mode in hydrogen transfer.

4. Hydrogen transfer represents a non-reversible reaction with a quantum yield close to 100%, i.e., there is no return to the enol geometry after the fast formation of the keto product. Calculations of wavepacket propagation in the S_1 state, assuming harmonic potentials for the modes contributing to

the initial wavepacket, demonstrate a substantial spreading of the wavepacket on the 30–50 fs time scale. In addition, intramolecular vibrational relaxation leads to a transfer of excitation into a multitude of other modes, corresponding to a multidimensional dephasing process. Both mechanisms stabilize the keto product, even though the directly excited low-frequency modes involved in the reaction continue to oscillate for periods much longer than the hydrogen transfer time. The weak 60 cm^{-1} frequency component present in the oscillations of Fig. 15.8 (e) is due to an underdamped mode which is not Raman active and, thus, not elongated upon electronic excitation. Instead, it is excited by vibrational redistribution on a time scale shorter than the vibrational period of 550 fs. At later times, transient populations of a larger manifold of modes also underlie the blue-shift of the fingerprint modes mediated via anharmonic couplings [8, 45].

15.5
Conclusions

The results discussed in this chapter demonstrate a prominent role of vibrational low-frequency quantum coherences for the structural and reactive dynamics of hydrogen bonds in the liquid phase. Underdamped oscillatory motions of modes directly affecting the hydrogen bonding geometry have been induced via vibrational excitation of the hydrogen donor stretching mode in the electronic ground state or via electronic excitation to the S_1 state of molecules undergoing ultrafast intramolecular hydrogen transfer. In the electronic ground state, femtosecond excitation of the O–H or O–D stretching vibrations generates a nonstationary coherent superposition of several quantum states of a low-frequency mode that couples anharmonically to the fast stretching vibration. In the intramolecular hydrogen bonds investigated, coherent motions are dominated by a single mode, whereas motions along several underdamped modes have been found in hydrogen-bonded dimers of acetic acid. The coherent low-frequency response is dominated by wavepackets in the $\nu_{OH} = 0$ state which are created through a Raman process resonantly enhanced by the O–H stretching transition dipole. The occurrence of wavepacket motions also confirms the much debated picture of vibrational low-frequency progressions within the strongly broadened O–H stretching bands, as introduced in the early theoretical literature on linear vibrational spectra. The picosecond decay of low-frequency coherences allows generation and manipulation of vibrational motion with phase-shaped infrared pulses. This may be of particular interest for reactive systems in which processes of hydrogen transfer along hydrogen bonds occur and may become accessible for optical control.

In excited state hydrogen transfer occurring on sub-100 fs time scales, Raman-active low-frequency modes that couple strongly to the electronic S_0–S_1 transition, are part of the reaction coordinate. Quantum coherent propagation along such coordinates sets the time scale for the intramolecular transfer of the hydrogen, creating new molecular structure without significant excitation of high-frequency vibrational modes. The excess energy released in the reaction is contained in low-frequency vibrations of the product species. Spreading of the vibronic wavepacket and vibrational relaxation are crucial for the stabilization of the reaction product. This qualitative picture describes hydrogen transfer along a pre-existing intramolecular hydrogen bond for a larger class of molecular systems. A quantitative description, however, requires a more detailed analysis of nuclear motions and anharmonic couplings in the electronically excited state.

Acknowledgements

I would like to acknowledge the important contributions of my present and former coworkers Jens Stenger, Dorte Madsen, Nils Huse, Karsten Heyne, Jens Dreyer, Peter Hamm, and Erik Nibbering to the work reviewed in this chapter. It is my pleasure to thank Casey Hynes for many interesting discussions. I also thank the Deutsche Forschungsgemeinschaft and the Fonds der Chemischen Industrie for financial support.

References

1 P. Schuster, G. Zundel, C. Sandorfy (Eds.), *The Hydrogen Bond: Recent Developments in Theory and Experiment, Vol. I–III*, North Holland, Amsterdam, **1976**.

2 T. Elsaesser, H. J. Bakker (Eds.), *Ultrafast Hydrogen Bonding Dynamics and Proton Transfer Processes in the Condensed Phase*, Kluwer, Dordrecht, **2002**.

3 S. Mukamel, *Principles of Nonlinear Optical Spectroscopy*, Oxford University Press, Oxford, **1995**.

4 M. D. Fayer (Ed.), *Ultrafast Infrared and Raman Spectroscopy*, Marcel Dekker, New York, **2001**.

5 E. T. J. Nibbering, T. Elsaesser, *Chem. Rev.* **2004**, *104*, 1887–1914.

6 T. Elsaesser, W. Kaiser, W. Lüttke, *J. Phys. Chem.* **1986**, *90*, 2901–2905.

7 T. Elsaesser, W. Kaiser, *Chem. Phys. Lett.* **1986**, *128*, 231–237.

8 M. Rini, J. Dreyer, E.T.J. Nibbering, T. Elsaesser, *Chem. Phys. Lett.* **2003**, *374*, 13–19.

9 C. Sandorfy, *Bull. Pol. Acad. Sci: Chem.* **1995**, *43*, 7–24.

10 W. Mikenda, S. Steinbock, *J. Mol. Struct.* **1996**, *384*, 159–163.

11 B. I. Stepanov, *Nature* **1946**, *157*, 808–810.

12 Y. Maréchal, A. J. Witkowski, *J. Chem. Phys.* **1968**, *48*, 3697–3705.

13 Y. Maréchal, *J. Chem. Phys.* **1987**, *87*, 6344–6353.

14 O. Henri-Rousseau, P. Blaise, D. Chamma, *Adv. Chem. Phys.* **2002**, *121*, 241–309.

15 D. Chamma, O. Henri-Rousseau, *Chem. Phys.* **1999**, *248*, 53–70.

16 K. Heyne, N. Huse, J. Dreyer, E. T. J. Nibbering, T. Elsaesser, S. Mukamel, *J. Chem. Phys.* **2004**, *121*, 902–913.

17 M. Lim, R. M. Hochstrasser, *J. Chem. Phys.* **2001**, *115*, 7629–7643.

18 G. M. Florio, T. S. Zwier, E. M. Myshakin, K. D. Jordan,

E. L. Sibert III, *J. Chem. Phys.* **2003**, *118*, 1735–1746.

19 C. Emmeluth, M. A. Suhm, D. Luckhaus, *J. Chem. Phys.* **2003**, *118*, 2242–2255.

20 N. Huse, B. D. Bruner, M. L. Cowan, J. Dreyer, E. T. J. Nibbering, R. J. D. Miller, T. Elsaesser, *Phys. Rev. Lett.*, submitted.

21 J. Dreyer, *J. Chem. Phys.*, submitted.

22 D. W. Oxtoby, *Adv. Chem. Phys.* **1979**, *40*, 1–48.

23 N. Rösch, M. Ratner, *J. Chem. Phys.* **1974**, *61*, 3344–3351.

24 S. Bratos, *J. Chem. Phys.* **1975**, *63*, 3499–3509.

25 G. N. Robertson, J. Yarwood, *J. Chem. Phys.* **1978**, *32*, 267–282.

26 R. Rey, K. B. Moller, J. T. Hynes, *Chem. Rev.* **2004**, *104*, 1915–1928.

27 J. Stenger, D. Madsen, J. Dreyer, E. T. J. Nibbering, P. Hamm, T. Elsaesser, *J. Phys. Chem. A* **2001**, *105*, 2929–2932.

28 D. Madsen, J. Stenger, J. Dreyer, P. Hamm, E. T. J. Nibbering, T. Elsaesser, *Bull. Chem. Soc. Jpn.* **2002**, *75*, 909–917.

29 H. Naundorf, G. A. Worth, H. D. Meyer, O. Kühn, *J. Phys. Chem. A* **2002**, *106*, 719–724.

30 O. Kühn O, H. Naundorf, *Phys. Chem. Chem. Phys.* **2003**, *5*, 79–86.

31 D. Madsen, J. Stenger, J. Dreyer, E. T. J. Nibbering, P. Hamm, T. Elsaesser, *Chem. Phys. Lett.* **2001**, *341*, 56–62.

32 G. Seifert, T. Patzlaff, H. Graener, *Chem. Phys. Lett.* **2001**, *333*, 248–254.

33 K. Heyne, N. Huse, E. T. J. Nibbering, T. Elsaesser, *Chem. Phys. Lett.* **2003**, *369*, 591–596.

34 K. Heyne, N. Huse, E. T. J. Nibbering, T. Elsaesser, *J. Phys: Condens. Matter* **2003**, *15*, S129–S136.

35 O. F. Nielsen, P. A. Lund, *J. Chem. Phys.* **1983**, *78*, 652–655.

36 T. Nakabayashi, K. Kosugi, N. Nishi, *J. Phys. Chem. A* **1999**, *103*, 8595–8603.

37 M. Rini, A. Kummrow, J. Dreyer, E. T. J. Nibbering, T. Elsaesser, *Faraday Discuss.* **2002**, *122*, 27–40.

38 T. Elsaesser, in: *Ultrafast Hydrogen Bonding Dynamics and Proton Transfer Processes in the Condensed Phase*, T. Elsaesser, H.J. Bakker (Eds.), Kluwer, Dordrecht, **2002**, pp. 119–153.

39 M. Pfeiffer, K. Lenz, A. Lau, T. Elsaesser, T. Steinke, *J. Raman Spectrosc.* **1997**, *28*, 61–72.

40 F. Laermer, T. Elsaesser, W. Kaiser, *Chem. Phys. Lett.* **1988**, *148*, 119–124.

41 C. Chudoba, E. Riedle, M. Pfeiffer, T. Elsaesser, *Chem. Phys. Lett.* **1996**, *263*, 622–628.

42 S. Lochbrunner, A. J. Wurzer, E. Riedle, *J. Chem. Phys.* **2000**, *112*, 10699–10702.

43 S. Lochbrunner, A. J. Wurzer, E. Riedle, *J. Phys. Chem. A* **2003**, *107*, 10580–10590.

44 D. Marks, H. Zhang, P. Borowicz, A. Grabowska, M. Glasbeek, *Chem. Phys. Lett.* **1999**, *309*, 19–25.

45 P. Hamm, S. M. Ohline, W. Zinth, *J. Chem. Phys.* **1997**, *106*, 519–529.

16
Proton-Coupled Electron Transfer: Theoretical Formulation and Applications

Sharon Hammes-Schiffer

16.1
Introduction

The coupling between proton and electron transfer plays an important role in a wide range of chemical and biological processes, including photosynthesis [1–7], respiration [8, 9], and numerous enzyme reactions [10]. The coupled transfer of protons and electrons is denoted proton-coupled electron transfer (PCET) [11–14]. In general, the electron and proton can transfer between different centers and can transfer either in the same direction or in different directions. A variety of model systems have been investigated experimentally to elucidate the general mechanisms of PCET reactions [15–20]. The theoretical description of these types of PCET reactions is challenging due to the quantum nature of the transferring electron and proton, the importance of nonadiabatic effects, and the wide range of timescales associated with the reaction. A number of theoretical approaches have been developed to address these challenges [21–36].

This chapter presents a general theoretical formulation for PCET and summarizes the applications of this theory to a wide range of experimentally relevant systems. Section 16.2 reviews the fundamental physical concepts of PCET reactions and discusses approaches for inclusion of the proton donor–acceptor motion, explicit molecular solvent and protein, and the corresponding dynamical effects. Section 16.3 provides an overview of theoretical studies of PCET reactions in solution and in proteins. General conclusions are given in Section 16.4.

Hydrogen-Transfer Reactions. Edited by J. T. Hynes, J. P. Klinman, H. H. Limbach, and R. L. Schowen
Copyright © 2007 WILEY-VCH Verlag GmbH & Co. KGaA, Weinheim
ISBN: 978-3-527-30777-7

16.2
Theoretical Formulation for PCET

16.2.1
Fundamental Concepts

PCET systems involve a wide range of timescales associated with the active electrons (i.e., the transferring electron and the bonding electrons in the proton transfer interface), transferring proton(s), donor and acceptor groups, and solvent electrons and nuclei. In our theoretical formulation [26–28], the active electrons and transferring proton are treated quantum mechanically. The solvent is described with either a dielectric continuum or an explicit molecular representation. The Born–Oppenheimer approach [37], which assumes that the solvent electronic degrees of freedom are infinitely fast relative to all other degrees of freedom, is adopted. The electron donor and acceptor are assumed to be fixed in space, which is a reasonable approximation for systems in which the electron donor and acceptor consist of heavy groups such as metal complexes. The motion of the proton donor and acceptor has been included at a number of different levels [30]. The effects of additional intramolecular vibrations of the solute have also been included [27, 38–40].

In our theoretical formulation for PCET [26, 27], the electronic structure of the solute is described in the framework of a four-state valence bond (VB) model [41]. The most basic PCET reaction involving the transfer of one electron and one proton may be described in terms of the following four diabatic electronic basis states:

$$
\begin{aligned}
(1a)\ &D_e^- \cdots\cdots D_p H^+ \cdots\cdots A_p \cdots\cdots A_e \\
(1b)\ &D_e^- \cdots\cdots D_p \cdots\cdots {}^+HA_p \cdots\cdots A_e \\
(2a)\ &D_e \cdots\cdots D_p H^+ \cdots\cdots A_p \cdots\cdots A_e^- \\
(2b)\ &D_e \cdots\cdots D_p \cdots\cdots {}^+HA_p \cdots\cdots A_e^-
\end{aligned}
\tag{16.1}
$$

where 1 and 2 denote the electron transfer (ET) state, and a and b denote the proton transfer (PT) state. Given these four VB states, PT processes can be described as $1a{\rightarrow}1b$ and $2a{\rightarrow}2b$ transitions, ET processes as $1a{\rightarrow}2a$ and $1b{\rightarrow}2b$ transitions, and EPT processes as $1a{\rightarrow}2b$ and $1b{\rightarrow}2a$ transitions. Here EPT processes refer to synchronous electron and proton transfer.

The general formulation for PCET can be represented in terms of a dielectric continuum environment or an explicit molecular environment. In both representations, the free energy of the PCET system can be expressed in terms of the solute coordinates r_p and R and two scalar solvent coordinates z_p and z_e corresponding to the PT and ET reactions, respectively [26, 42, 43]. In the dielectric continuum model for the environment, the solvent or protein is represented as a dielectric continuum characterized by the electronic (ε_∞) and inertial (ε_0) dielectric constants. The scalar solvent coordinates z_p and z_e represent the differences in electrostatic interaction energies of the charge densities ρ_i corresponding to the VB basis states involved in the PT and ET reactions, respectively, with the inertial polarization potential $\Phi_{in}(\mathbf{r})$ of the solvent:

$$z_p = \int d\mathbf{r}(\rho_{1b} - \rho_{1a})\Phi_{in}(\mathbf{r})$$
$$z_e = \int d\mathbf{r}(\rho_{2a} - \rho_{1a})\Phi_{in}(\mathbf{r})$$

(16.2)

In general, these solvent coordinates depend on the solute coordinates r_p and R, but this dependence is usually very weak and can be neglected. In the molecular description of the solvent, the scalar coordinates z_p and z_e are functions of the solvent coordinates ξ and can be defined in terms of the solute–solvent interaction potential $W_s\left(r_p, R, \xi\right)$ as

$$z_p(\xi) = \langle \psi_{1b}^{el} | W_s | \psi_{1b}^{el} \rangle - \langle \psi_{1a}^{el} | W_s | \psi_{1a}^{el} \rangle$$
$$z_e(\xi) = \langle \psi_{2a}^{el} | W_s | \psi_{2a}^{el} \rangle - \langle \psi_{1a}^{el} | W_s | \psi_{1a}^{el} \rangle$$

(16.3)

where the ψ_i^{el} are the wavefunctions corresponding to the VB states defined in Eq. (16.1).

For many PCET systems, the single PT reaction is electronically adiabatic and the single ET reaction is electronically nonadiabatic. Here electronically adiabatic refers to reactions occurring in a single electronic state, and electronically non-adiabatic refers to reactions involving multiple electronic states. The electronically adiabatic (or nonadiabatic) limit corresponds to strong (or weak) electronic coupling between the charge transfer states. Even for cases in which the single ET reaction is electronically adiabatic, the overall PCET reaction is usually nonadiabatic, because the coupling between the reactant and product vibronic states is small due to averaging over the reactant and product proton vibrational wavefunctions (i.e., due to the small overlap factor, analogous to the Franck–Condon factor in theories for single ET [40, 44]). In this case, the ET diabatic free energy surfaces corresponding to ET states 1 and 2 are calculated as mixtures of the a and b PT states. The reactants (I) are mixtures of the $1a$ and $1b$ states, and the products (II) are mixtures of the $2a$ and $2b$ states. The proton vibrational states are calculated for both the reactant (I) and product (II) ET diabatic surfaces, resulting in two sets of two-dimensional vibronic free energy surfaces that may be approximated as paraboloids. In this theoretical formulation, the PCET reaction is described in terms of nonadiabatic transitions from the reactant (I) to the product (II) ET diabatic surfaces. Thus, the ET diabatic states I and II, respectively, may be viewed as the reactant and product PCET states.

The unimolecular rate expression derived in Ref. [27] for a fixed proton donor–acceptor distance is

$$k = \frac{2\pi}{\hbar} \sum_{\mu}^{\{I\}} P_{\mu}^{I} \sum_{\nu}^{\{II\}} |V_{\mu\nu}|^2 \left(4\pi\lambda_{\mu\nu}k_B T\right)^{-1/2} \exp\left(\frac{-\Delta G_{\mu\nu}^{\dagger}}{k_B T}\right) \qquad (16.4)$$

where the summations are over the reactant and product vibronic states, P_{μ}^{I} is the Boltzmann probability for state Iμ, and $\Delta G_{\mu\nu}^{\dagger}$ is the free energy barrier defined as

$$\Delta G_{\mu\nu}^{\dagger} = \frac{\left(\Delta G_{\mu\nu}^{o} + \lambda_{\mu\nu}\right)^2}{4\lambda_{\mu\nu}} \qquad (16.5)$$

In this expression the free energy of reaction is defined as

$$\Delta G_{\mu\nu}^{0} = \varepsilon_{\nu}^{II}\left(\bar{z}_{p}^{IIv}, \bar{z}_{e}^{IIv}\right) - \varepsilon_{\mu}^{I}\left(\bar{z}_{p}^{I\mu}, \bar{z}_{e}^{I\mu}\right) \qquad (16.6)$$

where $\left(\bar{z}_{p}^{I\mu}, \bar{z}_{e}^{I\mu}\right)$ and $\left(\bar{z}_{p}^{IIv}, \bar{z}_{e}^{IIv}\right)$ are the solvent coordinates for the minima of the ET diabatic free energy surfaces $\varepsilon_{\mu}^{I}\left(z_p, z_e\right)$ and $\varepsilon_{\nu}^{II}\left(z_p, z_e\right)$, respectively. Moreover, the outer-sphere reorganization energy is defined as

$$\lambda_{\mu\nu} = \varepsilon_{\mu}^{I}\left(\bar{z}_{p}^{IIv}, \bar{z}_{e}^{IIv}\right) - \varepsilon_{\mu}^{I}\left(\bar{z}_{p}^{I\mu}, \bar{z}_{e}^{I\mu}\right) = \varepsilon_{\nu}^{II}\left(\bar{z}_{p}^{I\mu}, \bar{z}_{e}^{I\mu}\right) - \varepsilon_{\nu}^{II}\left(\bar{z}_{p}^{IIv}, \bar{z}_{e}^{IIv}\right) \qquad (16.7)$$

The free energy difference and outer-sphere reorganization energy are indicated in Fig. 16.1. The coupling $V_{\mu\nu}$ in the PCET rate expression is defined as

$$V_{\mu\nu} = \left\langle \phi_{\mu}^{I} \left| V\left(r_p, z_p^{\dagger}\right) \right| \phi_{\nu}^{II} \right\rangle_p \qquad (16.8)$$

where the subscript of the angular brackets indicates integration over r_p, $z_p^{\dagger}$ is the value of z_p in the intersection region, $V\left(r_p, z_p\right)$ is the electronic coupling between states I and II, and ϕ_{μ}^{I} and ϕ_{ν}^{II} are the proton vibrational wavefunctions for the reactant and product vibronic states, respectively. For many systems [45, 46], the coupling is approximately proportional to the overlap between the reactant and product proton vibrational wavefunctions:

$$V_{\mu\nu} \approx V^{el}\left\langle \phi_{\mu}^{I} | \phi_{\nu}^{II} \right\rangle_p \qquad (16.9)$$

where V^{el} is a constant effective electronic coupling. The effects of inner-sphere solute modes have also been included in this theoretical formulation for several different regimes [27, 38–40]. In the high-temperature approximation for uncoupled solute modes, the inner-sphere reorganization energy is added to the outer-sphere reorganization energy in Eq. (16.7) [45, 47].

Despite the similarity in form, the rate expression given in Eq. (16.4) for PCET is fundamentally different than the conventional rate expression for single electron transfer with uncoupled intramolecular solute modes [40, 44]. The most fundamental difference is that the reorganization energies, equilibrium free energy

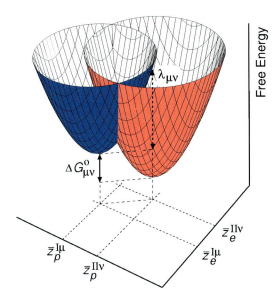

Figure 16.1 Two-dimensional vibronic free energy surfaces as functions of two collective solvent coordinates, z_p and z_e, for a PCET reaction. The lowest energy reactant and product free energy surfaces are shown. The minima for the reactant and product surfaces, respectively, are $\left(\bar{z}_p^{I\mu}, \bar{z}_e^{I\mu}\right)$ and $\left(\bar{z}_p^{II\nu}, \bar{z}_e^{II\nu}\right)$. The free energy difference $\Delta G_{\mu\nu}^0$ and outer-sphere reorganization energy $\lambda_{\mu\nu}$ are indicated.

differences, and couplings in Eq. 4 are defined in terms of two-dimensional paraboloids instead of one-dimensional parabolas. Another important difference is that the reorganization energies in Eq. 4 are different for each pair of intersecting ET diabatic surfaces due to varying positions of the minima within the reactant and product states, whereas in conventional single electron transfer theory the reorganization energy is the same for all pairs of intersecting parabolas. The final difference is that the coupling in Eq. 4 cannot be expressed rigorously as the product of a constant coupling and an overlap of the reactant and product vibrational wavefunctions because the electronic coupling depends on the proton coordinate. As mentioned above, however, this separation of the coupling is a reasonable approximation for many PCET reactions.

16.2.2
Proton Donor–Acceptor Motion

The proton donor–acceptor motion plays an important role in PCET reactions. This motion modulates the proton tunneling distance and therefore the overlap between the reactant and product proton vibrational wavefunctions. Thus, the nonadiabatic coupling between the reactant and product vibronic states for PCET reactions depends strongly on the proton donor–acceptor distance. We have derived nonadiabatic rate expressions that include the effects of the proton donor–acceptor vibrational motion in a number of different regimes [30].

In the simplest case, the R mode is characterized by a low frequency and is not dynamically coupled to the fluctuations of the solvent. The system is assumed to maintain an equilibrium distribution along the R coordinate. In this case, we can exclude the R mode from the dynamical description and consider an equilibrium ensemble of PCET systems with fixed proton donor–acceptor distances. The electrons and transferring proton are assumed to be adiabatic with respect to the R coordinate and solvent coordinates within the reactant and product states. Thus, the reaction is described in terms of nonadiabatic transitions between two sets of intersecting free energy surfaces $\varepsilon_{\mu}^{I}\left(R, z_{\mathrm{p}}, z_{\mathrm{e}}\right)$ and $\varepsilon_{v}^{II}\left(R, z_{\mathrm{p}}, z_{\mathrm{e}}\right)$ corresponding to the reactant and product electron–proton vibronic states for fixed R. For each fixed value of R, we recover the nonadiabatic rate expression given in Eq. (16.4). In this regime, all of the quantities in the rate expression, including the Boltzmann factors, depend explicitly on the coordinate R. Since the PCET system is assumed to be in equilibrium along the coordinate R, the total rate constant can be calculated by integrating Eq. (16.4) with a renormalized distribution function over all R distances:

$$
k^{\mathrm{aver}} = \frac{1}{\hbar} \int_{0}^{\infty} \mathrm{d}R \sum_{\mu}^{\{I\}} P_{\mu}^{I}(R) \sum_{v}^{\{II\}} \left| V_{\mu v}^{\neq}(R) \right|^2 \sqrt{\frac{\pi \beta}{\lambda_{\mu v}(R)}} \exp\left\{ -\frac{\beta \left[\Delta G_{\mu v}^{0}(R) + \lambda_{\mu v}(R) \right]^2}{4 \lambda_{\mu v}(R)} \right\}
$$

(16.10)

The above rate expression does not follow rigorously from the Golden Rule general expression. Nevertheless, it provides a physically reasonable method for estimating the rate constant in cases for which the dynamical coupling of the slow R mode to the solvent fluctuations is negligible.

In another limit, the R mode is characterized by a high frequency Ω and a relatively low reduced mass M. In this case, the motion along the R mode occurs on a much faster timescale than the timescale associated with the dominant solvent fluctuations, so the R mode fluctuations are dynamically uncoupled from the solvent fluctuations. In contrast to the previous case of the slow dynamically uncoupled R mode, however, the quantum character of this motion becomes important, especially at low temperatures where $\beta \hbar \Omega \gg 1$. To include these quantum effects, the R mode can be treated quantum mechanically on the same level as the electron and proton coordinates. The electrons, transferring proton, and R mode are assumed to be adiabatic with respect to the solvent coordinates within the reactant and product states. Thus, the reaction can be described in terms of nonadiabatic transitions between two sets of intersecting free energy surfaces $\varepsilon_{k}^{I}\left(z_{\mathrm{p}}, z_{\mathrm{e}}\right)$ and $\varepsilon_{l}^{II}\left(z_{\mathrm{p}}, z_{\mathrm{e}}\right)$ corresponding to the reactant and product electron–proton–R-mode states. The resulting nonadiabatic rate expression in the high-temperature limit of a Debye solvent is

$$k^{\text{quant}} = \frac{1}{\hbar} \sum_{k}^{\{I\}} P_k^I \sum_{l}^{\{II\}} |V_{kl}^{\neq}|^2 \sqrt{\frac{\pi\beta}{\lambda_{kl}}} \exp\left\{ -\frac{\beta(\Delta G_{kl}^0 + \lambda_{kl})^2}{4\lambda_{kl}} \right\} \tag{16.11}$$

This expression formally resembles the expression in Eq. (16.4) except that the quantities are calculated for pairs of mixed electron–proton–R-mode vibronic free energy surfaces.

16.2.3
Dynamical Effects

Recently, we derived rate expressions that include the dynamical effects of both the R mode and the solvent, as well as the quantum character of the R mode [30]. As mentioned above, one of the most important effects of the R coordinate motion in PCET systems is the modulation of the proton tunneling distance and thereby the nonadiabatic coupling between the reactant and product vibronic states. The fluctuations of the nonadiabatic coupling due to the R motion can be dynamically coupled to the fluctuations of the solvent degrees of freedom, which are responsible for bringing the system into the degenerate state required for nonadiabatic transitions. Here we consider the case in which the electron and transferring proton are adiabatic with respect to the R mode and solvent within the reactant and product states. The R mode is treated dynamically on the same level as the solvent modes in order to include the effects of the dynamical coupling between the R mode and the solvent.

To facilitate the derivation of analytical rate expressions, the R dependence of the overall coupling $V_{\mu\nu}$ is approximated by a single exponential:

$$V_{\mu\nu} \approx V_{\mu\nu}^{(0)} \exp\left[-a_{\mu\nu}\left(R - \bar{R}^{I\mu} \right) \right] \tag{16.12}$$

where $\bar{R}^{I\mu}$ is the equilibrium value of the R coordinate on the reactant surface Iμ, $V_{\mu\nu}^{(0)}$ is the value of the coupling at $\bar{R}^{I\mu}$, and $a_{\mu\nu}$ can be calculated from the R dependence of the coupling. The justification for this approximation is that the nonadiabatic coupling can be approximated as the product of a constant electronic coupling and a Franck–Condon overlap of the reactant and product proton vibrational wavefunctions, as given in Eq. (16.9). For PCET reactions, typically this overlap depends only weakly on the solvent coordinates but depends very strongly on the proton donor–acceptor separation R. For a simple model based on two ground state harmonic oscillator wavefunctions with centers separated by R, the overlap increases exponentially with decreasing R. The approximation in Eq. (16.12) has been shown to be reasonable for model PCET systems and was also used previously for nonadiabatic proton transfer systems [48–50].

In this case, the nonadiabatic rate constant can be expressed as

$$k^{\text{dyn}} = \sum_{\mu}^{\{I\}} P_{\mu}^{I} \sum_{\nu}^{\{II\}} k_{\mu\nu}^{\text{dyn}} \tag{16.13}$$

where the partial rate constant $k_{\mu\nu}^{\text{dyn}}$ describes nonadiabatic transitions between the quantum states for the pair of electron–proton vibronic surfaces Iμ and IIν and can be written as an integral of the time-dependent probability flux correlation function $j_{\mu\nu}(t)$:

$$k_{\mu\nu}^{\text{dyn}} = \frac{1}{\hbar^2} \int_{-\infty}^{\infty} j_{\mu\nu}(t)\mathrm{d}t \tag{16.14}$$

We have used this formulation to derive rate expressions for both a dielectric continuum and a molecular representation of the environment.

16.2.3.1 Dielectric Continuum Representation of the Environment

For a dielectric continuum environment, the probability flux correlation function is given by:

$$j^{\text{cont}}(t) = \left| V_{\mu\nu}^{(0)} \right|^2 \exp\left[\frac{\mathrm{i}}{\hbar}\left(\Delta G^0 + \lambda_z + \lambda_R \right)t \right]$$

$$\times \exp\left\{ \frac{2M\lambda_a}{\hbar^2}[C_R(0) + C_R(t)] - \frac{4\mathrm{i}}{\hbar^2} M\Omega\sqrt{\lambda_a\lambda_R} \int_0^t C_R(\tau)\mathrm{d}\tau \right.$$

$$\left. - \frac{1}{\hbar^2} \int_0^t \mathrm{d}\tau_1 \int_0^{\tau_1} \mathrm{d}\tau_2\, C_Z(\tau_1 - \tau_2) - \frac{2M\Omega^2\lambda_R}{\hbar^2} \int_0^t \mathrm{d}\tau_1 \int_0^{\tau_1} \mathrm{d}\tau_2\, C_R(\tau_1 - \tau_2) \right\} \tag{16.15}$$

Although the indices μ and ν are omitted in Eq. (16.15) for simplicity, the quantities in this expression are defined in terms of a pair of reactant and product free energy surfaces $\varepsilon_{\mu}^{I}\left(R, z_p, z_e \right)$ and $\varepsilon_{\nu}^{II}\left(R, z_p, z_e \right)$. The time correlation functions for the R mode and the solvent variables are given by $C_R(t) = \langle \delta R(0)\delta R(t)\rangle$, where $\delta R(t) = R(t) - \langle R \rangle$, and $C_Z(t) = \langle \delta Z(0)\delta Z(t)\rangle$. In the continuum solvent representation,

$$\delta Z(t) = \tilde{A}_p\left(z_p(t) - \bar{z}_p^{I\mu} \right) + \tilde{A}_e\left(z_e(t) - \bar{z}_e^{I\mu} \right) \tag{16.16}$$

$$\tilde{A}_p = A_{z_p z_p}\left(\bar{z}_p^{I\mu} - \bar{z}_p^{II\nu} \right) + A_{z_p z_e}\left(\bar{z}_e^{I\mu} - \bar{z}_e^{II\nu} \right) + A_{z_p R}\left(\bar{R}^{I\mu} - \bar{R}^{II\nu} \right) \tag{16.17}$$

$$\tilde{\Lambda}_e = \Lambda_{z_p z_e}\left(\bar{z}_p^{I\mu} - \bar{z}_p^{II\nu}\right) + \Lambda_{z_e z_e}\left(\bar{z}_e^{I\mu} - \bar{z}_e^{II\nu}\right) + \Lambda_{z_e R}\left(\bar{R}^{I\mu} - \bar{R}^{II\nu}\right) \tag{16.18}$$

$$\Lambda_{xy} = \left.\frac{\partial^2 \varepsilon_\mu^I\left(R, z_p, z_e\right)}{\partial x \partial y}\right|_{\bar{z}_p^{I\mu},\bar{z}_e^{I\mu},\bar{R}^{I\mu}} \quad , \quad x, y = R, z_p, z_e \tag{16.19}$$

In Eq. 16.16, ΔG^0 is the reaction free energy defined as the difference in the equilibrium free energies for the reactant and product surfaces:

$$\Delta G_{\mu\nu}^0 = \varepsilon_\nu^{II}\left(\bar{R}^{II\nu}, \bar{z}_p^{II\nu}, \bar{z}_e^{II\nu}\right) - \varepsilon_\mu^I\left(\bar{R}^{I\mu}, \bar{z}_p^{I\mu}, \bar{z}_e^{I\mu}\right) \tag{16.20}$$

The R mode reduced mass M and frequency Ω are related to the reactant surface $\varepsilon_\mu^I\left(R, z_p, z_e\right)$ through the second derivative at the minimum:

$$M\Omega^2 = \left.\frac{\partial^2 \varepsilon_\mu^I\left(R, z_p, z_e\right)}{\partial R^2}\right|_{\bar{R}^{I\mu},\bar{z}_p^{I\mu},\bar{z}_e^{I\mu}} \tag{16.21}$$

λ_z and λ_R are the solvent and R-mode reorganization energies, respectively, defined as

$$\lambda_z = \varepsilon_\mu^I\left(\bar{R}^{I\mu}, \bar{z}_p^{II\nu}, \bar{z}_e^{II\nu}\right) - \varepsilon_\mu^I\left(\bar{R}^{I\mu}, \bar{z}_p^{I\mu}, \bar{z}_e^{I\mu}\right) \tag{16.22}$$

$$\lambda_R = \varepsilon_\mu^I\left(\bar{R}^{II\nu}, \bar{z}_p^{I\mu}, \bar{z}_e^{I\mu}\right) - \varepsilon_\mu^I\left(\bar{R}^{I\mu}, \bar{z}_p^{I\mu}, \bar{z}_e^{I\mu}\right) \approx \frac{1}{2}M\Omega^2 \Delta R^2 \tag{16.23}$$

where $\Delta R = \bar{R}^{II\nu} - \bar{R}^{I\mu}$ is the difference between the equilibrium R coordinates on the reactant and product surfaces. Finally, the quantum coupling term λ_α is defined as

$$\lambda_\alpha = \frac{\hbar^2 a_{\mu\nu}^2}{2M} \tag{16.24}$$

Previously, expressions similar to Eq. 16.15 were derived and analyzed by Borgis, Lee, and Hynes in the context of vibrationally nonadiabatic proton transfer reactions in solution [49, 51, 52]. Note that the time correlation function for the solvent variables is fundamentally different for PCET reactions because PCET reactions involve two correlated solvent coordinates z_p and z_e rather than a single solvent coordinate z_p. Moreover, the physical meaning of the quantum coupling term λ_α is fundamentally different for PCET reactions because the coupling for PCET reactions involves ET nonadiabatic coupling as well as PT coupling. The most interesting features of the expression in Eq. (16.15) are the terms in the exponential that are proportional to the quantum coupling term λ_α and to $\sqrt{\lambda_\alpha \lambda_R}$.

These terms reflect the dynamical correlations between the R mode and the non-adiabatic coupling fluctuations.

The rate expression in Eq. (16.15) can be simplified significantly under a series of well-defined approximations. In the short-time high-temperature approximation for a Debye solvent, the solvent dynamics is negligible on the time scale of the probability flux correlation function, leading to the following simplification:

$$\exp\left[-\frac{1}{\hbar^2}\int_0^t d\tau_1 \int_0^{\tau_1} d\tau_2\, C_Z(\tau_1 - \tau_2)\right] = \exp\left[-\frac{\lambda_z t^2}{\beta\hbar^2}\right] \tag{16.25}$$

Moreover, when the probability flux correlation function decays on a time scale shorter than the time scale of the solvent effects on the R motion, the solvent effects on the R motion can be neglected. In this case, $C_R(t)$ can be approximated by the standard analytical expression for the time correlation function of an undamped quantum mechanical harmonic oscillator [53]:

$$C_R(t) = \frac{\hbar}{2M\Omega}\left[\coth\left(\frac{1}{2}\beta\hbar\Omega\right)\cos\Omega t + i\sin\Omega t\right] \tag{16.26}$$

These approximations lead to the following closed analytical expression for the nonadiabatic partial rate $k_{\mu\nu}^{\mathrm{dyn}}$:

$$k_{\mu\nu}^{\mathrm{dyn}} = \frac{\left|V_{\mu\nu}^{(0)}\right|^2}{\hbar^2\Omega}\exp\left[\frac{2\lambda_a\zeta}{\hbar\Omega}\right]\int_{-\infty}^{\infty} d\tau\exp\left[-\frac{1}{2}\chi\tau^2 + p(\cos\tau - 1) + i(q\sin\tau + \theta\tau)\right] \tag{16.27}$$

where the dimensionless parameters are defined as

$$\zeta = \coth\left(\frac{1}{2}\beta\hbar\Omega\right); \quad \chi = \frac{2\lambda_z}{\beta\hbar^2\Omega^2}; \quad \theta = \frac{\Delta G^0 + \lambda_z}{\hbar\Omega}$$

$$\tag{16.28}$$

$$p = \zeta\frac{\lambda_R + \lambda_a}{\hbar\Omega} - 2\frac{\sqrt{\lambda_R\lambda_a}}{\hbar\Omega}; \quad q = \frac{\lambda_R + \lambda_a}{\hbar\Omega} - 2\zeta\frac{\sqrt{\lambda_R\lambda_a}}{\hbar\Omega}$$

Again the indices μ and ν are omitted for simplicity. Note that the imaginary part of the rate in Eq. (16.27) is identically zero. The PT analog of this expression is given in Refs. [49, 51, 52].

The real part of the integrand in Eq. (16.27) is a damped oscillating function of the frequency-scaled time $\tau = \Omega t$. The strength of the damping factor $\exp[-\chi\tau^2/2]$ depends on the temperature, the solvent reorganization energy λ_z, and the R mode quantum energy $\hbar\Omega$. Note that in the regime considered here, the solvent fluctuations are the key element responsible for damping the quantum coherent electron–proton tunneling. In the absence of the solvent reorganization (i.e., $\lambda_z = 0$), the time integral in Eq. (16.27) is divergent and the reaction rate constant is not defined. Ref. [50] provides an alternative approach that avoids these divergences by performing a transformation of a model Hamiltonian corresponding to proton transfer. Since the thermal average of the transformed perturbation is zero, the integral of the time correlation function is convergent for the entire range of system parameters.

In certain limiting regimes, the time integral in Eq. (16.27) can be simplified to obtain closed analytical rate expressions. In the high temperature (low frequency) limit for the R mode ($\beta\hbar\Omega \ll 1$) the partial rate can be expressed as

$$
k_{\mu\nu}^{\text{high}-\text{T}} = \frac{\left|V_{\mu\nu}^{(0)}\right|^2}{\hbar} \exp\left[\frac{4\lambda_a}{\beta\hbar^2\Omega^2}\right]
$$

$$
\times \sqrt{\frac{\pi\beta}{\lambda_{\text{tot}}\left(1 - \frac{4\lambda_a\lambda_R}{\lambda_{\text{tot}}^2}\right)}} \exp\left[-\frac{\beta\left(\Delta G^0 + \lambda_z - \frac{4}{\beta\hbar\Omega}\sqrt{\lambda_a\lambda_R}\right)^2}{4\lambda_{\text{tot}}}\left(2 - \frac{\lambda_z}{\lambda_{\text{tot}}}\right)\right]
$$

(16.29)

where $\lambda_{\text{tot}} = \lambda_z + \lambda_R + \lambda_a$. This expression closely resembles the analogous expression in Eq. (16.4) for fixed R with a few important distinctions. First, it has an additional temperature-dependent exponential prefactor $\exp\left[4\lambda_a/(\beta\hbar^2\Omega^2)\right]$, which results in non-Arrhenius behavior of the rate constant at high temperatures. Second, the conventional Marcus activation barrier $(\Delta G^0 + \lambda_z)^2/(4\lambda_z)$ is modified due to the presence of the cross term $4\sqrt{\lambda_a\lambda_R}/(\beta\hbar\Omega)$, which is a quantum effect that is related to the dynamical correlation between the R mode and the nonadiabatic coupling fluctuations. Note that if $\lambda_a = \lambda_R = 0$ (i.e., the coupling is independent of R and the equilibrium value of R is the same for reactant and product free energy surfaces), this expression is identical to the previously derived rate expression for PCET reactions with fixed R, as evident by a comparison to Eq. (16.4).

In the low temperature (high frequency) limit for the R mode in the strong solvation regime (i.e., $|\Delta G^0| < \lambda_z$), the partial rate can be expressed as:

$$
k_{\mu\nu}^{\text{low}-\text{T}} = \frac{\left|V_{\mu\nu}^{(0)}\right|^2}{\hbar} \exp\left[\frac{2\lambda_a}{\hbar\Omega}\right] \exp\left[-\frac{\left(\sqrt{\lambda_a} - \sqrt{\lambda_R}\right)^2}{\hbar\Omega}\right] \sqrt{\frac{\pi\beta}{\lambda_z}} \exp\left[-\frac{\beta(\Delta G^0 + \lambda_z)^2}{4\lambda_z}\right]
$$

(16.30)

which is similar to the previously derived PCET rate expressions except for the additional exponential prefactors that modify the nonadiabatic coupling. In this quantum limit for the R mode, the rate expression corresponds qualitatively to the expression in Eq. (16.11) for the case of a fast quantized R mode. The square of the coupling in conjunction with the two additional exponential prefactors in Eq. (16.30) can be interpreted in terms of the square of the coupling between the reactant and product R mode vibrational ground states. Transitions involving R mode vibrationally excited states are not significant at low temperatures. Note that typical PCET reactions are expected to be in the strong solvation regime, in which the solvent reorganization energy exceeds the absolute reaction free energy. The rate expression can also be derived for the weak solvation regime, following the procedure in Ref. [54]. Moreover, the rigorous derivation of the low-temperature rate expressions for proton transfer reactions given in Ref. [50] could also be applied to PCET reactions.

16.2.3.2 Molecular Representation of the Environment

In the molecular representation of the environment, the solvent/protein coordinates ξ are treated explicitly. In this case, the probability flux correlation function is:

$$j_{\mu\nu}^{MD}(t) = \left|V_{\mu\nu}^{(0)}\right|^2 \exp\left[-2a_{\mu\nu}\langle\delta R_\mu\rangle\right]\exp\left[\frac{i}{\hbar}\left(\langle\mathcal{E}_{\mu\nu}\rangle + \langle\tilde{D}_{\mu\nu}\rangle\langle\delta R_\mu\rangle\right)t\right]$$

$$\times \exp\left\{a_{\mu\nu}^2[C_R(0) + C_R(t)] - \frac{2ia_{\mu\nu}}{\hbar}\langle\tilde{D}_{\mu\nu}\rangle\int_0^t C_R(\tau)d\tau \right. \tag{16.31}$$

$$\left. -\frac{1}{\hbar^2}\int_0^t d\tau_1\int_0^{\tau_1} d\tau_2 C_{\mathcal{E}}(\tau_1 - \tau_2) - \frac{1}{\hbar^2}\int_0^t d\tau_1\int_0^{\tau_1} d\tau_2 C_D(\tau_1 - \tau_2)C_R(\tau_1 - \tau_2)\right\}$$

where the time evolution on the reactant vibronic surface is described in terms of the energy gap

$$\mathcal{E}_{\mu\nu}(t) = \Delta\varepsilon_{\mu\nu}\left(\bar{R}^{I\mu}, \xi(t)\right) = \varepsilon_\nu^{II}\left(\bar{R}^{I\mu}, \xi\right) - \varepsilon_\mu^{I}\left(\bar{R}^{I\mu}, \xi\right) \tag{16.32}$$

the derivative of the energy gap

$$\tilde{D}_{\mu\nu}(\xi(t)) = \left.\frac{\partial\Delta\varepsilon_{\mu\nu}}{\partial R}\right|_{R=\bar{R}^{I\mu}} \tag{16.33}$$

and the R mode. The time correlation function for the energy gap is defined as $C_{\mathcal{E}}(t) = \langle\delta\mathcal{E}_{\mu\nu}(0)\delta\mathcal{E}_{\mu\nu}(t)\rangle$, where $\delta\mathcal{E}_{\mu\nu}(t) = \mathcal{E}_{\mu\nu}(t) - \langle\mathcal{E}_{\mu\nu}\rangle$. The time correlation functions for the other variables are defined analogously. Note that the probability flux expression in Eq. (16.31) is analogous to the probability flux expression given in Ref. [55] for vibrationally nonadiabatic PT reactions occurring on a single adiabatic electronic state. In PCET reactions, however, the energy gap and coupling are defined for pairs of electron–proton vibronic surfaces corresponding to different electronic states.

The above quantities can be evaluated with molecular dynamics simulations of the full solute–solvent system on the reactant vibronic surface $\varepsilon_\mu^I(R, \xi)$. Specifically, the quantities $C_R(t)$, $\langle\delta R^2\rangle$, and $\langle\delta R_\mu\rangle = \langle R(t) - \bar{R}^{I\mu}\rangle$ can be calculated from a molecular dynamics simulation with an unconstrained R coordinate, and the quantities $\langle\mathcal{E}_{\mu\nu}\rangle$, $\langle\tilde{D}_{\mu\nu}\rangle$, $C_{\mathcal{E}}(t)$, and $C_D(t)$ can be calculated from molecular dynamics simulations of the full solute–solvent system with the R coordinate fixed to $R = \bar{R}^{l\mu}$. Alternatively, the time correlation function $C_R(t)$ can be calculated in terms of a frequency shift correlation function, which is obtained from a simulation of the system with the R coordinate fixed to $R = \bar{R}^{l\mu}$, and the calculation of the correlation function for an undamped quantum mechanical harmonic oscillator [55]. Typically the solvent effects on the R mode occur on a much longer time scale than the decay of the probability flux correlation function, and the solvent effects on the R mode can be neglected. In this case, the correlation functions involving the R coordinate can be calculated directly from the analytical expressions for an undamped quantum mechanical harmonic oscillator, as given in Eq. (16.26).

The rate expression in Eq. (16.31) can be simplified under the following well-defined conditions: the coupling between the R coordinate and the solvent coordinates is neglected, the surfaces are approximated to be harmonic along the R coordinate, and the R mode frequency is assumed to be the same for both the reactant and product surfaces. In this case, $\langle \delta R_{\mu} \rangle = \langle \delta R \rangle = 0$ and the energy gap derivative defined in Eq. (16.31) becomes

$$\bar{D}_{\mu v} = \tilde{A}_{\mu v} \equiv -M\Omega^2 \Delta R_{\mu v} \tag{16.34}$$

where the R mode frequency Ω is given by

$$M\Omega^2 = \left. \frac{\partial^2 \varepsilon_{\mu}^{l}(R, \xi)}{\partial R^2} \right|_{R=\bar{R}^{l\mu}} \tag{16.35}$$

Under these conditions, the expression in Eq. (16.31) simplifies to:

$$j_{\mu v}^{\mathrm{MD,harm}}(t) = \left| V_{\mu v}^{(0)} \right|^2 \exp\left[\frac{i}{\hbar} \langle \mathcal{E}_{\mu v} \rangle t \right]$$

$$\times \exp\left\{ a_{\mu v}^2 [C_R(0) + C_R(t)] - \frac{2ia_{\mu v}\tilde{A}_{\mu v}}{\hbar} \int_0^t C_R(\tau)d\tau \right.$$

$$\left. - \frac{1}{\hbar^2} \int_0^t d\tau_1 \int_0^{\tau_1} d\tau_2 C_{\mathcal{E}}(\tau_1 - \tau_2) - \frac{\tilde{A}_{\mu v}^2}{\hbar^2} \int_0^t d\tau_1 \int_0^{\tau_1} d\tau_2 C_R(\tau_1 - \tau_2) \right\} \tag{16.36}$$

In Eq. (16.36), the quantities $\langle \mathcal{E}_{\mu v} \rangle$ and $C_{\mathcal{E}}(t)$ can be evaluated with molecular dynamics simulations of the full solute–solvent system on the constrained reactant vibronic surface $\varepsilon_{\mu}^{l}(R = \bar{R}^{l\mu}, \xi)$. Since the effects of the solvent on the R mode are neglected in this limit, the quantities pertaining to the R coordinate can be calculated from analytical expressions for an undamped quantum mechanical harmonic oscillator with frequency Ω and reduced mass M, as given in Eq. (16.26). In this manner, the quantum character of the R coordinate is included in the rate expression. Thus, all of the quantities required for the calculation of the nonadiabatic flux can be calculated from a single molecular dynamics simulation on the constrained reactant vibronic surface. Note that the form of Eq. (16.36) is the same as the form of Eq. (16.15), which was derived for a continuum representation of the environment.

We have used this theoretical formulation to analyze the dynamical aspects of model PCET reactions in solution with molecular dynamics simulations [56, 57]. For these model systems, the time dependence of the probability flux correlation function is dominated by the solvent damping term, and only the short-time equilibrium fluctuations of the solvent impact the rate. The proton donor–acceptor motion does not impact the dynamical behavior of the reaction but does influence the magnitude of the rate.

Although this dynamical treatment is analogous to the previous treatment of vibrationally nonadiabatic PT reactions [49, 51, 52], a number of important new issues arise for PCET reactions. In general, PCET reactions are described by at least a four-state model representing the four charge transfer states rather than the two-state model used for single PT reactions. As a result, general PCET reactions are described in terms of at least two scalar solvent coordinates corresponding to ET and PT rather than the single scalar solvent coordinate used for single PT reactions. The use of two scalar solvent coordinates leads to fundamentally different solvent terms in the rate expressions based on a dielectric continuum representation of the solvent. In addition, the previous work on vibrationally nonadiabatic PT reactions assumed that the reaction was electronically adiabatic (i.e., the reaction occurs on the electronic ground state). In contrast, typically PCET reactions are electronically nonadiabatic and occur on two different electronic surfaces corresponding to the two diabatic ET states. Thus, the coupling for PCET reactions involves ET nonadiabatic coupling as well as PT coupling.

16.3
Applications

We have applied this theoretical formulation [26–28] to a series of PCET reactions. The systems were chosen based on the availability of experimental data that had not yet been fully explained. The systems that will be discussed in this section are iron bi-imidazoline complexes, ruthenium polypyridyl complexes, amidinium-carboxylate interfaces, DNA–acrylamide complexes, tyrosine oxidation, and the enzyme lipoxygenase. In all cases, the solvent was treated as a dielectric continuum [58, 59].

16.3.1
PCET in Solution

A comparative experimental study of single ET and PCET reactions in the iron bi-imidazoline complexes shown in Fig. 16.2 indicated that the rates of ET and PCET

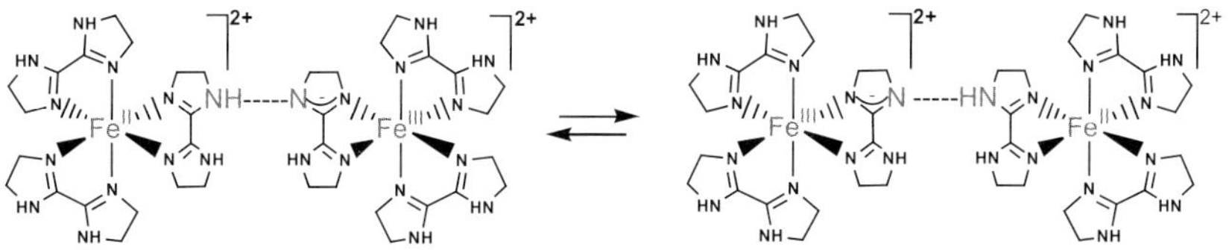

Figure 16.2 PCET reaction between iron bi-imidazoline complexes [45]. For the corresponding single electron transfer reaction, all nitrogen atoms are protonated.

are similar [15]. Previously this result was explained in the context of adiabatic Marcus theory, and the PCET reaction was viewed as a hydrogen atom transfer involving negligible solute charge rearrangement, leading to zero solvent reorganization energy [15]. The similarity of the ET and PCET rates was thought to be due to the compensation of the larger solvent reorganization energy for ET by a larger solute reorganization energy for PCET. The kinetic isotope effect (KIE) for PCET was measured to be a moderate value of 2.3. Our calculations, which were based on nonadiabatic rate expressions for ET and PCET, provided an alternative explanation for the experimental results [45]. The fundamental PCET mechanism for this reaction is illustrated in Fig. 16.3. In our calculations, the inner-sphere reorganization involving the Fe–N bonds was assumed to be the same for both ET and PCET. The solvent reorganization energies $\lambda_{\mu\nu}$ for the dominant contributions to the PCET reaction were found to be substantial and were ≈ 1–3 kcal mol^{-1} lower than the solvent reorganization energy for single ET. The overall coupling for PCET was found to be smaller than the coupling for ET due to averaging over the reactant and product hydrogen vibrational wavefunctions (i.e., multiplying by the vibrational overlap factor in Eq. (16.9)). The calculations indicated that the similarity of the rates for ET and PCET is due mainly to the compensation of the larger solvent reorganization energy for ET by the smaller coupling for PCET. The moderate KIE was determined to arise from the relatively large overlap factor and the significant contributions from excited vibronic states.

An experimental study [16,17] of PCET in the ruthenium polypyridyl complexes shown in Fig. 16.4 revealed that the CompB rate is nearly one order of magnitude larger than the CompA rate, and the CompA KIE of 16.1 is larger than the CompB KIE of 11.4. As shown in Fig. 16.5, our density functional theory calculations [46] illustrated that the steric crowding near the oxygen proton acceptor is significantly greater for CompA than for CompB. Consistent with this observation, our nonadiabatic rate calculations [46] implied that the proton donor–acceptor distance is larger for CompA than for CompB, leading to a larger overlap between the reactant and product hydrogen vibrational wavefunctions for CompB than for CompA, as shown in Fig. 16.6. The rate for CompB is larger than the rate for CompA because the rate increases as this overlap factor increases. The KIE for CompB is smaller than the KIE for CompA because the KIE decreases as this overlap factor increases. Both of these KIEs are larger than the KIE for the iron bi-imidazoline complexes described above because the vibrational overlap factor is smaller for the ruthenium systems.

(a)

(b)

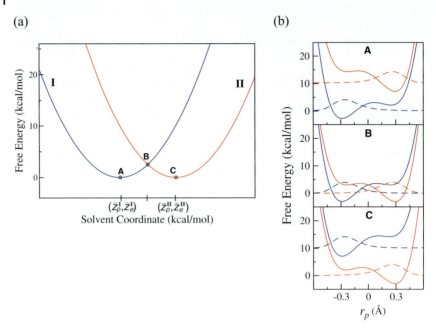

Figure 16.3 Illustration of the fundamental mechanism for the PCET reaction depicted in Fig. 16.2. (a) Slice of the two-dimensional ET diabatic free energy surface along the line connecting the two minima. The lowest energy reactant (I) and product (II) free energy surfaces are shown in blue and red, respectively. Points A, B, and C represent the equilibrium reactant configuration, the intersection point, and the equilibrium product configuration, respectively. (b) Proton potential energy curves and corresponding ground state proton vibrational wavefunctions as functions of the proton coordinate r_p for the solvent coordinates associated with points A, B, and C indicated in (a). The proton potential energy curves and vibrational wavefunctions denote the reactant (or product) ET diabatic free energy surface. Reproduced from Ref. [45].

In addition, this theory has been applied to PCET through amidinium-carboxylate salt bridges, in which the ET reaction is coupled to the motion of two protons at the proton transfer interface [60]. In this case, the reaction is described in terms of eight valence bond states to include all possible charge transfer states, two hydrogen nuclei are treated quantum mechanically, and the free energy surfaces depend on three solvent coordinates corresponding to the electron and two proton transfer reactions. Experimental studies of photoinduced PCET in analogous systems revealed that the rate for the donor–(amidinium-carboxylate)–acceptor system is substantially slower than the rate for the switched interface donor–(carboxylate-amidinium)–acceptor system [18]. The calculations illustrated that this difference in rates is due mainly to the opposite dipole moments at the proton transfer interfaces for the two systems, leading to an endothermic reaction for the donor–(amidinium-carboxylate)–acceptor system and an exothermic reaction for the switched interface system.

CompA

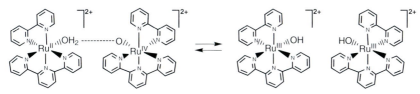

CompB

Figure 16.4 PCET comproportionation reactions in ruthenium polypyridyl complexes [46].

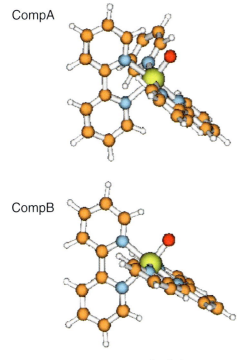

CompA

CompB

Figure 16.5 Structures optimized with density functional theory for the acceptor complexes in Fig. 16.4. Reproduced from Ref. [46].

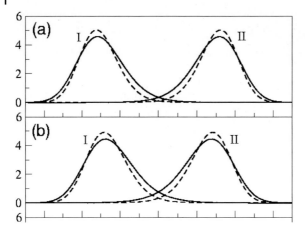

Figure 16.6 Reactant (I) and product (II) vibrational wavefunctions for H (solid) and D (dashed) for the reactions shown in Fig. 16.4. The overlap is smaller for CompA than for CompB because the O–O distance is larger for CompA due to greater steric crowding near the acceptor oxygen. Reproduced from Ref. [46].

We have also applied this theory to biologically relevant systems, such as PCET in DNA–acrylamide complexes [61]. Experiments implied that PCET may occur in such complexes [62]. The influence of neighboring DNA base pairs was determined theoretically by studying both solvated thymine–acrylamide and solvated DNA–acrylamide models. The calculations indicated that the final product corresponds to single ET for the solvated thymine–acrylamide complex but to a net PCET reaction for the solvated DNA–acrylamide models. This difference is due to a decrease in solvent accessibility in the presence of DNA, which alters the relative free energies of the ET and PCET product states. Thus, the balance between ET and PCET in the DNA–acrylamide system is highly sensitive to the solvation properties of the system.

Another recent application [63] of this theory was to the compound depicted in Fig. 16.7, which was designed to model tyrosine oxidation in Photosystem II [3–7]. Upon photoexcitation of the complex to produce Ru(III), an electron is transferred to the ruthenium from the tyrosine, which is concurrently deprotonated. The dependence of the rates on pH and temperature was measured experimentally [20]. The mechanism was determined to be PCET at pH<10 when the tyrosine is initially protonated and single ET for pH>10 when the tyrosine is initially deprotonated. As shown in Fig. 16.8, the PCET rate increases monotonically with pH, whereas the single ET rate is independent of pH and two orders of magnitude faster than the PCET rate. The calculations reproduced these experimentally observed trends. The pH dependence for the PCET reaction resulted from the decrease in the PT and PCET reaction free energies with pH. The calculations indicated that the larger rate for single ET arises from a combination of factors, including the greater exoergicity for ET, the smaller solvent reorganization energy

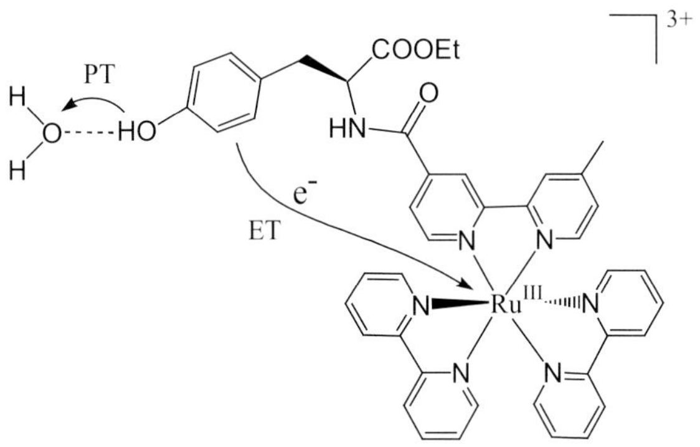

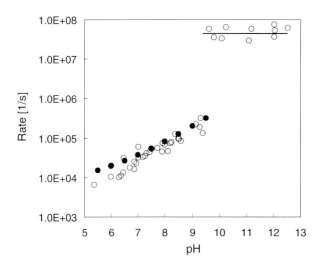

Figure 16.7 PCET reaction in a model for tyrosine oxidation in photosystem II. In the first step of the experiment, the ruthenium-tris-bipyridine portion absorbs light, and the excited electron is transferred to an external methyl viologen acceptor. In the second step, which is shown here, the tyrosine portion transfers an electron to the ruthenium and is deprotonated. Reproduced from Ref. [63].

Figure 16.8 Experimental and theoretical data for the pH dependence of the rates for single electron transfer and PCET in the tyrosine oxidation model shown in Fig. 16.7. The experimental values are denoted with open circles. The theoretical PCET rates are denoted with filled circles, and the theoretical ET rate is represented by a solid line because it is independent of pH. The ET and PT couplings were fit to the experimental ET rate at pH>10 and PCET rate at pH 7. Reproduced from Ref. [63].

for ET, and the averaging of the coupling for PCET over the reactant and product hydrogen vibrational wavefunctions (i.e., the vibrational overlap factor). The calculated temperature dependence of the rates and the deuterium kinetic isotope effects were also consistent with the experimental results.

16.3.2
PCET in a Protein

Recently we applied this theory to the PCET reaction catalyzed by soybean lipoxy-genase and investigated the role of the proton donor–acceptor motion for this enzyme reaction [47]. This investigation was motivated by experimental measurements of unusually large KIEs at room temperature in conjunction with a weak temperature dependence of the rates [64]. Lipoxygenases catalyze the oxidation of unsaturated fatty acids and have a wide range of biomedical applications [65, 66]. Kinetic studies have shown that the hydrogen transfer from the carbon atom C11 of the linoleic acid substrate to the Fe(III)–OH cofactor is rate limiting above 32 °C for SLO [64]. The deuterium KIE on the catalytic rate has been measured experimentally to be as high as 81 at room temperature [64, 67–71]. The rates for hydrogen and deuterium transfer were found to depend only weakly on temperature [64]. Various theoretical models have been invoked to analyze the temperature dependence of the KIEs [36, 47, 64, 67, 68, 72, 73].

In our calculations [47], we treated the net hydrogen transfer reaction catalyzed by lipoxygenase as a PCET mechanism, as illustrated in Fig. 16.9. Quantum mechanical calculations indicate that the electron transfers from the π system of the linoleic acid to an orbital localized on the Fe(III) center, and the proton transfers

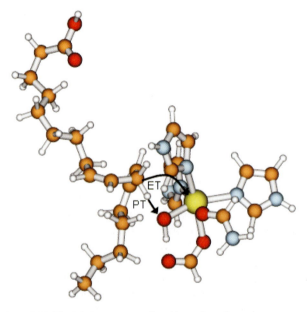

Figure 16.9 The PCET reaction catalyzed by soybean lipoxygenase. This reaction entails a net hydrogen atom transfer from the linoleic acid substrate to the Fe(III)–OH cofactor. This conformation was obtained from docking calculations that included the entire soybean lipoxygenase protein. Reproduced from Ref. [47].

from the donor carbon to the oxygen acceptor [74]. Moreover, analysis of the thermodynamic properties of the single PT and ET reactions, as well as the concerted PCET mechanism, indicates that the single PT and ET reactions are significantly endothermic, whereas the PCET reaction is exothermic [47, 64, 74]. These analyses imply that the mechanism is PCET, where the electron and proton transfer simultaneously between different sites.

As shown in Fig. 16.10, the temperature dependence of the rates and KIEs predicted by the multistate continuum theory is in remarkable agreement with the experimental data [47]. The calculations indicate that the weak temperature dependence of the rates is due to the relatively small free energy barrier arising

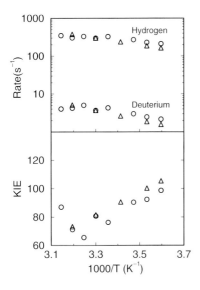

Figure 16.10 Temperature dependence of the rates and KIEs for the PCET reaction catalyzed by soybean lipoxygenase. The theoretical results are denoted with open triangles, and the experimental data are denoted with closed circles. The theoretical results were generated with the multistate continuum theory including the proton donor–acceptor vibrational motion. Reproduced from Ref. [47].

from a balance between the reorganization energy and the reaction free energy. The unusually high KIE of 81 arises from the small overlap of the reactant and product proton vibrational wavefunctions and the dominance of the lowest energy reactant and product vibronic states in the tunneling process. The proton donor–acceptor vibrational motion was included in our calculations. The dominant contribution to the overall rate was found to correspond to a proton donor–acceptor distance that is significantly smaller than the equilibrium donor–acceptor distance. This dominant distance is determined by a balance between the larger coupling and the smaller Boltzmann factor as the distance decreases. Thus, the proton donor–acceptor vibrational motion plays an important role in decreasing the dominant donor–acceptor distance relative to its equilibrium value to facilitate the PCET reaction. The quantum mechanical and nonequilibrium dynamical aspects of the proton donor–acceptor vibrational motion are not essential for the description of the experimentally observed temperature dependence of the rates and KIES within the framework of this model.

16.4
Conclusions

This chapter describes a general theoretical formulation for PCET and summarizes the results of applications to a wide range of different types of PCET reactions in solution and enzymes. This theoretical formulation treats the active electrons and transferring proton(s) quantum mechanically and includes the interactions among the electrons, proton(s), and solvent or protein environment. Moreover, this formulation allows the inclusion of the proton donor–acceptor motion, explicit molecular solvent and protein, and dynamical effects. The theory described here is directly applicable to nonadiabatic PCET reactions accompanied by substantial solute charge redistribution and solvent reorganization.

Within this framework, hydrogen atom transfer is considered as a special type of PCET in which the electron and proton are transferred between the same sites [31, 75]. Note that this definition of hydrogen atom transfer is not rigorous due to the delocalized nature of electrons and protons. Nevertheless, this terminology is used throughout the literature. An important characteristic of hydrogen atom transfer reactions is the lack of significant solute charge redistribution and solvent reorganization. As a result, hydrogen atom transfer reactions tend to be dominated by solute reorganization. Moreover, hydrogen atom transfer reactions are typically electronically adiabatic (i.e., occur on the electronic ground state), although they may be vibrationally nonadiabatic. Currently we are extending this theoretical formulation to study hydrogen atom transfer reactions.

Theoretical calculations on model PCET systems have assisted in the interpretation of experimental data and have provided insight into the underlying fundamental principles of PCET reactions. The theoretical framework described in this chapter enables the prediction of experimentally testable trends in rates and kinetic isotope effects. The interplay between experiment and theory will be vital to further progress in the field.

Acknowledgments

This work was supported by NIH grant GM56207 and NSF grant CHE-0096357. I am grateful to Alexander Soudackov, Nedialka Iordanova, and Elizabeth Hatcher for useful discussions and generation of the figures.

References

1 Babcock, G. T.; Barry, B. A.; Debus, R. J.; Hoganson, C. W.; Atamian, M.; McIntosh, L.; Sithole, I.; Yocum, C. F., *Biochemistry*, **1989**, *28*, 9557–9565.

2 Okamura, M. Y.; Feher, G., *Annu. Rev. Biochem.*, **1992**, *61*, 861–896.

3 Tommos, C.; Tang, X.-S.; Warncke, K.; Hoganson, C. W.; Styring, S.; McCracken, J.; Diner, B. A.; Babcock, G. T., *J. Am. Chem. Soc.*, **1995**, *117*, 10325–10335.

4 Hoganson, C. W.; Babcock, G. T., *Science*, **1997**, *277*, 1953–1956.

5 Hoganson, C. W.; Lydakis-Simantiris, N.; Tang, X.-S.; Tommos, C.; Warncke, K.; Babcock, G. T.; Diner, B. A.; McCracken, J.; Styring, S., *Photosynth. Res.*, **1995**, *47*, 177–184.

6 Blomberg, M. R. A.; Siegbahn, P. E. M.; Styring, S.; Babcock, G. T.; Akermark, B.; Korall, P., *J. Am. Chem. Soc.*, **1997**, *119*, 8285–8292.

7 Diner, B. A.; Babcock, G. T.; in *Oxygenic Photosynthesis: The Light Reactions*; Ort, D. R., Yocum, C. F., Eds., Kluwer: Dordrecht, 1996, pp. 213–247.

8 Babcock, G. T.; Wikstrom, M., *Nature*, **1992**, *356*, 301–309.

9 Malmstrom, B. G., *Acc. Chem. Res.*, **1993**, *26*, 332–338.

10 Siegbahn, P. E. M.; Eriksson, L.; Himo, F.; Pavlov, M., *J. Phys. Chem. B*, **1998**, *102*, 10622–10629.

11 Chang, C. J.; Chang, M. C. Y.; Damrauer, N. H.; Nocera, D. G., *Biochim. Biophys. Acta - Bioenergetics*, **2004**, *1655*, 13–28.

12 Cukier, R. I., *Biochim. Biophys. Acta-Bioenergetics*, **2004**, *1655*, 37–44.

13 Hammes-Schiffer, S.; Iordanova, N., *Biochim. Biophys. Acta-Bioenergetics*, **2004**, *1655*, 29–36.

14 Mayer, J. M., *Annu. Rev. Phys. Chem.*, **2004**, *55*, 363–390.

15 Roth, J. P.; Lovel, S.; Mayer, J. M., *J. Am. Chem. Soc.*, **2000**, *122*, 5486–5498.

16 Binstead, R. A.; Meyer, T. J., *J. Am. Chem. Soc.*, **1987**, *109*, 3287–3297.

17 Farrer, B. T.; Thorp, H. H., *Inorg. Chem.*, **1999**, *38*, 2497–2502.

18 Kirby, J. P.; Roberts, J. A.; Nocera, D. G., *J. Am. Chem. Soc.*, **1997**, *119*, 9230–9236.

19 Huynh, M. H. V.; Meyer, T. J., *Angew. Chem. Int. Ed.*, **2002**, *41*, 1395–1398.

20 Sjodin, M.; Styring, S.; Akermark, B.; Sun, L.; Hammarstrom, L., *J. Am. Chem. Soc.*, **2000**, *122*, 3932–3936.

21 Cukier, R. I., *J. Phys. Chem.*, **1994**, *98*, 2377–2381.

22 Cukier, R. I., *J. Phys. Chem.*, **1996**, *100*, 15428–15443.

23 Cukier, R. I.; Nocera, D. G., *Annu. Rev. Phys. Chem.*, **1998**, *49*, 337–369.

24 Cukier, R. I., *J. Phys. Chem. A*, **1999**, *103*, 5989–5995.

25 Cukier, R. I., *J. Phys. Chem. B*, **2002**, *106*, 1746–1757.

26 Soudackov, A.; Hammes-Schiffer, S., *J. Chem. Phys.*, **1999**, *111*, 4672–4687.

27 Soudackov, A.; Hammes-Schiffer, S., *J. Chem. Phys.*, **2000**, *113*, 2385–2396.

28 Hammes-Schiffer, S., *Acc. Chem. Res.*, **2001**, *34*, 273–281.

29 Hammes-Schiffer, S., in *Electron Transfer in Chemistry Vol I. Principles, Theories, Methods, and Techniques*, Balzani, V., Ed., Wiley-VCH, Weinheim, 2001, pp. 189–214.

30 Soudackov, A.; Hatcher, E.; Hammes-Schiffer, S., *J. Chem. Phys.*, **2005**, *122*, 014505.

31 Mayer, J. M.; Hrovat, D. A.; Thomas, J. L.; Borden, W. T., *J. Am. Chem. Soc.*, **2002**, *124*, 11142–11147.

32 Mincer, J. S.; Schwartz, S. D., *J. Chem. Phys.*, **2004**, *120*, 7755–7760.

33 Georgievskii, Y.; Stuchebrukhov, A. A., *J. Chem. Phys.*, **2000**, *113*, 10438–10450.

34 Moore, D. B.; Martinez, T. J., *J. Phys. Chem. A*, **2000**, *104*, 2367–2374.

35 Siegbahn, P. E. M.; Blomberg, M. R. A.; Crabtree, R. H., *Theoret. Chem. Acc.*, **1997**, *97*, 289–300.

36 Siebrand, W.; Smedarchina, Z., *J. Phys. Chem. B*, **2004**, *108*, 4185–4195.

37 Kim, H. J.; Hynes, J. T., *J. Chem. Phys.*, **1992**, *96*, 5088–5110.

38 Kestner, N. R.; Logan, J.; Jortner, J., *J. Phys. Chem.*, **1974**, *78*, 2148.

39 Ulstrup, J.; Jortner, J., *J. Chem. Phys.*, **1975**, *63*, 4358.

40 Bixon, M.; Jortner, J., *Adv. Chem. Phys.*, **1999**, *106*, 35–202.

41 Warshel, A. *Computer Modeling of Chemical Reactions in Enzymes and Solutions*; John Wiley & Sons, New York, 1991.

42 Basilevsky, M. V.; Chudinov, G. E.; Newton, M. D., *Chem. Phys.*, **1994**, *179*, 263–278.

43 Basilevsky, M. V.; Vener, M. V., *J. Mol. Struct.: THEOCHEM*, **1997**, *398*, 81.

44 Barbara, P. F.; Meyer, T. J.; Ratner, M. A., *J. Phys. Chem.*, **1996**, *100*, 13148–13168.

45 Iordanova, N.; Decornez, H.; Hammes-Schiffer, S., *J. Am. Chem. Soc.*, **2001**, *123*, 3723–3733.

46 Iordanova, N.; Hammes-Schiffer, S., *J. Am. Chem. Soc.*, **2002**, *124*, 4848–4856.

47 Hatcher, E.; Soudackov, A. V.; Hammes-Schiffer, S., *J. Am. Chem. Soc.*, **2004**, *126*, 5763–5775.

48 Trakhtenberg, L. I.; Klochikhim, V. L.; Pshezhetsky, S. Y., *Chem. Phys.*, **1982**, *69*, 121–134.

49 Borgis, D. C.; Lee, S. Y.; Hynes, J. T., *Chem. Phys. Lett.*, **1989**, *162*, 19–26.

50 Suarez, A.; Silbey, R., *J. Chem. Phys.*, **1991**, *94*, 4809–4816.

51 Borgis, D.; Hynes, J. T., *Chem. Phys.*, **1993**, *170*, 315–346.

52 Borgis, D.; Hynes, J. T., *J. Chem. Phys.*, **1991**, *94*, 3619–3628.

53 Chandler, D., in *Classical and Quantum Dynamics in Condensed Phase Simulations*; Berne, B. J., Ciccotti, G., Coker, D. F., Eds.; World Scientific, Singapore, 1998, pp. 3–23.

54 Borgis, D.; Hynes, J. T., *Chem. Phys.*, **1993**, *170*, 315–346.

55 Borgis, D.; Hynes, J. T., *J. Chem. Phys.*, **1991**, *94*, 3619–3628.

56 Hatcher, E.; Soudackov, A.; Hammes-Schiffer, S., *Chem. Phys.*, **2005**, *319*, 93–100.

57 Hatcher, E.; Soudackov, A.; Hammes-Schiffer, S., *J. Phys. Chem. B*, **2005**, *109*, 18565–18574.

58 Basilevsky, M. V.; Rostov, I. V.; Newton, M. D., *Chem. Phys.*, **1998**, *232*, 189–199.

59 Newton, M. D.; Basilevsky, M. V.; Rostov, I. V., *Chem. Phys.*, **1998**, *232*, 201–210.

60 Rostov, I.; Hammes-Schiffer, S., *J. Chem. Phys.*, **2001**, *115*, 285–296.

61 Carra, C.; Iordanova, N.; Hammes-Schiffer, S., *J. Phys. Chem. B*, **2002**, *106*, 8415–8421.

62 Taylor, J.; Eliezer, I.; Sevilla, M. D., *J. Phys. Chem. B*, **2001**, *105*, 1614–1617.

63 Carra, C.; Iordanova, N.; Hammes-Schiffer, S., *J. Am. Chem. Soc.*, **2003**, *125*, 10429–10436.

64 Knapp, M. J.; Rickert, K. W.; Klinman, J. P., *J. Am. Chem. Soc.*, **2002**, *124*, 3865–3874.

65 Samuelsson, B.; Dahlen, S.-E.; Lindgren, J.; Rouzer, C. A.; Serhan, C. N., *Science*, **1987**, *237*, 1171–1176.

66 Holman, T. R.; Zhou, J.; Solomon, E. L., *J. Am. Chem. Soc.*, **1998**, *120*, 12564–12572.

67 Rickert, K. W.; Klinman, J. P., *Biochemistry*, **1999**, *38*, 12218–12228.

68 Jonsson, T.; Glickman, M. H.; Sun, S.; Klinman, J. P., *J. Am. Chem. Soc.*, **1996**, *118*, 10319–10320.

69 Glickman, M. H.; Wiseman, J. S.; Klinman, J. P., *J. Am. Chem. Soc.*, **1994**, *116*, 793–794.

70 Hwang, C.-C.; Grissom, C. B., *J. Am. Chem. Soc.*, **1994**, *116*, 795–796.

71 Lewis, E. R.; Johansen, E.; Holman, T. R., *J. Am. Chem. Soc.*, **1999**, *121*, 1395–1396.

72 Olsson, M. H. M.; Siegbahn, P. E. M.; Warshel, A., *J. Biol. Inorg. Chem.*, **2004**, *9*, 96–99.

73 Olsson, M. H. M.; Siegbahn, P. E. M.; Warshel, A., *J. Am. Chem. Soc.*, **2004**, *126*, 2820–2828.

74 Lehnert, N.; Solomon, E. L., *J. Biol. Inorg. Chem.*, **2003**, *8*, 294–305.

75 Hammes-Schiffer, S., *ChemPhysChem*, **2002**, *3*, 33–42.

17

The Relation between Hydrogen Atom Transfer
and Proton-coupled Electron Transfer in Model Systems

Justin M. Hodgkiss, Joel Rosenthal, and Daniel G. Nocera

17.1
Introduction

A hydrogen atom transfer (HAT) reaction encompasses a proton transfer (PT) and electron transfer (ET). To this end, the HAT reaction is a subclass of the more general reaction of proton-coupled electron transfer (PCET). In PCET, the thermodynamics for electron and proton transport mandate that they couple. This is most easily elucidated by the square scheme shown in Fig. 17.1 [1]. The four corners of the scheme comprise the thermodynamic limiting (and measurable) species for any PCET event that involves the transfer of a single electron and a single proton. The edges of the square represent the fundamental ET (horizontal lines) and PT steps (vertical lines) that connect the four potential PCET species. The diagonal line represents a HAT where the PCET event is *concerted* and synchronous with a single transition state (i.e., no detectable intermediates). The thermochemistry of the total PCET reaction is path-independent; therefore the free energy of the diagonal PCET reaction is equivalent to the sum of either ET/PT or PT/ET stepwise pathways around the edges of the square. It is important to realize that the two

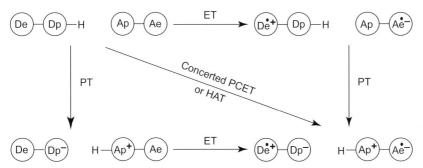

Figure 17.1 The generalized square-scheme describing the PCET reaction space. PCET reactions coordinate ET and PT at four distinct sites; De and Dp are the electron and proton donors respectively, while Ae and Ap are the corresponding acceptors. The thermodynamics associated with PCET or HAT is equivalent to the sum of the stepwise ET and PT steps around the edges.

Hydrogen-Transfer Reactions. Edited by J. T. Hynes, J. P. Klinman, H. H. Limbach, and R. L. Schowen
Copyright © 2007 WILEY-VCH Verlag GmbH & Co. KGaA, Weinheim
ISBN: 978-3-527-30777-7

sets of ET and PT equilibria are both distinct from one another and contain different thermodynamic information. The two ET equilibria are defined by the reduction potentials of the reactants in each protonation state. The two PT equilibria are
defined by the pK_as of the protonated species in each oxidation state.

With the limiting thermodynamics for a PCET event defined, one can often
make inferences about the mechanism by which the coupled transfer takes place.
In all cases, the driving force for concerted PCET is more favorable than for the
competing initial ET or PT steps of the stepwise processes. If this were not true,
these ET-only or PT-only states at the corners of Fig. 17.1 (versus the diagonal
PCET state) would be the thermodynamic product [1]. In some cases, either or
both of the intermediate states are so far uphill that their existence can be ruled
out. The thermochemical properties (i.e. pK_a or E_{red}) measured for the isolated
species can be significantly perturbed when considering the interaction of species
undergoing a PCET reaction. Since PCET reactions usually take place within an
encounter or precursor complex with a preferred orientation, it may be necessary
to account for hydrogen bonding or steric interactions that can favor or disfavor
certain intermediates [2].

Many definitions and descriptions of HAT, prior to the emergence of PCET as a
field of study, did not adequately take into account the complexity embodied by
Fig. 17.1. HAT is traditionally defined as the transfer of an electron and a proton
from one location to another along a spatially coincidental pathway. In this case,
the electron and proton are donated from one atom and they are accepted by
another atom. These transfers are well described mechanistically as the diagonal
pathway of Fig. 17.1 and they have been treated formally by a number of investiga-
tors [3, 4]. However, many reactions treated within a formalism of HAT are more
complex as ET and PT are site-differentiated either along uni- or bidirectional
pathways. As will be discussed in this chapter, traditional descriptions of HAT do
not address how the electron and proton transfer events are coordinated mechanistically in these more complicated reactions and more general treatments of
PCET are warranted.

17.1.1
Formulation of HAT as a PCET Reaction

A PCET reaction is described by four separate transfer sites derived from a donor
and an acceptor for both an electron *and* a proton [5]. This four state description
of PCET gives rise to two important considerations. A *geometric* aspect to PCET
arises when considering the different possible spatial configurations of the four
transfer sites. A HAT reaction comprises just one possible arrangement – where
the electron and proton transfer sites are coincidental – however this need not be
the case for PCET in general. In addition, the two-dimensional reaction space
spanned by the four PCET states shown in Fig. 17.1 encompasses infinite *mechanistic* possibilities (i.e., pathways) for the coordinated transfer of an electron and a
proton. These two issues of *geometry* and *mechanism* must be taken into account

in a description of a PCET reaction and consequently for the specific reaction subset of HAT.

Stepwise PT and ET reactions occur along the edges of Fig. 17.1, while PCET includes the entire space within the square. The stepwise and PCET mechanisms (including HAT) are clearly distinct. The PCET mechanism is defined by a *single* transition state in which the proton and electron both transfer in one step, with no intermediate states populated along the reaction coordinate; PCET is thus concerted but the electron and proton events can be asynchronous as opposed to their synchronous transfer for a HAT reaction. In a stepwise mechanism, an intermediate is formed and there are two distinct rate constants for the forward reaction and two separate transition states. Stepwise ET/PT or PT/ET can, in principle, be broken down and treated experimentally and theoretically as separate ET and PT events. Like any series of reactions, the rate-limiting rule applies,

$$k_{ET/PT} = k_{ET}^{-1} + k_{PT}^{-1} \qquad (17.1)$$

The overall reaction is thus described by conventional treatments of ET [6–12] and PT [12–15]. The stepwise reaction mechanism is tangential to HAT but nevertheless is important to consider in a discussion of HAT. If the second step is fast, the intermediate state may not be observed and the stepwise mechanism may not be recognized. We will return to this point in Section 17.2.

Concerted PCET provides a mechanistic framework for HAT [16, 17]. Cukier, Hammes-Schiffer and their coworkers have theoretically described the physics of negotiating this two-dimensional electron–proton tunneling space [18–28]. A detailed theoretical account of PCET as it pertains to HAT may be found elsewhere in this volume, but a few general comments are warranted. The description of the two-dimensional transfer event is dictated by coordinates describing the electron and proton, as shown in Fig. 17.2. Both are charged particles whose transfers occur within a medium and separate solvent coordinates z_e and z_p are required for the electron and the proton, respectively. These solvent coordinates are analogous to the single solvent coordinates employed in ET or PT reactions and as such they are parametric in the distance coordinates of the electron and proton. In this representation, a family of pathways may be used to describe PCET, depending on how closely the reaction approaches the limiting stepwise pathways on the edges. Regions of this space are designated as displaying more ET-like or more PT-like character in their transition-states and reaction kinetics. The degree of collective solvent motion required for the ET versus PT components for a given PCET geometry may dictate that the one solvent coordinate is much more likely to dominate the transition state. For this reason, the different pathways available may be of significant consequence to the overall rate of the reaction. For instance, in a nonpolar environment, a diagonal path that keeps charge separation minimized may be kinetically preferred in view of the large energetic cost associated with creating a dipole in a low dielectric environment. In this case, the reaction is driven by inner-sphere modes rather than solvent fluctuations. On the other hand, in a high dielectric environment, thermodynamic parameters such as pK_as or reduction

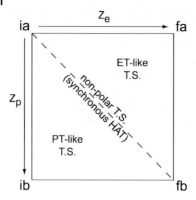

Figure 17.2 The four-state PCET reaction in a solvent coordinate system. The four states are abbreviated with labels according to the initial and final states of the electron (i and f, respectively), and the initial and final states of the proton (a and b, respectively). The coordinates z_e and z_p refer to the collective solvent coordinates that are coupled to ET and PT, respectively. A concerted PCET reaction can have a trajectory anywhere within this space with a single transition state. The synchronicity of the reaction reflects the nature of this trajectory; the synchronous HAT reaction is defined by the strictly diagonal line, whereas deviation from this line reflects asynchronous PCET with varying degrees of ET or PT character dominating the transition state.

potentials may drive the charge to separate via a more PT-like or ET-like transition state as long as the medium can support the increased polarization.

Within the context of Fig. 17.2, HAT is the special case of the diagonal path. The electron and proton can transfer more or less synchronously along the same physical coordinate (r_e and r_p), in which case there is no development of polarization to couple to the solvent. The reaction then follows the diagonal path in Fig. 17.2, which is only available in the HAT geometry where ET and PT coordinates are coincidental. All other PCET geometries require some degree of charge-separation (and hence divergence of z_e and z_p) in order to connect their initial and final states. It is important to reiterate that all paths within the square of Fig. 17.2 will be described by a single transition state and may be thought of as concerted, despite the asynchronous nature of the electron and proton transfer. The height of the barrier and its position relative to reactants and products depend on the nature of the pathway. Notwithstanding, because all paths are described by a single transition state, it is difficult to use linear free energy relations to glean detailed mechanistic information as to the precise nature of the PCET event. Ingenious system design, broad substrate scope, temperature dependences of rate constants, kinetic isotope effects, and new time-resolved laser methods that correlate proton and electron are needed to pin down the precise nature of the HAT-PCET mechanism.

17.1.2
Scope of Chapter

Several recent reviews have explored PCET [1, 5, 23, 24, 29, 30] and specifically HAT [1, 16, 17] from reaction chemistry and theoretical perspectives. This chapter will not exhaustively re-examine this material, but rather introduce a descriptive framework for the electron and proton that adequately depicts both the *geometric* and *mechanistic* complexities of PCET and its relation to HAT. Most examples will be restricted to systems in which kinetics have been measured and discussed within a PCET framework. Accordingly, more classical topics, such as radical organic photochemistry (e.g., Norrish Type I and II reactions) will not be considered.

This chapter will specifically consider the cases schematically represented in Fig. 17.3.

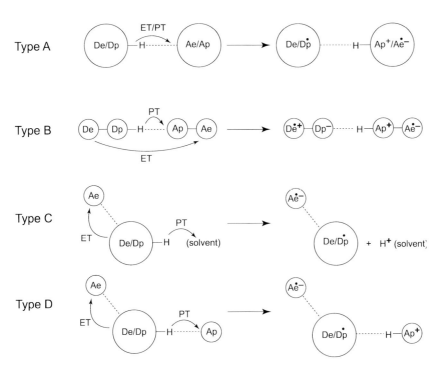

Figure 17.3 General geometries of PCET reactions defined by the spatial configuration of the four transfer elements, De, Dp, Ae, and Ap. In the Type A reaction, ET and PT coordinates are unidirectional with little or no site-differentiation. The latter case is formally an HAT reaction. Type B also represents unidirectional PCET, but with significant differentiation between the ET and PT sites. Type C represents a bidirectional PCET reaction where the electron is transferred to Ae and the proton is dissociated to the bulk. The opposite reaction can occur with ET from De and proton association from the bulk. Type D is the case of bidirectional PCET that employs specific Ae and Ap sites.

17.1.2.1 Unidirectional PCET

The electron and the proton transfer along collinear coordinates. The Type A case of Fig. 17.3 illustrates a limiting scenario for the electron and proton originating from the same group (i.e., as a hydrogen atom) and transferring along collinear coordinates to their respective acceptor sites. When the electron and proton acceptor sites are not differentiated, the traditional definition of HAT is realized. However, in the most general description of unidirectional PCET, the electron and proton can be site-differentiated on both the donor and acceptor sides of the reaction (Type B in Fig. 17.3). In the Type B configuration, the ET coordinate can be considerably longer than the PT coordinate, which remains confined to a hydrogen bonding length scale. Nevertheless, the thermodynamic relations for each limiting case are still defined by the square scheme of Fig. 17.1. Coupling between the electron and proton transfer sites ensures that E_{red} values for the ET sites depend on the position of the proton, while pK_a values for the PT sites depend on the oxidation state of the redox sites. This is not meant to imply that the mechanisms for all cases of unidirectional PCET are indistinguishable. For the site-differentiated situation where the ET coordinate extends beyond the PT coordinate, considerable charge separation may accompany the reaction, and the ET reaction becomes nonadiabatically coupled through the PT interface. For true HAT, the reaction can proceed without charge separation and in a strongly coupled adiabatic manner.

17.1.2.2 Bidirectional PCET

The theoretical treatments of PCET confirm that proton motion can affect electron transport even when the electron and proton do not move along a collinear path. All that is required for direct coupling is that the kinetics (and thermodynamics) of electron transport depend on the position of a specific proton or set of protons at any given time. The two cases highlighted in Fig. 17.3 describe a majority of experimental bidirectional PCET systems studied to date. In nonspecific 3-point PCET (Type C in Fig. 17.3), one molecule donates both an electron and a proton. The electron is transferred to an oxidant with no proton accepting ability, while the proton dissociates and is received by bulk solvent. Although the PT coordinate is less well defined than in other PCET reactions, there can still be tight and specific coupling between ET and PT. The microscopic reverse – where an electron is accepted from a reductant and a proton is taken from the bulk solvent – may also occur. In site-specified 3-point PCET (Type D in Fig. 17.3), a donor is again the nexus of ET and PT events, donating both an electron and a proton in a coordinated manner. Again, the electron is transferred to an oxidant with no proton accepting ability, but rather than being released to the solvent, the proton is transferred to a specific site of another component (usually via a pre-formed hydrogen bond). One can also imagine the microscopic reverse of this reaction, where a reductant and an acid group transfer an electron and a proton, respectively, to an electron/proton acceptor in a coordinated fashion.

This chapter is organized about the four PCET geometries illustrated in Fig. 17.3. Section 17.2 will describe the experiments applied to gain mechanistic information from PCET reactions, and how their interpretation can often have peculiarities specific to PCET. Sections 17.3 and 17.4 detail the characteristics found in each of the uni- and bidirectional PCET geometries, respectively, from an experimental point of view. We will see that relationships do exist between geometry and mechanism, however they do not necessarily exhibit a one-to-one correspondence and hence the need to discuss both *geometry* and *mechanism*. Section 17.5 will discuss biological examples of PCET reactions, and Section 17.6 will introduce emerging coherent laser methods to probe correlated structural and solvent dynamics.

17.2
Methods of HAT and PCET Study

The complexity of PCET and its relation to HAT suggests that rate constants for the process are best measured directly using time-resolved methods that are capable of detecting intermediates, if present. Due to the disparate timescales and the diverse molecular nature of the species involved, different methods of reaction initiation and spectroscopic detection are needed to probe the various PCET reactions of Fig. 17.3. Many PCET reactions occur from a pre-organized and well-defined precursor complex; such designs are simplifying inasmuch as the diffusional component of the PCET reaction is eliminated. These types of diffusion-free reactions invariably rely on fast to ultrafast laser methods to photoinitiate and monitor the PCET transformation directly. Slower PCET reactions are often kinetically resolved in solution without the need for excited states. However, inherent PCET kinetics can be masked by diffusion terms and the kinetics of attaining the PCET precursor geometry. In these cases, the transformation of a substrate is monitored with the implicit assumption that the PCET or HAT step is rate determining.

For fast reactions, the simplest kinetics experiment is to resolve the disappearance of the reactants, for example, by transient emission if the reaction can be photoinitiated and a reactant is luminescent. If there are multiple reaction channels available for reactant decay, the kinetics are described by a mono-exponential decay according to the sum of these rates of which PCET is only one. A more powerful experiment is to observe the disappearance of PCET reactants and growth of PCET products directly. In photoinitiated optical experiments, this means probing by transient absorption (TA) spectroscopy rather than transient emission. If PCET proceeds in a concerted fashion then concomitant mono-exponential disappearance of reactant and growth of product will be observed. If a stepwise mechanism operates, the growth of the products will be delayed (and fit by a bi-exponential function), however, this observation does not reveal the sequence in which the electron and proton were transferred. Moreover, in the limit where one of the steps is significantly faster then the other, the bi-exponential character of the kinetics trace will not be discernible, and the reaction may appear as if it were concerted.

Direct observation of intermediates (or lack thereof) provides credence to any mechanistic assignment. Integrated rate expressions for the intermediates will generally be less convoluted than the products since they are further upstream in the kinetic cascade. However, it is often difficult to observe independent spectroscopic signatures for each of the four PCET states. This is partly a consequence of the inherent coupling between electronic states and protonic states in PCET systems. In addition, PCET systems have not incorporated design elements for independent spectroscopic signatures of the proton and the electron.

If it is determined that PCET occurs in a concerted step, no further information can be obtained about the two-dimensional trajectory from kinetics alone. It is conceivable that a concerted PCET reaction could be dominated by either the proton or the electron solvent coordinate (thus approaching the edges of the four-state diagram in what is termed asynchronous PCET); however if no intermediate states are transiently populated, ensemble kinetics measurements only reveal the time-dependent statistical populations of the initial and final states rather than the reaction coordinate itself. For example, consider a concerted PCET reaction where $k_{PCET} = 10^9$ s^{-1}. The reactant population is depleted on a nanosecond time-scale, yet an individual electron/proton tunneling event could still occur on a sub-picosecond timescale, where the remainder of the one nanosecond is spent in inconsequential fluctuations of the solvent coordinates. Even if there is a strong spectral signature to characterize the actual transfer event, it will contribute a mere 'blip' in the time-course of the reaction whose ensemble kinetics are defined on a longer timescale. This blip will be lost in an ensemble measurement because molecules are not negotiating the reaction coordinate in unison. A deeper understanding of the two-dimensional trajectory can be gained by correlating the concerted PCET rate constant to the effect of the various other parameters that probe the nature of the transition state.

17.2.1
Free Energy Correlations

A general understanding as to how changes in reaction free energy affect kinetics has been developed for selected reaction types. In such reactions, a relative displacement of reactant and product surfaces ($\Delta\Delta G^\circ$) is reflected in a proportional displacement of the barrier between them ($\Delta\Delta G^\ddagger$). This leads to linear correlations between reaction free energies and the activation energies (or natural log of rate constants) over a narrow range of reactants and energy differences, assuming factors like $\Delta S^\ddagger$ and electronic coupling are constant. These principles are responsible for the success of Hammett free energy correlations, Brönsted plots for PT reactions, Marcus theory for ET reactions and Evans–Polanyi relations for HAT reactions involving organic radicals [31–34]. In concerted PCET reactions, insight as to the nature of the transition state may be garnered from free energy correlations. For example, an asynchronous transfer that is dominated by the ET coordinate will have a polarized transition state and exhibit a rate correlation with E_{red} (or ionization energy). Conversely, a genuinely synchronous HAT reaction will have a neutral transition state and exhibit a rate correlation with BDE. This issue is discussed in depth in Section 17.3.1.

17.2.2
Solvent Dependence

The solvent dielectric constant is a measure of how effectively the solvent medium screens the force between separated charges [35]. All PCET geometries except HAT involve some degree of charge separation, therefore the surface that defines the free energies of the four states and the barriers between them is a function of the solvent dielectric constant. Indeed, fluctuations in the electron and proton solvent coordinates drive these PCET reactions [5, 23, 24]. The extent to which electron and proton solvent coordinates contribute to the character of the transition state(s) is subject to the geometric constraints of the system, namely the origin and destination of the proton relative to the electron. For example, if PT is confined to a relatively short distance within a long-distance ET coordinate, as in the collinear systems described in Section 17.3.2, the solvent dependence is more ET-like.

17.2.3
Deuterium Kinetic Isotope Effects

The deuterium kinetic isotope effect (KIE) for a PCET reaction is defined as the ratio of rate constants for the reactions involving a protonated species and the deuterated analogue (k_H/k_D) [36–38]. This ratio reflects the degree of proton motion required to reach the transition state, and is one of the most tangible kinetic benchmarks through which theoretical PCET rate expressions can connect with experimental observations [16, 23, 28]. Classically, a reaction whose rate-determining step is breaking a bond to hydrogen will proceed faster than the deuterium case because the reaction coordinate must involve a hydrogen (deuterium) vibrational mode. The heavier deuterium atom has a lower zero-point energy in this mode, which is manifest in a higher activation energy to reach the transition state. Large KIEs can result when the proton quantum-mechanically tunnels, because tunneling rates scale according to the overlap between reactant and product wavefunctions. Vibrational wavefunctions involving the heavier deuterium atom are less diffuse than those of hydrogen, leading to less favorable tunneling overlap for deuteron transfer. This differentiation is amplified when the transfer distance is increased. Finally, it is important to note that the observation of KIE $\neq 1$ does not necessarily prove formal PT in the rate-determining step. For example, proton vibrational motion or displacement that does not result in PT can still be important to the reaction transition state geometry thus inducing a small KIE [39–41].

KIEs are frequently measured to determine whether PCET reactions proceed via stepwise or concerted mechanisms. Small KIEs are frequently cited as evidence for an asynchronous PCET or stepwise ET/PT reaction whereas large KIEs often are attributed to HAT [42–48]. However, KIEs exhibit pronounced distance dependences that can compromise such generalities. A large KIE can result, even for asynchronous PCET, if the transfer of the proton occurs over a large distance [28]. Conversely, a HAT can have a modest KIE if the transfer is occurring over a very compressed length scale [16]. Inverse KIEs (< 1) are encountered under rare circumstances, for example, when there are more thermally accessible deuteron states contributing to PCET [49]. KIEs are often complemented with other kinetic correlations to resolve ambiguities arising from intermediate magnitudes of KIE.

17.2.4
Temperature Dependence

Reaction rate expressions can be partitioned into two general terms: (i) an exponential thermal activation energy term that typically dominates the temperature dependence of reactions; and (ii) a pre-exponential term representing the probability of crossing from the reactant to the product surface in the thermally activated complex. Measuring the temperature dependence of reaction rates and fitting to the appropriate rate equation allows these terms to be separated experimentally, which is particularly valuable for interpreting PCET kinetics. For example, temperature dependence measurements reveal the activation barrier of a concerted PCET reaction for comparison with those of ET or PT steps. Comparison of pre-exponential terms exposes the effect that a PT network has on mediating electronic coupling. Temperature dependent rate measurements can be used to complement deuterium KIE measurements and determine whether the isotope effect originates from differences in activation energies or coupling.

17.3
Unidirectional PCET

17.3.1
Type A: Hydrogen Abstraction

HAT between X and Y is described as follows,

$$X–H + {}^{\bullet}Y \rightarrow X^{\bullet} + H–Y \tag{17.2}$$

The electron comes from the X–H bond (typically sigma), and transfers collinearly with the proton to become part of the new H–Y bond. Eq. (17.2) aptly describes the hydrogen abstraction reactions of organic radicals. Figure 17.4(a) shows the traditional arrow-pushing scheme for the reaction where a donor atom X provides both the electron and proton to atom Y, which accepts them both. The reaction is driven by hydrogen atom acceptors that exhibit a high affinity for an electron and a proton together. For example, ${}^{\bullet}$Cl is an exceptionally reactive HAT participant as opposed to its more inert periodic table relative, ${}^{\bullet}$I, because ${}^{\bullet}$Cl hydrogen abstrac-

A. Synchronous PCET (HAT)

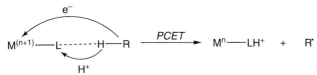

electron and proton localized as part of the same bond

B. Asynchronous PCET

$M^{(n+1)}$——L- - - - - - -H——R $\xrightarrow{PCET}$ M^n——LH$^+$ + R$^\bullet$

electron and proton ultimately are transferred to different locations

Figure 17.4 Schematic representation of the differences between HAT involving radicals and PCET involving transition metal complexes.

tions are more exothermic by roughly 32 kcal mol^{-1} compared to that of $^\bullet$I. Similarly, radical reactions employing $^\bullet$Br are typically much more selective than the corresponding reactions involving $^\bullet$Cl, due to the fact that the reaction with $^\bullet$Br is 15 kcal mol^{-1} less exothermic [50].

Based on Eq. (17.2), early treatments of HAT emphasize the orbital contributions of the singly occupied HOMOs of the donor and acceptor and bond strengths. From the perspective of PCET models [16], the close-range linear orbital pathway makes it reasonable that the electron and proton transfer adiabatically in a synchronous manner along the diagonal of Fig. 17.2, avoiding a polarized transition state. Indeed, such reactions do contain signatures of synchronous transfer through a neutral transition state. For example, they exhibit a linear correlation of log k with BDE, lack of dependence on solvent dielectric constant, and KIEs indicative of bond breaking in the transition state.

Equation (17.2) has been extended by substituting the organic oxidant $^\bullet$Y with a transition metal complex. The case for HAT begins with a discussion of the thermochemistry presented in Fig. 17.1. Mayer and coworkers have defined the thermochemistry of proton-coupled oxidation of dihydroanthracene (DHA) by a Fe^{III} bis-imidazole complex (Fe^{III}Hbim). This PCET reaction is shown in Table 17.1(a), with the deprotonated Fe^{III}Hbim and DHA reacting to give the fully protonated and reduced Fe^{II}H$_2$bim and monohydroanthracenyl radical (HA$^\bullet$) [51]. The thermodynamics of the individual PCET, ET and PT steps have been measured for this system and are shown in Fig. 17.5 [51–54]. ET from DHA to Fe^{III}Hbim is uphill by $\Delta G^\circ = 55$ kcal mol^{-1} and PT is uphill by $\Delta G^\circ = 32$ kcal mol^{-1}. By comparison, the concerted PCET process is only uphill by ~2 kcal mol^{-1}. It is therefore concluded that the oxidation of DHA by Fe^{III}Hbim has a large thermochemical bias that favors the concerted PCET process over either stepwise pathway. Furthermore, the experimentally determined activation barrier ($\Delta G^\ddagger = 22$ kcal mol^{-1}) is significantly smaller than the energy required to populate either of the potential intermediate states, ruling out both stepwise mechanisms.

Tab. 17.1 Various mechanistic studies of PCET involving transition metal complexes.

PCET mechanistic studies

a

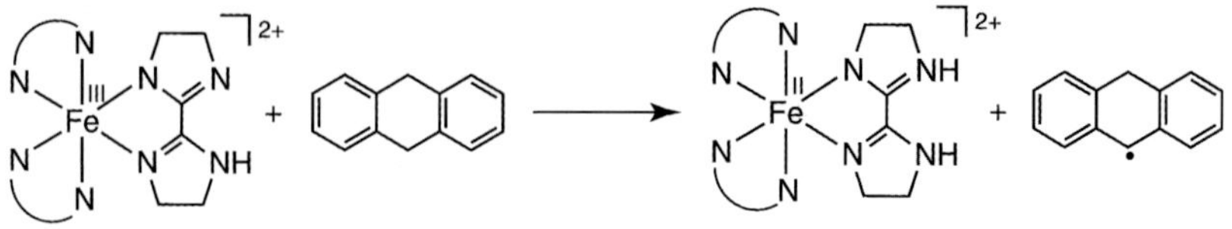

b $\quad$ $MnO_4^- + R{-}H \longrightarrow MnO_4H^- + R^\bullet$

c $\quad$ $[(N_4Py)Fe^{IV}{=}O]^{2+} + R{-}H \longrightarrow [(N_4Py)Fe^{III}{-}OH]^{2+} + R^\bullet$

d $\quad$ $[(bpy_2)(py)Ru^{IV}{=}O]^{2+} + R{-}H \longrightarrow [(bpy_2)(py)Ru^{III}{-}OH]^{2+} + R^\bullet$

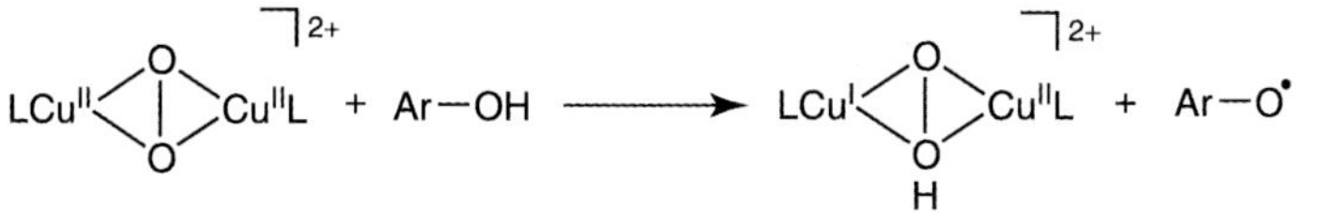

e

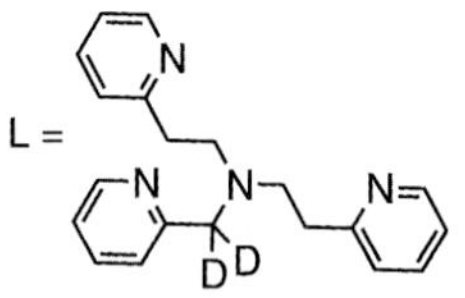

f

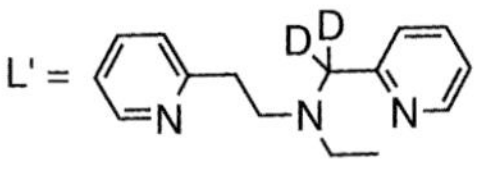

The evidence for authentic HAT reactions involving reactions in which $^\bullet Y =$ metal complex is derived from the correlation of the log of the observed rate constant, $\log k$, to the X–H bond strength. The treatment is effectively encompassed by an Evans–Polanyi relationship [31–34], which relates the driving force of a series of hydrogen atom abstraction reactions to the activation energy (E_a) [55–57]. The connection of E_a to enthalpic quantities such as bond strength is most robust when similar hydrogen atom donors and hydrogen atom acceptors are compared, based on the assumption that the transition state entropy and geometry are constant across the series [58]. The oxidation of typical hydrocarbons of varying hydrogen atom donor ability by permanganate [59] is exemplary of this approach. Table 17.1(b) illustrates that the rate constants and activation enthalpies of hydrogen atom abstraction correlate well with the change in enthalpy for the abstraction process. This change in enthalpy is a measure of the difference between the

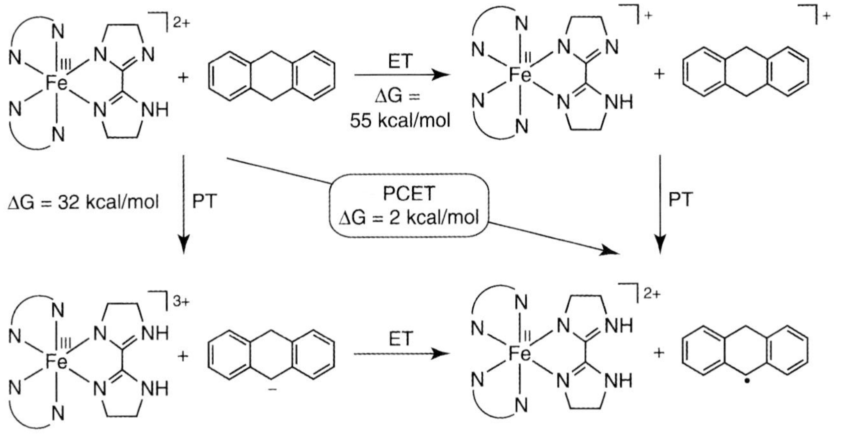

Figure 17.5 Thermochemical scheme for reaction between $Fe^{III}Hbim$ and dihydroanthracene. Figure adapted from Ref. [51].

strengths of the substrate C–H bond which is broken and that of the O–H bond which is being formed. The successful application of Evans–Polanyi correlations of the type shown in Fig. 17.6 for transition metal oxidants has led to the generalization that these reactions proceed by a synchronous PCET process that is mechanistically identical to hydrogen abstraction by an organic radical oxidant [52, 54].

We note however that the reactivity summarized in Table 17.1 for transition metal acceptors fundamentally differs from their organic counterparts in one important aspect – the proton and electron accepting sites for the reactions in Table 17.1 are distinct. It has been well documented that transition metal complexes that are capable of abstracting hydrogen atoms from substrates do not need to have unpaired spin density at the abstracting atom [52]. With the unpaired spin residing mainly at the transition metal center, upon completion of a PCET event, the electron is transferred to the metal ($M^{(n+1)} \rightarrow M^n$) while the proton comes to rest at the ligand. The same is true for a variety of hydrogen abstraction reactions accomplished by metal oxo complexes. Hydrogen is transferred to the oxo ligand with concomitant reduction of the metal. More generally, HAT is distinguished from the organic radical reactivity as illustrated in Fig. 17.4(b). Because the sites for electron and proton transfer are site-differentiated, the possibility for charge separation between the electron and proton is more likely and a PCET description of the overall kinetics, as opposed to a more classical HAT one, is more accurate. This has been theoretically described for the specific case of Table 17.1(a) [60].

The challenge of pinning down the exact nature of the reaction mechanism, whether it is a synchronous PCET or not, from BDEs is highlighted by the oxidation of hydrocarbons by the nonheme $Fe^{IV}{=}O$ complex [61–64] shown in Table 17.1(c). Figure 17.7(a) plots the measured second order rate constants for the oxidation of hydrocarbons by $[(N_4Py)Fe^{IV}{=}O]^{2+}$ against the BDEs of the hydrocarbons

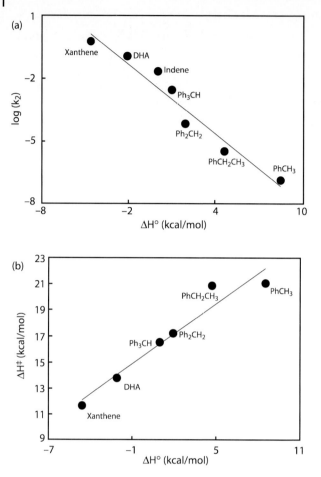

Figure 17.6 (a) Log k versus $\Delta H°$ for PCET between various hydrocarbons and Bu$_4$NMnO$_4$ (Table 3.1b); (b) relation between $\Delta H^{\ddagger}$ and $\Delta H°$ for PCET between various hydrocarbons and Bu$_4$NMnO$_4$ (Table 3.1b). Figure adapted from Ref. [59].

[64]. A good correlation holds between the substrate BDE and log k for oxidation. We note, however, that the reaction is performed in acetonitrile as a solvent. The BDE of acetonitrile (BDE$_{C-H}$ = 94.8 kcal mol^{-1}) [65, 66] is less than that of cyclohexane (BDE$_{C-H}$ = 99.5 kcal mol^{-1}), which is oxidized at an appreciable rate. The inertness of acetonitrile is all the more striking inasmuch as its concentration as a neat solvent is 10–10^2 times greater than the hydrocarbon substrate. The oxo complex does decompose slowly over time at a rate of 5.8 × 10^{-5} s^{-1}. If one assumes that this decomposition rate is essentially equivalent to the rate of hydrogen atom abstraction of acetonitrile by the [(N$_4$Py)Fe^{IV}=O]$^{2+}$, then the Polanyi correlation for BDE is compromised (Fig. 17.7(a)). Conversely, acetonitrile as a substrate for oxi-

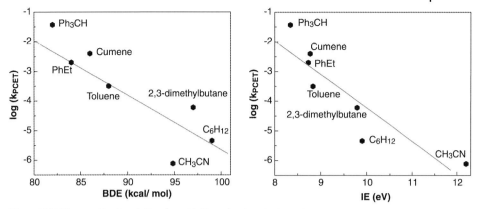

Figure 17.7 Thermo-kinetic relation for oxidation of various hydrocarbons by a $Fe^{IV}=O$ complex (Table 3.1c). Figure adapted from Ref. [64].

dation is described by a plot of $\ln k_2$ (in M^{-1} s^{-1}) vs. the ionization potential (IE) of the substrate. An equally strong correlation is observed for all the data, including acetonitrile (Fig. 17.7(b)) [67]. The IE correlation might suggest a PCET mechanism that is asynchronous, with an ET-like transition state in the upper right section of Fig. 17.2. Alternatively, the Evans–Polanyi relation may breakdown for acetonitrile owing to different entropic contributions to the transition state and thus the HAT correlation prevails for $[(N_4Py)Fe^{IV}=O]^{2+}$ reactivity. At the same time, the system exhibits high KIEs for substrate oxidation [64], ostensibly supporting a HAT mechanism. But it has been suggested that asynchronous PCET can lower the frequencies of transition state vibrations that contribute heavily to the KIEs [67–70]. In such cases, anomalously high kinetic isotope effects can be observed for an asynchronous PCET as well. The point here is that a simple BDE correlation becomes obscured in the light of the results of Fig. 17.7(b).

Similar discrepancies between BDE and IE correlations appear in other oxidations carried out by transition metal oxo compounds. The reaction of $[Ru^{IV}(bpy)_2$ $(py)O]^{2+}$ with hydrocarbons of varying reducing ability (i.e., varying hydrogen atom donor ability) is shown in Table 17.1(d) [71, 72]. The organic substrates that were studied are listed in Table 17.2 along with the respective BDE, IE and hydrogen abstraction rate data in acetonitrile. Again, the inability to oxidize acetonitrile is surprising given that toluene reacts at $k = 6.4 \times 10^{-3}$ M^{-1} s^{-1}; both acetonitrile ($BDE_{C-H} = 94.8$ kcal mol^{-1}) and toluene ($BDE_{C-H} = 90$ kcal mol^{-1}) [73] have similar enthalpies for hydrogen atom abstraction. In the absence of a reported self decay rate, it is reasonable to assume that the rate of hydrogen atom abstraction involving acetonitrile must be of the order of $k \leq 10^{-7}$ M^{-1} s^{-1}. This estimate is based on the concentration difference of acetonitrile (~19 M) versus substrate (0.1 M) and an expected decay rate that would be ~10^{-2} slower than that recorded for the least reactive substrate studied $k \sim 10^{-3}$ M^{-1} s^{-1}. A plot of the data shown in Table 17.2 with the exclusion of acetonitrile and anthracene shows a very strong correlation between the log k for oxidation of aliphatic hydrocarbons with BDE (Fig. 17.8(a));

the correlation is compromised when acetonitrile and anthracene are included on the plot. As shown in Fig. 17.8(b), if log k is plotted against substrate IE, a good correlation is observed. Although the correlation of Fig. 17.8(b) is somewhat inferior to that of Fig. 17.8(a) with respect to the typical hydrocarbons such as toluene and DHA, all the substrates are accommodated by the IE plot.

Tab. 17.2 Thermodynamic and kinetic data for substrate oxidation by $[(bpy)_2(py)Ru^{IV}=O]^{2+}$, adapted from Ref. [71].

Substrate	k $(M^{-1}\,s^{-1})$	BDE (C–H) (kcal mol^{-1})	IE (eV)
xanthene	5.77×10^2	75.5	7.65
DHA	1.25×10^2	78	–
indene	10.8	78.9	8.14
fluorine	21.9	80	7.91
cyclohexene	0.92	81.6	8.95
cumene	0.033	84.8	8.73
ethylbenzene	0.022	87	8.77
toluene	6.4×10^{-3}	90	8.83
anthracene	0.27	~111	7.44

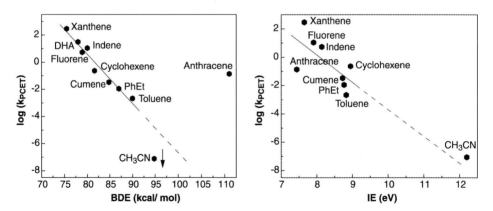

Figure 17.8 Thermo-kinetic relation for oxidation of various hydrocarbons by a $Ru^{IV}=O$ complex (Table 3.1d). Figure adapted from Ref. [71].

The ambiguity in the data presented in Figs. 17.7 and 17.8 arises from the relation between IE and BDE shown in Fig. 17.9. As illustrated by the figure, the BDE_{C-H} and IE for most simple hydrocarbon substrates correlate with one another. This is especially true of the hydrocarbons used in the aforementioned thermo-kinetic correlations (shown within the small box in Fig. 17.9). Even methane, which is difficult to oxidize in a controlled fashion, fits the overall trend. Given the parallel between IE and BDE for most typical hydrocarbon substrates lacking polar functional groups, correlations between log k vs. BDE, and log k vs. IE produce very similar plots and cannot uniquely distinguish between synchronous (HAT) and asynchronous PCET. For this reason, outliers to the correlation in Fig. 17.9 are the most interesting from a mechanistic perspective. These substrates are either aromatic (i.e., benzene and anthracene) or contain either polar functional groups (acetonitrile, formaldehyde and ethanol) or double and triple bonds (i.e., acetylene and cyclohexadiene).

The discrepancies highlighted by Figs. 17.7 and 17.8 may be due to a breakdown of the Evans–Polanyi relation for a HAT reaction or alternatively the contribution of asynchronous PCET induced by the site-differentiation inherent to the metal oxidations as presented in Fig. 17.4. Studies that comprehensively treat the IE vs. BDE issue are therefore valuable. A penetrating study [74, 75] by Fukuzumi, Itoh and coworkers has attempted to address this issue by undertaking a comparative study of the oxidation of a series of phenols by a metal complex vs. an organic radical.

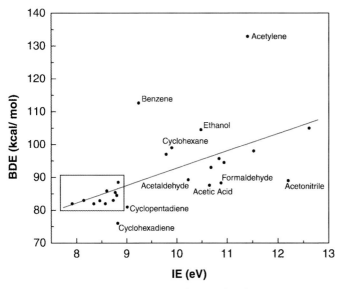

Figure 17.9 Thermodynamic relationship between bond dissociation energy (BDE) and ionization energy (IE) for a variety of hydrocarbons. The small box contains data for hydrocarbon substrates that have been typically used for BDE correlations.

Phenol oxidations by metal complexes employ the $(\mu\text{-}\eta^2\text{:}\eta^2\text{-peroxo})$dicopper(II) and bis$(\mu\text{-oxo})$-dicopper(III) complexes shown in Table 17.1(e) and (f), respectively as the oxidants [75]. In this study, the rate of phenol oxidation was correlated with phenol oxidation potential (E°_{ox}). We note that the use of the oxidation potential is more appropriate than the IEs of Figs. 17.7 and 17.8, since the latter approximates a gas phase internal energy for oxidation whereas the oxidation potential is the free energy for the solution reaction. Reaction of the substrates shown in Table 17.3 with the complexes in Table 17.1(e) and (f) allowed the determination of a Marcus driving force dependence for the oxidation reaction. As performed in the original work, Figure 17.10 presents the rate constants for oxidation vs. the E°_{ox} of the various phenols. In addition the figure includes a plot of the rate data vs. BDE values [66] of the substrate. Whereas there is no discernible correlation between the oxidation kinetics and substrate BDE, the correlation with E°_{ox} is very strong. The possibility of a stepwise ET/PT or PT/ET is ruled out by the slope of the line. A process in which ET is rate determining is predicted to yield a Marcus slope of –0.5 [6]; alternatively, a rate-determining PT step is predicted to yield a slope of

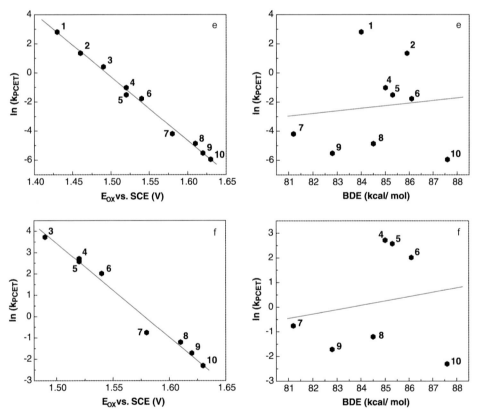

Figure 17.10 Thermo-kinetic relation for oxidation of various phenols of Table 17.3 by a set of dicopper-dioxygen complexes (e and f, see Table 17.1). Figure adapted from Ref. [75].

Tab. 17.3 Thermodynamic and kinetic data for substrate oxidation by $\mu\text{-}\eta^2{:}\eta^2$-peroxo)dicopper(II) (A) and bis(μ-oxo)-dicopper(III) (B) complexes and cumylperoxide radical (C).

ArOH	E^0_{ox} / V (vs. SCE)	k_2 / M^{-1} s^{-1}		
		A	B	C
1 4-OMe phenol	1.43	16.6	–	70
2 2,6-di-tBu phenol	1.46	3.84	–	28
3 4-OPh phenol	1.49	1.51	41.4	8.6
4 4-Ph phenol	1.52	0.36	15.1	8.3
5 4-tBu phenol	1.52	0.22	13.1	9.0
6 4-CH3 phenol	1.54	0.17	7.5	11
7 2,4,6-tri-tBu phenol	1.58	0.015	0.47	61
8 4-F phenol	1.61	0.0077	0.30	–
9 2,6-di-tBu phenol	1.62	0.0040	0.18	17
10 4-Cl phenol	1.63	0.0026	0.10	–

−1.0 [76]. When, however, the rates of electron and proton transfer are comparable and coupled to each other, intermediate slopes are obtained (between −0.5 and −1.0) [76–78]. Accordingly, the measured Marcus slopes of −0.72 and −0.71 support the assignment of a concerted PCET process for the oxidation of the substituted phenols by the Cu metal complexes. Additionally, KIEs of 1.21 to 1.56 are also consistent with an asynchronous PCET process in which there is some proton motion but a large ET component in the transition state.

Oxidation of the same set of substituted phenols listed in Table 17.3 was also undertaken employing the organic oxidant, cumylperoxyl radical (Cum•). Unlike the metal based oxidation, Fig. 17.11 shows that the phenol oxidation rate constants do not correlate well with the substrate E°_{ox} anny better than with substrate BDE. This behavior implicates a more synchronous HAT mechanism for the organic radical oxidant.

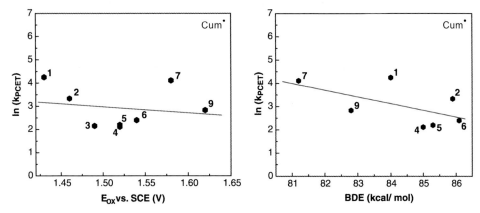

Figure 17.11 Thermo-kinetic relation for oxidation of various phenols of Table 17.3 by cumylperoxide radical (Cum•). Figure adapted from Ref. [75].

We emphasize that all of the hydrogen abstraction reactions described and listed in Table 17.1 are concerted because they are described by a single transition state within the square of Fig. 17.2, rather than proceeding stepwise around the edges of Fig. 17.1. The paramount mechanistic issue that we have highlighted in this section is whether the two-dimensional trajectories are best described by a true HAT (concerted and synchronous) reaction along the diagonal of Fig. 17.2, or a PCET (concerted and asynchronous) reaction with a nondiagonal trajectory. The energy of the transition state can vary remarkably between the two mechanisms depending on the charge separation in the transition state. The PCET reaction may be induced when the electron and proton acceptors are site-differentiated. Additionally, this mechanism may be favored when the hydrogen donor has polar groups (e.g., acetonitrile) that can stabilize a more polar transition state associated with an asynchronous trajectory. To resolve this issue, the relative energy barriers for concerted PCET reactions need to be assessed as a function of substrate. In this regard, PCET/HAT studies taking advantage of broadened substrate scope, solvent effects, and temperature dependences in addition to studies of isotope effects may be able to deliver a unified mechanistic picture.

17.3.2
Type B: Site Differentiated PCET

PCET can occur when the electron and proton are site-differentiated on *both* the donor and acceptor sides of the reaction. The PT coordinate must still be constrained to a hydrogen bond length scale, however, it is feasible for the ET coordinate to span an extended distance [79–81]. Nevertheless, coupling between the electron and proton may be strong since the redox potentials depend on the protonation state and the pK_as depend on the redox state. Consequently, the square scheme of Fig. 17.1 must be used to evaluate the attendant thermodynamics. From a geometric perspective, Type B PCET reactions begin to look less like HAT, and are more reminiscent of a nonadiabatic ET reaction through a hydrogen-bonded bridge. The electrostatic energy terms are dominated by the longer ET coordinate. In this description, the proton occupies a position in the electron tunneling pathway so the electronic coupling term becomes parametric in the coordinate of the proton. This means that ET through the hydrogen bonding bridge can depend on proton fluctuations, regardless of whether or not the proton is formally transferred. Such reactions are especially important in biology since ET in many proteins and enzymes is supported along pathways exhibiting hydrogen bond contacts between amino acid residues and polypeptide chains [82–84].

In model systems used to study Type B PCET reactions, the PCET complex is provided by self-assembly, which entails the noncovalent interaction of two or more molecular subunits to form a complex structure whose properties are determined by the nature and positioning of the molecular components [85–91]. The strengths and geometries of hydrogen bonds through which ET occurs can profoundly affect the PCET kinetics [92, 93].

17.3.2.1 PCET across Symmetric Hydrogen Bonding Interfaces
Type B PCET has been examined at a mechanistic level following the strategy outlined in Fig. 17.12. In this scheme, PCET is photoinduced between an electron donor (De) and acceptor (Ae) spanned by a hydrogen bonding interface ([H$^+$]), which contains the Dp–Ap pair. The first De–[H$^+$]–Ae construct to be studied was supramolecule 1, which is formed from the association of a Zn^{II} porphyrin donor with a 3,4-dinitrobenzoic acid acceptor by a carboxylic acid dimer interface [94]. In nonpolar, nonhydrogen bonding solvents such as CH_2Cl_2, 1 assembles with an association constant of $K_A = 552$ M^{-1}, as determined by IR shifts upon titration of the two components. This result is in good agreement with the association constants measured by static fluorescence quenching ($K_A = 698$ M^{-1}). Static quenching of 1 was not observed when the carboxylic acid dimer was disrupted by the addition of polar hydrogen bonding solvents or upon esterification of the donor and acceptor pair.

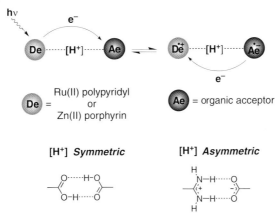

Figure 17.12 General strategy for the study of photoinduced PCET in model DeDp–ApAe systems.

The charge transfer properties of **1** were monitored optically by following the formation and decay of Zn^{II} porphyrin cation on picosecond time scales. The forward and reverse PCET rate constants of protonated **1** were $k_{fwd} = 5.0 \times 10^{10}$ s^{-1} and $k_{back} = 1.0 \times 10^{10}$ s^{-1}. The deuterated analog of **1** gave a rate constant of $k_{fwd} = 3.0 \times 10^{10}$ s^{-1} and $k_{back} = 6.2 \times 10^9$ s^{-1}, yielding KIEs of 1.7 and 1.6 for the forward and reverse PCET reactions, respectively. As theoretically elaborated [20, 95], it is this observed deuterium isotope effect that reveals coupling between electron and proton.

The results obtained for **1** provided the first direct experimental measurement of PCET in a model system. Comparison of the kinetics recorded for **1** with those of covalently linked model systems with similar donor–acceptor separation and driving force [96], reveals that the rates of ET for these systems are of a similar order of magnitude. These results confirmed that hydrogen-bonding pathways for ET can in fact compete with electronic tunneling through covalent bonds [82, 97] as predicted by tunneling models [98].

Attempts to probe the electronic coupling through hydrogen bonds more quantitatively were made using supramolecular dyad **2** [99]. This dyad is similar to **1**, inasmuch as a Zn^{II} porphyrin functions as De and a carboxylic acid dimer functions as the hydrogen bonding interface. The main distinction between **1** and **2** lies in the identity of the Ae. For **2**, it is a Fe^{III} chloride porphyrin as opposed to the dinitrobenzoate. Time-resolved fluorescence quenching of the Zn^{II} porphyrin across the hydrogen bonding interface yields a quenching rate of $k = 8.1 \times 10^9$ s^{-1},

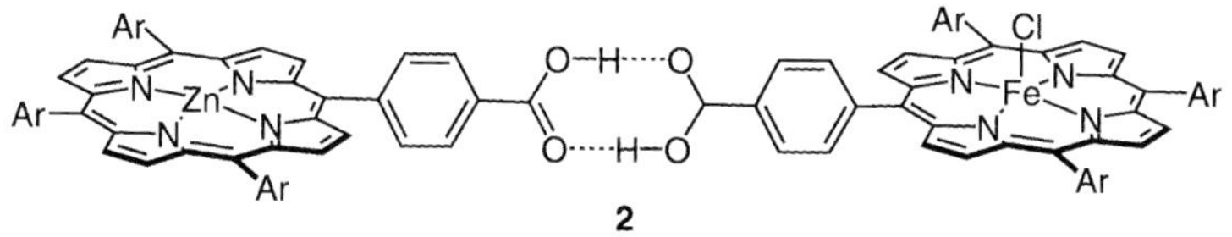

2

which was attributed to ET. This result was compared to the rate of decay of the Zn^{II} porphyrin singlet excited state of **3** and **4**, which juxtapose the same two redox sites between covalent σ (**3**) and π (**4**) bridges containing the same number of bonds as **2**. The rates measured for **3** and **4** were $k = 4.3 \times 10^9$ s^{-1} and $k = 8.8 \times 10^9$ s^{-1}, respectively [99]. The fact that the ET rate constant is nearly doubled when comparing the carboxylic acid bridged **2** with the system bridged by the saturated carbon bridge (**3**) was attributed to greater electronic coupling mediated by the hydrogen bonding scaffold. This result opposes *ab initio* analysis of **2**–**4** [83] that predicts a weaker electronic coupling through the hydrogen bond interface of **2**. The discrepancy may be resolved once the driving forces of the systems are considered. In the experimental study, redox potentials were deduced from porphyrin precursors rather than the electron transfer reactants themselves. **2** may have a higher driving force for ET and hence the free energy, as opposed to the electronic coupling, is the reason for the faster observed rate of the hydrogen bonded complex. Furthermore, energy transfer, and not electron transfer, from the singlet excited state of the Zn^{II} porphyrin to the Fe^{III} porphyrin may be the primary contribution to the quenching reaction mechanism. Recent TA studies have established that energy transfer prevails between Zn^{II} and Fe^{III} porphyrins for five-coordinate high-spin heme centers [100], as is the case for **2**–**4**. Only when the heme is coordinated by two strong-field ligands does the quenching mechanism switch from an energy transfer to an electron transfer.

Type B PCET systems of Fig. 17.3 may also be assembled using the three-point hydrogen bond of Watson–Crick base pairs such as guanine (G) and cytosine (C). Sessler and coworkers provided the first example of this assembly with **5**, for which only energy transfer is observed [101, 102]. A Type B PCET is realized when the cytosine of the GC base pair is appended with an Ae functional group. In **6**, a Zn^{II} porphyrin serves as De and *p*-benzoquinone as Ae [103]. Time resolved fluorescence quenching experiments reveal that the rate of ET across the GC interface

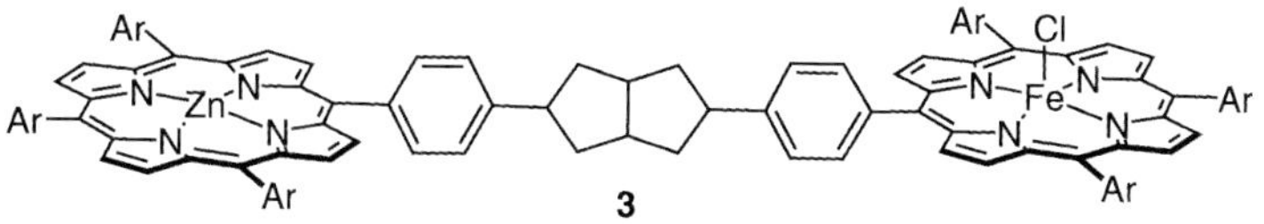

3

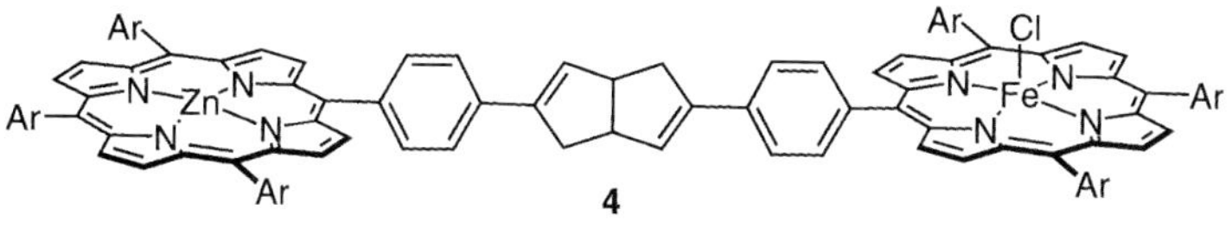

4

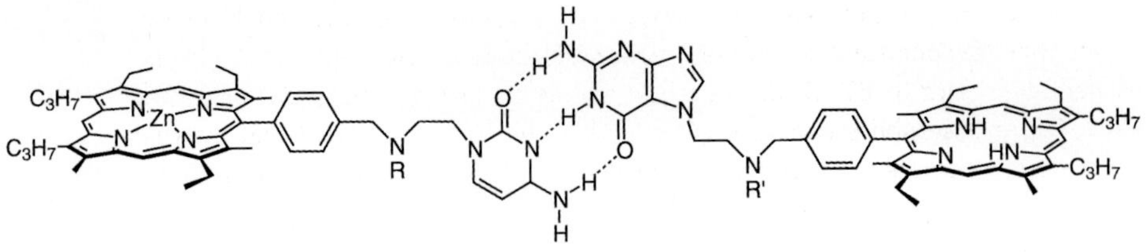

5

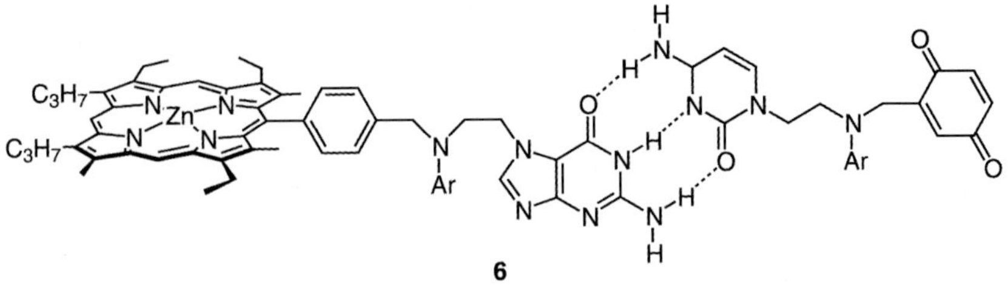

6

is k_{ET} = 4.2 ×10⁸ s⁻¹. Control experiments carried out with porphyrin and benzo-quinone subunits that lack the Watson–Crick base pair functionality show no fluorescence quenching at similar concentrations, clearly indicating that the proton interface mediates the charge transfer event.

Although the distance from the porphyrin De of **6** to the benzoquinone Ae is on the order of ~20 Å, the complex as a whole is highly flexible with a myriad of possible conformations that can allow ET to occur through space. To address this concern, more rigid systems such as **7** have been developed [104]. The greater rigidity of **7** as compared to **6** may be evident from its higher association constant (K_A = 1.6 × 10⁴ M⁻¹). Time-resolved emission experiments reveal that the rate of ET (k_{ET} ~ 8 × 10⁸ s⁻¹) for the rigid system was nearly double that for the flexible complex. This is a noteworthy result, considering that the driving force for ET in **7** (ΔG° ~ -43 kJ mol⁻¹) is substantially less than that for **6** (ΔG° ~ -96 kJ mol⁻¹). The complexities of photoinducing a PCET reaction become apparent when W-band

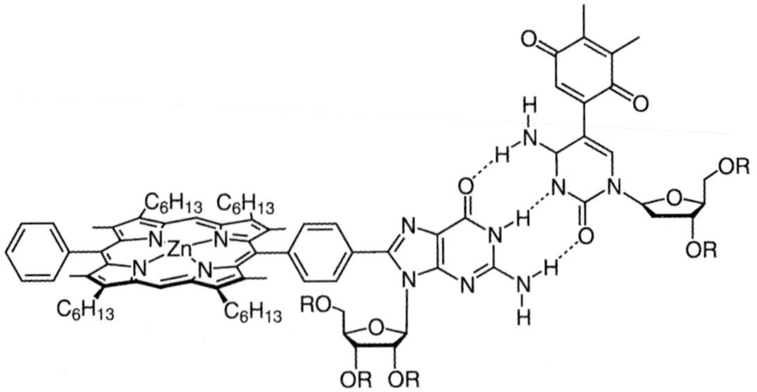

7

time-resolved EPR experiments are performed on **8** [105, 106]. EPR signals are observed that are consistent with intersystem crossing from the singlet to the triplet excited state prior to ET. Unlike covalent De–Ae ET systems, the PT interface of DeDp–ApAe assemblies can strongly modulate the ET to the extent that internal conversion processes can compete kinetically with the charge transfer event.

8

The dicarboxylic acid dimers and Watson–Crick base-paired hydrogen bonding interfaces described above are similar, in that the protonic networks are uncharged and very little charge redistribution occurs within the interface upon ET from donor to acceptor. For the former, proton displacement from one side of the interface is compensated by the concomitant displacement of a proton from the opposite side owing to the inability of a carboxylic acid to support two protons. The same is true for the Watson–Crick base pairs, which are not easily ionized [107]. Since charge redistribution within these types of interfaces is negligible, the only mechanism available to couple electron and proton arises from the dependence of electronic coupling on the position of the protons within the interface [19, 20, 95]. As a result, the effect of the proton on the ET kinetics for symmetric hydrogen bonding interfaces is small and derived mostly from the electronic coupling matrix element as opposed to Franck–Condon terms.

17.3.2.2 PCET across Polarized Hydrogen Bonding Interfaces

A more pronounced role of the proton in the photoinduced PCET event can be imposed in model systems that contain a hydrogen bond interface composed of Dp and Ap pairs possessing pK_as that can support the transfer of a proton. We have designed PCET networks assembled from asymmetric amidinium–carboxylate salt bridges. In the solid state, the amidinium–carboxylate interface is ionic in nature and combines the dipole of an electrostatic ion-pair interaction within a hydrogen bonding network [108, 109]. The amidinium–carboxylate salt bridge interaction is similar to the arginine–asparatate (Arg–Asp) salt bridge, which is an important structural element in countless biological systems including dihyrofolate reductases [110], cyctochrome c oxidase [111–115] and nitric oxide synthase [116–118] but at the same time is more amenable to PCET studies. As shown in Fig. 17.13, whereas the guanidinium group of arginine has multiple carboxylate binding interactions, the amidinium group can adopt only one two-point binding mode for carboxylate, thereby simplifying the supramolecular chemistry for donor–acceptor association.

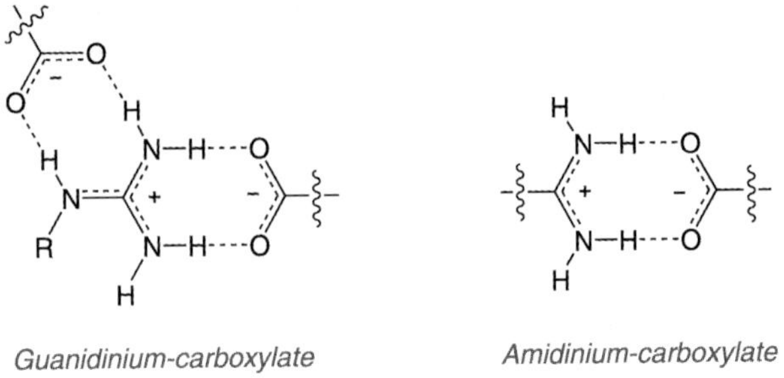

Guanidinium-carboxylate   Amidinium-carboxylate

Figure 17.13 Two-point binding modes for guandinium–carboxylate and amidinium–carboxylate salt-bridges.

The installation of an amidinium moiety on the periphery of De units permits the formation of supramolecular De–[H+]–Ae systems for the study of PCET. Interaction of an amidinium group with a carboxylate yields a unidirectional and remarkably stable two-point hydrogen bond, making it an effective association motif for the formation of such supramolecular complexes [119–122]. The location of the proton within the [H+] interface and the nature of its potential energy surface is of primary importance when considering the proton's effect on ET kinetics. One important factor to consider is the relative pK_as of Dp and Ap. Until recently, the pK_a of a donor amidinium in organic solutions had not been discerned owing to the absence of a simple experimental observable. To confront this issue, purpurin **9** was prepared, in which conjugation between amidinium functionality and chromophore permits the protonation state of Dp to be ascertained simply by monitoring shifts of the Soret and Q bands of the porphyrin framework [123]. Titration of **9** with bases of known pK_a reveals that the amidinium purpurin (pK_a

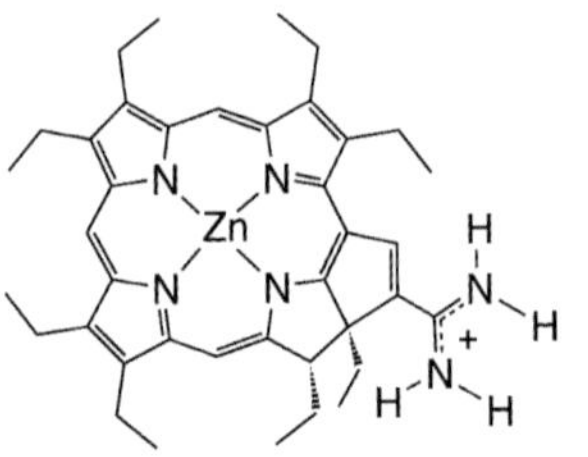

9

= 9.55 ± 0.10 in CH$_3$CN) is considerably more acidic than carboxylic acids such as benzoic acid and acetic acid in organic solvents (pK_a values of 20.1 and 22.3 in CH$_3$CN, respectively [124]). These relative pK_a values imply the prevalence of the amidine–carboxylic acid tautomer (Fig. 17.14, left) in low dielectric environments as opposed the amidinium–carboxylate salt bridge (Fig. 17.14, right), which is expected on the basis of aqueous pK_a values (amidinium salts, pK_a ~ 11–12 and carboxylic acids, pK_a ~ 5–6 [125]).

The precise nature of the tautomeric form that dominates the PCET assembly in low dielectric solvents has been elucidated by examining the optical spectrum of **9** bound to various anionic conjugates. Figure 17.15 shows that the spectrum of **9:benzoate** shifts relative to that of the purpurin amidinium **9**. The observed shift matches the shift seen upon deprotonation of the amidinium to amidine with DMAP, suggesting that Dp within the **9:benzoate** complex is more amidine-like than amidinium-like. As a comparison, the **9:phenylsulfonate** complex was also assembled and the optical spectrum in this case (Fig. 17.15) shows the definite presence of amidinium, and not the amidine. The phenylsulfonate is much more acidic than benzoate in organic solvents, thus explaining the differences between the two associated pairs. Comparison of these spectra leads to the conclusion that the [H$^+$] interface is on the razor's edge of the two tautomeric forms of Fig. 17.14. That such a delicate balance exists between tautomers is not an obvious prediction based solely on ΔpK_as. The ~10 pK_a unit difference in **9:benzoate** favors the amidine–carboxylic acid form by ~ 0.6 eV. However, a simple electrostatic calculation [123] for a positive and negative charge at a salt bridge distance of 3.8 Å translates into a stabilization energy of –0.50 eV in the solvent THF, which nearly offsets the stabilization of the amidine-carboxylic acid tautomer derived from the ΔpK_a. Electron-rich carboxylates such as benzoate are sufficiently basic that the amidine–carboxylic acid hydrogen bond interaction prevails while the interface retains its ionic nature for more acidic carboxylic acids and various sulfonic acids.

A case in point for the ionic tautomer comes from kinetics studies of supramolecules **10** and **11** (see Table 17.4). Both compounds juxtapose a modified [Ru(bpy)$_3$]$^{2+}$, De, and a 3,5-dinitrobenzene, Ae, between amidinium-carboxylate and carboxylate-amidinium hydrogen bonding interfaces [126, 127]. The geometry of each of the assemblies was constant, affording a pair of model compounds that directly probe the effect of interface directionality on PCET kinetics. Since PCET

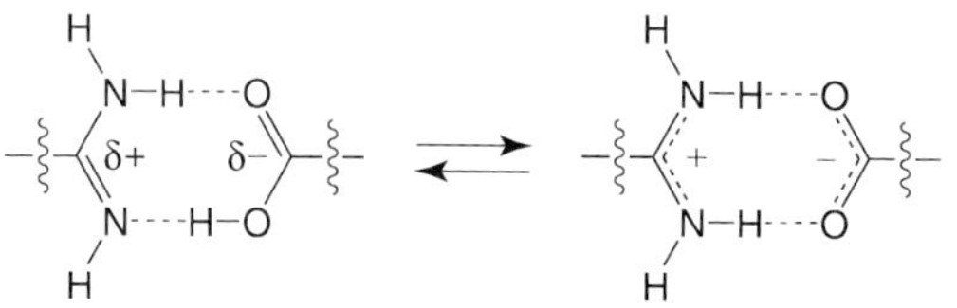

Figure 17.14 The two possible tautomers for assembly of amidines with carboxylic acids. The neutral interface is favored for basic carboxylates in low dielectric solvents. The ionic interface is favored for electron-poor carboxylates and in polar solvents.

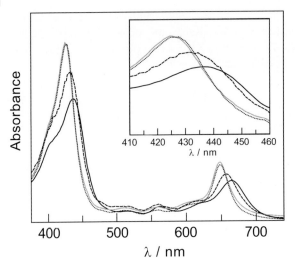

Figure 17.15 Absorption spectra of amidinium purpurin **9** (0.63 × 10⁻⁶ M) in protonated amidinium form (black, solid line), deprotonated amidine form (gray, solid line) and with a large excess of benzoate (8.5 × 10⁻² M) (black, dotted) and a large excess of phenyl sulfonate (black, dashed) in CH_2Cl_2.

is photoinitiated from the long-lived triplet metal-to-ligand charge transfer (^{3}MLCT) state of the Ru^{II} polypyridyl center, intermolecular ET reactions can compete with the desired intramolecular PCET reaction of the self-assembled complex. The intra- versus inter- charge transfer pathways can be deciphered from the concentration dependence of the observed rate constants. As is shown in Table 17.4, the directionality of the asymmetric hydrogen bonding interface has a pronounced effect on the rate of intramolecular ET from De to Ae. The rate constant for PCET in the De-[amidinium-carboxylate]-Ae system (**10**) is roughly 40 times slower than that for the De-[carboxylate-amidinium]-Ae complex (**11**). The large difference in PCET kinetics which is observed upon reversal of the asymmetric hydrogen bonding interface is due in part to the direction of ET relative to the orientation of the salt bridge. For **10**, the direction of the salt-bridge dipole is in the direction of PCET, however, for complex **11** the salt-bridge dipole opposes the direction of charge transfer. The salt-bridge orientation difference also contributes favorably to

10

11

the ET driving force for **11** and unfavorably for **10**. Additionally, differences in hydrogen bond strengths for **10** and **11** may also play a role in the disparate kinetics data since these bonds modulate the electronic coupling through the salt-bridge. Together, these results strongly implicate an ionized amidinium-carboxyl-ate salt bridge, as shown in Fig. 17.14 (right). The result is consistent with the increased acidity of an Ae owing to the presence of the electron-withdrawing groups on the aromatic ring.

Tab. 17.4 Rates for unimolecular and bimolecular electron transfer for DeDp–ApAe complexes with amidinium–carboxylate interfaces.

PCET model system	ΔG°/eV	k_{ET} (M^{-1} s^{-1})[a]	k_{PCET} (s^{-1})[b]
10	−0.14	1.2×10^9	8.4×10^6
11	−0.23	3.3×10^9	3.1×10^8

a The bimolecular reaction of the respective consitituents (non-hydrogen bonded) as determined by Stern–Volmer quenching kinetics.
b Unimolecular electron transfer of the associated DeDp–ApAe pair.

The direction of the PCET reaction may be switched by catenating the Ru^{II} poly-pyridyl photocenter to an electron donor in place of an electron acceptor. In **12** and **13**, the photoexcited Ru^{II} center is reduced by dimethylaniline. In order to accomplish these studies, a redesign of the Ru^{II} polypyridyl complexes was under-taken, such that the excited state electron of the photooxidant could be removed from the charge transfer pathway upon excitation into the MLCT state. This was achieved by modifying the bpy ligands of the Ru^{II} complex with strongly electron withdrawing diethylcarboxyl groups [128]. The rate of intramolecular PCET was determined by time-resolved emission for **12** ($k_{PCET} = 1.7 \times 10^9$ s^{-1}), whereas the switched-interface congener, **13**, showed no evidence of PCET within the ^{3}MLCT excited state lifetime. This comparison is consistent with the comparative kinetic studies of the switched interface systems described above (**10** and **11**) in that the directionality of the bridging interface has a profound effect on PCET kinetics. In both comparisons, charge transfer for De-[carboxylate-amidinium]-Ae assemblies can be very fast, while PCET is attenuated in De-[amidinium-carboxylate]-Ae complexes. As for **10** and **11**, the dominant source of the kinetics disparity in the comparison of **12** and **13** is undoubtedly the amplified thermodynamic bias induced by the orientation of the salt-bridge, which retards PCET when the electron traverses an ionic salt bridge oriented such that the amidinium is on De.

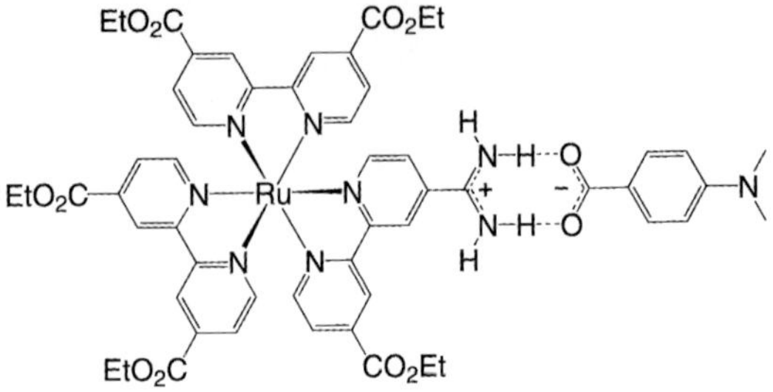

12

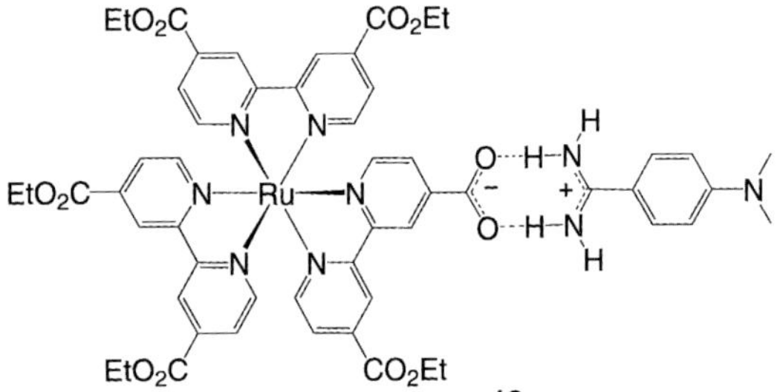

13

The large changes in optical density for ET [129–132] and PT [133–138] products of porphyrins make them ideal candidates for the study of PCET by transient absorption spectroscopy. In the three cases listed in Table 17.5, the porphyrin–amidinium conjugates **14** [139] and **15** [140] employ 3,5-dinitrobenzoic acid as the acceptor. The flexibility allowed by the vinyl bridge of **14** is eliminated by fusing the amidinium functionality directly onto the porphyrin macrocycle at the β-position of the ring. Compound **15**, which provided the first X-ray structural characterization of a De–[amidinium-carboxylate]–Ae assembly, reveals that the amidinium group and the macrocyclic ring can be in electronic structural communication. This conjugation may be slightly attenuated due to amidinium–porphyrin canting. The PCET rates listed in Table 17.5 show that the salt-bridge (as mentioned above, this tautomer should dominate for a 3,5-dinitrobenzoic acid as Ae) attenuates the charge transport rate by $\sim10^2$ when compared to that of covalently linked Zn^{II} porphyrin-[spacer]-quinone (spacer = rigid polycyclic bridge) systems of nearly equivalent driving forces [96]. The attenuated rate for the hydrogen bonding system is even more pronounced given the 3 Å shorter donor acceptor distance for **15** as compared to the covalent system. Indeed, the notable effect that the asymmetric hydrogen bonding interface has on the kinetics of PCET for **15**

Tab. 17.5 Rates for unimolecular and bimolecular electron transfer for porphyrin-based DeDp–ApAe complexes with amidinium–carboxylate interfaces.

PCET model system[a]	k_{PCET} (s^{-1})
14-DNB	7.5×10^8
15-DNB	6.4×10^7

a DNB represents 3,5-dinitrobenzoate.

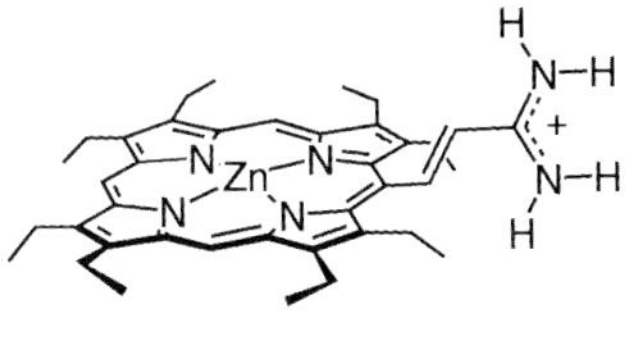

14

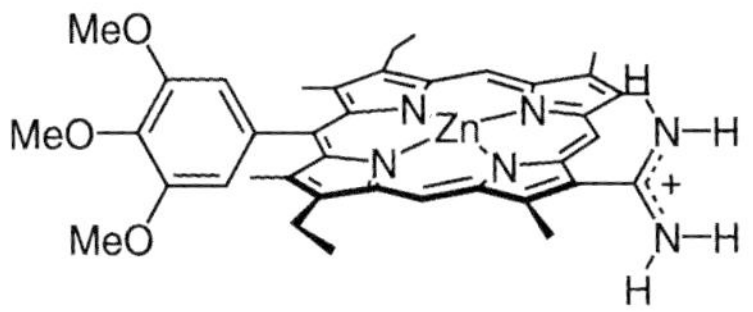

15

suggests that strong coupling of the protonic interface to the electron transfer pathway is enhanced by the direct attachment of the amidinium functionality to the porphyrin redox cofactor.

An alternative approach for the study of PCET across amidinium–carboxylate interfaces relies on the formation of ternary complexes such as that shown for **16** [141]. A N-propylisonicotinamidine serves as a ditopic spacer between a Zn^{II} tetraphenylporphyrin donor and 3,4-dinitrobenzoate acceptor. The PCET assembly is afforded by the axial coordination of the pyridyl group to the Zn^{II} porphyrin ($K_a = 1.9 \times 10^8$ M^{-2} in CH$_2$Cl$_2$). A benefit of this approach is a significantly reduced synthetic investment required for the incorporation of Dp/Ap into the assembly. Moreover, the system is highly modular and the De porphyrin can be

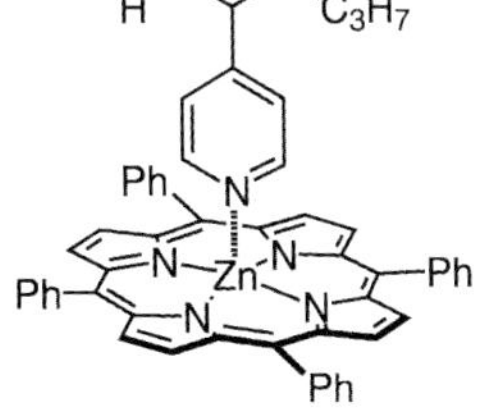

16

changed with relative ease. Fluorescence quenching of the Zn^{II} porphyrin in the presence of the ditopic isonicotinamidine spacer and dinitrobenzoate yields a $k = 1.9 \times 10^9 \text{ s}^{-1}$, which is notably faster than that observed for **14** or **15**. The disparate rate constants for the two types of systems may reflect significantly different electronic coupling for the differing topologies.

In most cases, PCET is implicated without the direct detection of appropriate reaction products. The challenge to the direct kinetic measurement by TA spectroscopy for such systems arises as a direct consequence of the presence of the proton transfer network proximal to the electron transfer pathway. In tightly coupled networks such as those for **14–16**, the hydrogen bonding interface significantly retards the ET rate such that charge transport is slow with respect to the dynamics of the photoexcited porphyrin donor. Large changes in optical intensities associated with $S_1 \rightarrow T_1$ interconversions often overwhelm the absorption signatures of the distinct PCET intermediates. Consequently, the measurement of charge transport in the strongly coupled PCET networks described above has been confined to monitoring the disappearance of the S_1 excited state by time-resolved emission spectroscopy. As discussed previously, this approach can be problematic because the dynamics of fluorescence decay are not necessarily specific to charge transport. Furthermore, PCET can proceed from non-emitting states to which time resolved emission spectroscopy is blind [105, 141]. Accordingly, there is great benefit offered by systems that allow the PCET kinetics to be determined directly from observed charge transport intermediates.

Amidinium porphyrins featuring an amidinium functionality directly attached to the porphyrin framework at the *meso-* and *β*-positions [142], **17** and **18**, respectively, are systems that provide the desired optical signatures of PCET reactions. ^{1}H NMR spectroscopy shows [143], that association of **18** to the naphthalenediimide (NI) carboxylate to form **19** in THF ($K_a = (2.4 \pm 0.6) \times 10^4$) occurs by a two-point hydrogen bond and that π-stacking between the porphyrin macrocycle and diimide acceptor is not a prevalent association mechanism. Carboxylate binding experiments carried out with purpurin **9**, as described above, indicate that the NI carboxylate is basic enough to deprotonate porphyrin amidiniums in organic solvent. Accordingly, it is believed that for dyad **19**, an uncharged amidine–carboxylic acid interface is established by the relatively basic acceptor, as opposed to the ionic amidinium–carboxylate bridge that is observed for the more acidic dinitrobenzoic acid acceptor used in previous PCET studies. TA optical signatures of the PCET reaction products were expected in the 600–700 nm region, namely the one-electron reduced diimide ($\Delta\varepsilon \sim 5000$ at 610 nm) and the porphyrin cation radical ($\Delta\varepsilon \sim 5000$ at 660 nm). However, the growth of the PCET product could not be resolved at most wavelengths due to the dominance of spectral signatures asso-

17

18

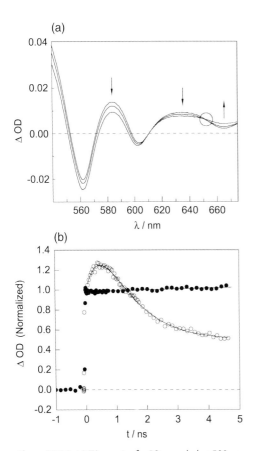

19

ciated with $S_1 \rightarrow T_1$ intersystem crossing in both bound and unbound porphyrins. [143]. This issue was remedied by performing single-wavelength kinetics at the S_1–T_1 isosbestic point ($\lambda_{probe} = 650$ nm) of unbound porphyrin **18**. Figure 17.16 (top) shows a collection of transient spectra for **18** at 500 ps, 1 ns and 2 ns. The small circle indicates the existence of the S_1–T_1 isosbestic point. Monitoring the PCET reaction kinetics at the unique wavelength, at which the dynamics of the

Figure 17.16 (a) TA spectra for **18** recorded at 500 ps, 1 ns and 2 ns. Circle denotes the S_1–T_1 isosbestic point at 650 nm. (b) Pump probe kinetics for **18** (solid circles) and **19** (open circles) at 650 nm.

excited states for unbound **18** are effectively nulled, allows the absorption features of the PCET products to be detected against a "flat" background. The transient absorption data recorded at the S_1–T_1 isosbestic point are shown in Fig. 17.16 (bottom). The solid circles show that the kinetics for **18** are invariant with time because the probe is at an isosbestic point. The same experiment for **19** results in a clear rise and decay that is associated with the formation of the porphyrin cation radical by PCET, followed by its subsequent decay. The kinetics were fit to yield forward and reverse PCET rates of $k_{PCET} = 9.3 \times 10^8$ s^{-1} and 1.4×10^9 s^{-1}, respectively [143]. This work has allowed the reverse rate of a PCET reaction across an amidinium-carboxylate salt bridge to be measured for the first time. In the light of these measurements, it can be seen that the forward PCET rate for **19** is nearly two orders of magnitude slower than those measured for covalently linked Zn^{II} porphyrin–naphthalenediimide dyads of comparable driving forces [144, 145].

The TA results obtained for **19** are noteworthy because this work outlines a method to detect PCET intermediates by transient optical spectroscopy. The propensity of PT networks to retard charge transfer rates has practical consequences for mechanistic studies of PCET reactions. Attenuated rates translate to low yields of PCET intermediates. For this reason, it is difficult to observe PCET intermediates directly by time-resolved methods. Assembly **19** shows, however, that PCET intermediates can be spectrally uncovered when the transient difference signal between S_1 and T_1 excited states is minimized. This procedure, which is similar to one previously exploited in studies of D–A dyads [146] and heme protein–protein complexes [147], opens the door to a host of future experiments designed to directly monitor rates of electron transfer that are strongly coupled to proton motion.

Variable temperature time-resolved experiments and KIE measurements on **19** have provided the most rigorous kinetics information of any Type B PCET system studied to date [49]. The slope and y-intercept of the modified Arrhenius plots (Fig. 17.17) permit the separation of nuclear and electronic contributions to the PCET rate. Weak electronic coupling matrix elements of 2.4 and 1.9 cm^{-1}, respectively, for the protonated and deuterated forms of **19** attest to a coupling bottleneck through the interface. As shown in Fig. 17.17, two different isotope effects are observed at high and low temperatures for **19** with a protonated and deuterated interface. A reverse isotope effect (i.e., $k_H/k_D < 1$) is observed as T approaches 120 K ($k_H/k_D = 0.9$, 120 K) whereas a normal isotope effect (i.e., $k_H/k_D > 1$) is recovered as the temperature is increased ($k_H/k_D = 1.2$, 300 K). The transition between these limits is smooth, with a crossover temperature of $T \sim 160$ K. This trend is interpreted in a model where fluctuations within the hydrogen bonding bridge dynamically modulate electronic coupling for ET, and consequently the rate of charge-separation becomes sensitive to the nature of proton modes within the bridge. Thermal population of vibrational states is the most likely cause of the reverse isotope effect in this system, where the low frequency mode is a localized vibration in the hydrogen bond. At low enough temperatures, the thermally-induced shift in the deuteron probability density contributes more to the PCET rate than the ^{1}H form of the interface (due to the difference in zero-point ener-

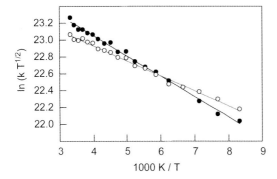

Figure 17.17 Temperature dependence of the rate of PCET in protonated **19** (solid circles) and deuterated **19** (open circles) in the solvent 2-MeTHF. Data are presented in a modified Arrhenius form with linear fits. Figure adapted from Ref. [49].

gies). The normal isotope effect is recovered with increasing temperature as the lowest lying excited states of the hydrogen bond vibration of interest begin contributing to the PCET rate. This microscopic insight into the role of mediating protons is very pertinent to collinear PCET in biology, where asymmetrical hydrogen bonding networks are frequently the bottlenecks for electron transport. However, this level of understanding can only be reached when the relevant dynamics are isolated in well-defined and spectroscopically-accessible model systems such as in **19**.

17.4
Bidirectional PCET

The square scheme in Fig. 17.1 applies equally well to the Types C and D bidirectional PCET schemes of Fig. 17.3. While that statement may seem trivial, it has the profound implication that the thermodynamics associated with a PT reaction can exert control over the rate and even the direction of an ET reaction along a spatially distinct coordinate. However, the thermodynamic square scheme conceals the major caveat to PCET; the transfer of the proton, as the heavier particle, is fundamentally limited to short distances whereas the electron, as the lighter particle, may transfer over very long distances [79–81]. The orthogonalization of ET and PT coordinates resolves the predicament of their disparate length scales while maintaining direct coupling. The electron can tunnel over a long distance and be coupled to a specific short-distance PT event that involves an additional acid or base group positioned nearby. Type C systems are relatively easy to study since the PT coordinate is not controlled. The Type D system is more challenging because a specific structure is imposed on the PT coordinate. In either case, the bidirectional approach is valuable since ET (redox potential) and PT (pK_a) parameters can be varied independently and correlated with resulting PCET rates.

17.4.1
Type C: Non-Specific 3-Point PCET

In this PCET configuration, a donor (De/Dp) transfers an electron to an acceptor (Ae) and loses a proton to the bulk (Eq. (C) in Fig. 17.3). Though there is no specific proton acceptor, the PCET event may be tightly coordinated and correlated to bulk properties of the medium, particularly the pH.

Hammarström and coworkers have conducted a thorough investigation of dissociative PCET from tyrosine (Y) using ruthenium-tyrosine model complexes. The work is motivated by the crucial role that tyrosine residues, play in charge-separation reactions in biology, particularly in Photosystem II (PS II). The model systems do not contain a specific PT coordinate as in PSII [148–156], yet the simplicity of the system permits the balance of stepwise versus concerted PCET to be studied in a controlled fashion.

In the initial model system ([Ru]–Y, **20**) [157], the oxidant is Ru^{III} tris(bipyridine) (designated [Ru]), which is generated from the starting Ru^{II} state by the flash-quench method. The Y moiety is covalently linked to one of the bipyridine ligands via an amide bond. The rate of ET from Y to the metal center of [Ru] as a function of pH was measured in water. These results are reproduced in Fig. 17.18. The data falls into two distinct pH regions, separated by the pK_a (~10) of Y. Above pH 10, the rate constant for charge transfer is ~100-fold greater, and independent of pH. Here, Y is initially deprotonated, and this region simply corresponds to ET from tyrosinate to Ru^{III}. In the low-pH region, the rate constant increases monotonically with increasing pH. Y is initially protonated, and the authors invoke a concerted PCET mechanism to account for this slope. A stepwise ET/PT mechanism can be ruled out because it does not give rise to pH-dependence in the rate of ET from Y to [Ru]. A gated stepwise PT/ET mechanism could give rise to pH-dependence, but this mechanism is also ruled out because the pK_a constrains the maximum rate via this mechanism to ~10 s^{-1}, which is too slow to account for the observed rate constants of >10^4 s^{-1}. The authors interpret their data for both pH regimes in the framework of Marcus' theory for ET [6, 7], which employs three parameters; driving force ($-\Delta G^o$), reorganization energy (λ), and electronic coupling (V). In the region above the pK_a of Y, $-\Delta G^o$ is greater (due to a lower oxidation potential of tyrosinate; $E_Y^- = 0.77$ V vs. NHE) [157] and is independent of pH. Below this pH, $-\Delta G^o$ increases with increasing pH, which is a consequence of the pH-dependent potential of the tyrosine/tyrosyl radical couple ($E_Y = (E_Y^o - [(RTln10/F) \times pH)$, where $E_Y^o = 1.34$ V vs. NHE) [157]. The Marcus treatment indeed predicts a positive corre-

20

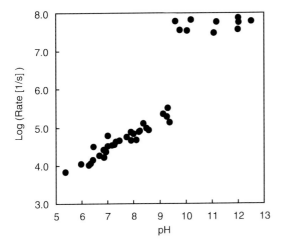

Figure 17.18 pH dependence of Ru^{III} reduction in system **20**.
The slope below pH 10 is indicative of a concerted Type C
PCET reaction. Figure adapted from Ref. [157].

lation between the rate constant and pH for concerted PCET, consistent with measurements in the low-pH region.

The physical nature of the concerted PCET reaction was further explored by measuring the temperature dependences of the rate constants in both the low-pH and high-pH regimes to afford an estimate of the Marcus reorganization energy, λ (after correction for the entropic contribution to ΔG° since the proton is dissociated to the bulk) [157, 158]. The reorganization energy associated with the concerted PCET reaction (pH = 6.5) was found to be 1.4 eV, compared with only 0.9 eV for the pure ET reaction at pH 12. The greater reorganization energy for the concerted PCET reaction was attributed to the contribution of O–H bond breaking to the reaction coordinate. The KIEs of 1.5–3 measured in the low-pH regime [158] also support the idea that an O–H stretching motion is implicated in the reaction coordinate.

Hammes-Schiffer and coworkers offer an alternative description of the same data [159] using their multi-state continuum theory developed specifically for PCET reactions. They also find that the PCET reaction carries a greater reorganization energy than the ET reaction. However, it is in the form of an outer-sphere (solvent) reorganization energy rather than an inner-sphere reorganization energy (associated with O–H bond cleavage). The increased solvent reorganization energy is justified because in a bidirectional geometry, the PT event adds to the charge-separation that the solvent must accommodate. The O–H bond cleavage is nevertheless an essential contribution to the attenuated PCET rate relative to ET in Hammes-Schiffer's model. In this case, it is in the form of a hydrogen vibrational wavefunction overlap factor. In other words, the PCET reaction has an attenuated coupling relative to ET due to averaging over the reactant and product vibronic states. Prior to this work, Cukier had also formulated a theory for dissociative

PCET, where the previous model is modified to include a repulsive surface for proton motion in the final state [21]. The PCET rates are found to be rather sensitive to the nature of the repulsive surface – which is coupled to the solvent.

Hammarström and coworkers developed related model compounds to explore the balance between concerted and consecutive PCET reactions. One such compound ([Ru]–W, **21**) retains the original Ru^{III} tris(bipyridine) core, however, the Y of **20** is replaced with tryptophan (W), which is easier to oxidize ($E_W^\circ = 1.13$ V vs. NHE) but harder to deprotonate ($pK_a(W) \sim 17$, $pK_a(W^{\bullet+}) \sim 4.7$) [160]. A pH-independent rate is observed in the region $4.7 < pH < 9$, beyond which the rate monotonically increases with increasing pH. The pH-independence is indicative of a stepwise ET/PT mechanism, an assignment that is also supported by resolving the ensuing PT step by TA spectroscopy. A third model compound (MeC(O)O–[Ru]–Y, **22**) differs from [Ru]–Y by the ethylester substituents added to the bipyridine ligands [160]. This increases the potential of the Ru^{II}/Ru^{III} couple by ~0.24 eV, while the oxidation potential and pK_a of the remote Y residue remain unperturbed. Interestingly, at pH < 6, the rate constant becomes independent of pH, despite there being a pH-dependent driving force for PCET. This is a signature of switching to a consecutive ET/PT mechanism, analogous to the behavior observed for [Ru]–W at low-pH and in contrast to [Ru]–Y. The temperature-dependence of Ru^{III} reduction in MeC(O)O–[Ru]–Y in the low-pH and high-pH regimes reveal that the reorganization energy for the concerted PCET process is twice that of the ET step in the consecutive mechanism (2.4 eV vs. 1.2 eV). When the measurements MeC(O)O–[Ru]–Y were performed in D_2O, the ET rate remained slow and independent of pH over the entire range studied. This indicates that the concerted PCET mechanism cannot compete with consecutive ET/PT. This leads to the estimate of a deuterium KIE > 10 for the concerted PCET reaction in this system.

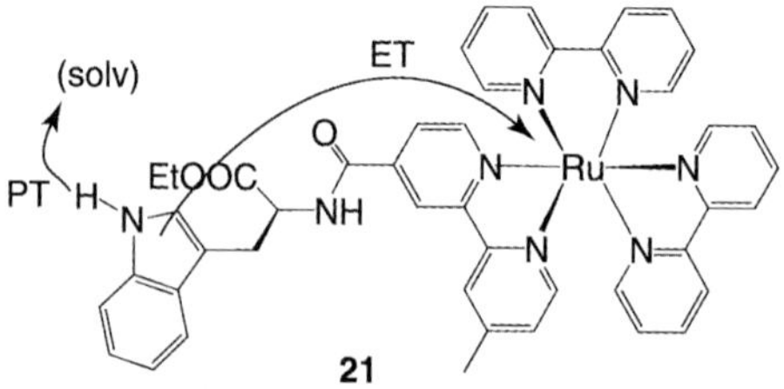

21

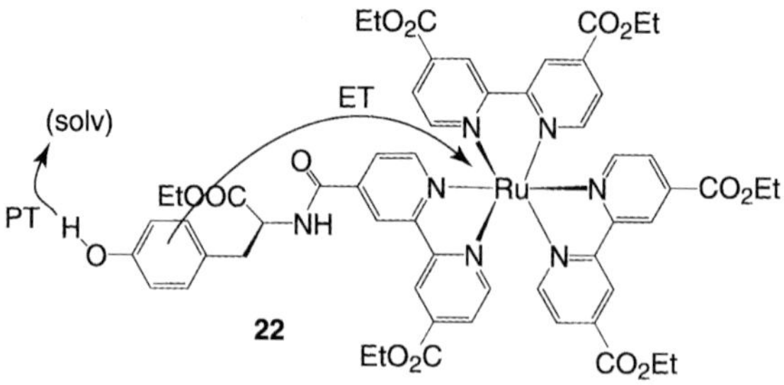

22

Hammarström and coworkers unify their data from the model compounds into the following picture. In a system that can undergo both concerted PCET and consecutive ET/PT to yield the same ultimate product, these two channels are in kinetic competition. The issue of which channel dominates depends largely on the balance of the driving force and the reorganization energy. The driving force is necessarily more favorable for a concerted PCET reaction (because it includes the thermodynamically favorable PT component), but on the other hand, concerted PCET carries the kinetic cost of a greater reorganization energy compared with consecutive ET/PT. Note that Hammes-Schiffer and coworkers' model invokes a vibrational overlap factor rather than reorganization energy as the main basis of the rate attenuation associated with concerted PCET, nevertheless, a parallel argument can be made. The problem reduces to the relative thermodynamics of the ET and PT components in the square scheme in Fig. 17.1. When the driving force for ET is large relative to PT, the small gain in driving force for concerted PCET is outweighed by the kinetic cost in terms of the activation energy. In this situation, the consecutive ET/PT mechanism is expected to dominate. This theory also explains the pH-independent region in [Ru]–W; the proton affinity of W means that PT does not contribute enough additional driving force to offset the attenuation inherent to the concerted mechanism. MeC(O)O–[Ru]–Y also proceeds via the consecutive ET/PT mechanism in the low-pH region, despite there being an appreciable driving force for PT (as in [Ru]–Y). However, the enhanced driving force for the ET component in this system relative to [Ru]–Y means that the additional driving force from the PT component is simply not required. In dissociative PCET systems, the driving force for PT increases with pH, which accounts for the eventual switch to a concerted PCET mechanism for [Ru]–W and MeC(O)O–[Ru]–Y at higher pH. This insight points to concerted PCET as an important mechanism in biology, where thermodynamic driving force for ET is typically conservative.

Type C PCET also describes several studies focused on the oxidation of guanine. Guanine (G) is the most easily oxidized of all nucleobases, and appears to be a trap for oxidative damage in DNA [161–167]. The pK_a of G shifts from 9.5 to 3.9 upon oxidation [161], thereby committing the deprotonation of N1 to accompany G oxidation in this pH regime. In fact, the strong coupling between ET and PT in nucleobases has been the source of substantial uncertainty in the determination of their redox potentials, due to the involvement of PT. Seidel and coworkers carried out a series of electrochemical measurements on nucleobases in aqueous and aprotic solvents [168]. They found that driving forces for ET in aqueous solution can be more favorable by 0.5–0.8 eV, due to coupled PT and hydrophobic interactions.

Thorp and coworkers examined the oxidation of G in duplex DNA using a series of metal-polypyridyl oxidants of varying driving force [78]. A Marcus plot yields a slope of ∼ –0.8. As discussed in Section 17.3.1, a stepwise mechanism with a rate-determining ET step produces a slope of –0.5 whereas a rate-determining PT step produces a slope of –1.0. The observations of an intermediate slope and attendant KIE of 2.1 signify that a PCET mechanism is operable. It is noted that in other studies, G has been oxidized directly to the radical cation, $G^{\bullet+}$, without concerted deprotonation [169–171]. Those systems employed much stronger oxidants, whereas ET-only reactions in Thorp's system are either endothermic or weakly exothermic. Accordingly, it is only necessary to couple to PT in a concerted step when there is otherwise insufficient driving force, consistent with the conclusions from the aforementioned Y and W oxidation studies.

Type C (associative) PCET in Fig. 17.3 describes the microscopic reverse of the nonspecific 3-point PCET in which a proton is lost to solvent. Instead, an acceptor is the recipient of an electron from a reductant and a proton from the solvent. This associative 3-point PCET has been found to occur in pyrimidine nucleobases, particularly cytosine (C), where N3 is protonated upon reduction. Geacintov and coworkers studied photoinduced ET in benzo[a]pyrenetetrol (BPT)–nucleoside complexes in aqueous solution and in the polar aprotic solvent DMSO [171]. Evaluation of the available driving forces for ET indicate that there is insufficient driving force for $^1BPT^*$ to reduce the pyrimidines C and thymine (T), unless reduction is coupled to the uptake of a proton. Accordingly, they find a lack of quenching by C and T in DMSO, however efficient quenching is observed in water. TA spectroscopy confirms the formation of $BPT^{\bullet+}$, and thermodynamics suggests that PT from water must accompany ET to the pyrimidine nucleobase. The authors find deuterium KIEs of 1.5–2.0 when the reactions are carried out in D_2O, which supports the proposed PCET mechanism.

Wagenknecht, Fiebig and their coworkers have used ultrafast spectroscopy to study reductive electron transport in pyrimidine nucleobases on the picosecond timescale. They developed a series of pyrene-modified nucleosides, in which the nucleosides are directly bound to pyrene, which acts as a photoreductant [172–174]. The first model compound (pyrene–dU) employed deoxyuridine (dU) as a model for the deoxythymidine nucleoside [172, 173]. The emission intensity from ^{1}(pyrene)* was measured as a function of pH in water. The emission is completely quenched in the low-pH range, and the fluorescence intensity follows a typical sigmoidal curve, with an inflection point at pH $\approx$ 5.5 reflecting the pK_a of the protonated pyrene$^{\bullet+}$–dU(H)$^\bullet$ biradical. However, TA spectroscopy reveals that the ET product (pyrene$^{\bullet+}$–dU$^{\bullet-}$) is initially formed in both pH regimes, therefore concerted protonation is not requisite for ET in this system, and the pH dependence of the emission intensity cannot simply be interpreted as quenching by concerted PCET in the low-pH regime. Instead, it is believed that pyrene$^{\bullet+}$–dU$^{\bullet-}$, is rapidly formed in a pure ET process regardless of pH. It is nonemissive, but is in equilibrium with the locally-excited state, (pyrene)*–dU, meaning that emission can still be observed. However, at pH < 5, pyrene$^{\bullet+}$–dU$^{\bullet-}$ is subsequently protonated, and efficient nonradiative decay prevents repopulation of the local-excited state, and thus emission is mitigated. In this system, there is sufficient driving force for ET without coupling to PT, and in the appropriate pH range, PT will ensue to complete a consecutive ET/PT mechanism. Similar behavior was observed when dU was replaced with deoxyadenosine (dA), and with dG, however in that case, the pH-dependent emission intensity curve was inverted because dG was oxidized (and deprotonated) [173]. When dC was used, fluorescence was efficiently quenched across the entire pH range of 1.5–12.5. This suggests that the pK_a of pyrene$^{\bullet+}$–dC(H)$^{\bullet}$ is greater than 13.

Further investigations on the pyrene–dC system (23) were undertaken in the polar and aprotic solvent acetonitrile (MeCN) in order to distinguish between pure ET and PCET reaction channels [174]. Fluorescence quenching was not observed. Moreover TA measurements reveal no ET dynamics, indicating that the pure ET reaction is endothermic. Similar measurements were undertaken in water at pH 5 and pH 11, where the formation of pyrene$^{\bullet+}$ is observed and time-resolved, along with the disappearance of ^{1}(pyrene)*. By comparison to the MeCN data, the authors contend that pyrene$^{\bullet+}$ must be formed by a concerted PCET mechanism (directly forming pyrene$^{\bullet+}$–dC(H$^{\bullet}$)), rather than via an energetically costly pyrene$^{\bullet+}$–dC$^{\bullet-}$ intermediate. The kinetics fit to a model that indicates that PCET is 3-fold faster at pH 5 than at pH 11, supporting the concerted PCET mechanism. The trend in this case is the reverse of that discussed above for Y oxidation in [Ru]–Y due to the opposite sense of the reaction. The issue of concerted versus stepwise PCET in the reduction of pyrimidine nucleobases also parallels the findings in 20–22. The nucleobase dU has a greater electron affinity compared with dC [168], but a weaker proton affinity. This is the basis for the consecutive ET/PT mechanism observed in pyrene–dU as compared with the concerted mechanism in pyrene–dC (23).

17.4.2
Type D: Site-Specified 3-Point PCET

Many ETs to an oxidant (or from a reductant) are accompanied by PT to a specific base (or from a specific acid), usually via a preformed hydrogen bond. As discussed in Section 17.5, Type D PCET often describes the transfer of a net hydrogen atom in Nature. It is the specific coordinate for PT that distinguishes Type D PCET from its Type C counterpart of Section 17.4.1.

Unlike proteins, which exert control over transfer distances by positioning amino acid residues according to the tertiary structure, most ternary PCET reactions studied in model systems to date are trimolecular reactions. This complicates kinetics measurements and analysis, and can mask the underlying physics. The PCET yield depends on both the association constant (K_{assoc}) to form the PCET precursor complex, and the subsequent pseudo-bimolecular PCET rate constant (k_{PCET}). It is imperative to decouple the measurement of K_{assoc} and k_{PCET} to allow rigorous comparisons among a carefully chosen series of reactants, since many variations in model compounds can affect both quantities.

Linschitz and coworkers pioneered the study of Type D PCET in model systems with extensive studies on **24** and derivatives. Ae is the triplet excited state of fullerene ($^3C_{60}$), De/Dp is a substituted phenol, and Ap is typically a substituted pyridine (py) [175, 176]. The pseudo-first-order rate constant of $^3C_{60}$ decay (k_{obs}) was measured as a function of [phenol] and [py], with k_0 representing the rate in the absence of phenol. Plots of ($k_{obs} - k_0$)/[phenol] vs. [py] were fit to yield both the association constant for the phenol:py complex ($K_{phenol:py}$), and the rate constant attributed to quenching from the bound complex. Independent measurements of the association constant using Mataga's titration treatment [177, 178] agree with the values for $K_{phenol:py}$ obtained from fitting to the kinetic model over a wide range of $K_{phenol:py}$. Their analysis reveals that the phenol:py complex dominates the quenching of $^3C_{60}$. The authors establish that the reaction proceeds according to the Type D process of Fig. 17.3 by using TA spectroscopy to detect PCET products directly; $C_{60}^{\bullet-}$, py–H$^+$, and the neutral phenoxyl radical are all generated concomitantly with the loss of $^3C_{60}$. As expected for a ternary PCET reaction that separates charge, the PCET quantum yield is found to increase as a function of solvent polarity. Furthermore, $^3C_{60}$ quenching is two orders of magnitude faster when trimethylpyridine is the proton acceptor, as compared to DMSO, which has comparable hydrogen bonding ability but cannot accept a proton from the oxidized phenol. The KIEs measured reflect complete PT coupled to ET in the case of the trimethylpyridine acceptor (k_H/k_D = 1.65), and ET coupled to a hydrogen bonding interaction for DMSO (k_H/k_D = 1.06). Moreover, the temperature dependence of the deuterium KIE in the former reaction reveals that its origin lies in the activation energy rather than electronic coupling. These results establish that, with the appropriately chosen proton acceptor, the electron and proton are both transferred to their respective acceptor sites concurrently.

The modularity of **24** permits PCET rates to be easily examined as a function of the pK_a of Ap. For a wide range of substituted phenols, and in various solvents, the concerted PCET rates consistently track the increasing basicity of the substituted pyridine. Moreover, these measurements, when taken together with their temperature dependence, reveal that PCET becomes more kinetically competent owing to a lower activation barrier as the PT component becomes more exergonic. It is interesting to note, that this conclusion is also obtained in Section 17.4.1 for effectively the same De/Dp. The difference between **20** and **24** is that for the latter, the hydrogen bonded Ae, rather than the bulk pH, determines the PT driving force.

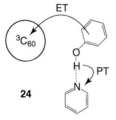

24

20 has been adapted to a Type D PCET model system (**25**) by the addition of two dipicolyamine ligands to Y [179]. The modification introduces an Ap site as an intramolecular hydrogen bond acceptor. The system was designed to mimic the interaction between Y_Z and H_{190} in PSII. Transient EPR measurements (coupled to a flash-quench experiment) in water revealed that $^\bullet Y$ is formed, confirming a PCET process. Flash-quench kinetics measurements reveal that **25** undergoes PCET at a rate two orders of magnitude faster than **20**. This acceleration in **25** is attributed to the intramolecular hydrogen bond, which decreases the oxidation potential of Y. Although the proton is ultimately lost to water, the dipicolyamine appears to acts as a primary proton acceptor (i.e., Ap) along a PT coordinate that is established by an intermolecular hydrogen bonding interaction. Similarly, Mayer and coworkers have studied the oxidation of a phenol that has a proton accepting amine group attached at the 2 position [180]. They carried out the reaction with a variety of oxidants to show that oxidation is coupled to concerted PT to the amine within an intramolecular hydrogen bond.

System **26** uses an extrinsic base to tune the reducing power of α-hydroxy radicals [181]. The photochemically generated diphenylketyl radical serves as De/Dp–H, Ae is 1,2,4,5-tetracyanobenzene (TCB), and Ap is a substituted pyridine (py). In the absence of the pyridine to facilitate PT, the ET reaction from the ketyl radical to TCB is endothermic by 0.4 eV. Consequently, the diphenylketyl radical decays slowly to reconstitute the starting reactants with no net reaction taking place. When coupled to PT by pre-associating the diphenylketyl radical with an Ap, the reaction can become thermodynamically favorable. For example, when lutidine is added as Ap, the driving force of the PT component increases by 0.8 eV, thus making the overall PCET reaction exergonic by 0.4 eV. TA spectroscopy confirms that addition of lutidine accelerates decay of the diphenylketyl radical, and a concomitant rise in the transient absorption of TCB$^{\bullet-}$ is observed. Note that direct PT from the diphenylketyl radical to the lutidine without coupling to ET is thermodynamically unfavorable and products of this reaction were not observed. Since experimental observations and thermodynamic arguments rule out direct ET and direct PT (per Fig. 17.1), a concerted PCET mechanism is implied, whereby the diphenylketyl radical transfers an electron to TCB and a proton to the py in a single step.

The PCET reaction rate for **26** has been slowed by using a less basic proton acceptor, thus reducing the driving force for the concerted PCET reaction. When lutidine is replaced with 2-Cl-py, the driving force diminishes from 0.4 eV to < 0.1 eV, and accordingly k_{PCET} decreases by a factor of four. A KIE of 3.2 was measured when the deuterated diphenylketyl radical was used and Ap was 2-Cl-py. This magnitude certainly supports the mechanism of concerted PCET.

Type D reactions can operate in reverse wherein an electron and proton converge at Ae/Ap. The development of **27** by Linschitz and coworkers [182] employs chloranil (CA) as Ae/Ap, $^3C_{60}$ as De, and an alcohol or carboxylic acid (R–OH) as Dp. This study was conducted in essentially the same manner as phenol oxidation for **24**, however the mechanisms were found to differ as a result of the proton's role in the PCET reaction. In the absence of R–OH, CA can quench $^3C_{60}$, albeit slowly. The instability of the charge-separated state results in a reverse ET reaction that rapidly depletes the radical yield. The rate of ET from $^3C_{60}$ to CA increases sharply upon addition of R–OH that is a strong hydrogen bond donor such as hexafluoro-2-propanol (HFIPA). Weaker R–OH hydrogen-bonding substrates such as trifluoroethanol (TFE) have a negligible effect on ET rates. Formal PT need not be invoked to rationalize these data; strong hydrogen bonding accelerates ET by shifting the reduction of the quinone to less negative potentials. Electrochemical measurements support this explanation. Quinone reduction potentials are found to track the hydrogen bonding ability of R–OH groups in aprotic solvents [183]. PCET products are only observed when R–OH is a very strong proton donor (in addition to being a strong hydrogen bond donor), such as trifluoroacetic acid (TFA, pK_a = 3.45 in DMSO). In such cases, the neutral semiquinone (CAH$^{\bullet}$), $C_{60}^{\bullet+}$, and R–O$^-$ are the PCET products. The first two of these were identified by TA spectroscopy; their absorption features grow concomitantly with the decay of

27

$^3C_{60}$. These observations lead to the conclusions: (i) ET is accelerated by strong hydrogen-bonding between CA and R–OH; (ii) rapid PT within the hydrogen-bonded complex ensues to generate the PCET products only when Dp is a sufficiently strong proton donor; and, (iii) if PT occurs, it prolongs the lifetime of $C_{60}^{\bullet+}$. In other words, coupling PT to an ET reaction can enhance the utility of charge-separated states, regardless of whether the mechanism of PCET is concerted or consecutive.

Moore, Moore, Gust and their coworkers have developed 3-point PCET systems that incorporate all the necessary ET and PT components within molecular dyads and triads [184]. In **28**, De is the singlet excited state of a free-base porphyrin, Ae/Ap is a naphthoquinone moiety that connects to the porphyrin via an amide linkage, and Dp is a carboxylic acid that is attached to the bridgehead of a norbornene system fused to the naphthoquinone. The carboxylic acid associates to the carbonyl group of the quinone by a hydrogen bond. The intramolecular hydrogen bonding interaction shifts the reduction potential of the quinones to more positive potential, thereby accelerating ET from the porphyrin to the quinone. The enhanced quenching rate is reflected by the three-fold shortening of the porphyrin fluorescence lifetime of **28** as compared with a control compound in which the norbornene and carboxylic acid are absent. Using TA spectroscopy, the fate of the charge-separated state was tracked by probing the TA signature of the porphyrin cation radical, $P^{\bullet+}$. An increase in the lifetime of $P^{\bullet+}$ from < 250 fs to ~4 ps for **28** (compared with the control compound) was ascribed to a fast PT (~1 ps), which followed ET. This is well justified considering that the quinone becomes more basic by >10 pK_a units upon reduction within the pre-established hydrogen bond network. Analogous to Linschitz' quinone system **27**, a hydrogen bonding and proton-donating group appropriately positioned with respect to a quinone leads to accelerated charge-separation and prolonged lifetimes.

Type D PCET clearly demonstrates the role of an extrinsic proton donor or acceptor in modulating the rate and direction of ET along a spatially separate coordinate, and in doing so stabilizes charged intermediates. In this way, the proton manages charge separation as long as the electronic states of Ae/Ap are tightly coupled to the protonic states of De/Dp. This idea is schematically depicted in Fig. 17.19. In the thermodynamic square scheme, this translates into large shifts in De/Ae redox potentials, depending on whether or not the proton is bound, and large shifts in Dp/Ap pK_a, depending on the accompanying oxidation state. For example, phenols (or tyrosine residues in biology) are found to be very prevalent in bidirectional PCET reactions that proceed according to Eq. (D) of Fig. 17.3 since

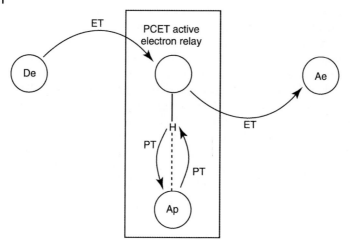

Figure 17.19 Schematic depiction of off-pathway PT as a means to assist long distance electron relaying. The coupled PT exerts control over the ET rate and direction by modulating the redox potential of the relay intermediate relative to De and Ae.

the pK_a of phenol shifts from 10 to –2 upon one-electron oxidation [185, 186]. Likewise, quinones (or other ketones) are frequently involved in orthogonal PCET reactions in the reverse direction on account of their pronounced pK_a-redox coupling. There are common features among these species (and others discussed such as guanine and cytosine) that give rise to the strong pK_a-redox coupling. They are aromatic molecules that have energetically accessible frontier electronic orbitals, thus allowing facile ET with minimal structural distortion. They also integrate hetero-atoms (typically oxygen or nitrogen) that are basic and thus able to support PT. Importantly, they are typically part of a small framework – typically based on a six-membered ring, ensuring that significant amplitudes of the frontier electronic orbitals reside on the hetero-atoms to increase pK_a-redox coupling. In other words, rather than extensively delocalizing charge, which is typically the paradigm for facile ET, focusing changes in charge onto acid–base functionalities amplifies pK_a-redox coupling, which can then be used as a control-handle when incorporated into a bidirectional PCET configuration.

17.5
The Different Types of PCET in Biology

Nature provides striking examples of each of the types of PCET discussed in this chapter. Enzymes often rely on PCET to affect primary metabolic steps involving charge transport and catalysis. Amino acid radical generation and transport is synonymous with PCET [187], as is the activation of substrate bonds at enzyme active sites [29]. PCET is especially prevalent for metallo-cofactors that activate substrates at carbon, oxygen, nitrogen, and sulfur atoms. In order to access these different types of reactivity, enzymes impart exquisite control over both ET and PT coordinates via the protein tertiary structure.

Type A PCET reactions describe amino acid radical generation steps in many enzymes, since the electron and proton transfer from the same site as a hydrogen atom [188]. Similarly, substrate activation at C–H bonds typically occurs via a Type A configuration at oxidized cofactors such as those in lipoxygenase [47, 48] galactose oxidase [189–191] and ribonucleotide reductase (Y oxidation at the di-iron cofactor, *vide infra*) [192]. Here, the "HATs" are more akin to the transition metal mediated reactions of Section 17.3.1 since the final site of the electron and proton are on site differentiated at Ae (redox cofactor) and Ap (a ligand).

Type B PCET represents a largely inevitable scenario in many proteins where electron transfer coordinates traverse the hydrogen bonding networks that are required to establish tertiary structure. In the case where the ET coordinate is long and no X–H bond is broken, the hydrogen bond modulates the kinetics for ET via the electronic coupling matrix element. Pathway models for ET in proteins have explicitly parametrized the effect of hydrogen bonding conduits in recognition of the ubiquity of Type B PCET in biology [82–84]. The coupling between the proton and electron is more pronounced and thermodynamically derived when X–H bond breaking is involved.

Type D PCET is utilized in Nature to effect a variety of catalytic processes because it allows enzymes to manage the disparate electron and proton length scales. The PT network is established by the tertiary structure of the protein matrix about a redox center. Figure 17.20 depicts a generalized scheme of PCET at a M–OH center. PCET reactions can affect oxidation or reduction at the metal center when an appropriate PT site is positioned nearby. The reactions are coupled to changes in the M–O bond order. Many oxidases also derive function from PT networks orthogonalized to redox-active M–O–O–H centers. Examples include peroxidases, catalases, and cytochrome P450 mono-oxygenases [193–198]. The heme-oxo intermediates, Compound I ($P^{\bullet+}Fe^{IV}=O$, P = porphyrin) and Compound II ($PFe^{IV}=O$), are generated upon protonation of ferric peroxy species at the unbound oxygen atom and loss of water in concert with oxidation at the metal redox center and increase in the M–O bond order. Proton-donating amino acid residues at well-defined distances from the porphyrin redox platform orchestrate

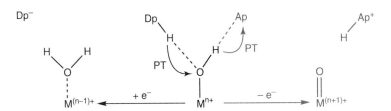

Figure 17.20 Schematic depiction of metal-centered bidirectional PCET. Oxidation (or reduction) at the M–OH center is coupled to loss (or gain) of a proton and an increase (or decrease) in the M–O bond order.

these metal-centered PCET reactions. The structure of cytochrome P450 [193] (Fig. 17.21) establishes the presence of a hard-wired water channel to direct protons to and from the heme redox cofactor. Compounds I and II are also photogenerated in modified microperoxidases upon oxidation of a ferric hydroxy species (PFe^{III}–OH) coupled to deprotonation to reveal the terminal oxo [199, 200].

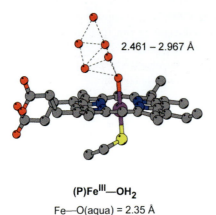

2.461 – 2.967 Å

(P)Fe^{III}—OH$_2$

Fe—O(aqua) = 2.35 Å

Figure 17.21 Crystal structure of cytochrome P450, displaying a water channel above the heme. Figure adapted from Ref. [193].

The so-called 'Hangman porphyrin' models [201] were developed as a simplified construct of the bidirectional PCET networks of heme enzymes. Compound **29** suspends a pendant acid group from a xanthene spacer at a fixed distance from the porphyrin platform. The structurally characterized Hangman porphyrin xanthene (HPX) (Fig. 17.22) shows that a water molecule (with a binding energy of 5.8 kcal mol^{-1}) [202] is suspended between the xanthene carboxylic acid hanging group and the hydroxide ligand [201]. This is the first synthetic redox-active site displaying an assembled water molecule as part of a structurally well-defined PT network. As in mono-oxygenases, PT from the acid–base hanging group in (HPX)M^{III}–O–O–H (HPX = Hangman porphyrin xanthene, M = Fe, Mn) peroxide complexes yields (HP$^{\bullet+}$X)M^{IV}=O [203]. In the presence of olefins, epoxidation occurs at high turnover [202] (M = Mn). In the absence of substrate, the Com-

^{t}Bu

O

OH

H

Mes

O

N—N

Mes

Fe

N—N

Mes

^{t}Bu

Mes

29

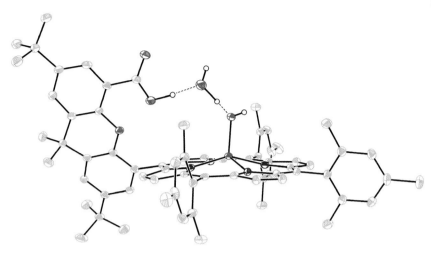

Figure 17.22 Crystal structure of the ferric hydroxide form of Hangman porphyrin **29** showing the water channel within the Hangman cleft. Figure adapted from Ref. [201].

pound I-type intermediate reacts with peroxide to generate oxygen and water in a catalase-like reaction, also at high turnover [204] (M = Fe). Mono-oxygenase and catalase activities are lost when the Hangman pillar is extended and the proton must transfer over long distance [202]. Activity is also severely reduced when the pK_a of the hanging acid–base group is increased [204].

The studies on these Hangman porphyrins and other macrocyclic Hangman platforms [205] clearly demonstrate that exceptional catalysis may be achieved when redox and PT properties of a cofactor are controlled independently. A key requirement is that the PT distance is kept short, which may be accomplished by orthogonalizing ET and PT coordinates. The benefits of incorporating PT functionality into redox catalysis can only be realized when a suitable geometry is established. Moreover, the Hangman platforms show that a multifunctional activity of a single metalloporphyrin-based scaffold is achieved by the addition of proton control to a redox platform. This observation is evocative of natural heme-dependent proteins that employ a conserved protoporphyrin IX cofactor to affect a myriad of chemical reactivities.

Other oxidases also derive function from bidirectional PCET pathways at the enzyme active site. The recent crystal structures of PSII [206, 207] support suggestions that as the oxygen evolving complex (OEC) steps through its various S-states [208, 209], substrate derived protons are shuttled to the lumen via a proton exit channel, the headwater of which appears to be the D_{61} residue hydrogen-bonded to Mn-bound water [210]. The protons are liberated with the proton-coupled oxidation of the Mn–OH_2 site. As shown by the structure reproduced in Fig. 17.23, D_{61} is diametrically opposite to Y_Z, which has long been known [148, 151, 152] to be the electron relay between the PS II reaction center and OEC. Notwithstanding,

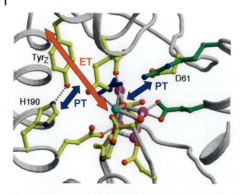

Figure 17.23 The 3.4 Å resolution structure of the oxygen evolving complex (OEC) and the immediate peptide environment. The directions of proposed PT and ET pathways are indicated with arrows. Figure adapted from Ref. [206].

Y_Z employs bidirectional PCET in its role as an electron relay. For the reasons described in Fig. 17.1, and consistent with the numerous related model systems described in Section 17.4, oxidation of Y_Z requires proton dissociation from the phenolic oxygen. Functional schemes of PSII prior to the crystal structure suggested that the proton was lost to a water channel via the primary acceptor H_{190} (i.e., a Type D PCET reaction) [155, 156] and hence Y_Z was the nexus for the requisite electron and proton transport to and from OEC. However, the more recently obtained crystal structure shows the Y_Z-H_{190} pair to be relatively isolated by α-helices. This potentially supports a 'rocking model', where the phenolic proton rocks back and forth between Y_Z and H_{190} concomitantly with the cycling of the Y_Z oxidation state as it relays holes [211, 212]. This case serves as a perfect example of the manner in which orthogonal PCET is used to control charge-transport, as illustrated in Fig. 17.19. The coupled PT occurs to a base (H_{190}) that is off-pathway from electron transport. Its function is to modulate reversibly the oxidation potential of an electron transport intermediate (Y_Z) that is on-pathway. It is crucial that H_{190} is positioned close enough to Y_Z to be effective in this role. As well as establishing the PT coordinate, the hydrogen bonding interaction presumably also relieves the activation energy associated with the PCET reaction.

Bidirectional PCET is also featured on the reduction side of the photosynthetic apparatus. In the bacterial photosynthetic reaction center, two sequential photo-induced ET reactions from the P680* excited state to a quinone molecule (Q_B) are coupled to the uptake of two protons to form the hydroquinone [213–215]. This diffuses into the inter-membrane quinone pool and is re-oxidized at the Q_0 binding site of the cytochrome bc_1 and coupled to translocation of the protons across the membrane, thereby driving ATP production. These PCET reactions are best described by a Type D mechanism because the PCET of Q_B appears to involve specifically engineered PT coordinates among amino acid residues [215]. In this case PT ultimately takes place to and from the bulk solvent. Coupling remains tight in the thermodynamic sense because the reduction potential of the quinone depends on the chemical potential of the coupled proton. In other words, ET is coupled to the local pH via a PT network and this coupling is manifest in the proton-pump that drives ATP production.

Bidirectional PCET also manifests itself in reductases. Crystal structures of hydrogenases [216–218] indicate that the mechanism for hydrogen production occurs by transporting protons into the active site along pathways distinct from those traversed by the electron equivalents. Electrons are putatively injected into the active site via a chain of [FeS] clusters, while proton channels and acid–base residues at the active site manage the substrate inventory.

Class I E. *coli* ribonucleotide reductase (RNR) exploits all the PCET variances of Fig. 17.3 in order to catalyze the reduction of nucleoside diphosphates to deoxynucleoside diphosphates. This reaction demands radical transport across two subunits and over a remarkable 35 Å distance [187, 188, 219]. The crystal structures of both R1 and R2 subunits have been solved independently [220–222] and a docking model has been proposed [220]. R2 harbors the diferric tyrosyl radical (•Y122) cofactor that initiates nucleotide reduction by generating a transient thiyl radical (•C439) in the enzyme active site located >35 Å away in R1 [223]. Substrate conversion is initiated by a hydrogen atom abstraction (Type A PCET) at the 3′ position of the substrate by •C439 [192].

Figure 17.24 presents the current model for radical transport in RNR [30]. Beginning at the Y122 cofactor, a bidirectional PCET step (Type D) involving PT between Y122 and the di-iron oxo/hydroxo cofactor [223] is suggested to lead to •Y122 radical generation. Oxidation of Y356, the redox terminus of the R2 pathway, demands a PCET reaction takes place, but this too appears to be bidirectional

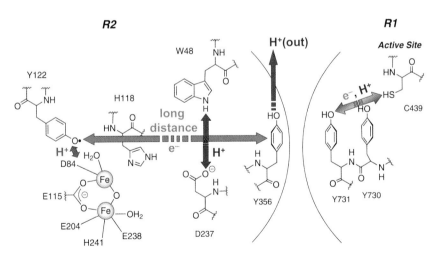

Figure 17.24 Proposed model for radical transport in RNR. The model employs both uni- and bidirectional PCET steps. The mode of transport at the interface (between Y356 and 731) is undefined. Figure adapted from Ref. [30].

with the PT occurring off pathway as determined from R2 mutants in which the pK_a and redox potential of Y356 is systematically altered with non-natural amino acids [224–228]. The proton of the tyrosine at 356 appears to be lost to the bulk (i.e., a Type C reaction). By moving the protons at Y122 and Y356 off pathway, the radical transport in R2 involves a long distance ET coupled to short PT hops at the tyrosine endpoints. In setting up the radical transport pathway in this fashion, the very different PT and ET length scales are managed in RNR. Dipeptide W–Y studies suggest that the direction of ET along the pathway may be controlled by an off-pathway PT step between W48 and D236 [229].

PCET in R1 has been studied in detail using photo-generated •Y sources in place of Y356 [30, 230], hence circumventing the R2 subunit and paving the way for photochemical activity studies (and potentially time-resolved kinetics measurements). Additional studies have been undertaken with various mutants incorporating fluorinated Y residues that perturb the thermodynamic parameters (E_{red} and pK_a) in a controlled fashion [30, 224, 227, 228]. Within R1, the combination of mutant activity studies suggests a unidirectional (Type A) PCET pathway through R1 in which both the electron and proton are transferred between Y731-Y730-C439. The close proximity of these three residues lends credence to this proposition.

17.6
Application of Emerging Ultrafast Spectroscopy to PCET

Direct kinetics measurements of PCET and HAT reactions to date have primarily focused on detecting the ET component of the transformation by time-resolved optical measurements. But this is only half the story in a PCET and HAT reaction. Insight into PCET and HAT reaction mechanisms will be expanded considerably by experimental methods that directly probe the structural changes accompanying the PT component of the reaction. Optical changes associated with PT can be designed into Dp or Ap components of the system [123] but generally they are overwhelmed in ultraviolet and visible spectral regions by those associated with ET.

A more promising line of inquiry for the purposes of directly detecting PT will likely involve transient spectroscopies that probe the IR spectral region. PT inherently involves motion of nuclei and changes in bonding. As a result, PT is accompanied by distinctive signatures in the IR region. At the very least, a proton shift or complete transfer from X–H to Y will cause the loss of X–H vibrations concomitantly with the gain of H–Y vibrations. Depending on the nature of the proton donor and acceptor groups, distinctive IR shifts may also propagate several bonds away. For example, the amidinium–carboxylate salt bridge was found to be a particularly effective interface for investigating collinear PCET reactions (see Section 17.3.2). Deprotonation of the amidinium on the Dp side of the interface to produce the corresponding amidine is reflected in a red-shift of the C–N stretching frequency by ~30 cm^{-1} [231]. This counter-intuitive shift is in line with observations for the protonation/deprotonation of other Schiff base complexes, and is explained by the mixing of anti-bonding character of the nitrogen lone-pair of the C–N bond [232]. It is emphasized here that the relevant amidine/amidinium vibration lies outside the fingerprint region of the porphyrins that are frequently employed as electron donors, enabling easy observation with IR ultrafast probe sources. Furthermore, PT can result in shifts of C–O stretching frequencies on the Ap side of the amidinium–carboxylate interface, in addition to changes in N–H and O–H stretches. The IR wavelengths required to probe the evolution of these signatures on an ultrafast timescale have recently become easily accessible using difference frequency generation methods [233]. HAT and PCET reactions of metal oxos also exhibit large changes in frequency owing to a formal change in bond order from two to one. Accordingly, we expect IR probe TA spectroscopy to become a routine tool for future investigations of PCET kinetics. With ET and PT kinetics in hand, a correlation of the rate constants to parameters such as solvent dielectric constant and reaction free energy can help to describe the nature of the transition state and the two-dimensional trajectory of a concerted PCET reaction.

The coupling of the charge shift resulting from electron and proton motion to the polarization of the surrounding environment is a distinguishing characteristic of PCET reactions of Type B to D, as well as many of the Type A reactions as discussed in Section 17.3.1. Thus, new experimental tools that probe the solute and solvent modes of the PCET reaction coordinate will be valuable. Of these, multi-dimensional optical and IR spectroscopies hold the most promise as a powerful tool to study molecular structure and chemical dynamics [234–238]. A 2D IR spectrum is obtained via a sequence of pulses in the time domain, and it characterizes how transient excitation of molecular vibrations at one frequency effect vibrations at a different frequency. Such experiments reveal vibrational couplings as off-diagonal peaks, and correlate the motion of different molecular entities on femtosecond to picosecond time scales, providing a characterization of transient and time-evolving structure. This type of information will undoubtedly be valuable for PCET, where correlated nuclear motion gets to the heart of how a PCET reaction proceeds on the most detailed level.

Application of 2D IR spectroscopy to PCET models of Section 17.3.2 is a logical starting point for this type of investigation. 2D methods can unravel the correlated nuclear motion in a PCET reaction and in principle decipher how vibrational coupling in the Dp/Ap interface couples to the ET event between the Ae/De sites. These data can identify the structural dynamics within the interface that promote PCET reactions in much the same way that local hydrogen bonding structure and dynamics mediate excited state PT reactions [239, 240]. In these experiments, the PCET reaction can be triggered by an ultrafast resonant visible laser pulse (as in a standard TA experiment) and a sequence of IR pulses may be employed to build a transient 2D IR spectrum. These experiments demand that systems be chosen so that the ET and PT events occur on an ultrafast timescale.

Another novel ultrafast methodology that has rapidly matured and is poised for PCET applications is time-domain nonresonant third-order Raman (TOR) spectroscopy [241–243], which can directly detect the low-frequency response of liquids. Blank and coworkers have successfully applied this technique to probe the role of the solvent in excited state intramolecular PT reactions and ultrafast solvation dynamics [244–247]. Prior to this work, inferences regarding the solvent response were made by comparing the observed kinetics to known timescales of solvent response, and by interpreting solvent-induced spectral shifts in the solute (reporter) molecules [248–251]. The more direct experiments of Blank revealed a nonequilibrium solvent response in the excited state PT reaction [245, 246]. The solvent response lagged behind progress along the PT reaction coordinate, leading to the breakdown of linear response and inducing time-dependent evolution of the reaction free energy surface.

PCET rate formalisms are cast primarily in terms of solvent coordinates for both the electron and proton, since both are charged particles that couple to the solvent polarization [5, 24]. In a concerted PCET reaction, the coupled transfer must occur via a common transition state and a common solvent configuration on both solvent coordinates. An ultrafast PCET reaction could be photoinitiated with resonant excitation, and a TOR probe would subsequently reveal the evolution of the two-dimensional reaction coordinate via the solvent response. Working in concert, these experiments would offer a powerful means to evaluate the coupling between the two coordinates in different types of PCET reactions and thus enable the PCET trajectories within the 2D space of Fig. 17.2 to be determined with much greater clarity.

Acknowledgment

J.R. thanks the Fannie and John Hertz Foundation for a pre-doctoral fellowship. The work on PCET and the activation of small molecules by PCET has been supported by grants from the NIH (GM47274) and the DOE DE-FG02-05ER15745.

References

1 J. M. Mayer, *Annu. Rev. Phys. Chem.* **2004**, *55*, 363.

2 J. D. Soper, J. M. Mayer, *J. Am. Chem. Soc.* **2003**, *125*, 12217.

3 J. S. Jaworski, *Trends Phys. Chem.* **2003**, *9*, 51.

4 J. Robertson, P. Jayasheela, R. K. Lush, *Chem. Soc. Rev.*, **2001**, *30*, 94.

5 R. I. Cukier, D. G. Nocera, *Annu. Rev. Phys. Chem.* **1998**, *49*, 337.

6 R. A. Marcus, N. Sutin, *Biochim. Biophys. Acta* **1985**, *811*, 265.

7 R. A. Marcus, *J. Chem. Phys.* **1956**, *24*, 966.

8 N. S. Hush, *Trans. Faraday Soc.* **1961**, *57*, 557.

9 G. L. Closs, J. R. Miller, *Science* **1988**, *240*, 440.

10 *Electron Transfer in Chemistry*, V. Balzani (Ed.), Wiley-VCH, Weinheim, 2001–2002, Vol. 1–5.

11 P. F. Barbara, T. J. Meyer, M. A. Ratner, *J. Phys. Chem.* **1996**, *100*, 13148.

12 *Electron and Proton Transfer In Chemistry and Biology*, A. Muller (Ed.), Elsevier, Amsterdam, 1992.

13 *Proton-transfer Reactions*, E. Caldin, V. Gold (Eds.), Wiley, New York, 1975.

14 P. M. Kiefer, J. T. Hynes, *Solid State Ionics*, **2004**, *168*, 219.

15 K. Ando, J. T. Hynes, *Adv. Chem. Phys.* **1999**, *110*, 381.

16 R. I. Cukier, *J. Phys. Chem. B.* **2002**, *106*, 1746.

17 S. Hammes-Schiffer, *ChemPhysChem*, **2002**, *3*, 33.

18 R. I. Cukier, *J. Phys. Chem.* **1996**, *100*, 15428.

19 R. I. Cukier, *J. Phys. Chem.* **1994**, *98*, 2377.

20 X. G. Zhau, R. I. Cukier, *J. Phys. Chem.* **1995**, *99*, 945.

21 R. I. Cukier, *J. Phys. Chem. A* **1999**, *103*, 5989.

22 R. I. Cukier, *Biochim. Biophys. Acta* **2004**, *1655*, 37.

23 S. Hammes-Schiffer, in *Electron Transfer in Chemistry*, V. Balzani (Ed.), Wiley-VCH: Weinheim, 2001, Vol. 1.1.5, p. 189.

24 S. Hammes-Schiffer, *Acc. Chem. Res.* **2001**, *34*, 273.

25 S. Hammes-Schiffer, N. Iordanova, *Biochim. Biophys. Acta* **2004**, *1655*, 29.

26 A. Soudackov, S. Hammes-Schiffer, *J. Chem. Phys.* **1999**, *111*, 4672.

27 A. Soudackov, S. Hammes-Schiffer, *J. Chem. Phys.* **2000**, *113*, 2385.

28 H. Decornez, S. Hammes-Schiffer, *J. Phys. Chem. A.* **2000**, *104*, 9370.

29 C. J. Chang, M. C. Y. Chang, N. H. Damrauer, D. G. Nocera, *Biochim. Biophys. Acta* **2004**, *1655*, 13.

30 S. Y. Reece, J. M. Hodgkiss, J. Stubbe, D. G. Nocera, *Philos. Trans. R. Soc. London, Ser. B* **2006**, *361*, in press.

31 O. Exner, in *Correlation Analysis in Chemistry*, N. B. Chapman, J. Shorter (Eds.), Plenum, New York, 1978, p. 439.

32 R. H. Lowry, K. S. Richardson, *Mechanism and Theory in Organic Chemistry*, 3rd Edn. Harper & Row, New York, 1987.

33 F. A. Carroll, *Perspectives on Structure and Mechanism in Organic Chemistry*, 4th Edn. Brooks/Cole, Pacific Grove, CA, 1998.

34 B. Miller, *Advanced Organic Chemistry*, Pearson Education, New York, 1997.

35 *Techniques of Chemistry Vol. II, Organic Solvents: Physical Properties and Methods of Purification*, 3rd Edn., p. 57.

36 K. B. Wiberg, *Chem. Rev.* **1955**, *55*, 713.

37 F. H. Westheimer, *Chem. Rev.* **1961**, *61*, 265.

38 S. E. Scheppele, *Chem. Rev.* **1972**, *72*, 511.

39 E. Buhks, M. Bixon, J. Jortner, *J. Phys. Chem.* **1981**, *85*, 3763.

40 N. Sutin, *Acc. Chem. Res.* **1982**, *15*, 275.

41 T. Guarr, E. Buhks, G. McLendon, *J. Am. Chem. Soc.* **1983**, *105*, 3763.

42 R. A. Binstead, T. J. Meyer, *J. Am. Chem. Soc.* **1987**, *109*, 3287.

43 R. A. Binstead, M. E. McGuire, A. Dovletoglou, W. K. Seok, L. E. Roecker, T. J. Meyer, *J. Am. Chem. Soc.* **1992**, *114*, 173.

44 M.-H. V. Huynh, T. J. Meyer, P. S. White, *J. Am. Chem. Soc.* **1999**, *121*, 4530.

45 M.-H. V. Huynh, P. S. White, T. J. Meyer, *Angew. Chem. Int. Ed.* **2000**, *39*, 4101.

46 M.-H. V. Huynh, T. J. Meyer, *Angew. Chem. Int. Ed.* **2002**, *41*, 1395.

47 A. Kohen, J. P. Klinman, *Acc. Chem. Res.* **1998**, *31*, 397.

48 Z.-X. Liang, J. P. Klinman, *Curr. Opin. Struct. Biol.* **2004**, *14*, 648.

49 J. M. Hodgkiss, N. H. Damrauer, S. Pressé, J. Rosenthal, D. G. Nocera, *J. Phys. Chem. B.* **2006**, in press.

50 J. M. Mayer, Thermodynamic Influences on C-H Bond Oxidation, in *Biomimetic Oxidations Catalyzed by Transition Metal Complexes*, Imperial College Press, London, **2000**, p. 1.

51 J. P. Roth, J. M. Mayer, *Inorg. Chem.* **1999**, *38*, 2760.

52 J. M. Mayer, *Acc. Chem. Res.* **1998**, *31*, 441.

53 J. P. Roth, S. Lovell, J. M. Mayer, *J. Am. Chem. Soc.* **2000**, *122*, 5486.

54 J. M. Mayer, I. J. Rhile, *Biochim. Biophys. Acta.* **2004**, *1655*, 51.

55 J. K. Kochi, *Free Radicals*, Wiley, New York, 1973.

56 J. M. Tedder, *Ang. Chem. Int. Ed. Engl.* **1982**, *21*, 401.

57 K. R. Korzekwa, J. P. Jones, J. R. Gillette, *J. Am. Chem. Soc.* **1990**, *112*, 7042.

58 J. H. Knox, *Oxidation of Organic Compounds*; Vol. 2, F. R. Mayo (Ed.), Adv. Chem. Series, **1968**, *76*, 11.

59 K. A. Gardner, L. L. Kuehnert, J. M. Mayer, *Inorg. Chem.* **1997**, *36*, 2069.

60 N. Iordanova, H. Decornez, S. Hammes-Schiffer, *J. Am. Chem. Soc.* **2001**, *123*, 3723.

61 J. U. Rohde, J. H. In, M. H. Lim, W. W. Brennessel, M. R. Bukowski, A. Stubna, E. Münck, W. Nam, L. Que Jr., *Science* **2003**, *299*, 1037.

62 M. H. Lim, J. U. Rohde, A. Stubna, M. R. Bukowski, M. Costas, R. Y. N. Ho, E. Münck, W. Nam, L. Que Jr., *Proc. Natl. Acad. Sci.U.S.A* **2003**, *100*, 3665.

63 A. Decker, J. U. Rohde, L. Que Jr., E. I. Solomon, *J. Am. Chem. Soc.* **2004**, *126*, 5378.

64 J. Kaizer, E. J. Klinker, N. Y. Oh, J. U. Rohde, W. J. Song, A. Stubna, J. Kim, E. Münck, W. Nam, L. Que Jr., *J. Am. Chem. Soc.* **2004**, *126*, 472.

65 J. Berkowitz, G. B. Ellison, D. Gutman, *J. Phys. Chem.* **1994**, *98*, 2744.

66 R. Y. Luo, *Handbook of Bond Dissociation Energies in Organic Compounds*, CRC Press: Boca Raton, 2003.

67 O. L. Sydora, J. I. Goldsmith, T. P. Vaid, A. E. Miller, P. T. Wolczanski, H. D. Abruña, *Polyhedron*, **2004**, *23*, 2841.

68 B. K. Carpenter, *J. Am. Chem. Soc.* **1995**, *117*, 6336.

69 B. J. Bahnson, J. P. Klinman, *Methods Enzymol.* **1995**, *249*, 373.

70 L. M. Slaughter, P. T. Wolczanski, T. R. Klinckman, T. R. Cundari, *J. Am. Chem. Soc.*, **2000**, *122*, 7953.

71 J. R. Bryant, J. M. Mayer, *J. Am. Chem. Soc.* **2003**, *125*, 10351.

72 J. R. Bryant, T. Matsuo, J. M. Mayer, *Inorg. Chem.* **2004**, *34*, 1587.

73 V. Bierbaum, C. DePuy, G. Davico, B. Ellison, *Int. J. Mass Spectrom. Ion Phys.* **1996**, *156*, 109.

74 S. Itoh, H. Kumei, M. Taki, S. Nagatomo, T. Kitagawa, S. Fukuzumi, *J. Am. Chem. Soc.* **2001**, *123*, 6708.

75 T. Osako, K. Ohkubo, M. Taki, Y. Tachi, S. Fukuzumi, S. Itoh, *J. Am. Chem. Soc.* **2003**, *125*, 11027.

76 M. S. Ran, J. T. Hupp, *J. Phys. Chem.* **1990**, *94*, 2378.

77 Y. Goto, Y. Watanabe, S. Fukuzumi, J. P. Jones, J. P. Dinnocenzo, *J. Am. Chem. Soc.* **1998**, *120*, 10762.

78 S. C. Weatherly, I. V. Yang, H. H. Thorp, *J. Am. Chem. Soc.* **2001**, *123*, 1236.

79 C. C. Moser, J. M. Keske, K. Warncke, R. S. Farid, P. L. Dutton, *Nature*, **1992**, *355*, 796.

80 J. R. Winkler, H. B. Gray, *Chem. Rev.* **1992**, *92*, 369.

81 H. B. Gray, J. R. Winkler, *Annu. Rev. Biochem.* **1996**, *65*, 537.

82 D. N. Beratan, J. N. Onuchic, J. J. Hopfield, *J. Chem. Phys.* **1987**, *86*, 4488.

83 D. N. Beratan, J. N. Onuchic, J. R. Winkler, H. B. Gray, *Science*, **1992**, *258*, 1740.

84 J. Lin, I. A. Balabin, D. N. Beratan, *Science* **2005**, *310*, 1311.

85 J.-M. Lehn, *Supramolecular Chemistry: Concepts and Perspectives*, VCH, Weinheim, 1995.

86 F. Vögtle, *Supramolecular Chemistry: An Introduction*, Marutzen, Tokyo, 1995.

87 D. Philp, J. F. Stoddart, *Angew. Chem. Int. Ed.* **1996**, *35*, 1154.

88 J. R. Fredericks, A. D. Hamilton, In *Comprehensive Supramolecular Chemistry*, Y. Murakami (Ed.), Pergamon, Oxford, 1996, Vol. 3.

89 G. M. Whitesides, J. P. Mathias, C. T. Seto, *Science* **1991**, *254*, 1312.

90 M. M. Conn, J. Rebek, Jr., *Chem. Rev.* **1997**, *97*, 1647.

91 J. W. Steed, J. L. Atwood, *Supramolecular Chemistry*, John Wiley & Sons, West Sussex, UK, 2000.

92 G. A. Jeffrey, *An Introduction to Hydrogen Bonding*, Oxford University Press, New York, 1997.

93 M. C. Etter, *Acc. Chem. Res.* **1990**, *23*, 120.

94 C. Turró, C. K. Chang, G. E. Leroi, R. I. Cukier, D. G. Nocera, *J. Am. Chem. Soc.* **1992**, *114*, 4013.

95 E. Cukier, S. Daniels, E. Vinson, R. J. Cave, *J. Phys. Chem. A* **2002**, *106*, 11240.

96 M. R. Wasielewski, M. P. Niemszyk, W. A. Svec, E. B. Pewitt, *J. Am. Chem. Soc.* **1985**, *107*, 1080.

97 J. N. Onuchic, D. N. Beratan, *J. Chem. Phys.* **1990**, *92*, 722.

98 D. N. Beratan, J. N. Betts, J. N. Onuchic, *Science* **1991**, *252*, 1285.

99 P. J. F. de Rege, S. A. Williams, M. J. Therien, *Science*, **1995**, *269*, 1409.

100 K. Pettersson, K. Kilså, J. Mårtensson, B. Albinsson, *J. Am. Chem. Soc.* **2004**, *126,* 6710.

101 A. Harriman, D. J. Magda, J. L. Sessler, *J. Chem. Soc.,Chem. Commun.* **1991**, 245.

102 A. Harriman, D. J. Magda, J. L. Sessler. *J. Phys. Chem.* **1991**, *95,* 1530.

103 A. Harriman, Y. Kubo, J. L. Sessler, *J. Am. Chem. Soc.* **1992**, *114,* 388.

104 J. L. Sessler, B. Wang, A. Harriman, *J. Am. Chem. Soc.* **1993**, *115,* 10418.

105 A. Berman, E. S. Israeli, H. Levanon, B. Wang, J. L. Sessler, *J. Am. Chem. Soc.* **1995**, *117,* 8252.

106 M. Asano-Someda, H. Levanon, J. L. Sessler, R. Wang, *Mol. Phys.* **1998**, *95,* 935.

107 J. L. Sessler, B. Wang, S. L. Springs, C. T. Brown, in *Comprehensive Supramoleuclar Chemistry,* Y. Murakami (Ed.), Pergamon, Oxford, 1997, Vol. 4, p. 311.

108 W. L. Jorgenson, J. Pranata, *J. Am. Chem. Soc.* **1990**, *112,* 2008.

109 J. Pranata, S. G. Wierschke, W. L. Jorgensen, *J. Am. Chem. Soc.* **1991**, *113,* 2810.

110 E. H. Howell, J. E. Villafranca, M. S. Warren, S. J. Oatley, J. Kraut, *Science* **1986**, *231,* 1123.

111 C. Ostermeier, A. Harrenga, U. Ermler, H. Michel, *Proc. Natl. Acad. Sci. U.S.A.* **1997**, *94,* 10547.

112 S. Iwata, C. Ostermeier, B. Ludwig, H. Michel, *Nature* **1995**, *376,* 660.

113 S. Yoshikawa, K. Shinzawa-Itoh, R. Nakashima, R. Yaono, E. Yamashia, N. Inoue, M. Yao, M. J. Fei, C. P. Libeu, T. Mizushima, H. Yamaguchi, T. Tomizaki, T. Tsukihara, *Science* **1998**, *280,* 1723.

114 T. Tsukihara, H. Aoyama. E. Yamashia, T. Tomizaki, H. Yamaguchi, K. Shinzawa-Itoh, R. Nakashima, R. Yaono, S. Yoshikawa, *Science* **1996**, *272,* 1136.

115 T. Tsukihara, H. Aoyama. E. Yamashia, T. Tomizaki, H. Yamaguchi, K. Shinzawa-Itoh, R. Nakashima, R. Yaono, S. Yoshikawa, *Science* **1995**, *269,* 1069.

116 B. R. Crane, A. S. Arvai, D. K. Ghosh, C. Wu, E. D. Getzoff, D. J. Stuehr, J. A. Tainer, *Science,* **1998**, *279,* 2121.

117 T. O. Fischmann, A. Hruza, X. Da Niu, J. D. Fosseta, C. A. Lunn, E. Dolphin, A. J. Prongay, P. Reichert, D. J. Lundell, S. K. Narula, P. C. Weber, *Nat. Struct. Biol.* **1999**, *6,* 233.

118 C. S. Raman, H. Li, P. Martasek, V. Kral, B. S. S. Masters, T. L. Poulos, *Cell* **1998**, *95,* 939.

119 D. Papoutsakis, J. P. Kirby, J. E. Jackson, D. G. Nocera, *Chem. Eur. J.* **1999**, *5,* 1474.

120 J. Otsuki, K. Iwasaki, Y. Nakano, M. Itou, Y. Araki, O. Ito, *Chem. Eur. J.* **2004**, *10,* 3461.

121 S. Camiolo, P. A. Gale, M. I. Ogden, B. W. Skelton, A. H. White, *J. Chem. Soc., Perkin Trans. 2* **2001**, 1294.

122 A. Terfort, G. von Kiedrowski, *Angew. Chem. Int. Ed. Engl.* **1991**, *5,* 654.

123 J. Rosenthal, J. M. Hodgkiss, E. Young, D. G. Nocera, *J. Am. Chem. Soc.* **2006**, in press.

124 K. Izutsu, *Acid–Base Dissociation Constants in Dipolar Aprotic Solvents,* Blackwell Scientific, Cambridge, USA, 1990.

125 M.B. Smith, J. March, *March's Advanced Organic Chemistry,* 5th Edn., John Wiley and Sons, New York, 2001.

126 J. P. Kirby, J. A. Roberts, D. G. Nocera, *J. Am. Chem. Soc.* **1997**, *119,* 9230.

127 J. A. Roberts, J. P. Kirby, D. G. Nocera, *J. Am. Chem. Soc.* **1995**, *117,* 8051.

128 J. A. Roberts, J. P. Kirby, S. T. Wall, D. G. Nocera, *Inorg. Chim. Acta* **1997**, *263,* 395.

129 R. H. Felton, D. Dolphin, D. C. Borg, J. Fajer, *J. Am. Chem. Soc.* **1969**, *91,* 196.

130 J. H. Fuhrhop, P. Wasser, D. Riesner, D. Mauzerall, *J. Am. Chem. Soc.* **1974**, *94,* 7996.

131 W. A. Oertling, A. Salehi, Y. C. Chung, G. E. Leroi, C. K. Chang, G. T. Babcock, *J. Phys. Chem.* **1987**, *91,* 5887.

132 D. Daulton, R. H. Felton, *Acc. Chem. Res.* **1974**, *7,* 26.

133 R. I. Walter, *J. Am. Chem. Soc,* **1953**, *75,* 3860.

134 B. Ward, P. M. Callahan, R. Young, G. T. Babcock, C. K. Chang, *J. Am. Chem. Soc.* **1983**, *105,* 634.

135 J. D. Petke, G. M. Maggiora, *J. Am. Chem. Soc.* **1984**, *106*, 3129.

136 B. Ward, C. K. Chang, R. Young, *J. Am. Chem. Soc.* **1984**, *106*, 3943.

137 L. K. Hanson, C. K. Chang, B. Ward, P. M. Callahan, G. T. Babcock, J. D. Head, *J. Am. Chem. Soc.* **1984**, *106*, 3950.

138 E. C. A. Ojadi, H. Linschitz, M. Gouterman, R. I. Walter, J. S. Lindsey, R. W. Wagner, P. R. Droupadi, W. Wang, *J. Phys. Chem.* **1993**, *97*, 13192.

139 J. P. Kirby, N. A. van Dantzig, C. K. Chang, D. G. Nocera, *Tetrahedron Lett.* **1995**, *36*, 3477.

140 Y. Deng, J. A. Roberts, S.-M. Peng, C. K. Chang, D. G. Nocera, *Angew. Chem. Int. Ed.* **1997**, *36*, 2124.

141 J. Otsuki, M. Takatsuki, M. Kaneko, H. Miwa, T. Takido, M. Seno, K. Okamoto, H. Imahori, M. Fujitsuka, Y. Araki, O. Ito, S. Fukuzumi, *J. Phys. Chem. A* **2003**, *107*, 379.

142 C.-Y. Yeh, S. E. Miller, S. D. Carpenter, D. G. Nocera, *Inorg. Chem.* **2001**, *40*, 3643.

143 N. H. Damrauer, J. M. Hodgkiss, J. Rosenthal, D. G. Nocera, *J. Phys. Chem. B* **2004**, *108*, 6315.

144 A. Osuka, R. Yoneshima, H. Shiratori, S. Okada, S. Taniguchi, N. Mataga, *J. Chem. Soc., Chem. Commun.* **1998**, 1567.

145 J. L. Sessler, M. Sathiosatham, C. T. Brown, T. A. Rhodes, G. Wiederrecht, *J. Am. Chem. Soc.* **2001**, *123*, 3655.

146 S. R. Greenfield, W. A. Svec, D. Gosztola, M. R. Wasielewski, *J. Am. Chem. Soc.* **1996**, *118*, 6767.

147 N. Liang, A. G. Mauk, G. J. Pielak, J. A. Johnson, M. Smith, B. M. Hoffman, *Science* **1988**, *240*, 311.

148 B. Barry, G. T. Babcock, *Proc. Natl. Acad. Sci. U.S.A.* **1987**, *84*, 7099.

149 H.-A. Chu, A. P. Nguyen, R. J. Debus, *Biochemistry* **1995**, *34*, 5839.

150 A.-M. A. Hays, I. R. Vassiliev, J. H. Golbeck, R. J. Debus, *Biochemistry* **1999**, *38*, 11851.

151 R. J. Debus, B. A. Barry, G. T. Babcock, L. McIntosh, *Proc. Natl. Acad. Sci. U.S.A.* **1988**, *85*, 427.

152 R. J. Debus, B. A. Barry, I. Sithole, G. T. Babcock, L. McIntosh, *Biochemistry* **1988**, *27*, 9071.

153 B. A. Diner, *Biochim. Biophys. Acta* **2001**, *1503*, 147.

154 R. J. Debus, *Biochim. Biophys. Acta* **2001**, *1503*, 164.

155 C. Tommos, G. T. Babcock, *Acc. Chem. Res.* **1998** *31*, 18.

156 C. Tommos, G. T. Babcock, *Biochim. Biophys. Acta* **2000**, *1458*, 199.

157 M. Sjödin, S. Styring, B. Åkermark, L. Sun, L. Hammarström, *J. Am. Chem. Soc.* **2000**, *122*, 3932.

158 M. Sjödin, R. Ghanem, T. Polivka, J. Pan, S. Styring, L. Sun, V. Sundström, L. Hammarström, *Phys. Chem. Chem. Phys.* **2004**, *6*, 4851.

159 C. Carra, N. Iordanova, S. Hammes-Schiffer, *J. Am. Chem. Soc.* **2003**, *125*, 10429.

160 M. Sjödin, S. Styring, H. Wolpher, Y. Xu, L. Sun, L. Hammarström, *J. Am. Chem. Soc.* **2005**, *127*, 3855.

161 L. P. Candeias, S. Steenken, *J. Am. Chem. Soc.* **1989**, *111*, 1094.

162 E. M. Boon, J. K. Barton, *Curr. Opin. Struct. Biol.* **2002**, *12*, 320.

163 B. Giese, *Annu. Rev. Biochem.* **2002**, *71*, 51.

164 B. Giese, *Acc. Chem. Res.* **2000**, *33*, 631.

165 F. D. Lewis, in *Electron Transfer in Chemistry*, Balzani, V. (Ed.), Wiley-VCH, Weinheim, 2001, Vol. 3.1.5, p. 105.

166 F. D. Lewis, R. L. Letsinger, M. R. Wasielewski, *Acc. Chem. Res.* **2001**, *34*, 159.

167 G. B. Schuster, *Acc. Chem. Res.* **2000**, *33*, 253.

168 C. A. M. Siedel, A. Schulz, M. H. M. Sauer, *J. Phys. Chem.* **1996**, *100*, 5541.

169 F. D. Lewis, T. Wu, Y. Zhang, R. L. Letsinger, S. R. Greenfield, M. R. Wasielewski, *Science* **1997**, *277*, 673.

170 F. D. Lewis, X. Liu, J. Liu, S. E. Miller, R. T. Hayes, M. R. Wasielewski, *Nature* **2000**, *406*, 51.

171 V. Y. Shafirovich, S. H. Courtney, N. Ya, N. E. Geacintov, *J. Am. Chem. Soc.* **1995**, *117*, 4920.

172 N. Amann, E. Pandurski, T. Fiebig, H.-A. Wagenknecht, *Angew. Chem. Int. Ed.* **2002**, *41*, 2978.

173 R. Huber, T. Fiebig, H.-A. Wagenknecht, *Chem. Commun.* **2003**, 1878.

174 M. Ratchev, E. Mayer, N. Amann, H.-A. Wagenknecht, T. Fiebig, *ChemPhysChem* **2004**, *5*, 706.

175 L. Biczók, H. Linschitz, *J. Phys. Chem.* **1995**, *99*, 1843.

176 L. Biczók, N. Gupta, H. Linschitz, *J. Am. Chem. Soc.* **1997**, *119*, 12601.

177 H. Miyasaka, A. Tabata, S. Ojima, N. Ikeda, N. Mataga, *J. Phys. Chem.* **1993**, *97*, 8222.

178 N. Mataga, S. Tsuno, *Bull. Chem. Soc. Jpn.* **1957**, *30*, 368.

179 L. Sun, M. Burkitt, M. Tamm, M. K. Raymond, M. Abrahamsson, D. LeGourriérec, Y. Frapart, A. Magnuson, P. H. Kenéz, P. Brandt, A. Tran, L. Hammarström, S. Styring, B. Åkermark, *J. Am. Chem. Soc.* **1999**, *121*, 6834.

180 I. J. Rhile, J. M. Mayer, *J. Am. Chem. Soc.*, **2004**, *126*, 12718.

181 D. Shukla, R. H. Young, S. Farid, *J. Phys. Chem. A* **2004**, *108*, 10386.

182 L. Biczók, H. Linschitz, *J. Phys. Chem. A* **2001**, *105*, 11051.

183 N. Gupta, H. Linschitz, *J. Am. Chem. Soc.* **1997**, *119*, 6384.

184 S.-C. Hung, A. N. MacPherson, S. Lin, P. A. Liddel, G. R. Seely, A. Moore, T. Moore, D. Gust, *J. Am. Chem. Soc.* **1995**, *117*, 1657.

185 E. J. Land, G. Porter, E. Strachan, *Trans. Faraday Soc.* **1961**, *57*, 1885.

186 W. T. Dixon, D. Murphy, *J. Chem. Soc., Faraday Trans. II* **1976**, *72*, 1221.

187 J. Stubbe, D. G. Nocera, C. S. Yee, M. C. Y. Chang, *Chem. Rev.* **2003**, *103*, 2167.

188 J. Stubbe, W. A. van der Donk, *Chem. Rev.* **1998**, *98*, 705.

189 A. Maradufu, G. M. Cree, A. S. Perlin, *Can. J. Chem.* **1971**, *49*, 3429.

190 J. W. Whittaker, *Arch. Biochem. Biophys.* **2005**, *433*, 227.

191 N. Ito, S. E. V. Phillips, K. D. S. Yadav, P. F. Knowles, *J. Mol. Biol.* **1994**, *238*, 794.

192 J. Stubbe, M. Ator, T. Krenitsky, *J. Biol. Chem.* **1983**, *258*, 1625.

193 T. L. Poulos, B. C. Finzel, A. J. Howard, *Biochemistry* **1986**, *25*, 5314.

194 M. Vidakovic, S. G. Sligar, H. Li, T. L. Poulos, *Biochemistry* **1998**, *37*, 9211.

195 A. E. Pond, A. P. Ledbetter, M. Sono, D. B. Goodin, J. H. Dawson, in *Electron Transfer in Chemistry*, V. Balzani (Ed.), Wiley-VC, Weinheim, 2001, Vol. 3.1.4; p. 56.

196 B. Meunier, S. P. de Visser, S. Shaik, *Chem. Rev.* **2004**, *104*, 3947.

197 S. L. Newmyer, P. R. Ortiz de Montellano, *J. Biol. Chem.* **1995**, *270*, 19430.

198 M. Tanaka, K. Ishimori, M. Mukai, T. Kitagawa, I. Morishima, *Biochemistry* **1997**, *36*, 9889.

199 D. W. Low, J. R. Winkler, H. B. Gray, *J. Am. Chem. Soc.* **1996**, *118*, 117.

200 J. Berglund, T. Pascher, J. R. Winkler, H. B. Gray, *J. Am. Chem. Soc.* **1997**, *119*, 2464.

201 C.-Y. Yeh, C. J. Chang, D. G. Nocera, *J. Am. Chem. Soc.* **2001**, *123*, 1513.

202 C. J. Chang, L. L. Chng, D. G. Nocera, *J. Am. Chem. Soc.* **2003**, *125*, 1866.

203 J. L. Dempsey, A. J. Esswein, D. R. Manke, J. Rosenthal, J. D., Soper, D. G. Nocera, *Inorg. Chem.* **2005**, *44*, 6879.

204 L. L. Chng, C. J. Chang, D. G. Nocera, *Org. Lett.* **2003**, *5*, 2421.

205 S.-Y. Liu, D. G. Nocera, *J. Am. Chem. Soc.* **2005**, *127*, 5278.

206 K. N. Ferreira, T. M. Iverson, K. Maghlaoui, J. Barber, S. Iwata, *Science* **2004**, *303*, 1831.

207 B. Loll, J. Kern, W. Saenger, A. Zouni, J. Biesiadka, *Nature* **2005**, *438*, 1040.

208 C. W. Hoganson, C. Tommos, *Biochim. Biophys. Acta* **2004**, *1655*, 116.

209 J. S. Vrettos, G. W. Brudvig, *Philos. Trans. R. Soc. London, Ser. B* **2002**, *357*, 1395.

210 J. Barber, K. Ferreira, K. Maghlaoui, S. Iwata, *Phys. Chem. Chem. Phys* **2004**, *6*, 4737.

211 M. Haumann, W. Junge, *Biochim. Biophys. Acta* **1999**, *1411*, 86.

212 W. Junge, M. Haumann, R. Ahlbrink, A. Mulkidjanian, J. Clausen, *Phil. Trans. R.l Soc. Lond. B* **2002**, *357*, 1407.

213 W. A. Cramer, D. B. Knaff, *Energy Transduction in Biological Membranes*, Springer-Verlag, New York, 1990.

214 J. Deisenhofer, J. R. Norris Jr., *The Photosynthetic Reaction Center*, Vol. I and II. Academic Press, San Diego, CA, 1993.

215 M. Y. Okamura, M. L. Paddock, M. S. Graige, G. Feher, *Biochim. Biophys. Acta* 2000, *1458*, 148.

216 J. W. Peters, W. N. Lanzilotta, B. J. Lemon, L. C. Seefeldt, *Science* 1998 *282*, 1853.

217 Y. Nicolet, C. Piras, P. Legrand, C. E. Hatchikian, J. C. Fontecilla-Camps, *Structure* 1999, *7*, 13.

218 A. Volbeda, J. C. Fontecilla-Camps, *Coord. Chem. Rev.* 2005, *249*, 1609.

219 A. Jordan, P. Reichard, *Annu. Rev. Biochem.* 1998, *67*, 71.

220 U. Uhlin, H. Eklund, *Nature* 1994, *370*, 533.

221 P. Nordlund, B.-M. Sjöberg, H. Eklund, *Nature* 1990, *345*, 593.

222 M. Högbom, M. Galander, M. Andersson, M. Kolberg, W. Hofbauer, G. Lassmann, P. Nordlund, F. Lendzian, *Proc. Natl. Acad. Sci. U.S.A.* 2003, *100*, 3209.

223 J. Stubbe, P. Riggs-Gelasco, *Trends Biochem. Sci.* 1998, *23*, 438.

224 C. S. Yee, M. C. Y. Chang, J. Ge, D. G. Nocera, J. Stubbe, *J. Am. Chem. Soc.* 2003, *125*, 10506.

225 C. S. Yee, M. R. Seyedsayamdost, M. C. Y. Chang, D. G. Nocera, J. Stubbe, *Biochemistry* 2003, *42*, 14541.

226 M. C. Y. Chang, C. S. Yee, D. G. Nocera, J. Stubbe, *J. Am. Chem. Soc.* 2004, *126*, 16702.

227 M. R. Seyedsayamdost, S. Y. Reece, D. G. Nocera, J. Stubbe, *J. Am. Chem. Soc.* 2006, *128*, 1569.

228 M. R. Seyedsayamdost, C. S. Yee, S. Y. Reece, D. G. Nocera, J. Stubbe, *J. Am. Chem. Soc.* 2006, *128*, 1562.

229 S. Y. Reece, J. Stubbe, D. G. Nocera, *Biochim. Biophys. Acta* 2005, *1706*, 232.

230 M. C. Y. Chang, C. S. Yee, J. Stubbe, D. G. Nocera, *Proc. Nat. Acad. Sci. USA* 2004, *101*, 6882.

231 Y.-Q. Deng, D. G. Nocera, unpublished results.

232 J. J. Lopez-Garriga, G. T. Babcock and J. F. Harrison, *J. Am. Chem. Soc.* 1986, *108*, 7241.

233 P. Hamm, R. A. Kaindl, J. Stenger, *Opt. Lett.* 2000, *25*, 1798.

234 S. Mukamel, *Annu. Rev. Phys. Chem.* 2000, *51*, 691.

235 M. Khalil, N. Demirdöven, A. Tokmakoff, *J. Phys. Chem. A* 2003, *107*, 5258.

236 M. T. Zanni, R. M. Hochstrasser, *Curr. Opin. Struct. Biol.* 2001, *11*, 516.

237 J. C. Wright, *Int. Rev. Phys. Chem.* 2002, *21*, 185.

238 S. Woutersen, P. Hamm, *J. Phys. Condens. Matter* 2002, *14*, R1035.

239 R. I. Cukier, J. Zhu *J. Chem. Phys.* 1999, *110*, 9587.

240 O. Vendrell, M. Moreno, J. M. Lluch, S. Hammes-Schiffer, *J. Phys. Chem. B* 2004, *108*, 6616.

241 J. D. Simon, *Acc. Chem. Res.* 1988, *21*, 128.

242 M. Cho, M. Du, N. F. Scherer, G. R. Fleming, S. Mukamel, *J. Chem. Phys.* 1993, *99*, 2410.

243 S. Ruhman, B. Kohler, A. G. Joly, K. A. Nelson, *Chem. Phys. Lett.* 1987, *141*, 16.

244 D. F. Underwood, D. A. Blank, *J. Phys. Chem. A* 2003, *107*, 956.

245 S. J. Schmidtke, D. F. Underwood, D. A. Blank, *J. Am. Chem. Soc.* 2004, *126*, 8620.

246 S. J. Schmidtke, D. F. Underwood, D. A. Blank, *J. Phys. Chem. A* 2005, *109*, 7033.

247 D. F. Underwood, D. A. Blank, *J. Phys. Chem. A* 2005, *109*, 3295.

248 P. Barbara, G. Walker, T. Smith, *Science*, 1992, *256*, 975.

249 T. Smith, K. Zaklika, K. Thakur, P. Barbara, *J. Am. Chem. Soc.* 1991, *113*, 4035.

250 R. Jimenez, G. R. Fleming, P. V. Kumar, M. Maroncelli, *Nature* 1994, *369*, 471.

251 M. L. Horng, J. A. Gardecki, A. Papazyan, M. Maroncelli, *J. Phys. Chem.* 1995, *99*, 17311.

Part V
Hydrogen Transfer in Organic and Organometallic Reactions

Part V is devoted to the study of H transfers in organic and organometallic reactions and systems. In Ch. 18 Koch describes kinetic studies of proton abstraction from CH groups by methoxide anion, of the reverse proton transfer from methanol to hydrogen bonded carbanion intermediates, and of proton transfer associated with methoxide promoted dehydrohalogenation reactions. Substitutent effects, kinetic isotope effects and *ab initio* calculations are treated. Of great importance is the extent of charge delocalization in the carbanions formed which determine the "kinetic" and "thermodynamic" acidities.

In Ch. 19 Williams describes theoretical simulations of free-energy-relationships in proton transfer processes. Both linear and non-linear relations are observed, usually described in terms of Brønsted coefficients or Marcus intrinsic barriers. Derived from empirical data, the phenomenological parameters of themselves do not lead to satisfying explanations at a fundamental molecular level. Theoretical simulations can fill in this gap.

In 1983 Kubas et al. found that transition metals can contain hydrogen not only in atomic (hydride) but also in molecular (H_2) form in side-on coordination. This field is reviewed in Ch. 20 by Kubas. The structure and dynamics of hydrogen containing transition metal complexes is studied mainly by neutron diffraction, neutron scattering, NMR, NMR relaxometry, IR and other spectroscopic. Naturally, quantum-mechanical treatments give rise to very detailed insights. Dihydrogen molecules bound to metals exhibit rotational tunnel splittings of the order of 10^{11} Hz observed by inelastic neutron scattering. When the rotational barriers are larger values of about 10^2 Hz are observed as "exchange couplings" by NMR. On top of the coherent rotational tunneling process incoherent exchange is observed as well. Proton donors can form hydrogen bonds to transition metal hydrides, forming complexes of the type M-H$\cdots$H-A. The authors shows the implication of these unusual features for the elementary steps of homogeneous and heterogeneous catalysis.

In Ch. 21 Buntkowsky and Limbach review recent NMR work on the dynamics of dihydrogen and dideuterium in the coordination sphere of transition metals. In addition to inelastic neutron scattering and liquid state NMR, the effects of coherent (exchange couplings) and incoherent rotational tunneling of D_2 pairs in transi-

Hydrogen-Transfer Reactions. Edited by J. T. Hynes, J. P. Klinman, H. H. Limbach, and R. L. Schowen
Copyright © 2007 WILEY-VCH Verlag GmbH & Co. KGaA, Weinheim
ISBN: 978-3-527-30777-7

tion metal hydrogen complexes are described, providing information about the corresponding dynamic H/D isotope effects. It is shown that the size of the rotational tunnel splittings as compared to those of magnetic nuclear interaction is the main control parameter which governs the transition metal catalyzed magnetic ortho-para hydrogen spin conversion. Applications in chemistry and medicine are discussed.

18

Formation of Hydrogen-bonded Carbanions as Intermediates in Hydron Transfer between Carbon and Oxygen

Heinz F. Koch

This chapter is divided into three sections: Proton transfer from carbon acids to methoxide ion; proton ttransfer from methanol to carbanion intermediates; proton transfer associated with methoxide promoted dehydrohalogenation reactions.

18.1
Proton Transfer from Carbon Acids to Methoxide Ion

Knowledge of equilibrium pK_a values of carbon acids is important for an understanding of organic chemistry; however, they are not always reliable indicators of relative rates for proton transfer reactions. Ritchie [1] predicted in 1969 that carbon acids whose conjugate bases have localized charge will show "kinetic acidities" greater than their thermodynamic acidities. This is illustrated by the weaker acid pentafluorobenzene-t [PFB-t], $pK_a = 25.8$ [2], with a methanolic sodium methoxide catalyzed protodetritiation rate, $k = 2.57 \times 10^{-2}$ M^{-1} s^{-1} [3], that is 15 times faster at 25 °C than that for 9-phenylfluorene-9-t [9-PhFl-9-t], $pK_a = 18.5$ and $k = 1.54 \times 10^{-3}$ M^{-1} s^{-1} [4]. Primary kinetic isotope effects [PKIE] can be used to determine C–H bond breaking in the rate limiting steps of any reaction mechanism. Normal hydrogen isotope effects are largely due to zero-point energy differences of the stretching frequencies of C–H vs. C–D bonds. If only stretching frequencies are considered the value for k^H/k^D would be 6.2 for the C–^{i}H bond. With allowance for bending vibrations this could result in a $k^H/k^D = 10$, $k^H/k^T = 27$ and $k^D/k^T = 2.7$ at 25 °C [5]. Melander [6] and Westheimer [7] suggested that lower values of k^H/k^D could be due to residual zero-point energy in an asymmetric transition structure. For this reason the magnitude of k^H/k^D was often used to assign early or late transition structures for hydron-transfer reactions. Smaller values due to asymmetric transition structures should still obey the Swain–Schaad relationship [8], $k^H/k^T = (k^H/k^D)^{1.442}$. The $k^D/k^T = 1.0$ associated with the exchange reactions of PFB^{i}H suggest that the C–H bond is not broken in the rate limiting step and this differs significantly from the k^D/k^T of 2.54 measured from the reactions of 9-PhFl-9-^{i}H.

Hydrogen-Transfer Reactions. Edited by J. T. Hynes, J. P. Klinman, H. H. Limbach, and R. L. Schowen
Copyright © 2007 WILEY-VCH Verlag GmbH & Co. KGaA, Weinheim
ISBN: 978-3-527-30777-7

In a 1975 Account, Kresge addressed "fast" vs. "slow" proton transfer [9]: "It is common experience that proton transfer between electronegative atoms such as oxygen and nitrogen is very fast whereas that involving carbon is usually quite slow. This would seem to be related to the fact that the electron pair which receives the proton onto an oxygen or nitrogen base is generally localized on a single atom, as in ammonia or amines. The corresponding pair of a carbon base, on the other hand, except in unusual circumstances, is strongly delocalized away from the atom to which the proton becomes attached; this is so, for example, in nitronate and enolate ions, the reprotonation of which are classic examples of slow proton transfer." The study of proton transfer from C–H bonds to a base ["kinetic acidity"] in aqueous media was largely carried out on compounds that generated nitronate or enolate anions and owed their stability to π-delocalization of the negative charge. Since the sp orbital of phenylacetylide ion is orthogonal to the π-bonds, Kresge chose the aqueous base catalyzed hydron exchange of phenylacetylene to study hydron-transfer reactions of localized carbanions [10]. The near unity PKIE associated with these exchange reactions is similar to the results obtained for PFB, where the $C_6F_5^-$ anion [PFB$^-$] is localized in an sp^2 orbital that is orthogonal to the π-electrons of the benzene ring. On the other hand, the 9-phenylfluorenyl anion [9-PhFl $^-$] is a highly π-delocalized aromatic anion. If a C–H bond is not broken in a rate-limiting step, it is not possible to know the rate of proton transfer.

Cram first suggested that near unity isotope effects for hydron exchange reactions are due to an internal-return mechanism with hydron transfer occurring prior to the rate-limiting step in the reaction mechanism [11]. The seven-step process given below would be for an internal-return mechanism applied to deuterium exchange of a carbon acid using methanolic methoxide as the base. The process starts as one of the three methanols associated with the lone pair electrons of a methoxide ion joins the bulk solvent and is replaced by the carbon acid forming the encounter complex, EC-d.

$$
\begin{array}{c}
\text{HOMe} \\
\ce{\backslash \; \overset{|}{C}-D \quad MeOH\cdots\overset{..}{\underset{..}{O}}-CH_3} \\
\text{HOMe}
\end{array}
\rightleftharpoons
\begin{array}{c}
\text{HOMe} \\
\ce{\backslash \; \overset{|}{C}-D\cdots\overset{..}{\underset{..}{O}}-CH_3} \\
\text{HOMe} \\
\text{EC-}d
\end{array}
\tag{18.1}
$$

The system is now ready to have the deuterium transfer from the carbon to methoxide ion, k_1^D, and generate a hydrogen-bonded carbanion, HB-d. The internal return step, k_{-1}^D, can compete with any forward reaction and would regenerate the encounter complex EC-d.

$$
\begin{array}{c}
\text{HOMe} \\
\ce{\backslash \; \overset{|}{C}-D\cdots\overset{..}{\underset{..}{O}}-CH_3} \\
\text{HOMe} \\
\text{EC-}d
\end{array}
\underset{k_{-1}^D}{\overset{k_1^D}{\rightleftharpoons}}
\begin{array}{c}
\text{HOMe} \\
\ce{\backslash \; \overset{|}{C}\cdots D-\overset{..}{\underset{..}{O}}-CH_3} \\
\text{HOMe} \\
\text{HB-}d
\end{array}
\tag{18.2}
$$

The next forward step occurs when the hydrogen bond breaks, k_2^D, to form a free carbanion, FC-d, that is not stabilized by any direct contact with the $DOCH_3$ that is still in the best position to interact with the lone pair of the carbanion, k_{-2}^D.

$$(18.3)$$

The exchange, k_{exc}, occurs when a CH_3OH replaces the CH_3OD in the most favorable position.

$$(18.4)$$

To complete the reaction there are three steps using CH_3OH in place of CH_3OD in steps 3, 2 and 1. When k_{exc} is greater than k_{-2}^H, the rate law for this mechanism is:

$$k_{obs} = \frac{k_1 k_2}{k_{-1} + k_2} \qquad (18.5)$$

There are two extremes for this rate law: (i) when $k_{-1} \gg k_2$, then $k_{obs} = [\, k_1 / k_{-1} \,] k_2$, and (ii) when $k_2 \gg k_{-1}$, then $k_{obs} = k_1$. For case (i) second order kinetics and near unity experimental PKIE values are expected. On the other hand for case (ii) second order kinetics are measured with normal experimental values of the PKIE that obey the Swain–Schaad relationship.

A major contribution to the analysis of isotope effects associated with cases where there is no single rate-limiting step was made by the Streitwieser group in 1971 [4, 12]. Using single temperature rate constants for all three hydrogen isotopes and the Swain–Schaad relationship, they calculate an internal-return parameter, $a = k_{-1}/k_2$, associated with the experimental rate constants for each of the isotopic exchange measurements.[1] The rate constant for the actual hydron transfer step can now be calculated:

$$k_1 = k_{obs}\,(a + 1) \qquad (18.6)$$

1) The original Swain–Schaad relationship[8] $(k^H/k^T) = (k^H/k^D)^{1.442}$ becomes $(k^H/k^T) = (k^D/k^T)^{3.26}$. For hydron exchange reactions the later relationship is more useful. The Streitwieser treatment uses 3.344 instead of 3.26 as the exponent [12]. We use the 3.344 value for our calculations.

This analysis can be applied to the methoxide-catalyzed rates of the exchange reactions for 9-PhFl-9-^{i}H that result in $k^D/k^T = 2.53$ and $k^H/k^T = 15.9$.[2] To satisfy Swain–Schaad, the value of k^H/k^T should be 20.6 or 22.3 depending on the value of the exponent used.[1] The deviations from the Swain–Schaad relationship predict that a small amount of internal return is associated with the hydron exchange reactions of 9-PhFl-9-^{i}H. The internal return is negligible for 9-PhFl-9-t ($a^T = 0.016$) and 9-PhFl-9-d ($a^D = 0.050$), but cannot be ignored for 9-PhFl ($a^H = 0.49$).[3] Experimental isotope effects that are near unity in magnitude do not allow the calculation of any a values; however, the return step must be at least 50 times faster than the forward step. Therefore, after making a correction for internal return, the hydron exchange reaction from the weaker acid, PFB-t is at least 750 times faster than that from the stronger acid, 9-PhFl-9-t.

Since fluorine-containing organic compounds are more acidic than the hydrocarbon analogs, we started to study benzylic compounds that have trifluoromethyl groups and chlorine atoms to increase their acidity. Two compounds with rates of protodetritiation similar to that for 9-PhFl-9-t were of interest: $C_6H_5CH(CF_3)_2$ **1** and p-$CF_3C_6H_4CHClCF_3$ [p-CF_3-**2**]. Measurement of the pK_a values for **1** and p-CF_3-**2** are not possible in solution due to elimination of a β-fluoride. However, Mishima has been able to obtained their gas phase acidities, $\Delta G^0_{acid}=$ 335.3 kcal mol^{-1} for **1** and $\Delta G^0_{acid} = 337.4$ kcal mol^{-1} for p-CF_3-**2**, and these are close to that for 9-PhFl, $\Delta G^0_{acid} = 335.6$ kcal mol^{-1} [13]. Although the protodetritiation and gas phase acidities are similar for these three compounds, the near unity experimental k^D/k^T values obtained from the reactions of **1** and p-CF_3-**2** differ significantly from that for 9-PhFl and are similar to the value for PFB. This suggests a significant amount of internal return associated with the hydron transfer reactions of **1** and p-CF_3-**2**. The effects of substituents on the phenyl ring of 9-PhFl are also very different from those on the phenyl ring of **1**, Table 18.1. The negative charge for the 9-PhFl$^-$ anion would not be π-delocalized into the phenyl ring, and a Hammett plot using the four 9-YPhFl compounds results in a rho of 2.1, which is similar to that obtained for the hydrolysis of methyl benzoates. A plot with the six Y-I compounds results in a rho value of 4.9, which is similar that obtained for the protodetritiation reactions of ring-substituted toluene-a-t's with lithium cyclohexylamide in cyclohexylamine [14].

Although the gas phase acidity of $C_6H_5CHClCF_3$ **2**, $\Delta G^0_{acid}= 348.7$ kcal mol^{-1}, is about the same as that for PFB, $\Delta G^0_{acid}= 349.2$ kcal mol^{-1}, the reaction of **2** is too slow in methanolic sodium methoxide and was carried out in ethanolic sodium ethoxide. Reactions in methanolic sodium methoxide are about 20 times slower than those in ethanol [see m-$ClC_6H_4CTClCF_3$ and m-$CF_3C_6H_4CTClCF_3$ in Table 18.1]. When the ethoxide rate for the protodetritiation of **2**-t is corrected for the change in base systems the reaction is 66,000 times slower at 25 °C than that for

2) Reference [4] has values of k^D / k^T of 2.54±0.11 to 2.46±0.20 and k^H / k^T of 16.1±0.6 to 15.9±0.6. Our values are from our Arrhenius plots for all three isotopes of 9-PhFl-9-^{i}H, see Table 18.1.

3) The equation for a^T is given in Ref. 12 p. 5099. We thank Professor Andrew Streiwieser for supplying the equations to calculate a^H and a^D.

Table 18.1. Rate constants and activation parameters for methanolic sodium methoxide promoted hydron exchange reactions.

Compound	ΔG^0_{Acid}	k, M⁻¹s⁻¹ (25 °C)	$\Delta H^{\ddagger}$, kcal / Tmol	$\Delta S^{\ddagger}$, eu (25 °C)	Temp. range,°C [no. of points]
m-CF₃C₆H₄CD(CF₃)₂	326.8	4.12 E-1	20.79 ± 0.18	9.4 ± 0.7	−20 to −5 [4]
m-CF₃C₆H₄CT(CF₃)₂		4.74 E-1	21.60 ± 0.10	12.4 ± 0.4	−19 to 0 [4]
m-CF₃C₆H₄CT(CF₃)₂ [MeOD]		1.33	20.63 ± 0.02	11.2 ± 0.1	−29 to −10 [3]
m-CF₃C₆H₄CH(CF₃)₂ [MeOD]		1.32	20.34 ± 0.12	10.2 ± 0.5	−25 to −5 [4]
m-FC₆H₄CD(CF₃)₂	331.5	8.50 E-2	21.55 ± 0.08	8.8 ± 0.3	−10 to 15 [4]
p-ClC₆H₄CD(CF₃)₂	331.4	3.45 E-2	23.10 ± 0.16	12.2 ± 0.6	0 to 20 [5]
C₆H₅CD(CF₃)₂	335.3	2.13 E-3[a]	24.42 ± 0.12	11.2 + 0.4	0 to 50 [7]
C₆H₅CT(CF₃)₂		2.09 E-3	24.26 ± 0.07	10.6 ± 0.2	5 to 50 [7]
C₆H₅CT(CF₃)₂ [MeOD]		5.49 E-3	23.85 ± 0.07	11.1 ± 0.2	5 to 50 [9]
C₆H₅CH(CF₃)₂ [MeOD]		6.14 E-3	23.86 ± 0.13	11.4 ± 0.4	10 to 50 [5]
m-CH₃C₆H₄CD(CF₃)₂	336.3	1.04 E-3	24.97 ± 0.15	11.5 ± 0.5	20 to 40 [5]
p-CH₃C₆H₄CD(CF₃)₂	337.0	4.70 E-4	25.53 ± 0.20	11.9 ± 0.7	25 to 45 [5]
p-NO₂C₆H₄CDClCF₃	329.8	4.40	16.62 ± 0.05	0.1 ± 0.2	−50 to −30 [4]
p-NO₂C₆H₄CHClCF₃ [MeOD]		4.04 E+1	16.23 ± 0.19	3.2 ± 0.8	−55 to −40 [3]
3,5-(CF₃)₂C₆H₃CDClCF	332.4	7.51 E-2	22.48 ± 0.07	11.7 ± 0.3	−5 to 25 [6]
3,5-F₂C₆H₃CDClCF₃	340.5	1.82 E-3	25.36 ± 0.10	14.0 ± 0.4	10 to 30 [3]
p-CF₃C₆H₄CDClCF₃	337.4	8.58 E-4	25.90 ± 0.08	14.3 ± 0.3	20 to 45 [6]
p-CF₃C₆H₄CTClCF₃		7.91 E-4[c]	25.42 ± 0.08	12.5 ± 0.3	20 to 55 [8]
p-CF₃C₆H₄CTClCF₃ [MeOD]		2.06 E-3[c]	24.86 ± 0.13	12.5 ± 0.4	20 to 45 [6]
m-CF₃C₆H₄CDClCF₃	340.1	1.09 E-4	26.36 ± 0.11	11.7 ± 0.3	30 to 55 [7]
m-CF₃C₆H₄CTClCF₃		1.03 E-4	26.67 ± 0.12	12.7 ± 0.4	30 to 60 [5]
m-CF₃C₆H₄CTClCF₃ [MeOD]		2.77 E-4	25.81 ± 0.33	11.8 ± 1.1	30 to 50 [3]
[EtOH]		2.20 E-3	25.25 ± 0.16	14.0 ± 0.5	10 to 40 [7]
m-CF₃C₆H₄CHClCF₃ [MeOD]		3.03 E-4	25.51 ± 0.14	10.9 ± 0.5	30 to 50 [3]
m-ClC₆H₄CTClCF₃	343.2	4.16 E-5	26.96 ± 0.08	11.8 ± 0.3	35 to 59 [5]
[EtOH]		7.74 E-4	25.31 ± 0.16	12.1 ± 0.5	25 to 50 [5]
m-FC₆H₄CDClCF₃	344.6	2.36 E-5	27.85 ± 0.06	13.7 ± 0.2	40 to 65 [5]

Table 18.1 Continued.

Compound	ΔG^0_{Acid}	k, M^{-1}s^{-1} (25 °C)	$\Delta H^{\ddagger}$, kcal / Tmol	$\Delta S^{\ddagger}$,eu (25 °C)	Temp. range,°C [no. of points]
C$_6$H$_5$CTClCF$_3$ [EtOH]	348.7	9.03 E-6	26.54 ± 0.01	7.4 ± 0.1	50 to 80 [3]
[EtOD]		2.08 E-5	26.41 ± 0.14	8.6 ± 0.4	40 to 80 [5]
9-(p-CF$_3$C$_6$H$_4$)-fluorene-9-t	326.9	2.03 E-2	15.78 ± 0.11	−13.4 ± 0.4	−10 to 25 [4]
9-(m-CF$_3$C$_6$H$_4$)-fluorene-9-t	327.4	1.58 E-2	17.07 ± 0.06	−9.5 ± 0.2	0 to 20 [3]
[MeOD]		3.38 E-2	16.29 ± 0.14	−10.6 ± 0.5	−10 to 10 [3]
9-m-FC$_6$H$_4$)-fluorene-9-t	331.3	8.24 E-3	17.84 ± 0.05	−8.2 ± 0.2	0 to 25 [4]
9-Phenylfluorene-9-d	335.6[b]	3.89 E-3	17.22 ± 0.04	−11.8 ± 0.1	5 to 35 [3]
9-Phenylfluorene-9-t		1.54 E-3	18.42 ± 0.13	−9.6 ± 0.4	25 to 45 [5]
9-Phenylfluorene-9-t [MeOD]		3.25 E- 3	18.13 ± 0.07	−9.1 ± 0.2	25 to 40 [3]
9-Phenylfluorene-9-h [MeOD]		5.16 E-2	15.18 ± 0.07	−13.5 ± 0.3	−25 to 5 [6]
CDCl$_2$CF$_3$	348.2	1.93 E-2[d]	21	5	0 to 20 [2]
C$_6$F$_5$D	349.2	3.07 E-2	19.52 ± 0.19	0.0 ± 0.7	0 to 21 [5]
C$_6$F$_5$T		2.96 E-2[e]	19.76 ± 0.07	0.7 ± 0.2	−15 to 15 [6]
C$_6$F$_5$T [MeOD]		5.93 E-2	19.45 ± 0.05	1.1 ± 0.2	−20 to 15 [4]
C$_6$F$_5$H [MeOD]		6.53 E-2	19.25 ± 0.07	0.6 ± 0.3	−15 to 10 [4]

a Reference [25]
b Lias, S. G.; Bartmess, J. E.; Leibman, J. F.; Holmes, J. L.; Levin,
   R. D.; Mallard, G. W. *J Phys. Chem., Ref. Data 1* **1988**, *17*,
   Suppl. 1.
c Ref. [37]
d Ref. [20a]
e Data in Ref. [3] gave the following: C$_6$F$_5$T, k = 2.57 E-2 (25 °C),
   $\Delta H^{\ddagger}$ = 19.92 ± 0.05 kcal/mol^{-1},$\Delta S^{\ddagger}$ = 1.0 ± 0.2 eu.

PFB-t. The much stronger acid 3,5-(CF$_3$)$_2$C$_6$H$_3$CDClCF$_3$, ΔG^0_{acid}= 332.4 kcal mol^{-1}, has a rate of deuterium exchange that is only 2.5 times faster than that for PFB-d. Do all the reactions with near unity isotope effects have the same balance of steps in an internal return mechanism, and should these compounds be compared with each other? The activation entropies, $\Delta S^{\ddagger}$, should be in the same range when reactions have a similar balance of steps that contribute to the value of k_{obs} in the internal-return mechanism.

The exchange reactions of 9-phenylfluorenes have experimental PKIE values that are normal in magnitude, but do not obey the Swain–Schaad relationship, suggesting that there is a small amount of internal return associated with these reactions. Reactions of methanolic sodium methoxide and the four 9-YPhFl-9-t in Table 18.1 result in experimental $\Delta S^{\ddagger}$ with values between – 8 and –13 eu. Such values are consistent with the second order kinetics of a bimolecular reaction and suggest that an encounter complex, EC, could be the dominating step along the reaction path. The reactions of methanolic sodium methoxide with 15 $YC_6H_4CH=CF_2$ have experimental $\Delta S^{\ddagger}$ values that range from –12 to –20 eu [15]. The rate-limiting step for the reaction of CH_3O^- with $CF_2=CHC_6H_5$ generates a carbanion intermediate, $\{CH_3OCF_2CHC_6H_5\}^-$, and would result in bimolecular kinetics. This reaction will be discussed in Section 18.2.

Reactions that have large amounts of internal return are ones with near unity experimental isotope effects, and these normally have activation entropies of around +10 eu. This is true for the hydron exchange reactions of the six YC_6H_4-$C^iH(CF_3)_2$ and four $YC_6H_4C^iHClCF_3$ listed in Table 18.1 that resulted in experimental $\Delta S^{\ddagger}$ values from +12 to +14 eu. An exception to this is p-$NO_2C_6H_4$-C^iHClCF_3 where the $\Delta S^{\ddagger}$ values are between 0 and +2 eu; however, this compound also differs from the others with $k^H/k^D \approx 3$ after correcting for a solvent isotope effect k^{OD}/k^{OH} of 2.6. This can be due to the more π-delocalized carbanion having much less internal return than the other ring substituted hydrogen-bonded carbanion intermediates. Another exception to the higher activation entropy is PFB with $\Delta S^{\ddagger}$ values of about +1 eu and a near unity PKIE. To get a better understanding of these reactions density functional calculations using B3LYP/6-31+G(d,p) were carried out on the energetics of some of the intermediates associated with the reactions of 9-PhFl, **1**, **2** and PFB.

Zuilhof first performed calculations to evaluate the stability of any possible intermediates for the reaction of $C_6H_5CH(CF_3)_2$ **1**, and CH_3O^- to $\{C_6H_5C(CF_3)_2\}^-$ and CH_3OH. An encounter complex, EC, that is not in an energy minimum is 30.1 kcal mol^{-1} more stable than the two reactants, and proceeds by a barrierless process to form the hydrogen-bonded intermediate that is 13.1 kcal mol^{-1} more stable [16]. More recent calculations for $C_6H_5CHClCF_3$ have an encounter complex at a definite energy minimum that is 19.1 kcal mol^{-1} more stable than reactants and a hydrogen-bonded carbanion that is 6.5 kcal mol^{-1} more stable than the encounter complex. It is not surprising that the process is highly exothermic since calculations are for gas phase species and the ΔG^0_{acid} of 375.5 kcal mol^{-1} for methanol [17] makes it a weaker acid than toluene. Since these calculations did not agree with the experimental observations, our current calculations start with three methanols solvating the methoxide ion, and follow the steps outlined in Eqs. (18.1)–(18.3).

Calculations for the reaction of **1** and $CH_3O^-(HOCH_3)_3$ to $\{C_6H_5CH(CF_3)_2\cdots$ $OCH_3(HOCH_3)_2\}^-$ and CH_3OH now requires + 4.7 kcal mol^{-1}, see Fig. 18.1. The energy used in the calculation for the released CH_3OH is one third of the energy of $HOCH_3(HOCH_3)_2$. Proton transfer to form $\{C_6H_5C(CF_3)_2\cdots HOCH_3$-$(HOCH_3)_2\}^-$, requires another + 4.1 kcal mol^{-1} and then +1.1 kcal mol^{-1} is needed to generate $\{C_6H_5C(CF_3)_2\}^-$ and $HOCH_3(HOCH_3)_2$. We have not yet calculated transition structures associated with these reactions; however, the energetics are reasonable for a mechanism that has the proton transfer with significant internal return dominating the rate of reaction. A big difference occurs when comparing the calculations for **1** and 9-PhFl. The 9-PhFl requires +11.1 kcal mol^{-1} to form the corresponding encounter complex, and the formation of the hydrogen-bonded carbanion is then favorable by −3.7 kcal mol^{-1}. The formation of the 9-PhFl$^-$ and HOCH$_3$(HOCH$_3$)$_2$ requires +1.2 kcal mol^{-1}. Therefore, the calculations for 9-PhFl can support the measurement of negative activation entropies that are attributed to a bimolecular reaction that forms an encounter complex as the major factor in the overall rate process [18].

When calculations are carried out to compare the reactions of **2** with those of **PFB**, the formation of the two encounter complexes are similar to those calculated for **1**, with values of +5.5 kcal mol^{-1} (**2**) and +5.1 kcal mol^{-1} (**PFB**). Differences between the reactions of **2** and **PFB** occur for proton transfer to generate the corresponding hydrogen-bonded carbanions with +7.8 kcal mol^{-1} for **2** compared to only +1.3 kcal mol^{-1} for the reactions of **PFB**. This difference is offset in the last

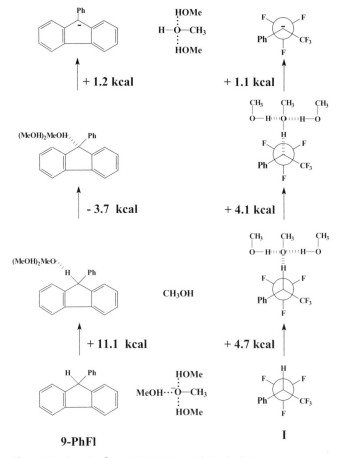

Figure 18.1 Energies from B3LYP/6-31 + G(d,p) calculations.

step to form $\{C_6H_5CClCF_3\}^-$ and $(HOCH_3)_3$, +7.3 kcal mol^{-1}, compared to the formation of PFB$^-$ and $(HOCH_3)_3$, +15.3 kcal mol^{-1}. Since the exchange reactions of PFB are 66000 times faster than those for **2**, it is apparently not necessary to form PFB$^-$. The exchange must be able to occur from the hydrogen-bonded intermediate, and this is the same conclusion as from the $\Delta S^{\ddagger}$ of +1 eu for the reactions of PFB compared to values of > +10 eu for the reactions of methanolic sodium methoxide with the **Y-2** compounds. Calculations are still needed to assess the actual mechanism of the exchange process for PFB as well as to obtain the energetics for barriers associated with the exchange processes.

18.2
Proton Transfer from Methanol to Carbanion Intermediates

In 1948 Miller reported the reactivity of 1,1-difluoroalkenes toward nucleophiles by the rapid methoxide ion catalyzed addition of methanol to $CF_2=CCl_2$ to give a saturated ether, $CH_3OCF_2CHCl_2$ [19]. Hine's group carried out kinetic studies of the methoxide-promoted dehydrofluorination of $CHCl_2CF_3$ [55 °C and 70 °C] with $CH_3OCF_2CHCl_2$ and $CH_3OCF=CCl_2$ as the isolated products [20]. The formation of the saturated ether is expected since the reaction of methanolic sodium methoxide and $CF_2=CCl_2$ has a half life of 40 min at 0 °C. Since the elimination of $CH_3OCF_2CHCl_2$ is twice as fast as that of $CHCl_2CF_3$, the formation of $CH_3OCF=CCl_2$ is a secondary reaction product. The rate of methoxide-catalyzed exchange of $CH_3OCF_2CDCl_2$ is over 7000 times faster than the extrapolated rate of dehydrofluorination for $CH_3OCF_2CHCl_2$ at 20 °C, and it is not surprising that the carbanion, $\{CH_3OCF_2CCl_2\}^-$, generated from the reaction of methoxide and $CF_2=CCl_2$ adds a proton to form the saturated ether rather than eliminating a fluoride ion to generate a vinyl ether. The two types of reactions are an excellent method to generate the same carbanion intermediate by different approaches.

Our first use of this method was the detailed study of the reactions of $C_6H_5C(CF_3)=CF_2$ with ethanolic sodium ethoxide. Much to our surprise, the carbanion intermediate, $\{C_6H_5C(CF_3)CF_2OC_2H_5\}^-$, generated during the reaction at -78 °C favors elimination of fluoride ion to give 85% of two vinyl ethers, C_6H_5-$C(CF_3)=CFOC_2H_5$, and only 15% of the saturated ether, $C_6H_5CH(CF_3)CF_2OC_2H_5$ [21]. When the reaction is carried out in C_2H_5OD the amount of saturated ether remains almost constant at 13% and suggests hydron transfer to the carbanion occurs with a near unity isotope effect. Elimination and exchange kinetics required higher temperatures. The ethoxide-promoted dehydrofluorination of $C_6H_5CH(CF_3)CF_2OCH_3$ also occurs with a near unity value for k^H/k^D at 25 °C. Reaction of $C_6H_5CH(CF_3)CF_2OCH_3$ in C_2H_5OD results in 3–4% deuterium incorporation at 20% elimination that matches the partitioning of carbanions generated from the reaction of $C_6H_5C(CF_3)=CF_2$ in C_2H_5OD at 25 °C. Since there is no loss of the methoxide group, the nucleophilic addition of alkoxide, k_N, is not reversible. This suggests a mechanism similar to that for reactions occurring with internal return, Scheme 18.1 [22].

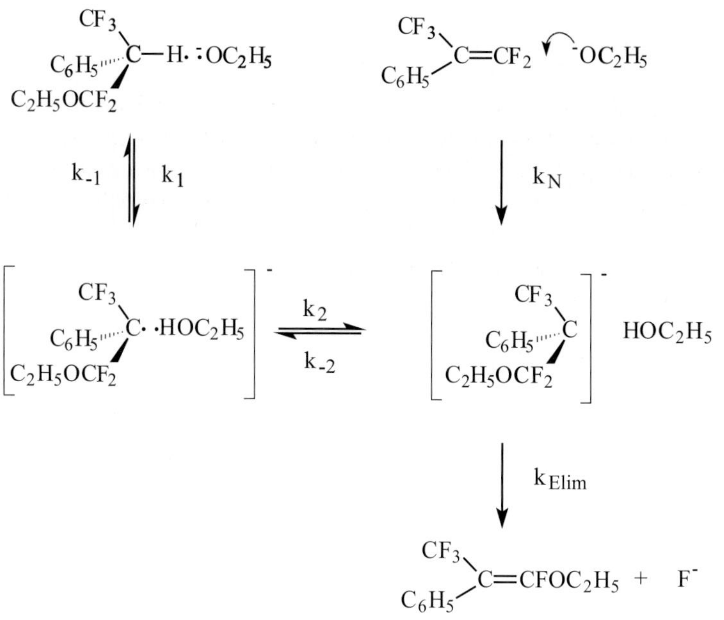

Scheme 18.1

The encounter complex of the saturated ether can transfer the proton, k_1, to form a hydrogen-bonded carbanion intermediate. Internal return, k_{-1}, regenerates the encounter complex and can wash out the experimental PKIE. The free carbanion is generated by breaking the hydrogen bond, k_2. The ejection of fluoride ion to form the vinyl ether requires formation of the free carbanion. The reaction of ethoxide with $C_6H_5C(CF_3)=CF_2$ directly forms the free carbanion, $\{C_6H_5C(CF_3)-CF_2OC_2H_5\}^-$, and it partitions between ejecting fluoride ion, k_{Elim}, and forming the hydrogen-bonded carbanion, k_{-2}. Since $k_{-1} \gg k_2$, the rate-limiting step for making the saturated ether would be the formation of the hydrogen-bonded carbanion, k_{-2}, and this does not involve a hydron transfer.

Eaborn and coworkers reported PKIE values for the protonation of various benzylic anions by methanol [23]. Anions are generated *in situ* by the reaction of methanolic sodium methoxide with $YC_6H_4CH_2Si(CH_3)_3$.

$$CH_3O^- + (CH_3)_3SiCH_2C_6H_4Y \rightarrow CH_3OSi(CH_3)_3 + {}^-CH_2C_6H_4Y$$
$$CH_3O^iH + {}^-CH_2C_6H_4Y \rightarrow CH_3O^- + {}^iHCH_2C_6H_4Y \qquad (18.7)$$

Isotope effects were calculated from the known initial ratio of $CH_3OH:CH_3OD$ and the measured H:D ratio in ${}^iHCH_2C_6H_4Y$. This resulted in $k^H/k^D \approx 1.2$ at 25 °C for the protonation of a variety of ${}^-CH_2C_6H_4Y$. The value of k^H/k^D became 10 for p-$NO_2C_6H_4CH_2^-$ and both fluorenyl and 9-methylfluorenyl anions. This suggested a study using benzylic anions generated by reactions of methanolic sodium methoxide with a series of $YC_6H_4CH=CF_2$ compounds [15].

$$CH_3O^- + CF_2=CHC_6H_4Y \rightarrow \{CH_3OCF_2CHC_6H_4Y\}^-$$

$$CH_3O^iH + \{CH_3OCF_2CHC_6H_4Y\}^- \rightarrow CH_3OCF_2CH^iHC_6H_4Y \quad (18.8)$$

Generating carbanions *in situ* by reaction of methoxide with either alkyltrimethyl-silanes or β,β-difluorostyrenes has several advantages. Reactions occur at lower temperatures and the problem of internal return is eliminated. The methanol sol-vent molecules are excellent trapping reagents for carbanions and only an intra-molecular trap appears to be more efficient. The carbanions generated from the β,β-difluorostyrenes have the added advantage of an intra-molecular trapping agent due to the elimination of a β-fluoride ion.

$$CH_3O^iH + \{CH_3OCF_2CHC_6H_4Y\}^- \rightarrow$$
$$CH_3OCF_2CH^iHC_6H_4Y + CH_3OCF=CHC_6H_4Y \quad (18.9)$$

The product ratios and isotope effects associated with the reactions of p-NO$_2$C$_6$-H$_4$CH=CF$_2$ can be compared to those of m-NO$_2$C$_6$H$_4$CH=CF$_2$, which has results similar to other YC$_6$H$_4$CH=CF$_2$ compounds [24]. Isotope effects for hydron trans-fer to neutralize $\{m$-NO$_2$C$_6$H$_4$CHCF$_2$OCH$_3\}^-$ increase slightly from $k^H/k^D = 1.20$ at $-50\,°C$ to $k^H/k^D = 1.39$ at $+50\,°C$, and this trend is the same for the reactions of other YC$_6$H$_4$CH=CF$_2$. The isotope effects differ significantly for the neutralization of $\{p$-NO$_2$C$_6$H$_4$CHCF$_2$OCH$_3\}^-$ with larger values that decrease with increasing temperature in a normal fashion: $k^H/k^D = 11.3$ at $-70\,°C$ and 6.44 at $+25\,°C$. Partitioning of the carbanions between hydron transfer and fluoride ion loss is also different. The $\{p$-NO$_2$C$_6$H$_4$CHCF$_2$OCH$_3\}^-$ gives 94% p-NO$_2$C$_6$H$_4$-CH$_2$CF$_2$OCH$_3$, and equal amounts of E- and Z-p-NO$_2$C$_6$H$_4$CH=CFOCH$_3$ at $-70\,°C$. Formation of the saturated ether decreases to 50% with equal amounts of the E- and Z-vinyl ethers at 25 °C. This differs from the reactions of $\{m$-NO$_2$C$_6$H$_4$CHCF$_2$OCH$_3\}^-$ where the amount of saturated ether increases slightly from 53% at $-50\,°C$ to 59% at $+50\,°C$ and the E:Z ratios decrease from 6:1 at $-50\,°C$ to 3:1 at $+50\,°C$.

The reactions of methanolic methoxide with CF$_2$=CCl$_2$ can be compared to those of CF$_2$=CClC$_6$H$_5$ where a benzene ring replaces a chlorine atom. Although the rates of reaction are similar at 0 °C, product distributions are different as CF$_2$=CCl$_2$ forms only CH$_3$OCF$_2$CHCl$_2$, and the reactions of CF$_2$=CClC$_6$H$_5$ result in 49% of the satu-rated ether, CH$_3$OCF$_2$CHClC$_6$H$_5$, and 51% of the vinyl ethers, E- and Z-CH$_3$OCF=CClC$_6$H$_5$. There is a similarity between the rates of dehydrofluorina-tions for CHCl$_2$CF$_3$, $k = 1.9 \times 10^{-5}$ M^{-1} s^{-1} at 70 °C in methanol [20b], and those of C$_6$H$_5$CHClCF$_3$ **2**, $k = 3.7 \times 10^{-4}$ M^{-1} s^{-1} at 75 °C in ethanol (a correction is necessary to account for rate enhancement by ethoxide over methoxide). In both dehydro-fluorinations the rates of exchange are faster than the elimination; however, the big difference is the ease of the exchange for CDCl$_2$CF$_3$, $k = 1.93 \times 10^{-2}$ M^{-1} s^{-1} at 25 °C in methanol, and that for C$_6$H$_5$CDClCF$_3$, $k = 6.73 \times 10^{-3}$ M^{-1} s^{-1} at 75 °C in ethanol [25]. The replacement of a chlorine with a phenyl ring dramatically alters the trans-fer of a hydron from an alcohol to a carbanion. The gas phase acidities and rates of methoxide-catalyzed exchange of $CDCl_2CF_3$ and PFB-d are similar, Table 18.1. The difference between $\{C_6H_5CClCF_3\}^-$, **2**$^-$, and either the PFB$^-$ or $(CCl_2CF_3)^-$ anions is that **2**$^-$ is a π-delocalized anion and both the PFB$^-$ and $(CCl_2CF_3)^-$ have their negative charge in localized sp^2 or sp^3 orbitals.

18.3
Proton Transfer Associated with Methoxide Promoted Dehydrohalogenation Reactions

The E2 mechanism for alkoxide-promoted dehydrohalogenations is generally thought to be concerted with the breaking of the C–H and C–X bonds and the formation of the O–H and π-bond occurring in the transition structure [26]:

Obtaining both an isotope effect, k^H/k^D, and an element effect, k^{HBr}/k^{HCl}, are the experimental evidence that the C–H and C–X bonds are breaking in the transition structure. Since measurement of heavy atom isotope effects requires special instrumentation, the element effect has taken the place of heavy atom isotope effects in most investigations. The element effect was first proposed by Bunnett in a 1957 paper dealing with the nucleophilic substitution reactions of activated aromatic compounds [27], and later applied to dehydrohalogenation mechanisms by Bartsch and Bunnett [28]. The lack of any incorporation of deuterium prior to elimination has also been used as experimental evidence favoring the concerted mechanism [29]. The stereochemistry should be a trans-elimination.

The consequences of hydrogen-bonded carbanions as the first intermediate for proton transfer from the carbon acid to methoxide have important mechanistic implications when considering dehydrohalogenation reactions. If a group on the beta carbon can leave from that hydrogen-bonded carbanion, HB-h, the experimental results would be similar to those expected for the concerted E2 mechanism. The hydrogen bond will inhibit exchange of that proton if the reaction is carried out in CH_3OD and will retain any stereochemistry at the carbanion site. This is illustrated by the methoxide-promoted dehydrochlorination in Scheme 18.2 where $k_{Elim} \gg k_2^H$.

Hydrogen isotope effects for methoxide-promoted dehydrohalogenations of $C_6H_5C^iHClCF_2Cl$, $k^H/k^D = 2.35$, and $C_6H_5CHBrCF_2Br$, $k^H/k^D = 4.10$, with a k^{HBr}/k^{HCl} of 66 at 0 °C could suggest a concerted mechanism with asymmetric

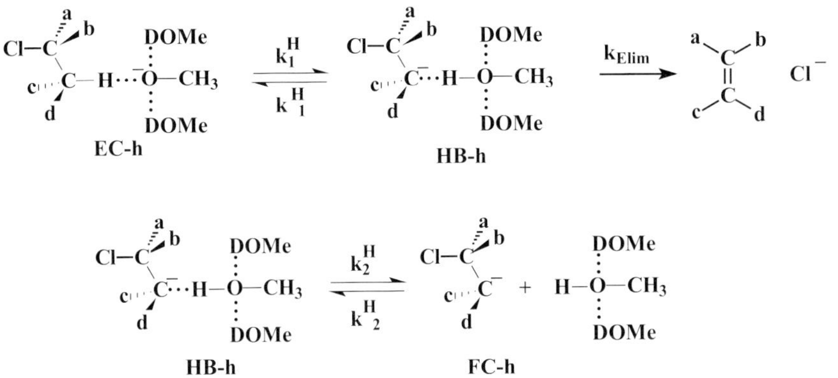

EC-h $\dfrac{k_1^H}{k_{-1}^H}$ HB-h $\xrightarrow{k_{Elim}}$ Cl^-

HB-h $\dfrac{k_2^H}{k_{-2}^H}$ FC-h

Scheme 18.2

transition structures. Since the C–Cl bond is stronger than the C–Br bond the smaller PKIE associated with the dehydrochlorination could suggest a later transition structure. The problem is that there are anomalous Arrhenius parameters with $E_a^D - E_a^H = 0.0$ and $A^H/A^D = 2.4$ for $C_6H_5C^iHClCF_2Cl$ and $E_a^D - E_a^H = 0.1$ and $A^H/A^D = 3.9$ for $C_6H_5C^iHBrCF_2Br$ [30]. Isotope effects come from zero-point energy differences and Arrhenius parameters should be due to differences in the activation energies, $E_a^D - E_a^H = RTln(k^H/k^D)$ with $A^H/A^D = 1$. Therefore, these anomalous Arrhenius parameters rule out the concerted E2 mechanism. What Arrhenius parameters would be expected from a reaction that occurs by an internal-return mechanism?

Dahlberg calculated the anticipated Arrhenius behavior for elimination reactions occurring by the mechanism shown in Scheme 18.2 [31]. Several surprising situations were obtained. Mid-range values of k^H/k^D could be modeled with normal Arrhenius behavior, $A^H/A^D \approx 1$. Also possible are normal values of k^H/k^D that have low values of A^H/A^D normally considered to come from reactions that feature quantum mechanical tunneling. Although there were situations that resulted in $A^H/A^D > 1$, no model explained the values of A^H/A^D and $E_a^D - E_a^H$ obtained in our dehydrohalogenations. However, the calculations are able to model a number of other systems.

Shiner and Smith obtained Arrhenius parameters for ethanolic sodium ethoxide-promoted dehydrobromination reactions of $C_6H_5CH(CH_3)CH_2Br$ over a 50 °C range. The k^H/k^D of 7.51 at 25 °C was normal, but the A^H/A^D of 0.4 was considered experimental evidence for reaction occurring with quantum mechanical tunneling [32]. Shiner and Martin later reported tritium rates and corrected the earlier data for small amounts of a substitution reaction that formed $C_6H_5CH(CH_3)$-$CH_2OC_2H_5$ [33]. Corrected values at 25 °C were $k^H/k^D = 7.82$ and $k^D/k^T = 2.65$. To satisfy the Swain–Schaad relationship the experimental $(k^H/k^T)_{obs}$ of 20.6 should be 17% to 26% greater with $(k^H/k^T)_{obs}$ of 23 or 26 [11]. Dahlberg was able to model their results with an internal-return mechanism [34].

The methoxide-promoted elimination reactions of four $YC_6H_4C^iHXCH_2X$ can be modeled by using all three hydrogen isotopes, Table 18.2. The results for $m\text{-}ClC_6H_4C^iHBrCH_2Br$ have such a small internal return parameter, $a^H = 0.045$ at 25 °C, that a concerted mechanism cannot be ruled out; however, a $\Delta S^{\ddagger}$ of −3 eu is in line with an elimination occurring from a hydrogen-bonded carbanion intermediate. The $a^H = 0.59$ for $m\text{-}ClC_6H_4C^iHClCH_2Cl$ and $m\text{-}CF_3C_6H_4C^iHClCH_2Cl$ as well as the $\Delta S^{\ddagger}$ values of 0 to −2 eu definitely favor the two-step mechanism over a concerted pathway. That $p\text{-}CF_3C_6H_4C^iHClCH_2F$, $a^H = 2.1$, reactions have more internal return is not surprising since breaking a C–F bond requires more energy than for a C–Cl bond. The rates of $p\text{-}CF_3C_6H_4C^iHClCH_2Cl$ and $p\text{-}CF_3C_6H_4\text{-}C^iHClCH_2F$ have an element effect of $k^{HCl}/k^{HF} = 54$. Can a two-step reaction with internal return have a large element effect associated with dehydrohalogenation?

Table 18.2. Internal return parameters associated with methanolic sodium methoxide-promoted promoted dehydrohalogenation reactions at 25 °C.[a]

Compound	k, M^{-1} s^{-1}	$\Delta H^{\ddagger}$, kcal mol^{-1}	$\Delta S^{\ddagger}$, eu	k^H/k^D or k^D/k^T	a^iH
$m\text{-}ClC_6H_4CHClCH_2Cl$	1.60×10^{-3}	21.1 ± 0.1	-0.5 ± 0.1	3.40	0.59
$m\text{-}ClC_6H_4CDClCH_2Cl$	4.71×10^{-4}	21.9 ± 0.1	-0.4 ± 0.5	1.83	0.14
$m\text{-}ClC_6H_4CTClCH_2Cl$	2.58×10^{-4}	22.3 ± 0.2	-0.0 ± 0.5		0.072
$m\text{-}CF_3C_6H_4CHClCH_2Cl$	3.34×10^{-3}	20.3 ± 0.1	-1.7 ± 0.2	3.49	0.59
$m\text{-}CF_3C_6H_4CDClCH_2Cl$	9.58×10^{-4}	21.1 ± 0.1	-1.7 ± 0.2	1.88	0.13
$p\text{-}CF_3C_6H_4CTClCH_2Cl$	5.10×10^{-4}	21.5 ± 0.1	-1.5 ± 0.3		0.068
$p\text{-}CF_3C_6H_4CHClCH_2Cl$	8.03×10^{-3}	20.1 ± 0.1	-0.7 ± 0.2	3.75	
$p\text{-}CF_3C_6H_4CDClCH_2Cl$	2.14×10^{-3}	20.7 ± 0.1	-1.3 ± 0.1		
$p\text{-}CF_3C_6H_4CHClCH_2F$	1.49×10^{-4}	23.4 ± 0.1	2.6 ± 0.3	2.19	2.1
$p\text{-}CF_3C_6H_4CDClCH_2F$	6.73×10^{-5}	23.9 ± 0.1	2.5 ± 0.3	1.63	0.50
$p\text{-}CF_3C_6H_4CTClCH_2F$	4.22×10^{-5}	24.2 ± 0.2	2.6 ± 0.6		0.27
$m\text{-}ClC_6H_4CHBrCH_2Br$	3.26×10^{-2}	18.6 ± 0.1	-2.8 ± 0.1	4.95	0.045
$m\text{-}ClC_6H_4CDBrCH_2Br$	6.60×10^{-3}	19.7 ± 0.2	-2.6 ± 0.2	2.02	0.0095
$m\text{-}ClC_6H_4CTBrCH_2Br$	3.27×10^{-3}	20.0 ± 0.1	-2.8 ± 0.4		0.0070

a Data from Ref. [37]

The element effect has been used to replace the need to measure a heavy atom isotope effect like k^{35}/k^{37} for chlorine; however, the chlorine isotope effect has the advantage that the nature of the leaving group does not change. Measuring k^{35}/k^{37} for both the protium and deuterium compounds can distinguish between a reaction occurring by a concerted E2 mechanism or by the two-step process with internal return [35]. For the concerted reaction, the values of $(k^{35}/k^{37})^{HCl}$ should equal those for $(k^{35}/k^{37})^{DCl}$; however, the two-step process will have a $(k^{35}/k^{37})^{HCl}$ larger than that for $(k^{35}/k^{37})^{DCl}$. This is due to the fact that $k^H_{-1} > k^D_{-1}$ and the elimination step, k_{Elim}, has a larger role in the experimental rate constant, $k_{obs} = [k_1 k_{Elim}] / [k_{-1} + k_{Elim}]$, for HCl than for DCl. There is a large difference in the values k^{35}/k^{37} obtained for methoxide-promoted dehydrochlorination of $C_6H_5C^iHClCH_2Cl$, $(k^{35}/k^{37})^{HCl} = 1.00978$–$0.00020$ compared to $(k^{35}/k^{37})^{DCl} = 1.00776$–$0.00020$. Similar differences are obtained for the reaction using ethanolic sodium methoxide, and for the alkoxide-promoted dehydrochlorinations of $C_6H_5C^iHClCF_2Cl$, Table 18.3. These are very large values for chlorine isotope effects. McLennan was able to explain these values by calculating that a considerable lengthening of the C–Cl bond during the formation of the hydrogen-bonded carbanion, k_1, would result in a (k^{35}/k^{37}) occurring for the proton-transfer step as well for the actual breaking of the C–Cl bond, k_{Elim} [36].

Table 18.3 Chlorine and hydrogen isotope effects, and element effects associated with alcoholic sodium alkoxide-promoted dehydrochlorination ractions.[a]

Compound	Solvent	k^{35}/k^{37} [°C]	k^H / k^D [°C]	kHBr / kHCl [°C]
$C_6H_5CHClCH_2Cl$	EtOH	1.00908 ± 0.00008 [24]	4.24 [25]	35 [25]
$C_6H_5CDClCH_2Cl$		1.00734 ± 0.00012 [24]		24 [25]
$C_6H_5CHClCH_2Cl$	McOH	1.00978 ± 0.00020 [21]	3.83 [25]	35 [25]
$C_6H_5CDClCH_2Cl$		1.00776 ± 0.00020 [21]		25 [25]
$C_6H_5CHClCF_2Cl$	EtOH	1.01229 ± 0.00047 [0]	2.77 [0]	66 [0]
$C_6H_5CDClCF_2Cl$		1.01003 ± 0.00024 [0]		39 [0]
$C_6H_5CHClCF_2Cl$	MeOH	1.01255 ± 0.00048 [20]	2.28 [25]	47 [25]
$C_6H_5CDClCF_2Cl$		1.01025 ± 0.00043 [20]		29 [25]

a Data from Ref. [35b].

Element effects for the hydrogen and deuterium compounds can also be used to distinguish between a concerted E2 mechanism and one with two steps and internal return, Scheme 18.2. Data in Table 18.2 can be used to calculate element effects comparing p-$CF_3C_6H_4C^iHClCH_2Cl$ and p-$CF_3C_6H_4C^iHClCH_2F$ where a

k^{HCl}/k^{HF} of 54 becomes a k^{DCl}/k^{DF} of 32. Similar results come from the ethoxide-promoted eliminations of $C_6H_5C^iHBrCH_2Br$, $C_6H_5C^iHBrCH_2Cl$ and $C_6H_5C^i$-$HBrCH_2F$: $k^{HBr}/k^{HCl} = 35$ and $k^{DBr}/k^{DCl} = 16$; $k^{HCl}/k^{HF} = 67$ and $k^{DCl}/k^{DF} = 35$ [37]. An element effect of $k^{Cl}/k^F = 68$ had been previously reported for the ethoxide-promoted dehydrohalogenataions of $C_6H_4CH_2CH_2X$ [38].

18.4
Conclusion

Ritchie's prediction that carbon acids whose conjugated bases have localized charge will show "kinetic acidities" greater than their thermodynamic acidities only needs to be changed slightly to read that the orbital of the carbanion is localized on that carbon atom. Density functional calculations using the natural population analysis allowed charge distributions to be calculated for the various anions. The negative charge for 9-PhFl⁻ is highly π-delocalized to form an aromatic species with only 18% of the negative charge in the 9 position and this increases to 25% in the hydrogen-bonded species. The negative charge of the pentafluorophenyl anion, PFB⁻, cannot undergo π-delocalization and is thought to be localized in an sp^2 orbital; however, the ipso carbon of PFB⁻ only has 41% of the negative charge, which increases to 44% for the hydrogen-bonded species. Since only 41% to 44% of the negative charge is on the ipso carbon of PFB⁻, the charge is also delocalized by the field effects of five fluorine atoms in the benzene ring.[4] The case of $CF_3CCl_2^-$ is similar with 45% to 50% of the charge on the carbanion carbon in a localized sp^3 orbital, and the remainder delocalized by field effects. The benzylic carbon atoms of Y-1⁻ and Y-2⁻ still retain 37% and 60% of the negative charge, but that charge is in a π-delocalized orbital.

Both pentafluorobenzene and $CHCl_2CF_3$ exchange their protons much faster than expected from their gas phase acidities. The exchange reactions occur with large amounts of internal return, and there must be a mechanism that allows the exchange process to occur without having to form a free carbanion. Although the benzylic carbanions have about the same amount of the negative charge as those two anions they have much slower exchange reactions that also occur with large amounts of internal return resulting in near unity hydrogen isotope effects. The exchange of the 9-hydrogen of 9-phenylfluorene has a similar rate of exchange as benzylic compounds with similar gas phase acidities. The difference is that the reaction occurs with only a small amount of internal return and has a sizable hydrogen isotope effect.

Gas-phase studies have been made to gain information about the intrinsic stability of hydrogen-bonded ionic intermediates of alcohols and carbon acids. Many anionic hydrogen-bonded complexes appear to be better when there is a good match of the gas-phase acidities of a neutral and the conjugate acid of the ion.

---

4) When this material was first presented at an international conference, Paul Schleyer stated that a pentafluorophenyl anion is not a localized carbanion, and this has certainly proven to be the case from our calculations.

However, it does not work for toluene and methanol which have comparable acidities, and Brauman's group were not successful in their attempts to form the hydrogen-bonded complex of methoxide and toluene [39]. They were successful in the study of CHF_3, which is slightly more acidic than toluene, and methanol complexes [40]. Studies of complexes between CHF_3 and ethanol or isopropanol were also made, and the energies range from -21.1 to -23.5 kcal mol^{-1}, which are larger than those that are typical ion-dipole complexes, $10–15$ kcal mol^{-1}. There is the same difference between the benzyl anion which is π-delocalized and $^-CF_3$ where the charge is in an sp^3 orbital.

The generality of concerted 1,2 eliminations was questioned by Bordwell [41], and supported by Saunders [42]. What it boils down to is what can be considered to be an intermediate during a chemical reaction. Jencks gives an excellent definition: "An intermediate is, therefore, defined as a species with a significant lifetime, longer than that of a molecular vibration of $\sim 10^{-13}$ s, that has barriers for its breakdown to both reactants and products." [43]. Bunnett has an excellent summation: "It is axiomatic that one cannot define a reaction mechanism absolutely. What one can do is to reject conceivable mechanisms that are not compatible with experimental evidence. When all conceivable mechanisms but one have been rejected, one is tempted to consider the mechanism to be established. However, such reasoning does not, and by definition cannot, take account of inconceivable mechanisms. At a later time, especially in the light of advances in the meanwhile, one or more further alternative mechanisms may be recognized, and the mechanism previously thought to be the only tenable one may be found to be incompatible with new experimental evidence. Such has happened many times in the history of mechanistic studies. The conservative scientist therefore refers to the better-understood mechanisms as 'generally accepted' or 'well recognizes' rather than as 'established' or 'proven'." [44].

References

1 C. D. Ritchie, *J. Am. Chem. Soc.* **1969**, *91*, 6749–6753.

2 A. Streitwieser, Jr., P. J. Scannon, N. M. Neimeyer, *J. Am. Chem. Soc.* **1972**, *94*, 7938–7937.

3 A. Streitwieser, Jr., J. A. Hudson, F. Mares, *J. Am. Chem. Soc.* **1968**, *90*, 648–650.

4 A. Streitwieser, Jr., W. B. Hollyhead, A. H. Pudjaatmake, P. H. Owens, T. L. Kruger, P. A. Rubenstein, R. A. MacQuarrie, M. L. Brokaw, W. K. C. Chu, H. M. Niemeyer, *J. Am. Chem. Soc.* **1971**, *93*, 5088–5096.

5 (a) R. P. Bell, *Chem. Soc. Rev.* **1974**, *3*, 513–544, (b) R. A. More O' Ferrall, *Proton-Transfer Reactions*, E. F. Caldin, V. Gold (Eds.), Chapman & Hall, London, 1975, pp. 216–227.

6 L. Melander, *Isotope Effects on Reaction Rates*, Ronald Press, New York, 1960, pp. 24–32.

7 F. H. Westheimer, *Chem. Rev.* **1961**, *61*, 265–273.

8 C. G. Swain, E. C. Stivers, J. F. Reuwer, Jr., L. J. Schaad, *J. Am. Chem. Soc.* **1958**, *80*, 5885–5893.

9 A. J. Kresge, *Acc. Chem. Res.* **1975**, *9*, 354–360.

10 A. J. Kresge, A. C. Lin, *J. Chem. Soc. Chem. Commun.* **1973**, 761.

11 D. J. Cram, D. A. Scott, W. D. Nielsen,
   J. Am. Chem. Soc. **1961**, *83*, 3696–3707.
12 A. Streitwieser Jr., W. B. Hollyhead,
   G. Sonnichsen, A. H. Pudjaatmake,
   C. J. Chang, T. L. Kruger, *J. Am. Chem.
   Soc.* **1971**, *93*, 5096–5102.
13 H. F. Koch, J. C. Biffinger, M. Mishima,
   Mustanir, G. Lodder, *J. Phys. Org. Chem.*
   1998, *11*, 614–616.
14 A. Streitwieser Jr., H. F. Koch, *J. Am.
   Chem. Soc.* **1964**, *86*, 404–409.
15 H. F. Koch, J. G. Koch, N. H. Koch,
   A. S. Koch, *J. Am. Chem. Soc.* **1983**, *105*,
   2388–2393.
16 H. F. Koch, M. Mishima, H. Zuilhof,
   Ber. Bunsenges. Phys. Che. **1998**, *102*,
   567–572.
17 K. M. Ervin, V. F. DeTuri, V. F., *J. Phys.
   Chem. A* **2002**, 9947–9956.
18 V. F. DeTuri, H. F. Koch, J. G. Koch,
   G. Lodder, M. Mishima, H. Zuilhof,
   N. M. Abrams, C. E. Anders,
   J. C. Biffinger, P. Han, A. R. Kurl;and,
   J. M. Nichols, A. M. Ruminski,
   P. R. Smith, K. D. Vasey, *J. Phys. Org.
   Chem.* **2006**, *19*, 308–317.
19 W. T. Miller, E. W. Fager, P. H. Griswold,
   J. Am. Chem. Soc. **1948**, *70*, 431–432.
20 (a) J. Hine, R. Wiesboeck,
   R. G. Ghirardelli, *J. Am. Chem. Soc.*
   1961, *83*, 1291–1222; (b) J. Hine,
   R. Wiesboeck, O. B. Ramsay, *J. Am.
   Chem. Soc.* **1961**, *83*, 1222–1226.
21 H. F. Koch, A. J. Kielbania Jr., *J. Am.
   Chem. Soc.* **1970**, *92*, 729–730.
22 H. F. Koch, J. G. Koch, D. B. Donovan,
   A. G. Toczko, A. J. Kielbania Jr., *J. Am.
   Chem. Soc.* **1981**, *103*, 5417–5423.
23 (a) C. Eaborn, D. R. M. Walton,
   G. Seconi, *J. Chem. Soc., Perkin Trans. 2*
   1976, 1857–1861; (b) D. Macciantelli,
   G. Seconi, C. Eaborn, *J. Chem. Soc.,
   Perkin Trans. 2* **1978**, 834–838.
24 H. F. Koch, A. S. Koch, *J. Am. Chem.
   Soc.* **1984**, *106*, 4536–4539.
25 H. F. Koch, D. B. Dahlberg, G. Lodder,
   K. S. Root, N. A. Touchette, R. L. Solsky,
   R. M. Zuck, L. J. Wagner, N. H. Koch
   and M. A. Kuzemko, *J. Am. Chem. Soc.*
   1983, 105, 2394–2398.
26 E. D. Hughes, C. K. Ingold,
   S. Masterman, B. J. McNulty, *J. Chem.
   Soc.* **1940**, 899–912.
27 J. F. Bunnett, E. W. Garbisch,
   K. M. Pruitt, *J. Am. Chem. Soc.* **1957**, *79*,
   385–391.
28 R. A. Bartsch, J. F. Bunnett, *J. Am.
   Chem. Soc.* **1968**, *90*, 408–417.
29 P. S. Skell, C. R. Hauser, *J. Am. Chem.
   Soc.* **1945**, *67*, 1661.
30 H. F. Koch, D. B. Dahlberg,
   M. F. McEntee, C. J. Klecha, *J. Am.
   Chem. Soc.* **1976**, *98*, 1060–1061.
31 H. F. Koch, D. B. Dahlberg, *J. Am.
   Chem. Soc.* **1980**, *102*, 6102–6107.
32 V. J. Shiner, M. L. Smith, *J. Am. Chem.
   Soc.* **1961**, *83*, 593–598.
33 V. J. Shiner, B. Martin, *Pure Appl. Chem.*
   1964, *8*, 371–378.
34 Reference [31], p. 6106.
35 (a) H. F. Koch, J. G. Koch, W. Tumas,
   D. J. McLennan, B. Dobson, G. Lodder,
   J. Am. Chem. Soc. **1980**, *102*, 7955–7956;
   (b) H. F. Koch, D. J. McLennan,
   J. G. Koch, W. Tumas, B. Dobson,
   N. H. Koch, *J. Am. Chem. Soc.* **1983**,
   105, 1930–1937.
36 Reference [35b], pp. 1934–1935.
37 H. F. Koch, G. Lodder, J. G. Koch,
   D. J. Bogdan, G. H. Brown,
   C. A. Carlson, A. B. Dean, R. Hage,
   P. Han, J. C. P. Hopman, L. A. James,
   P. M. Knape, E. C. Roos, M. L. Sardina,
   R. A. Sawyer, B. O. Scott, C. A. Testa,
   III, S. D. Wickham, *J. Am. Chem. Soc.*
   1997, 119, 9965–9974.
38 C. H. DePuy, C. A. Bishop, *J. Am.
   Chem. Soc.* **1960**, *82*, 2535–2537.
39 G. C. Gatev, M. Zhong, J. I. Brauman,
   J. Phys. Org. Chem. **1997**, *10*, 531–536.
40 M. L. Chabinyc, J. I. Brauman, *J. Am.
   Chem. Soc.* **1998**, *120*, 10863–10870.
41 F. G. Bordwell, *Acc. Chem. Res.* **1972**, *5*,
   374–381.
42 W. H. Saunders, Jr., *Acc. Chem. Res.*
   1976, *9*, 19–25.
43 W. P. Jencks, *Acc. Chem. Res.* **1980**, *13*,
   161–169.
44 J. F. Bunnett, in *Techniques of Chemistry,
   Volume VI. Investigation of Rates and
   Mechanisms of Reactions, Part I,* 3rd Edn.
   E. S. Lewis (Ed.), Wiley Interscience,
   New York, 1974, Ch. VIII, p. 369.

19
Theoretical Simulations of Free Energy Relationships in Proton Transfer

Ian H. Williams

19.1
Introduction

The observation of a rate equilibrium relationship, or equivalently a free energy relationship (FER), between logarithms of experimentally determined rate constants and equilibrium constants, raises questions concerning its validity, the possible significance of the slope of a linear correlation (LFER), or the origin of curvature, which are natural concerns for theoretical enquiry. Values of a Brønsted coefficient or a Marcus intrinsic barrier (see below) derived from empirical data are phenomenological parameters that of themselves do not provide satisfying explanations at a fundamental molecular level for observed or predicted reactivity trends. The subject of theoretical simulation involves the construction of FERs from the bottom up rather than from the top down: that is, it starts from microscopic models for chemical bonding and intermolecular interactions and proceeds, by way of explicit consideration of molecular structures for species undergoing chemical reaction, to evaluate energy changes governing the kinetic and thermodynamic aspects of a macroscopically observed process. However, since numerical computations alone rarely provide insight in a readily digestible form, it is common to seek interpretations of the results of theoretical simulations by means of simple analytical models involving parameters that may be understood in terms of fundamental molecular properties and chemical concepts.

We begin by reviewing briefly some of the qualitative models for FERs before surveying some examples of simulated FERs for proton transfer (PT) reactions. These include molecular orbital (MO)-based studies of potential energy surfaces (PES) for gas-phase reactions and valence bond (VB)-based studies of free energy changes for reactions in condensed phases.

Hydrogen-Transfer Reactions. Edited by J. T. Hynes, J. P. Klinman, H. H. Limbach, and R. L. Schowen
Copyright © 2007 WILEY-VCH Verlag GmbH & Co. KGaA, Weinheim
ISBN: 978-3-527-30777-7

19.2
Qualitative Models for FERs

Theory and experiment are inextricably linked in the development of models for chemical reactivity [1]. Since the 1930s, when the ideas of both transition-state (TS) theory [2] and of rate equilibrium relationships [3] were first promulgated, many distinguished scientists have contributed to the body of knowledge, concerning the relationships between barrier heights for simple group transfer processes and their reaction energies, sometimes referred to as the "Bema Hapothle" [4] – an acronym based upon the initials of a subset of these workers, namely Bell [5], Marcus [6], Hammond [7], Polanyi [8], Thornton [9] and Leffler [10]. Among other omissions from this short-list are the names of Jencks [11], More O'Ferrall [12], and Murdoch [13].

A general group transfer reaction (19.1a) may be considered as the sum of the two half-reactions 19.1b and 19.1c (in which homolysis or heterolysis is deliberately not specified).

$$A\text{–}X + B \rightarrow A + X\text{–}B \tag{19.1a}$$

$$A\text{–}X \rightarrow A + X \tag{19.1b}$$

$$X + B \rightarrow X\text{–}B \tag{19.1c}$$

Evans and Polanyi [2c, 8] considered the energetics of each step separately: step (19.1b) could be described by an energy curve for bond dissociation of A–X as a function of the interatomic distance between A and X, and step (19.1c) could be described by an energy curve for association of atoms X and B (Fig. 19.1). The overall reaction could avoid the energetic cost of complete bond dissociation (to A + X + B) in step (19.1b) if there were an electronic reorganisation occurring at a particular distance between A and B that allowed a cross-over from the rising energy curve for step (19.1b) to the falling energy curve for step (19.1c). The barrier to reaction in this model, which assumes that the bond-breaking and bond-making processes are independent of each other, is given simply by the point of intersection of the two energy curves relative to the energy of the reactants, A–X + B. Since variation in the nature of B would not be expected to affect the energy curve for the bond-breaking step (19.1b), but would at least alter the energy of the products A + X–B relative to the reactants, the effect of changes in B (e.g. B^1, B^2 and B^3 in Fig. 19.1) could be estimated by means of moving the curve for step (19.1c) up or down relative to that for step (19.1b). If the central part of each curve were reasonably linear, where the points of intersection occurred, then the change in the barrier height $\Delta E^{\ddagger}$ would be expected to be proportional to the change in the reaction energy ΔE_{rxn} for reaction (19.1a), according to Eq. (19.2).

$$\Delta E^{\ddagger} = a\,\Delta E_{rxn} + C \tag{19.2}$$

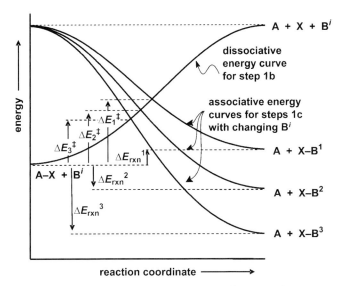

Figure 19.1 Intersecting curves to illustrate the Bell–Evans–Polanyi Principle for a general group-transfer reaction (19.1a). Three members of a reaction series involving different acceptor groups B^i ($i = 1, 2, 3$) give separate energy curves for step (19.1c) each intersecting the energy curve for step (19.1b).

Bell [5] considered the case of X = H, representing proton transfer between a Brønsted acid and a base, and noted that, if $\Delta E^\ddagger$ and ΔE_{rxn} were, respectively, proportional to log k and log K, Eq. (19.2) was equivalent to the Brønsted relation (Eq. (19.3)) where k is the rate constant and K is the equilibrium constant for the acid–base reaction of Eq. (19.4).

$$\log k = a \log K + C' \tag{19.3}$$

$$\text{A–H} + \text{B} \rightarrow \text{A} + \text{H–B} \tag{19.4}$$

These simple considerations yield several corollaries, sometimes known together as the Bell–Evans–Polanyi (BEP) principle [14]. First, there is an approximately linear relation between the barrier height and the reaction energy: this is the basis of the Brønsted relation (and other LFERs). Second, the proportionality constant a in Eq. (19.2) tends to be smaller for exothermic reactions (but larger for endothermic reactions). Third, the position of the crossing point between the curves lies closer to the reactants for more exothermic reactions: this is the basis of the Hammond postulate, that the TS for a more exothermic reaction more closely resembles the reactants (and that for a more endothermic reaction more closely resembles the products).

Murdoch [15] pointed out that the quantitative barrier expressions given by London–Eyring–Polanyi–Sato [16], Johnston and Parr [17], Marcus [6, 18], Murdoch

[13], Rehm and Weller [19], Agmon and Levine [20], Bell [21], le Noble [22], Lewis [23], Kurz [24], Thornton [9], Zavitsas [25], and Chatt and co-workers [26] all reduced to specific cases, or simple extensions, of Eq. (19.5).

$$\Delta E^{\ddagger} = \Delta E_{int}^{\ddagger} (1 - f_2) + {}^1\!/_2\, \Delta E_{rxn} (1 + f_1) \tag{19.5}$$

Here $\Delta E^{\ddagger}$ is the barrier for the reaction (or activation energy) in which the energy change (or reaction asymmetry) is ΔE_{rxn}, and $\Delta E_{int}^{\ddagger}$ is the intrinsic barrier for an identity reaction with $\Delta E_{rxn} = 0$. The functions f_1 and f_2 are functions involving, respectively, only odd or even powers of ΔE_{rxn} and govern the relative contributions of the kinetic term $\Delta E_{int}^{\ddagger} (1 - f_2)$ and the thermodynamic term ${}^1\!/_2\, \Delta E_{rxn} (1 + f_1)$ to the overall barrier. Very exoergic reactions, for which ΔE_{rxn} is large and negative, $f_1 \rightarrow -1$ and $f_2 \rightarrow +1$, leading to $\Delta E^{\ddagger} = 0$. on the other hand, very endoergic reactions, for which ΔE_{rxn} is large and positive, $f_1 \rightarrow +1$ and $f_2 \rightarrow +1$, leading to $\Delta E^{\ddagger} = \Delta E_{rxn}$. Near to thermoneutrality, reactions for which $\Delta E_{rxn} \approx 0$ have both f_1 and $f_2 \approx 0$, yielding $\Delta E^{\ddagger} = \Delta E_{int}^{\ddagger} + {}^1\!/_2\, \Delta E_{rxn}$. The functions f_1 and f_2 serve to interpolate between the limiting cases of very high exoergicity and endoergicity. Changes in the structure or composition of the reactant species may affect both the kinetic term and the thermodynamic term in Eq. (19.5). If changes in the thermodynamic term dominate, or the changes in both terms are in the same direction, the BEP principle is likely to be upheld, with the more stable product being formed more quickly. However, if the kinetic term dominates and opposes changes in the thermodynamic term, then the BEP principle is expected to fail and a less stable product may be formed more quickly than a more stable product. The kinetic term determines the relative barrier heights for reactions having similar thermodynamic terms.

If the interpolation functions in Eq. (19.5) take the values $f_1 = \Delta E_{rxn}/4\Delta E_{int}^{\ddagger}$ and $f_2 = (f_1)^2$, this relationship between the barrier and the overall energy change for reaction becomes the Marcus relation, Eq. (19.6), whose range of applicability is $|\Delta E_{rxn}/\Delta E_{int}^{\ddagger}| \leq 4$, corresponding to $-1 \leq f_1 \leq +1$ [15].

$$\begin{aligned}\Delta E^{\ddagger} &= \Delta E_{int}^{\ddagger} [1 - (\Delta E_{rxn}/4\Delta E_{int}^{\ddagger})^2] + {}^1\!/_2\, \Delta E_{rxn} (1 + \Delta E_{rxn}/4\Delta E_{int}^{\ddagger}) \\ &= \Delta E_{int}^{\ddagger} + {}^1\!/_2\, \Delta E_{rxn} + \Delta E_{rxn}{}^2/16\Delta E_{int}^{\ddagger}\end{aligned} \tag{19.6}$$

It is well known [6, 14a, 27] that the Marcus relation may be derived from a model of intersecting parabolas for the reactant and product energy curves (Fig. 19.2). A parabola provides an unrealistic description of the energy profile for extension of an A–H bond far from its equilibrium length and towards complete dissociation: a Morse curve offers a better description. It is perhaps not too widely appreciated [28, 29] that transformation of a Morse function (Eq. (19.7)) from bond length (r) to bond order (n) coordinates (Eq. (19.8)) yields a parabola (Eq. (19.9)).

$$E(r) = D_e \exp[-\beta(r - r_e)] \{\exp[-\beta(r - r_e)] - 2\} \tag{19.7}$$

$$n = \exp[-\beta(r - r_e)] \tag{19.8}$$

$$E(n) = D_e\, n\, (n - 2) \tag{19.9}$$

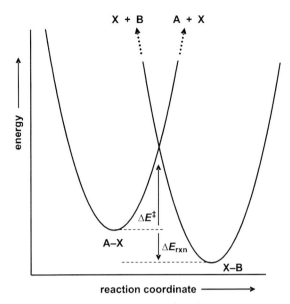

Figure 19.2 Intersecting parabolas of equal curvature.

Regardless of the choice of coordinates, consideration of an intersecting curves model in which the bond-breaking and bond-making processes are independent of each other implies that the reactant and product curves describe diabatic states of the system, i.e. there is no mixing between them that would allow for a smooth and gradual transformation from one to the other. Deformation of the reactant geometry without allowing for any electronic reorganisation from a reactant-like VB structure incurs an energetic penalty. The energy of the reactant diabatic state at the geometry corresponding to the energy minimum of the product is often known as the reorganisation energy, λ; in Marcus theory, $\lambda = 4\Delta E_{int}^{\ddagger}$. If the reactant and product diabatic states have equal curvature, then for a thermoneutral reaction λ is also the energy of the product at the reactant geometry (Fig. 19.3).

It is more realistic to allow the reactant-like and product-like states to interact with each other, rather than to consider them as being independent. Thus electronic reorganisation may accompany geometrical reorganisation along a reaction coordinate for proton transfer (PT). Mixing between the reactant and product diabatic states generates two new adiabatic states. The energy profile for PT follows the ground-state adiabatic curve that shows a smooth maximum passing through the geometry for which the diabatic curves intersect. Necessarily there is also an excited-state adiabatic curve containing an energy minimum with respect to the reaction coordinate. In the adiabatic model there is an avoided crossing between the reactant-like and product-like electronic arrangements. The degree of avoided crossing is given by the difference in energy E_{res} between the point of intersection of the diabatic curves and the maximum of the adiabatic curve (Fig. 19.3).

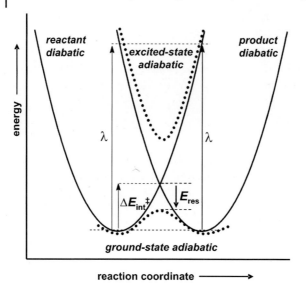

Figure 19.3 Intersection of parabolic diabatic curves for reactant-like and product-like states, together with adiabatic energy curves arising from mixing of the diabatic states.

19.2.1
What is Meant by "Reaction Coordinate"?

The horizontal axis in diagrams such as Figs. 19.1–19.3, often labelled as the "reaction coordinate", is usually not well defined. Figure 19.4 shows qualitatively a ground-state adiabatic PES for the PT (or hydrogen-atom transfer) reaction (19.4) as a function of two coordinates, the bond distances $A \cdots H$ and $H \cdots B$. Let us first follow the Evans–Polanyi argument, outlined above, to trace an energy profile across this surface. We start near point y, where the $H \cdots B$ distance has some finite value, and elongate the A–H bond by traversing parallel to the $A \cdots H$ axis towards point z. Now we freeze the partially broken A–H bond but let the $H \cdots B$ distance decrease: we move parallel to the $H \cdots B$ axis towards point x. This path (dashed lines in Fig. 19.4) is continuous but its energy profile contains a cusp at point z, rather like the lower parts of the intersecting curves in Fig. 19.1. In this case the "reaction coordinate" of Fig. 19.1 contains two distinct segments, described by changes in different interatomic distances. The energy minimum along the path yz occurs at point r, corresponding to a reactant-like species $A–H \cdots \cdots \cdot B$, which is not necessarily an energy minimum with respect to the $H \cdots B$ coordinate. Similarly, the product-like species $A \cdots \cdots \cdot H–B$ corresponding to point p (the energy minimum along path zx) which is not necessarily an energy minimum with respect to the $A \cdots H$ coordinate. A direct reaction path across the PES from r to p may be defined as a linear combination of the two coordinates,

(A $\cdots$ H) – (H $\cdots$ B). The energy profile along this path (diagonal dotted line in Fig. 19.4) is smooth and continuous but the species A $\cdots$ H $\cdots$ B corresponding to point s at the energy maximum is not a stationary point with respect to displacement in the perpendicular direction, defined by the linear combination (A $\cdots$ H) + (H $\cdots$ B). The energy minimum along the extension of the diagonal line zs occurs at point t, which is a saddle point with zero gradient in both degrees of freedom, corresponding to the true transition structure [A $\cdots$ H $\cdots$ B]‡. The points z, s and t lie, of course, in the dividing surface between reactants and products. The curved path passing through r, t and p (thick solid line in Fig. 19.4) corresponds to the minimum-energy reaction path for proton transfer between A and B. This path is described by some nonlinear combination of the two coordinates that cannot be simply predicted without prior knowledge of the shape of the PES. It is usually determined by following the "intrinsic reaction coordinate" from the saddle point towards both the reactants and products; obviously this requires the saddle point to have been located beforehand.

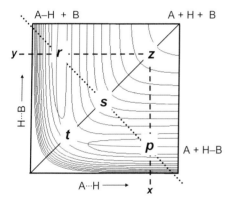

Figure 19.4 Schematic potential energy surface to illustrate alternative reaction coordinates.

19.2.2
The Brønsted a as a Measure of TS Structure

The Marcus relation, Eq. (19.6), is clearly not a linear relationship between the activation energy and the reaction asymmetry but a quadratic one. The first derivative of $\Delta E^\ddagger$ with respect to ΔE_{rxn} is equivalent to the Brønsted coefficient a in Eq. (19.2), but is itself a linear function of ΔE_{rxn}.

$$\mathrm{d}\Delta E^\ddagger / \mathrm{d}\Delta E_{rxn} = a = {}^1/_2 + \Delta E_{rxn}/8\Delta E_{int}^\ddagger \qquad (19.10)$$

Leffler [10] had proposed that the slope of an empirical linear correlation between $\log k$ and $\log K$ (i.e. of $\Delta G^\ddagger$ vs. ΔG_{rxn}) could be identified as a parameter measuring the degree of resemblance of the TS to the products as compared with its resemblance to the reactants. Thus for a PT from an acid to a series of related bases, the value of a obtained from the linear correlation $\Delta G^{\ddagger} = a\,\Delta G_{\mathrm{rxn}}$ would be interpreted as a measure of the degree of transfer of the proton in the TS. A value of $a = 0$ would indicate a reactant-like TS in which essentially no PT had occurred from the acid to the base, $a = 1$ would indicate a product-like TS in which PT essentially complete, and $a = 0.5$ would indicate a TS in which the proton was about half transferred [30].

The use of the Brønsted a (and also the slopes of other structure–reactivity correlations) as measures of TS structure has been severely criticised [31], but has also been thoughtfully defended with appropriate cautionary notes [32]. It has long been recognised that there is a fundamental incompatibility within the Bema Hapothle collection of ideas that arise from TS theory: if the slope of a (genuinely) linear rate-equilibrium relationship is interpreted as some index of TS structure, it must reflect some *constant* property of the TS for the entire family of reactions used to construct the correlation; however, the BEP [2c, 5, 8] and Leffler [10] principles, Hammond postulate [7], and related notions all suggest that changing the reaction energetics should cause *variation* in TS structure within this same family.

19.3
FERs from MO Calculations of PESs

Modern computational chemistry offers a means by which the paradox between the existence of linear FERs and variation in TS structure may be resolved. For a particular reaction of a series of substituted compounds, the energetics and geometries of reactants, transition structures and products may be obtained from calculated PESs. Rate-equilibrium relationships may be constructed from plots of calculated $\Delta G^{\ddagger}$ vs. ΔG_{rxn} (or $\Delta H^{\ddagger}$ vs. ΔH_{rxn}) values. The merit of this approach is that the TS structure, deduced indirectly from the slopes of these plots, may be compared with directly determined transition structures [33].

19.3.1
Energies and Transition States

Two comments are necessary before we proceed further. First, most of the preceding discussion of reaction energetics has been couched rather imprecisely in terms of undefined energy changes ΔE, whereas the logarithms of experimental rate constants or equilibrium constants are proportional to changes in Gibbs free energy ΔG. A linear rate-equilibrium relationship ($\Delta G^{\ddagger} = a\,\Delta G_{\mathrm{rxn}}$) might arise as the result of: (a) $\Delta H^{\ddagger}$ and ΔH_{rxn} each being constant for all members of a series of related reactions; (b) $\Delta S^{\ddagger}$ and ΔS_{rxn} each being constant throughout the same series; or (c) enthalpy and entropy changes being linearly related to each other. With careful design of a theoretical simulation it is often possible to ensure that condition (b) is met satisfactorily, so that changes in free energy may be modelled adequately by changes in enthalpy. Moreover, within a series of related reactions which differ, for example, in regard to the nature of a remote substituent upon one reactant, it is also reasonable to approximate enthalpy changes ΔH by potential energy changes ΔE. In the remainder of this chapter we will consider examples of theoretical simulations of rate equilibrium relationships based upon changes in potential energy, enthalpy, and free energy. The second point is to clarify our usage of the terms "transition state" and "transition structure". The latter term, transition structure, refers to the molecular geometry corresponding to a saddle point on a potential energy surface: this is a well-defined microscopic entity which may be located and characterized within a particular computational model. In contrast, the term transition state refers to the properties of an ensemble of molecular entities at a finite (nonzero) temperature: it may be associated with the maximum along a simulated free energy profile for an elementary step. We consider it useful to distinguish the two concepts in view of the significant difference between them: the one is microscopic and related to a feature of an (unobservable) PES; the other is macroscopic and related to phenomenological interpretation of empirical kinetic data. The connection between a transition structure and a TS involves statistical-mechanical averaging. Of course, the same distinction could be made for a stable molecule between its minimum energy structure and its equilibrium state at a given temperature and pressure.

Perhaps the first example of a computationally simulated Brønsted correlation was the semiempirical MNDO study of Anhede, Bergman and Kresge (ABK) [34] for a series of PT reactions involving fluoroethanols, Eq. (19.11), with $n = 0–3$.

$$CH_{(3-n)}F_nCH_2OH + {}^-OCH_2CH_3 \rightarrow CH_{(3-n)}F_nCH_2O^- + HOCH_2CH_3 \qquad (19.11)$$

The geometries of reactant and product ion–molecule complexes, together with transition structures, were completely optimized, and for each reaction the enthalpies of activation and of reaction were computed for the elementary step of PT between the reactant and product complexes. Each of the non-identity reactions ($n = 1, 2, 3$) was considered in both the forward and backward direction; together with the identity reaction ($n = 0$), this yielded seven pairs of values for a plot of computed $\Delta H^{\ddagger}$ vs. ΔH_{rxn} which showed noticeable curvature over a range $-100 < H_{rxn} < +100$ kJ mol^{-1}. A good fit to the Marcus equation was obtained with $\Delta H_{int}^{\ddagger} = 67$ kJ mol^{-1}. The slope of the Brønsted correlation, as determined by Eq. (19.10), varied between $a = 0.32$ for the most exothermic PT to 0.68 for the most endothermic PT. ABK did not comment upon the position of the proton in the transition structures for the non-identity reactions.

Scheiner and Redfern (SR) [35] performed *ab initio* HF/4-31G calculations for PT in (A-H······B)$^+$ where B = NH$_3$ and A = MeNH$_2$, EtNH$_2$, Me$_2$NH. Intramolecular PT along a hydrogen bond within a complex was modelled by holding fixed the distance between the donor and acceptor groups. This approach corresponds to following the straight dotted line *rsp* across the PES as in Fig. 19.4: consequently the maxima in the energy profiles do not correspond to genuine transition structures. Nonetheless, SR found that their directly calculated changes in barriers and reaction energies were quantitatively reproduced by the Marcus equation for all systems and that there was fair agreement between the Brønsted a computed as the slope of the Marcus curve and as the position of the proton along the reaction coordinate in the transition structure. An intrinsic barrier of ~18 kJ mol^{-1} may be obtained by fitting a quadratic to the published data for the N $\cdots$ N distance (2.75 Å) closest to that in the energy minimum of the double-well potential surface. The Brønsted plot for this transect is markedly curved, with a changing from 0.17 to 0.83 over a range of 93 kJ mol^{-1}. Increasing the fixed N $\cdots\cdots$ N distance to 3.10 Å raises the intrinsic barrier (cf. Ref. [36]) to 77 kJ mol^{-1} and decreases the degree of curvature, with a changing from 0.4 to 0.6 over a range of 124 kJ mol^{-1}. Scheiner and Redfern noted that the difficulties experienced a little previously, by Evleth and co-workers [37], in correlating similar *ab initio* SCF results for a range of PT reactions using the Marcus relation were due partly to the equation not being applied only to a single elementary step and partly to the estimation of intrinsic barriers from data on the symmetric systems AHA and BHB, where one of these was described by a single-well potential rather than a double well.

Hoz, Yang and Wolfe (HYW) [38] made ingenious use of the gas-phase PES for the concerted addition of water to formaldehyde in order to obtain a Brønsted correlation for PT between oxygen atoms. PT from water to the carbonyl oxygen is endothermic in the reactant complex (Fig. 19.5, top left) but very exothermic in the zwitterionic species (bottom left) formed by nucleophilic attack at the carbonyl carbon, as shown by the More O'Ferrall–Jencks diagram. There is a single true

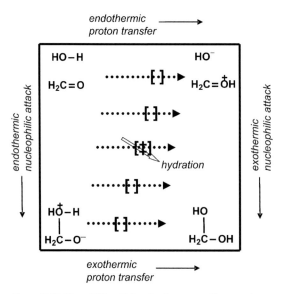

Figure 19.5 More O'Ferrall–Jencks diagram to illustrate how cross-sections of the potential energy surface for gas-phase hydration of formaldehyde yield a family of barriers from which a Brønsted correlation may be generated (cf. Ref. [38]).

transition structure on this surface for hydration, as shown by the open arrow, but there are many energy maxima for PT along the cross-sections indicated by dotted lines in Fig. 19.5. The activation energy $\Delta E^{\ddagger}$ and reaction aymmetries ΔE_{rxn} computed with both the semiempirical AM1 and *ab initio* HF/3-21G MO methods for each of several of these cross-sections were used to construct apparently very good linear Brønsted correlations. HYW argued that, although their computed energies fit the Marcus equation reasonably well, the linear fit was better. However, re-analysis of their published data actually suggests that a quadratic fit is slightly better than a simple linear correlation, with intrinsic barriers of ~96 kJ mol^{-1} (HF/3-21G) and ~143 kJ mol^{-1} (AM1).

A larger matrix of related PT reactions was considered by Williams and Austin (WA) [39] using the semiempirical AM1 method. They modulated the proton affinities of donor and acceptor ammonia molecules in the complex (A-H$\cdots$B)$^+$ where A = B = NH$_3$ by tuning the magnitude of external dipoles placed remotely along the N$\cdots$N axis. This allowed a range of 4-substituted pyridines to be mimicked as both acids YPyH$^+$ and bases PyX (X, Y = CN, Cl, H, Me, NMe$_2$) thereby yielding 25 distinct reactions; geometries for all ion-dipole reactant/product complexes and transition structures were located with complete optimization (subject to a three-fold symmetry constraint). Figure 19.6 shows the Brønsted correlation generated by plotting $-\Delta H_{rxn}$ against $-\Delta H^{\ddagger}$ for intra-complex PT for each combination of donor and acceptor; to a good approximation this is equivalent to a plot of log k against log K, since changes in ΔS_{rxn} and $\Delta S^{\ddagger}$ are likely to be essentially constant for the whole set of reactions. A least squares fit of a quadratic to the 25 data points gives an excellent correlation (r^2 = 0.999) with an intercept of 5.1 kJ mol^{-1}, but a Marcus equation with intrinsic barrier $\Delta H^{\ddagger}_{int}$ of 5.22 kJ mol^{-1} (= $\Delta H^{\ddagger}$ for X = Y = H) provides an almost equally satisfactory fit to the AM1 calculated data. (Note that the coefficient of the quadratic term in the Marcus equation is a function of the constant and the coefficient of the linear term, and is therefore less flexible than a generalised second-order polynomial.) The slope of the curved Brønsted correlation for PT changes from 0 for the most exothermic to 1 for the most endothermic reactions, in accord with expectation [13a, 40], over a range of

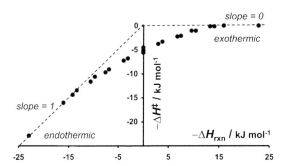

Figure 19.6 Curved Brønsted correlation for gas-phase proton transfer between mimicked pyridine donors and acceptors; AM1 results from Ref. [39].

46 kJ mol^{-1} corresponding to about $8\Delta H^{\ddagger}_{int}$. The transition structures varied sensibly with the thermodynamic reaction asymmetry, and it was found that the slope of the Brønsted correlation provided an approximate measure of both the degree of PT and of charge development along the reaction coordinate between the reactant and product ion–dipole complexes. An analogous simulation for methyl transfer [39] between the same sets of donors and acceptors (now electrophiles and nucleophiles rather than acids and bases) gave an imperceptibly curved Brønsted correlation for which the Marcus intrinsic barrier was very high (152 kJ mol^{-1}); this demonstrates that the apparent linearity of an LFER depends upon the fraction of $8\Delta H^{\ddagger}_{int}$ spanned by the range of thermodynamic reaction asymmetry covered by a simulation or by a series of experiments.

The examples discussed above all employ MO theory to simulate gas-phase PT, and all yield energy minima and maxima along defined reaction paths across adiabatic PESs, as represented qualitatively by Fig. 19.4. Progress from a reactant ion–dipole complex through a transition structure (either fully optimized or subject to some constraint) towards a product ion–dipole complex is accompanied smoothly by changes in the shapes and energies of the MOs, corresponding to the changing charge distributions. All give results that correlate well with the Marcus equation. All show variation in geometry of the transition structure with changes in reaction energy, and all show curved Brønsted correlations.

Closer examination of the WA data [39] is warranted in view of the fact that this simulation covered the whole range $0 \le a \le 1$. It has often been noted that different values of a are obtained from apparently linear Brønsted correlations according to whether the acid is kept constant and the nature of the base is changed, or the acid is varied while keeping the base fixed [41], and this effect has been attributed to variation in the intrinsic barrier within a series of reactions. However, the WA intrinsic barriers vary by only 1 kJ mol^{-1}. Selection from the WA data of the series with acid = $^{+}$HPyH and bases PyX (X = CN, Cl, H, Me, NMe$_2$) gives a linear correlation ($r^2 = 0.97$) of ΔH_{rxn} against $\Delta H^{\ddagger}$ with $a = 0.48$; selection of the related series with base = PyH and acids $^{+}$HPyY (Y = CN, Cl, H, Me, NMe$_2$) gives a very similar linear correlation ($r^2 = 0.98$) with $a = 0.52$. Each of these two series has a range of reaction enthalpies extending either side of the $\Delta H_{rxn} = 0$ by about the same amount. However, if one selects the series with acid = $^{+}$HPyCN and the same five bases, a fair linear correlation ($r^2 = 0.81$) of ΔH_{rxn} against $\Delta H^{\ddagger}$ is obtained with a slope of 0.24, whereas with base = PyCN and variation of the acid a good linear correlation ($r^2 = 0.98$) with a slope of 0.76 is obtained. This marked difference in the value of the Brønsted coefficient arises not from variation in the intrinsic barrier but because the former series covers an exothermic range of ΔH_{rxn} whereas the latter covers an endothermic range. To comment that neither $a = 0.24$ nor $a = 0.76$ can be taken as a measure of the position of the TS along the reaction coordinate does not add value beyond the understanding that the correlation is intrinsically curved in consequence of the continuously varying nature of the transition structure.

The four examples of PT simulations discussed above employed a variety of MO methods, giving rise to wide variation in the intrinsic barriers. The large intrinsic barriers obtained by ABK [34] for ROH/$^-$OR identity reactions ($\Delta H_{int}^{\ddagger} = 65, 79, 74$ and 69 kJ mol^{-1}, respectively, for $n = 0, 1, 2$ and 3 in $CH_{(3-n)}F_nCH_2OH/CH_{(3-n)}F_nCH_2O^-$) might arise purely from errors in the MNDO Hamiltonian, but they might also be due to overestimated O$\cdots$O distances; ABK did not report their geometries. We have already noted that an increase of 0.35 Å in the N$\cdots$N distance for PT in NH_4^+/RNH_2 complexes [35] raises the HF/4-31G intrinsic barrier by 59 kJ mol^{-1}; SR reported a barrier of 15.9 kJ mol^{-1} for the NH_4^+/NH_3 identity reaction at the HF/4-31G with constrained geometry (N$\cdots$N = 2.731 Å) [35], as compared with 7.7 kJ mol^{-1} for reaction through the true transition structure (N$\cdots$N = 2.594 Å) [37].

Intrinsic barriers are also very sensitive to the basis set and the degree of electron correlation. Table 19.1 collects values for MeOH/$^-$OMe computed for HF/6-31G(d) optimised geometries [42]: p polarisation functions on hydrogen atoms lower the intrinsic barrier, whereas s and p diffuse functions on carbon and oxygen atoms raise it; inclusion of electron correlation at the MP2 level markedly reduces the intrinsic barrier; the values vary over an 18 kJ mol^{-1} range. Table 19.1 also contains data for NH_3/$^-NH_2$ computed for MP2/6-31+G(d,p) optimised geometries [43]: extra polarisation functions may either raise or lower the intrinsic barrier, whereas improved methods for electron correlation give lower intrinsic barriers than Hartree–Fock theory; the values vary over a 24 kJ mol^{-1} range. Furthermore, inclusion of zero-point energy and the thermal energy correction for 298 K (using HF/6-31+G(d,p) frequencies scaled by 0.9) together serve to reduce the intrinsic barrier by about 14 kJ mol^{-1}. The "best" estimate for the NH_3/$^-NH_2$ intrinsic barrier [QCISD(T)/6-311+G(d,p)//MP2/6-31+G(d,p) + ΔE_{zpe} + $E_{thermal}$(298 K)] is about 5 kJ mol^{-1}. Wu, Shaik and Saunders [44] have used a VB method to analyse the factors contributing to intrinsic barriers for identity PT reactions between neutral acids A–H and anionic bases A$^-$: they note that the reactant- and product-like covalent VB structures A–H A$^-$ and A$^-$ H–A alone do not provide an adequate description, and point out the importance of triple-ion VB structures A$^-$ H$^+$ A$^-$.

In summary, these examples of MO simulations for gas-phase PT (and many more similar studies for hydrogen atom [45] or hydride transfer [46]) reactions clearly demonstrate the qualitative validity of quadratic FERs (such as the Marcus relation) over wide ranges of thermodynamic reaction asymmetry, and of approximately linear FERs over relatively narrow ranges. Quantitatively it remains a very demanding task for electronic structure methods to predict intrinsic barriers or kinetic activation energies with high accuracy: while feasible for single calculations, high-level calculations for families of PT reactions are not performed routinely even for gas-phase reactions, let alone in solution.

Finally it may be noted that simulations yielding Brønsted coefficients outside the normal range of $0 > \alpha > 1$ (*e.g.* hydrogen-atom transfer in R–H + Cl (R = Me, Et, ^{i}Pr, ^{t}Bu) [45a]) or failure to obtain any sort of sensible FER at all (*e.g.* nucleophilic addition to a carbonyl group concerted with PT [47], *cf.* HYW [38]) probably indicates that the TSs involve structural features not present in either the reactants or products [48]. This type of behavior sometimes manifests itself in terms of large variations in Marcus intrinsic barriers [6c].

Tab. 19.1 Method dependence of intrinsic barriers calculated with molecular orbital theory.

Energy method	Intrinsic barrier / kJ mol^{-1}
MeOH/$^-$OMe [a]	
HF/6-31G(d)	9.2
HF/6-31G(d,p)	6.4
HF/6-31+G(d)	12.2
MP2/6-31G(d)	−1.8
MP2/6-31G(d,p)	−5.5
MP2/6-31+G(d,p)	0.1
H$_2$NH/$^-$NH$_2$ [b]	
basis set	
MP2/6-311+G(d,p)	16.3
MP2/6-311+G(2df,p)	13.8
MP2/6-311+G(3df,2p)	17.2
electron correlation	
HF/6-311+G(d,p)	37.7
MP2/6-311+G(d,p)	16.3
MP3/6-311+G(d,p)	20.5
MP4/6-311+G(d,p)	18.4
QCISD(T)/6-311+G(d,p)	18.8
thermodynamic corrections[c]	
ΔE_{zpe}(0 K)	−10.0
$\Delta E_{\text{thermal}}$(298 K)	−3.8

a Ref. 42; geometries calculated at HF/6-31G(d).
b Ref. 43; geometries calculated at MP2/6-31+G(d,p).
c Ref. 43; geometries calculated at HF/6-31+G(d,p)

19.4
FERs from VB Studies of Free Energy Changes for PT in Condensed Phases

Warshel and coworkers have employed the empirical valence-bond (EVB) method [49] to simulate FERs for PT [50] and other reactions [51]. The PT step between two water molecules in the mechanism of the reaction catalysed by carbonic anhydrase was described as an effective two-state problem involving "reactant-like" $(HOH)(OH_2)$ and "product-like" $(HO^-)(HOH_2^+)$ VB structures [50a]. Diabatic energy curves for these two VB structures were calibrated to reproduce the experimental free energy change ΔG_{rxn} for autodissociation in water, and the mixing of the reactant-like and product-like states (cf. E_{res} in Fig. 19.3) was calibrated to reproduce the observed activation energy $\Delta G^{\ddagger}$ for the uncatalysed reaction in water. The magnitude of the off-diagonal matrix element E_{res} that coupled the two diabatics to yield the ground-state adiabatic energy surface depended upon the value of the reaction coordinate. Classical free energy surfaces for reactions in aqueous solution or in an enzyme were evaluated by a combination of EVB with a free-energy perturbation/umbrella sampling method [51a]: within each of many overlapping windows along the reaction coordinate, a series of molecular dynamics trajectories were propagated to sample many configurations of the surrounding water or protein. The reaction coordinate used was not a geometrical parameter but rather the difference in energy between the reactant and product diabatic states. This approach did not attempt to evaluate PESs of reacting fragments but instead fit them to known experimental (or theoretical) results; however, it did focus upon the important issue of how changing the environment (from water to enzyme) affected the quantum mechanical region of the chemical reaction.

It turns out that charge-transfer reactions, such as this, follow the linear response approximation in solution [51a, 52] and in enzymes [50a, 53], meaning that the diabatic energy curves are quadratic. Consequently, the (kinetic) activation free energy on the adiabatic surface is related to the (thermodynamic) reaction asymmetry according to Eq. (19.12).

$$\Delta G^{\ddagger} = \Delta G_{int}^{\ddagger} + {}^{1}/_{2}\,\Delta G_{rxn} + \Delta G_{rxn}^{2}/16\Delta G_{int}^{\ddagger} - E_{res}^{\ddagger} + (E_{res}^{R})^{2}/(\Delta G_{rxn} + 4G_{int}^{\ddagger})$$

(19.12)

The first three terms on the right-hand side correspond to the Marcus relation for the nonadiabatic case where there is no coupling between the diabatic energy states (i.e. $E_{res} = 0$ at all values of the reaction coordinate). The fourth and fifth terms reflect the effect of the adiabatic coupling of the two surfaces on the transition state and reactant state, respectively, and $E_{res}^{R} < {}^{1}/_{2}|\Delta G_{rxn} + 4\Delta G_{int}^{\ddagger}|$.

The initial work of Warshel, Hwang and Åqvist [50a] for PT in water and in carbonic anhydrase (a zinc metalloenzyme) reported FERs simulated in two ways. First, the autodissociation in water catalysed by a metal ion was studied using explicit Zn^{2+}, Mg^{2+}, Ca^{2+} and Na^{+} cations. Second, the energy of the product diabatic curve (for the ionic VB structure) was shifted vertically in order to change the value of ΔG_{rxn} for PT in both water and the enzyme active site. Both methods gave apparently linear Brønsted correlations, with the points for substitution of the catalytic metal ions lying on the same line as those for parametrically changing the reaction asymmetry in water. The slopes of both lines were similar ($a \approx 0.5$) but the intrinsic barriers at $\Delta G_{rxn} = 0$ were ~46 and ~21 kJ mol^{-1}, respectively, for the reactions in water and catalysed by carbonic anhydrase; the lower value for the enzyme reflects stabilisation of the transition state. However, the reaction asymmetry is less endoergic in the enzyme than with the Zn^{2+}-catalysed reaction in water, suggesting a thermodynamic component to the enzyme catalysis as well as a kinetic component. A later paper by Hwang and Warshel [50b] considered the role of quantum-mechanical nuclear motions using the quantized classical path approach: it was found that while these corrections could be quite large, they were not drastically different for PT in the enzyme than for PT in aqueous solution. This study found the reorganisation energies (Fig. 19.3) to be about 250 and 100 kJ mol^{-1}, respectively, for the enzymic and nonenzymic reactions. These values correspond to Marcus intrinsic barriers (as given by the intersection of the diabatic curves for $\Delta G_{rxn} = 0$) of ~60 and 25 kJ mol^{-1}, respectively. In contrast, the Marcus intrinsic barrier deduced by phenomenological fitting to experimental data for carbonic anhydrase gave a Marcus intrinsic barrier of only 6 kJ mol^{-1} [54]. The discrepancy was suggested to arise for two reasons: first, the reaction studied experimentally might involve more than an elementary step, and thus more than two intersecting parabolas; second, the conventional Marcus relation (Eq. (19.6)) might not be valid for reactions involving strong coupling between the reactant and product diabatic states, for which Eq. (19.12) would be more appropriate. Recently, Schutz and Warshel [50c] re-examined this problem and found that the FER reflected a much more complex situation than previously thought: they suggested the PT process involved three or more parabolic free energy surfaces rather than the two assumed in the simple Marcus treatment. Their analysis reproduced the strongly curved observed FER without using any adjustable parameters and, in contrast to the result obtained from application of a two-parabola Marcus treatment [54], they found the work function to have a negligible value. Their three-state EVB description considered PT from an active site Lys or His residue to the zinc-bound hydroxyl by means of a bridging water molecule, as in Eq. (19.13), with each of the elementary steps being treated by Eq. (19.12).

$$BH^+(H_2O)_b(OH^-)_aZn^{2+} \rightarrow B(H_3O^+)_b(OH^-)_aZn^{2+} \rightarrow B(H_2O)_b(H_2O)_aZn^{2+} \quad (19.13)$$

They argued that the Marcus-like behavior of the observed FER was not due to transition to the Marcus inverted region but to change in energy of the intermediate; as the overall PT becomes more exoergic, the reaction rate increases but then starts to decrease as the first step becomes rate limiting instead of the second step. Schutz and Warshel [50c] emphasised that their simulation approach was based on realistic molecular parameters obtained while starting from the the X-ray structure of the protein and reproducing the relevant pK_as and reorganisation energy.

Interpretation of numerical simulations in terms of a (modified) Marcus treatment provides valuable insight owing to the important distinction between kinetic and thermodynamics factors in reactivity and catalysis. Feierberg, Cameron and Åqvist [55] applied the EVB methodology to simulate the reaction catalysed by glyoxalase I. They proposed the rate-limiting step to be PT from carbon on the substrate to a glutamate residue, forming a high-energy enolate intermediate. Simulated LFERs for the PT step revealed that the effect of an active-site divalent metal cation was mainly to reduce ΔG_{rxn} with a rather smaller contribution to the activation energy reduction coming from a decrease in the reorganisation energy λ.

Kiefer and Hynes [56] have developed a quadratic FER for acid-base ionization PT reactions (Eq. (19.14)) in a polar environment in the proton adiabatic regime, in which the proton is treated quantum mechanically but does not tunnel.

$$\text{A–H} \cdots \text{B} \rightarrow \text{A}^- \cdots \text{X–B}^+ \tag{19.14}$$

Their approach is based on a two-state VB treatment involving neutral reactant and ionic product VB structures as in Eq. (19.14), and the reaction coordinate is a solvent coordinate. Strong electronic coupling between the VB states produces the ground electronically adiabatic surface on which the reaction occurs, similarly to Warshel's EVB method. The underlying picture of PT is very different from the conventional view implicit in the examples discussed in Section 19.3. A PT reaction is driven by configurational changes in its surrounding polar environment and the activation free energy is largely determined by the reorganisaton of this environment rather than by the height of any potential barrier in the coordinate of the transferring proton. The rapidly vibrating proton adiabatically follows the slower rearrangement of the environment: evolution of the solvent coordinate leads to an evolving proton potential pattern, in which the proton is initially bound to a donor in the reactant state, to a TS with the proton delocalised to a degree between the donor and acceptor groups, and finally to the product state with the proton bound to the acceptor. In the proton adiabatic regime the quantised proton vibrational level lies above the proton barrier at the TS along the coordinate for rearrangement of the solvent environment: at the TS the proton potential is a double well, but the free energy barrier for PT is located along the solvent coordinate, not the proton coordinate. The resulting FER is given by Eq. (19.15), where a_0 is the Brønsted coefficient at $G_{rxn} = 0$ and a_0' is the first derivative of the Brønsted coefficient also at $\Delta G_{rxn} = 0$.

$$\Delta G^{\ddagger} = \Delta G_{int}^{\ddagger} + a_0 G_{rxn} + {}^1\!/_2\, a_0' \Delta G_{rxn}{}^2 \tag{19.15}$$

In this model the intrinsic barrier is governed by the solvent reorganisation, modulated by certain electronic structural changes, and the change in the combined zero-point energy (ZPE) of the proton and the hydrogen-bond vibrations to reach the TS in the solvent coordinate. Involvement of the proton ZPE in the intrinsic barrier is one key difference between this model and the conventional view of PT; in the latter there is no such contribution because the proton coordinate is the reaction coordinate, whereas in the former it is transverse to the reaction coordinate. Although application of this new approach to specific examples of PT is still awaited, nonetheless it is of considerable interest that the FER of Eq. (19.15), which is clearly related to the Marcus equation, is obtained from the proton adiabatic picture that differs so fundamentally from the traditional view of PT. A valuable aspect of the new approach is that it yields an analytical expression for the intrinsic barrier in terms of its microscopic ingredients, whereas usually in the Marcus treatment it is merely a phenomenological parameter with no clear relation to features of molecular structure and bonding.

19.5
Concluding Remarks

Among the many topics not dealt with in this chapter are intrinsic barrier asymmetry [57], disparity/tightness and their relation to observed Brønsted coefficients [45a, 58], the nitroalkane anomaly and role of additional VB states dissimilar from reactants and products [59], and several other interesting models for FERs [60, 61]. This reviewer was surprised at the lack, as yet, of applications of hybrid quantum-mechanical/molecular-mechanical methodology (other than EVB-based methods) for simulations of FERs for any chemical reactions, let alone PT. However, the subject is alive and well, and it is to be expected that many more theoretical simulations of FERs for PT will be performed in the relatively near future.

References

1 I. H. Williams, *Chem. Soc. Rev.* **1993**, *22*, 277–283.

2 (a) H. Eyring, M. Polanyi, *Z. Phys. Chem.* **1931**, *12B*, 279–311; (b) R. A. Ogg, M. Polanyi, *Trans. Faraday Soc.* **1935**, *31*, 604–620; (c) M. G. Evans, M. Polanyi, *Trans. Faraday Soc.* **1935**, *31*, 875–894; (d) H. Eyring, *J. Chem. Phys.* **1935**, *3*, 107–115; *Chem. Rev.* **1935**, *17*, 65–77.

3 L. P. Hammett, *Chem. Rev.* **1935**, *17*, 125–136.

4 W. P. Jencks, *Chem. Rev.* **1985**, *85*, 511–527.

5 R. P. Bell, *Proc. R. Soc. London, Ser. A* **1936**, *154*, 414–429.

6 (a) R. A. Marcus, *J. Phys. Chem.* **1968**, *72*, 891–899; (b) A. O. Cohen, R. A. Marcus, *J. Phys. Chem.* **1968**, *72*, 4249–4256; (c) R. A. Marcus, *J. Am. Chem. Soc.* **1969**, *91*, 7224–7225.

7 G. S. Hammond, *J. Am. Chem. Soc.*, **1955**, *77*, 334–338.

8 (a) M. G. Evans, M. Polanyi, *Trans. Faraday Soc.* **1936**, *32*, 1333–1360; (b) M. G. Evans, M. Polanyi, *Trans. Faraday Soc.* **1938**, *34*, 11–24.

9 (a) E. R. Thornton, *J. Am. Chem. Soc.* **1967**, *89*, 2915–2927; (b) E. K. Thornton, E. R. Thornton, in *Transition States of Biochemical Processes*, Ed. R. D. Gandour, R. L. Schowen, Plenum, New York, 1978.

10 J. E. Leffler, *Science*, **1953**, *117*, 340–341; J. E. Leffler, E. Grunwald, *Rates and Equilibria of Organic Reactions*, Wiley, New York, 1963.

11 (a) W. P. Jencks, *Chem. Rev.*, **1972**, *72*, 705–718; (b) D. A. Jencks, W. P. Jencks, *J. Am. Chem. Soc.* **1977**, *99*, 7948–7960.

12 R. A. More O'Ferrall, *J. Chem. Soc. B* **1970**, 274–277.

13 (a) J. R. Murdoch, *J. Am. Chem. Soc.* **1972**, *94*, 4410–4418; (b) J. R. Murdoch, D. E. Magnoli, *J. Am. Chem. Soc.* **1982**, *104*, 3792–3800.

14 M. J. S. Dewar, D. A. Dougherty, *PMO Theory or Organic Chemistry*, Plenum Press, New York, 1975.

15 J. R. Murdoch, *J. Am. Chem. Soc.* **1983**, *105*, 2159–2164.

16 S. Sato, *J. Chem. Phys.* **1955**, *23*, 592–593.

17 (a) H. S. Johnston, C. Parr, *J. Am. Chem. Soc.* **1963**, *85*, 2544–2551; (b) H. S. Johnston, *Adv. Chem. Phys.* **1960**, *3*, 131–170.

18 R. A. Marcus, *J. Chem. Phys.* **1956**, *24*, 966–978.

19 D. Rehm, A. Weller, *Isr. J. Chem.* **1970**, *8*, 259–271.

20 N. Agmon, R. D. Levine, *J. Chem. Phys.* **1979**, *71*, 3034–3041.

21 R. P. Bell, *J. Chem. Soc., Faraday Trans. 2* **1976**, 2088–2094.

22 W. J. le Noble, A. R. Miller, S. D. Hamann, *J. Org. Chem.* **1977**, *42*, 338–342.

23 (a) E. S. Lewis, *Top. Curr. Chem.* **1978**, *74*, 31; (b) E. S. Lewis, C. C. Shen, R. A. More O'Ferrall, *J. Chem. Soc., Perkin Trans. 2* **1981**, 1084–1088.

24 J. L. Kurz, *Chem. Phys. Lett.* **1978**, *57*, 243–246.

25 (a) A. A. Zavitsas, *J. Am. Chem. Soc.* **1972**, *94*, 2779–2789; (b) A. A. Zavitsas, A. A. Melikian, *J. Am. Chem. Soc.* **1975**, *97*, 2757–2763.

26 S. Ahrland, J. Chatt, N. R. Davies, A. A. Williams, *J. Chem. Soc.* **1958**, 276–288.

27 V. G. Levich, R. R. Dogonadze, E. D. German, R. M. Kuznetsov, Y. I. Kharkats, *Electrochim. Acta* **1970**, *15*, 353–367.

28 (a) Footnote 44 in Ref. 14; (b) J. R. Murdoch, *J. Mol. Struct. (THEOCHEM)* **1988**, *40*, 447–476.

29 (a) R. B. Hammond, I. H. Williams, *J. Chem. Soc., Perkin Trans. 2* **1989**, 59–66; (b) G. D. Ruggiero, I. H. Williams, *J. Chem. Soc., Perkin Trans. 2* **2001**, 448–458.

30 A. J. Kresge, in *Proton-transfer reactions*, Ed. E. F. Caldin, V. Gold, Chapman and Hall, London, 1975, pp. 179–199.

31 A. Pross, S. Shaik, *Nouv. J. Chim.* **1989**, *13*, 427–433.

32 W. P. Jencks, *Bull. Soc. Chim. Fr.* **1988**, 218–224.

33 I. H. Williams, *Bull. Soc. Chim. Fr.* **1988**, 192–198.

34 B. Anhede, N.-Å. Bergman, A. J. Kresge, *Can. J. Chem.* **1986**, *64*, 1173–1178.

35 S. Scheiner, P. Redfern, *J. Phys. Chem.* **1986**, *90*, 2969–2974.

36 S. Scheiner, *Acc. Chem. Res.* **1985**, *18*, 174–180.

37 H. Z. Cao, M. Allavena, O. Tapia, E. M. Evleth, *J. Phys. Chem.* **1985**, *89*, 1581–1592.

38 S. Hoz, K. Yang, S. Wolfe, *J. Am. Chem. Soc.* **1990**, *112*, 1319–1321.

39 I. H. Williams, P. A. Austin, *Can. J. Chem.* **1999**, *77*, 830–841.

40 A. J. Kresge, *Chem. Soc. Rev.* **1973**, *2*, 475–503.

41 R. P. Bell, *Faraday Symp. Chem. Soc.* **1975**, *10*, 7–19.

42 S. Wolfe, S. Hoz, C.-K. Kim, K. Yang, *J. Am. Chem. Soc.* **1990**, *112*, 4186–4191.

43 S. Gronert, *J. Am. Chem. Soc.* **1993**, *115*, 10258–10266.

44 W. Wu, S. Shaik, W. H. Saunders, *J. Phys. Chem. A*, **2002**, *106*, 11616–11622.

45 (a) H. Yamataka, S. Nagase, *J. Org. Chem.* **1988**, *53*, 3232–3238; (b) A. K. Chandra, V. Sreedhara Rao, *Chem. Phys.* **1995**, *200*, 387–393; (c) W. T. Lee, R. I. Masel, *J. Phys. Chem. A* **1998**, *102*, 2332–2341; (d) S. Gronert, C. Kimura, *J. Phys. Chem. A* **2005**, *107*, 8932–8938.

46 E.-U. Würthwein, G. Lang, L. H. Schappele, H. Mayr, *J. Am. Chem. Soc.* **2002**, *124*, 4084–4092.

47 I. H. Williams, D. Spangler, G. M. Maggiora, R. L. Schowen, *J. Am. Chem. Soc.* **1985**, *107*, 7717–7723.

48 A. J. Kresge, *J. Am. Chem. Soc.* **1970**, *92*, 3210–3211.

49 A. Warshel, *Computer Modeling of Chemical Reactions in Enzymes and Solutions*, Wiley, Chichester, 1991.

50 (a) A. Warshel, J.-K. Hwang, J. Åqvist, *Faraday Discuss.* **1992**, *93*, 225–238; (b) J.-K. Hwang, A. Warshel, *J. Am. Chem. Soc.* **1996**, *118*, 11745–11751; (c) C. N. Schutz, A. Warshel, *J. Phys. Chem. B* **2004**, *108*, 2066–2075.

51 (a) J.-K. Hwang, G. King, S. Creighton, A. Warshel, *J. Am. Chem. Soc.* **1988**, *110*, 5297–5311; (b) A. Warshel, T. Schweins, M. Fothergill, *J. Am. Chem. Soc.* **1994**, *116*, 8437–8442; (c) Y. S. Kong, A. Warshel, *J. Am. Chem. Soc.* **1995**, *117*, 6234–6242.

52 (a) R. A. Kuharski, J. S. Bader, D. Chandler, M. Sprik, M. L. Klein, R. W. Impey, *J. Chem. Phys.* **1988**, *89*, 3248–3257; (b) G. King, A. Warshel, *J. Chem. Phys.* **1990**, *93*, 8682–8692.

53 (a) J. Åqvist, A. Warshel, *J. Am. Chem. Soc.* **1990**, *112*, 2860–2868; (b) A. Warshel, *Pontif. Acad. Sci. Scr. Var.* **1984**, *55*, 59–81; (c) A. Warshel, S. Russell, F. Sussman, *Isr. J. Chem.* **1987**, *27*, 217–224.

54 D. N. Silverman, C. Tu, X. Chen, S. M. Tanhauser, A. J. Kresge, P. J. Laipis, *Biochemistry* **1993**, *34*, 10757–10762.

55 I. Feierberg, A. D. Cameron, J. Åqvist, *FEBS Lett.* **1999**, *453*, 90–94.

56 (a) P. M. Kiefer, J. T. Hynes, *J. Phys. Chem. A* **2002**, *106*, 1834–1849; (b) P. M. Kiefer, J. T. Hynes, *J. Phys. Chem. A* **2002**, *106*, 1850–1861.

57 (a) R. A. Marcus, *Faraday Symp. Chem. Soc.* **1975**, *10*, 60–68; (b) G. W. Koeppl, A. J. Kresge, *J. Chem. Soc., Chem. Commun.* 1973, 371–373; (c) F. Wiseman, N. R. Kestner, *J. Phys. Chem.* **1984**, *88*, 4354–4358.

58 (a) M. M. Kreevoy, I.-S. H. Lee, *J. Am. Chem. Soc.* **1984**, *106*, 2550–2553; (b) E. Grunwald, *J. Am. Chem. Soc.* **1985**, *107*, 125–133.

59 (a) W. J. Albery, C. F. Bernasconi, A. J. Kresge, *J. Phys. Org. Chem.* **1988**, *1*, 29–31; (b) H. Yamataka, Mustanir, M. Mishima, *J. Am. Chem. Soc.* **1999**, *121*, 10223–10224; (c) J. R. Keefe, S. Gronert, M. E. Colvin, N. L. Tran, *J. Am. Chem. Soc.* **2003**, *125*, 11730–11745.

60 (a) S. J. Formosinho, *J. Chem. Soc. Perkin Trans. 2* **1987**, 61–66; (b) L. G. Arnaut, S. J. Formosinho, *J. Phys. Org. Chem.* **1990**, *3*, 95–109; (c) L. G. Arnaut, A. A. C. C. Pais, S. J. Formosinho, M. Barroso, *J. Am. Chem. Soc.* **2003**, *125*, 5236–5246.

61 (a) S. Shaik, H. B. Schlegel, S. Wolfe, *Theoretical aspects of Physical Organic Chemistry*, Wiley, New York, 1992; (b) L. Song, W. Wu, K. Dong, P. C. Hiberty, S. Shaik, *J. Phys. Chem. A* **2002**, *106*, 11361–11370.

20

The Extraordinary Dynamic Behavior and Reactivity of Dihydrogen and Hydride in the Coordination Sphere of Transition Metals

Gregory J. Kubas

20.1
Introduction

20.1.1
Structure, Bonding, and Activation of Dihydrogen Complexes

Transition metals can contain, within their coordination field, hydrogen in molecular (H_2) and/or atomic (hydride) form. These types of metal–ligand complexes are unquestionably the most dynamic molecular systems known in terms of structural variability and atom motion/exchange processes. Until about 20 years ago, metals were known to contain only atomically bound hydrogen, that is, metal hydrides and metal hydride complexes (L_nMH_x, where L is an ancillary ligand). However the discovery by Kubas and coworkers [1] in 1983 of side-on coordination of a nearly intact *dihydrogen molecule* (H_2) to a metal complex has led to a new paradigm in chemistry that is the subject of many review articles [2–10].

η^2-H_2 complex   dihydride complex

Molecules containing only strong "inert" σ bonds such as H–H in H_2 and C–H in alkanes had previously been believed to be incapable of stable binding to a metal. However, dihydrogen complexes (referred to as η^2-H_2 or H_2 complexes) that were only assumed to be unobservable intermediates in dihydride formation can be isolatable species, as exemplified by the first H_2 complex, $W(CO)_3(P^iPr_3)_2(H_2)$ (Fig. 20.1) [1, 2]. The H–H distance is elongated to 0.89 Å here, versus 0.75 Å in free H_2, indicating that the strong H–H bond is only partially broken. Hundreds of different H_2 complexes with nearly every transition metal have now been established, and in many cases the nearly intact H_2 ligand is reversibly bound, as in

Hydrogen-Transfer Reactions. Edited by J. T. Hynes, J. P. Klinman, H. H. Limbach, and R. L. Schowen
Copyright © 2007 WILEY-VCH Verlag GmbH & Co. KGaA, Weinheim
ISBN: 978-3-527-30777-7

physisorbed H_2. These species are part of a class of compounds called *σ complexes,* which refers to the side-on 3-center interaction of the bonding electron pair in H–H or other X–H bonds with a transition metal center M (X = B, C, Si, etc) [2, 10]. The dihydrogen complexes are remarkable in that, aside from elemental H_2 and materials containing weakly physisorbed H_2, no other stable compounds containing H_2 molecules had previously been known.

METAL–DIHYDROGEN COORDINATION

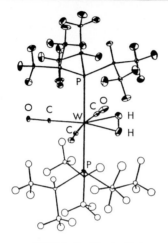

$W(CO)_3(P\text{-}i\text{-}Pr_3)_2(H_2)$

NEUTRON STRUCTURE, 30 K

Figure 20.1 Molecular structure of $W(CO)_3(P^iPr_3)_2(H_2)$ showing intact H–H bond elongated to 0.82(1) Å. Lower phosphine is disordered. The actual H–H distance is longer (0.89 Å from solid state NMR) because rapid rotation of the H_2 results in foreshortening of the neutron distance.

The metal-coordinated H_2 is "activated," a term used to describe the weakening of a chemical bond towards rupture when it is bound to a transition metal, as observed by the elongation of the H–H bond. This process, and indeed much of the structure/bonding/dynamic/exchange features of hydrogen in the coordination sphere of metals, can best be understood by examining the nature of the chemical bonding of H_2 to metals. The nonclassical 3-center 2-electron bonding in M–H_2 and other σ-bond complexes is stabilized by *backdonation* (*BD*), that is the retrodative donation of electrons from a filled metal d orbital to the σ* orbital of the H–H bond [2, 3]. This is analogous to the Dewar–Chatt–Duncanson model for olefin coordination in Scheme 20.1 [11]. BD is crucial in both increasing the strength of the M–H_2 bonding and activating the H–H bond towards homolytic cleavage to dihydride [12] ligands. If the backbonding becomes too strong, for example if more electron-donating co-ligands are put on M, the σ bond cleaves to form a dihydride because of overpopulation of H_2 σ^*. The entire reaction coordinate for the activation and ultimate splitting of H_2 on a metal can be mapped out and related to the degree of BD. This is dramatically demonstrated by the remarkable "stretching" of the H–H distance, d_{HH}, as displayed within the large regime of known complexes with H_2 bound to different metal–ligand fragments (Scheme 20.2) [2, 3]. A near *continuum* of d_{HH} ranging from 0.82 Å to nearly 1.6 Å is observed by crystallographic (X-ray and neutron diffraction) and NMR methods for the hundreds of known stable H_2 complexes. Much as in physisorbed H_2, the d_{HH} is relatively short (0.85–0.90 Å) and the H_2 is reversibly bound in "true" H_2 complexes, sometimes referred to as Kubas complexes, as best exemplified by $W(CO)_3(P^iPr_3)_2(H_2)$. The elongated (or stretched) H_2 complexes (d_{HH} = 1.0–1.2 Å) are the most dynamic species and the positions of the hydrogens are extremely delocalized [7], as will be shown below. Even longer d_{HH} have been observed, at which point the complex is best viewed as a "compressed dihydride" containing weak H$\cdots$H attractions. A type of complex originally synthesized in 1971 in Taube's group, $[Os(H_2)(ethylenediamine)_2(acetate)]^+$, shows a very long d_{HH} [13] and is on the verge of becoming a dihydride, which it was originally believed to be!

Complexes with d_{HH} > 1.6 Å are considered to be hydrides, but even this is a subjective boundary. There is little H–H bonding interaction remaining when d_{HH} becomes >1.1 Å, so intriguing questions arise such as "at what point is the bond broken?" By one criterion, the H–H bond can be considered "broken" when

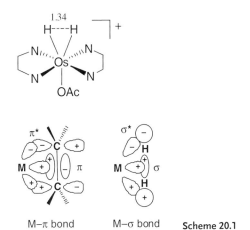

M–π bond   M–σ bond   **Scheme 20.1**

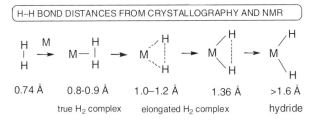

H–H BOND DISTANCES FROM CRYSTALLOGRAPHY AND NMR

Scheme 20.2

d_{HH} is twice the free molecule separation, 1.48 Å, at which point the bond order becomes zero [14]. Even before H_2 complexes were discovered, Xα calculations performed in 1978 by Ginsberg [15] showed that significant $H \cdots H$ interactions occur even at d_{HH} up to 1.9 Å and were believed to stabilize high coordination number polyhydrides such as $[ReH_9]^{2-}$ and $[ReH_8(PR_3)]^-$. Eisenstein also showed that weak long-range attractions exist between well-separated *cis*-hydride ligands [16]. The activation and splitting of H_2 is very sensitive to the nature of the metal, the ancillary ligands, and the overall charge, all of which can affect the amount of electron BD to H_2. Third-row metals, strongly donating ligands, and neutral charge favors elongation or splitting of the H–H bond (higher electron richness of metal increases BD), while first-row metals, electron-withdrawing ligands, and positive charge shortens the H–H distance. The binding site for hydrogen on a metal complex can thus be tuned electronically over a vast range, and the local ligand environment around the site can also be varied tremendously. Thus the activation process and associated hydrogen transfer processes on metal centers can feature extraordinary structural, bonding, and dynamic phenomena that are the subject of this chapter.

20.1.2
Extraordinary Dynamics of Dihydrogen Complexes

Long before the "nonclassical" dihydrogen complexes were discovered, classical polyhydride complexes had been known to be *stereochemically nonrigid* (also termed *fluxional*) in solution, as shown by investigations of complexes of the type $MH_n(PR_3)_4$ (n = 1, 2, 4; M = Group 6, 8, and 9 metals) by Meakin, Muetterties and coworkers in the early 1970s [12]. All of these systems showed fluxional behavior, and low barriers for n = 1 or 2 were rationalized by a "tetrahedral jump" mechanism for rapid ligand exchange. The study of solution nuclear magnetic resonance line shapes was critical to the determination of mechanisms for intramolecular hydride exchange [12c]. Before H_2 coordination was recognized, the fluxionality of polyhydrides was viewed as isolated H-atoms moving over the surface of the metal center. However, their association as H_2 ligands as intermediate steps is now much more attractive, as will be discussed in detail. For example, for hydride site exchange in polyhydrides such as ML_4H_4 (M = Mo, W; L = P-atom donor), transient intermediates with a geometry very much like $MH_2(H_2)L_4$ or trans-$M(H_2)_2L_4$ with elongated d_{HH} were considered possible, even in 1973 long before H_2 complexes were actually discovered (Scheme 20.3). Since the dihydrogen ligand nearly freely rotates, that is, has a relatively low barrier to rotation (1–10 kcal mol^{-1}), hydride ligand rearrangement could easily take place by rotating the intermediate H1–H2 ligand as shown. Many new examples of hydride fluxionality were later discovered, and the principle mechanistic aspects have been reviewed to include systems containing η^2-H_2 ligands [5a, 8, 17]. Fast exchange between terminal and bridging hydrides in dinuclear rhenium complexes has been shown calculationally to be facilitated by formation of dihydrogen-containing intermediates [18]. As will be shown, remarkably facile hydrogen site exchange between cis hydride and

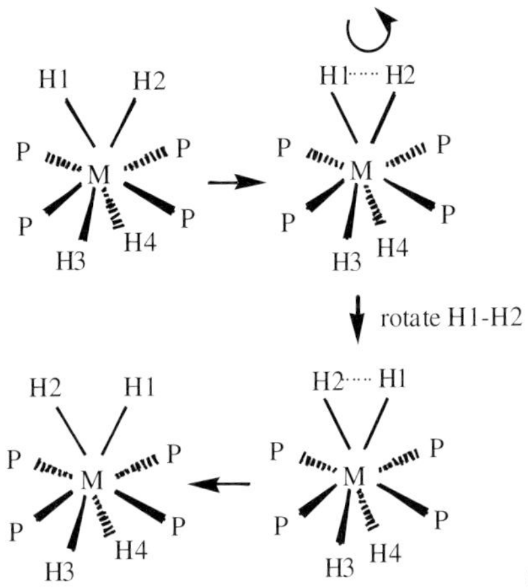

Scheme 20.3

H_2 ligands can occur, even in the *solid state* at temperatures below 77 K, with activation barriers as low as 1.5 kcal mol^{-1}!

For the H_2 ligand, the structure and dynamics are much more extensive and richer than for hydride ligands. These can include rotational/vibrational motion of η^2-H_2, binding and splitting of H_2 (including equilibria between η^2-H_2/dihydride tautomers), transfer of hydrogen to substrates, heterolytic cleavage of H_2, and σ bond metathesis processes (Scheme 20.4). Several of these processes can occur simultaneously on a metal center and all will be discussed in more detail below. The dihydrogen ligand by itself is remarkably dynamic. Except for rare cases, η^2-H_2 rotates rapidly as in a propeller (librational motion is more accurate), even in the solid state. One of the key diagnostics for coordination of *molecular* H_2 is in fact the observation by inelastic neutron scattering (INS) of rotational transitions for η^2-H_2 [19–22], which cannot exist for classical *atomic* hydrides. Most importantly, these extensive studies by Eckert and coworkers measure the barrier to rotation of H_2 ligands and, as will be shown, consequently offer direct experimental proof of M→H_2 backdonation. The rotational transitions and barrier are very sensitive to even minor changes in ligand environment about M and provide valuable insight into the reaction coordinate for splitting of H_2 on M and intramolecular interactions. Hydrogen reorientation among either identical or inequivalent sites is extremely complex because it can involve *quantum-mechanical* phenomena such as tunneling (in H_2 rotation) and *exchange coupling*. As will be shown, the latter is an NMR effect in certain polyhydrides that undergo exchange of chemically-inequivalent hydrogens.

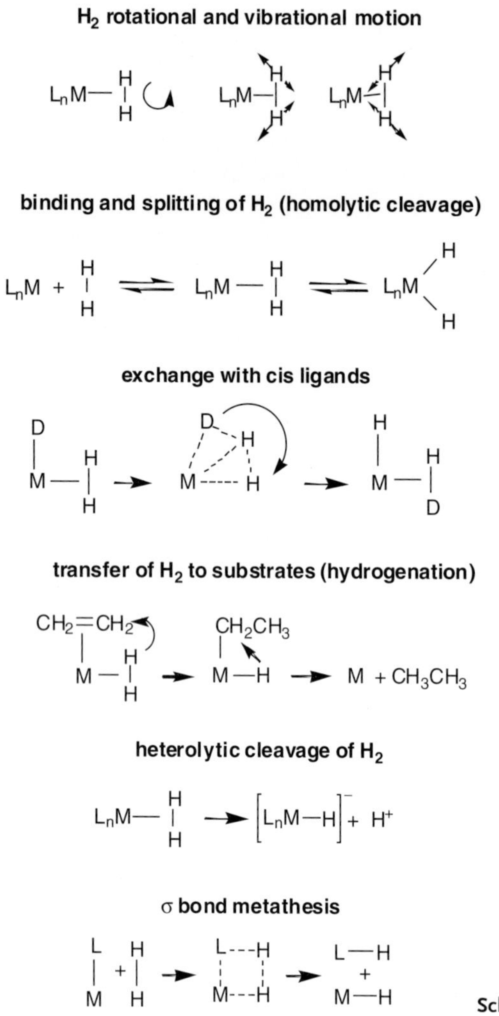

H$_2$ rotational and vibrational motion

$$L_nM—\overset{H}{\underset{H}{|}} \qquad L_nM—\overset{H}{\underset{H}{|}} \qquad L_nM\overset{H}{\underset{H}{\diagup}}$$

binding and splitting of H$_2$ (homolytic cleavage)

$$L_nM + \overset{H}{\underset{H}{|}} \;\rightleftharpoons\; L_nM—\overset{H}{\underset{H}{|}} \;\rightleftharpoons\; L_nM\overset{H}{\diagdown}_{H}$$

exchange with cis ligands

$$\overset{D}{\underset{M—\overset{H}{\underset{H}{|}}}{|}} \;\rightarrow\; M\text{-----}H \;\rightarrow\; \overset{H}{\underset{M—\overset{|}{D}}{|}}$$

transfer of H$_2$ to substrates (hydrogenation)

$$\overset{CH_2=CH_2}{\underset{M—\overset{H}{\underset{H}{|}}}{|}} \;\rightarrow\; \overset{CH_2CH_3}{\underset{M—H}{|}} \;\rightarrow\; M + CH_3CH_3$$

heterolytic cleavage of H$_2$

$$L_nM—\overset{H}{\underset{H}{|}} \;\rightarrow\; \left[L_nM—H\right]^- + H^+$$

σ bond metathesis

$$\overset{L}{\underset{M}{|}} + \overset{H}{\underset{H}{|}} \;\rightarrow\; \begin{matrix} L\text{---}H \\ M\text{---}H \end{matrix} \;\rightarrow\; \begin{matrix} L—H \\ + \\ M—H \end{matrix}$$

Scheme 20.4

20.1.2
Vibrational Motion of Dihydrogen Complexes

The vibrational motion of M–H$_2$ complexes has been analyzed in depth both spectroscopically and by calculation [2, 23]. The vibrational modes are completely different from those for metal hydrides, which typically have only two fundamental modes: M–H stretches in the range 1700–2300 cm^{-1} and M–H deformations at 700–950 cm^{-1}. When diatomic H$_2$ combines with a M–L fragment to form a η^2-H$_2$, five "new" vibrational modes in addition to ν_{HH} are created which are related to the "lost" translational and rotational degrees of freedom for H$_2$ (Scheme 20.5). ν_{HH} is still present, but shifted to much lower frequency and

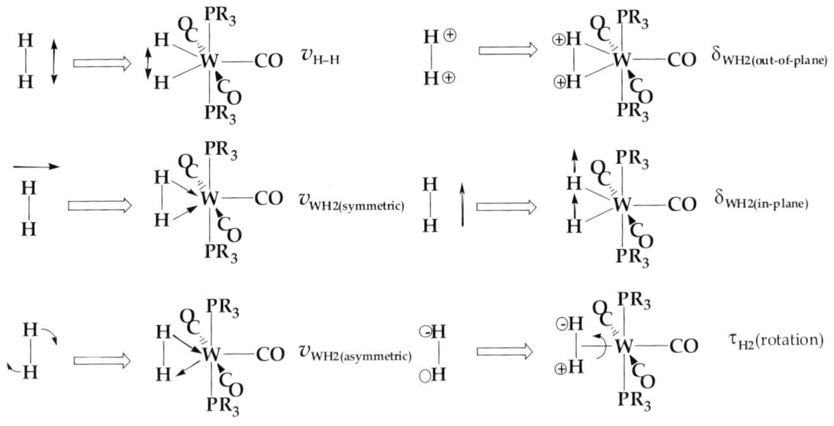

Scheme 20.5

becomes highly coupled with the $\nu(MH_2)$ modes. The bands shift hundreds of wavenumbers on isotopic substitution with D_2 or HD, which greatly facilitates their assignment. The entire set of bands has been identified only in the first H_2 complex, $W(CO)_3(PR_3)_2(H_2)$ (R= Cy, ^iPr) [23a]. All but $\nu_s(MH_2)$, observed in both the IR and Raman, are weak, and most of the bands tend to be obscured by other ligand modes. In the Nujol mull IR spectrum, four bands, $\nu(HH)$ at 2690 cm^{-1}, $\nu_{as}(MH_2)$ at 1575 cm^{-1}, $\nu_s(MH_2)$ at 953 cm^{-1}, and $\delta(MH_2)_{in\text{-}plane}$ at 462 cm^{-1}, can be observed and shift to lower frequency for the D_2 analog. The band at 442 cm^{-1} in the D_2 complex is assigned to $\delta(WD_2)_{out\text{-}of\text{-}plane}$. The modes for H_2 rotation about the M–H_2 axis, $\tau(H_2)$, and also $(MH_2)_{out\text{-}of\text{-}plane}$ near 640 cm^{-1} are observable only by inelastic neutron scattering (INS) methods (see below).

20.1.3
Elongated Dihydrogen Complexes

In elongated H_2 complexes, experimental and computational studies indicate that the η^2-H_2 ligand is greatly delocalized and cannot be envisaged as a fixed, rigid unit [7]. Rapid motion of two hydrogen atoms occurs on a flat potential energy surface with a shallow minimum at the neutron-diffraction determined position of 1.2 Å for $trans$-$[OsCl(H_2)(dppe)_2]^+$.

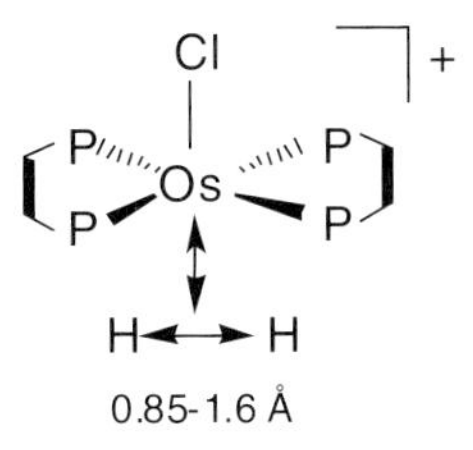

The potential energy surface for the H–H vibrational stretch is so flat for some complexes that the stretch of this bond can traverse the entire distance range from 0.85 Å to 1.60 Å with attendant variation in d_{MH} at an energy cost of merely 1 kcal mol^{-1}! The motion of the hydrogens is approximated in cartoon-like fashion in Scheme 20.6. It is astonishing that a bond as strong as H–H can be weakened so as to be lengthened by 0.8 Å without a significant rise in energy. The ν(HH) and ν(MH$_2$) vibrational stretches in fact lose their meaning and the "normal" stretching modes have to be redefined to one along the arrows shown in the middle drawing at the top of Scheme 20.4 (low-energy mode) and one orthogonal to it (top right, high-energy mode). Importantly, the soft vibrational mode parallels the reaction coordinate for metal-induced H–H bond splitting, an unprecedented situation in chemistry.

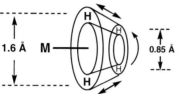

Scheme 20.6

20.1.4
Cleavage of the H–H Bond in Dihydrogen Complexes

Scheme 20.4 shows a set of dynamics involving the binding and splitting of H$_2$ that essentially represent the reaction coordinate for homolytic H–H bond cleavage or in the reverse process, the formation of H$_2$ from a dihydride species. Remarkably these can be *equilibrium processes* in certain cases. Solutions of the complex W(CO)$_3$(P^{i}Pr$_3$)$_2$(H$_2$) were observed by NMR spectroscopy to contain about a 4:1 ratio of dihydrogen to dihydride complex, proving that side-on bonded H$_2$ complexes were the first step in formation of hydride complexes (Eq. (20.1); P = P^{i}Pr$_3$) [1b, 2].

$$(20.1)$$

The structure of the dihydride tautomer could not be determined in the solid state because it is an equilibrium species that could not be isolated from solution (where it has a fluxional structure). However calculations indicated that a pentagonal bipyramid with distal hydrides is the lowest energy structure [24]. Thermodynamic and kinetic investigations of the equilibrium reactions of the tungsten complexes showed the following parameters [25].

$$t_{1/2} =$$

agostic   dihydrogen   dihydrogen

$$\Delta H = -10.1 \text{ kcal mol}^{-1} \qquad \Delta H = 1.2 \text{ kcal mol}^{-1}$$

(20.2)

The precursor for the dihydrogen complex here contains a so-called "agostic" interaction [26], that is an intramolecular interaction of a C–H bond with the metal that stabilizes the electronically (16-electron) and coordinatively unsaturated $W(CO)_3(P^iPr_3)_2$ species. It is noteworthy that the splitting/formation of the H–H bond is much slower here than its binding to and dissociation from the metal.

The above processes are directly relevant to catalysis of either hydrogenation reactions or the production of hydrogen, both of which have been well studied in industrial and biological systems. Metal-catalyzed transfer of hydrogen to substrates, both in heterogeneous and homogeneous (solution) systems, is a well-treated area of immense size and importance in chemistry and will not be discussed in detail here. Several books and reviews have been devoted to just homogeneous catalysis using inorganic and organometallic complexes, where the reaction mechanisms for hydrogen transfer can be best studied [27]. A review on catalytic processes that specifically involve dihydrogen complexes and other sigma complexes has recently been published [4]. Direct transfer from coordinated dihydrogen ligands to organic or other substrates has been shown to take place in certain systems, as illustrated in Scheme 20.4 for olefin hydrogenation. Conclusive evidence for direct transfer of the hydrogens in an H_2 ligand to co-bound substrates is limited, however, partly because it is difficult to prove that cleavage of the H–H bond to give a dihydride complex does not occur first [2, 4].

There are actually two different pathways for cleavage of H–H (and X–H bonds): homolytic cleavage to a dihydride as discussed above and also *heterolytic cleavage*, that is breaking the H–H bond into H^+ and H^- fragments (Scheme 20.4). Both paths have been identified in catalytic hydrogenation and are available for other σ bond activations. Heterolytic cleavage is one of the oldest and most widespread reactions of H_2 on metal centers [28, 29]. H_2 can bind in stable fashion to very electron-deficient cationic metal complexes that favor heterolysis nearly as well as to more electron-rich M. A proton can then split off from the H_2 ligand and migrate to either an external Lewis base B, a cis-ligand L, or an anion (A), as shown in Scheme 20.7 and discussed in detail in a recent review [30]. Especially on electron-poor cationic complexes, the H_2 ligand becomes highly acidic, i.e. polarized towards $H^{\delta-}$–$H^{\delta+}$ where the highly mobile H^+ readily transfers. Free H_2 is an extremely weak acid with a pK_a near 35 in THF, but when H_2 is bound to a

$$\left[\begin{array}{c} L \\ | \\ M{-}| \\ \end{array}\begin{array}{c} \\ H^{\delta+} \\ \\ H^{\delta-} \end{array}\right]^{+}A^{-} \longrightarrow \begin{array}{c} L \\ | \\ M{-}H\cdots HA \end{array} \xrightarrow{-HA} \begin{array}{c} L \\ | \\ M{-}H \end{array}$$

intramolecular

$$\left[\begin{array}{c} L \\ | \\ M{-}\square \end{array}\right]^{+}A^{-} \xrightarrow{H_2} \left[\begin{array}{c} L{:} \\ | \\ M{-}| \\ \end{array}\begin{array}{c} ^{\delta-} \\ \\ H^{\delta+} \\ H^{\delta-} \end{array}\right]^{+}A^{-} \longrightarrow \left[\begin{array}{c} LH \\ | \\ M{-}H \end{array}\right]^{+}A^{-} \xrightarrow{-LH} \left[\begin{array}{c} \square \\ | \\ M{-}H \end{array}\right]^{+}A^{-}$$

intramolecular

$$\left[\begin{array}{c} L \\ | \\ M{-}| \\ \end{array}\begin{array}{c} \\ H^{\delta+} \\ \\ H^{\delta-} \end{array}\right]^{+}A^{-} \xrightarrow{:B} \left[\begin{array}{c} L \\ | \\ M{-}| \\ \end{array}\begin{array}{c} \\ H\cdots B \\ \\ H \end{array}\right]^{+}A^{-} \xrightarrow{-[HB^+][A]^-} \begin{array}{c} L \\ | \\ M{-}H \end{array}$$

intermolecular

$\square$ = COORDINATIVELY UNSATURATED SITE (e.g. 16e) OR WEAK SOLVENT LIGAND

Scheme 20.7

highly electrophilic cationic metal center, the acidity of H_2 gas can be increased spectacularly, by up to 40 orders of magnitude. The pK_a of H_2 can become as low as −6 and thus the acidity of η^2-H_2 becomes as strong as that of sulfuric or triflic acid [31].

Intramolecular heterolysis involves proton transfer to a cis ligand L (e. g. H or Cl) or to the counteranion of a cationic complex. This can occur via the intermediacy of a so-called cis-interaction, which essentially is a hydrogen-bonding like interaction of H_2 with a cis ligand, such as a hydride, that has a partial negative charge (δ–) [2, 5a].

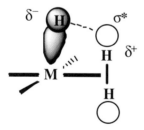

cis-interaction

This is essentially the reverse of the protonation reaction commonly used to synthesize H_2 complexes (all the reactions in Scheme 20.7 can be reversible). The intermediate here, M–H$\cdots$H–A, is held together by *dihydrogen-bonding*, also known as proton-hydride bonding, where the M–H hydrogen is hydridic and the HA hydrogen is protonic [5e, 32]. The unconventional hydrogen bonds here can be comparable in strength to classical X–H···(lone pair) hydrogen bonds ($3-7$ kcal mol^{-1}). Proton-hydride exchange has recently been observed in the reaction of CpRuH(PP) (PP =diphosphine) with CF_3CO_2H that led to a dihydrogen bonded complex with an extremely short RuH···HO_2CF_3 interaction [32e]. The proton-transfer process could occur via an intermediate state between a dihydrogen bond and a coordinated dihydrogen molecule.

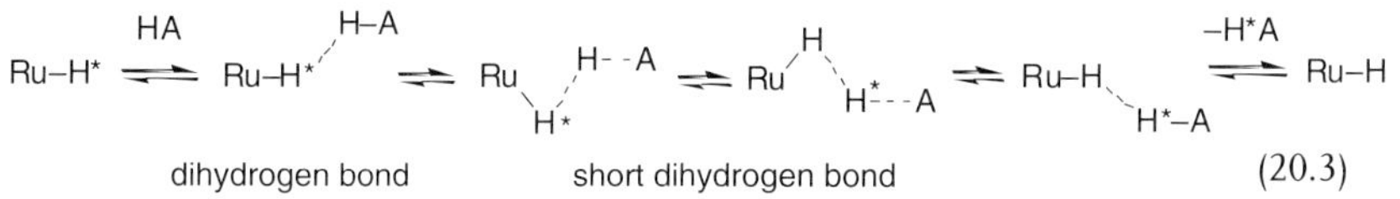

$$Ru-H^* \rightleftharpoons Ru-H^{*\overset{HA}{}}\rightleftharpoons Ru\overset{H-A}{\underset{H^*}{}} \rightleftharpoons Ru\overset{H}{\underset{H^*--A}{}} \rightleftharpoons Ru-H \rightleftharpoons Ru-H$$

dihydrogen bond   short dihydrogen bond   (20.3)

In this and a related Fe system [32h], the counteranion $CF_3CO_2^-$ (A) could change between the two hydrogen atoms of the dihydrogen bond, thus changing the nature of both in the manner represented here, which explains the observed proton-hydride exchange. In this way, this extremely short dihydrogen bond behaves as a dihydrogen molecule polarized by the counteranion. As will be shown below, interactions similar to dihydrogen bonds can also occur between H and H_2 ligands (cis interaction between H_b and H_c shown below) and other σ ligands (H···Si–H) and play an important role in hydrogen transfer/exchange processes.

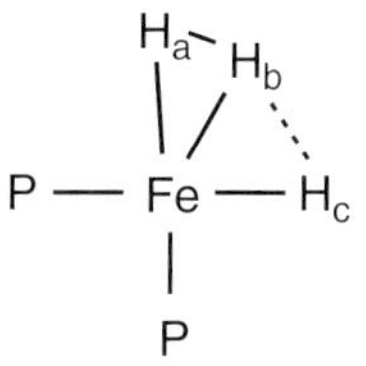

Reversible proton transfer to pendant basic sites can also occur (Eq. (20.4)) [32f] and is facilitated by dihydrogen bonding interactions, e.g. Eq. (20.4B) where the OH and IrH hydrogens scramble via rotation of the H_2 ligand.

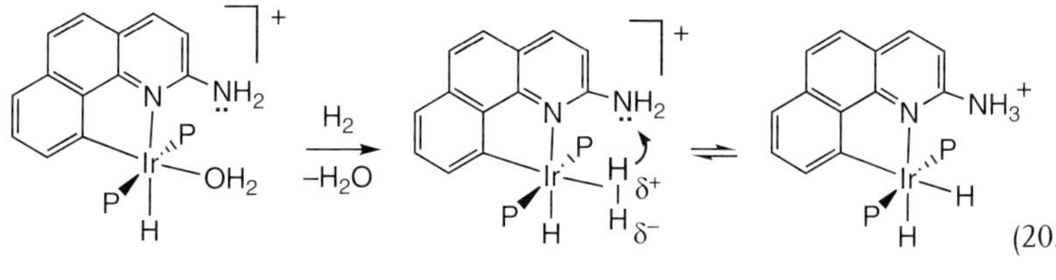

(20.4A)

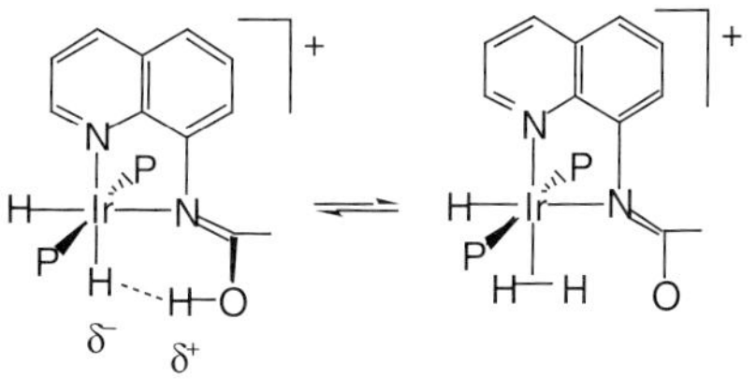

(20.4B)

H_2 heterolysis related to Eq. (20.4) also occurs via intramolecular proton transfer between nitrogens on Ru complexes containing phosphinopyridine ligands, $Cp^*RuH[PPh_2(pyridine)]_2$ [32i].

Intramolecular heterolysis of H_2 is closely related to the well-known σ-bond metathesis processes (Scheme 20.4) that generally occur on less electrophilic centers, especially d^0 systems [2, 5a]. Although the heterolytic process here is formally a concerted "ionic" splitting of H_2, as often illustrated by a four-center intermediate with partial charges, the mechanism does not have to involve such charge localization. In other words the two electrons originally present in the H–H bond do not necessarily both go into the newly-formed M–H bond while a bare proton transfers onto L or, at the opposite extreme, an external base. The term σ-bond metathesis is thus actually a better description and may comprise more transition states than the simple four-center intermediate shown in Scheme 20.4, for example initial transient coordination of H_2 to the metal cis to L and dissociation of transiently bound H—L as the final step.

Intermolecular heterolysis involves protonation of an external base B to give a metal hydride (H⁻ fragment) and the conjugate acid of the base, HB⁺. The bases can be as weak as organic ethers, and the [HB]⁺ formed can relay the proton to internal or external sites. Such intermolecular heterolysis can be the key step in catalytic processes on highly electrophilic metal centers [4, 9, 30], where the reaction mechanism must, almost by definition, be considered to involve direct transfer of hydrogen (as H⁺) from coordinated dihydrogen ligands. Finally, heterolysis of H_2 can occur via several mechanisms that may or may not involve H_2 complexes as stable intermediates.

Summarizing, the dynamics and transfer of hydrogen in the coordination sphere of metals is remarkably rich and complex, as illustrated above. The exact positions of the hydrogens, including critical parameters such as the H–H distance, cannot be determined accurately in many H_2 complexes because of the usual difficulty of observing hydrogen in the coordination sphere of metals by X-ray crystallography. Furthermore, the rapid rotation of the H_2 ligand can significantly foreshorten the observed d_{HH}, even in neutron diffraction crystal structures, requiring application of imprecise correction factors [33]. Because of the diminutive size and extraordinary dynamics of hydrogen, it is often difficult or impossible to distinguish the detailed mechanistic features for transfer of hydrogens, both intramolecularly and intermolecularly. Several complexes still defy attempts to determine whether the dihydrogen is molecularly or atomically bound because rapid intramolecular exchange hinders spectroscopic diagnostics such as NMR methodologies. The following sections will examine in more detail several of the dynamical and hydrogen transfer processes summarized in Scheme 20.4.

20.2
H₂ Rotation in Dihydrogen Complexes

The H_2 ligand undergoes rapid two-dimensional hindered rotation about the $M–H_2$ axis, that is it spins (librates) in propeller-like fashion with little or no wobbling. This phenomenon has been extensively studied by neutron scattering methods and computationally [19–22]. Significantly, there is always at least a small barrier to rotation, ΔE, brought about by $M{\rightarrow}H_2$ σ^* backdonation (BD) (Scheme 20.8). The σ-donation from H_2 to M cannot give rise to a rotational barrier since it is completely isotropic about the $M–H_2$ bond. The barrier actually

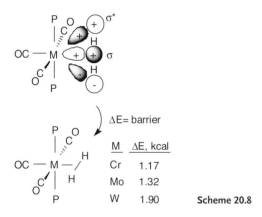

M	ΔE, kcal
Cr	1.17
Mo	1.32
W	1.90

$\Delta E =$ barrier

Scheme 20.8

arises from the disparity in the BD energies from the d orbitals when H_2 is aligned parallel to P–M–P versus parallel to OC–M–CO, where BD is less (though not zero). ΔE varies with M and other factors and can be analyzed in terms of the BD and other forces that lead to it, both by calculation and by a series of experiments where metal–ligand (M/L) sets are varied. In most "true" H_2 complexes with $d_{HH} < 0.9$ Å, the barrier is only a few kcal mol⁻¹ and observable only by neutron scattering methods. It can be as low as 0.5 kcal mol⁻¹ for symmetrical ligand sets, for example all cis L are the same, but has never been measured to be zero because minor geometrical distortions or crystal lattice-related effects are usually present. In the case of complexes with elongated H–H bonds or where rotation is sterically blocked as in $[Cp'_2M(H_2)(L)]^+$ (M = Nb, Ta), much higher barriers of 3–12 kcal mol⁻¹ are observed by INS (see below) or even solution NMR methods [22e, 34]. Interactions of η^2-H_2 with cis ligands can significantly lower the barriers as will be shown below. The hindered rotation of η^2-H_2 is thus governed by a variety of forces, which can be divided into bonded (electronic) and nonbonded interactions ("steric" effects). The direct electronic interaction between M and H_2 results from overlap of the appropriate molecular orbitals. Nonbonded interactions such as van der Waals forces between the η^2-H_2 atoms and the other atoms on the molecule may vary as η^2-H_2 rotates.

20.2.1
Determination of the Barrier to Rotation of Dihydrogen

The geometry and height of the barrier can be derived by fitting the rotational transitions observed by inelastic neutron scattering (INS) techniques to a model for the barrier. The simplest possible model for the rotations of a dumbbell molecule is one of planar reorientation about an axis perpendicular to the midpoint of the H–H bond in a potential of two-fold symmetry. Application of a barrier to rotation rapidly decreases the separation between the lowest two rotational levels, which may then be viewed as a split librational ground state. Transitions within this ground state as well as those to the excited librational state (often called torsions) may be observed by INS. The former occur by way of rotational tunneling [35] since the wave functions for the H_2 in the two wells 180° apart overlap. This rotational tunneling transition has an approximately exponential dependence on the barrier height, and is therefore extremely sensitive to the latter. It is this property that is exploited to gain information on the origin of this barrier.

Both the rotational tunneling transition and the transitions to the first excited librational state can readily be observed by INS techniques [19–22, 35]. Neutrons are extremely well suited as probes for molecular rotations when the motion involves mainly H atoms. The INS studies allow observation of low-lying transitions within the ground librational state of the η^2-H_2 (tunnel splitting), which corresponds to the para $(I = 0, J = 0)$ to ortho $(I = 1, J = 1)$ transition for free H_2 (120 cm^{-1} in liquid hydrogen). INS measurements are typically carried out at ~5 K using ~1 g of polycrystalline H_2 complex sealed under inert atmosphere in aluminum or quartz sample holders. This measurement can be performed without regard to other hydrogen-containing ligands, which do not have observable excitations at low temperatures in the energy range of those of the H_2. Typical intensities of the tunneling peaks in these complexes are in fact about 0.25% or less than that of the intense central elastic peak (Fig. 20.2). In most cases the ground-state rotational tunnel splitting, as well as the two transitions to the split excited librational state, are observed. Because the tunnel splittings (typically 1–10 cm^{-1}) can be measured with much better accuracy than the librational transitions, the value for the barrier height V_2 is usually extracted from the former. Prior to the discovery of H_2 complexes, the only systems known to contain hydrogen molecules were H_2 gas or H_2 that was barely affected by its surroundings (as in physisorbed H_2). The smallest splittings between the ortho- and para H_2 state that had previously been observed were 4.8–10.5 cm^{-1} for H_2 in K-intercalated graphite [36], and 30.6 cm^{-1} for H_2 in Co ion-exchanged NaA Zeolite [37]. In both of these cases H_2 is in all likelihood physisorbed as no indication of H–H bond activation could be found. However, for the $M(\eta^2$-$H_2)$ ground librational state, splittings between 17 and 0.6 cm^{-1} are observed at temperatures as high as 200 K. The signals shift to lower energy and broaden but remain visible into the quasielastic scattering region (Section 20.4.2). Observation of rotational tunneling, which is a quantum mechanical phenomenon, at such a high temperature is extraordinary: in all previous studies of this type involving CH_3 or $[NH_4]^+$ the transition to classical behavior occurs well below 100 K.

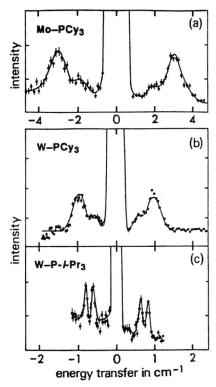

Figure 20.2 Rotational tunneling spectra for the closely related series of complexes $M(H_2)(CO)_3(PR_3)_2$ where M = Mo and R = Cy (top) and M = W and R = Cy (middle) or ^iPr (bottom). Note the change in energy scale between top and middle figures and the high sensitivity of the spectra to changes in metal and minor changes in phosphine ligand. The sharp splitting in the peaks in the lower spectrum is believed to be due to the disordered phosphine in the crystal structure (see Fig. 20.1).

20.3
NMR Studies of H₂ Activation, Dynamics, and Transfer Processes

20.3.1
Solution NMR

The solution and solid-state ^1H NMR spectra of H_2 complexes are normally quite distinctive compared to hydride complexes. In complexes containing both H_2 and hydride ligands, facile exchange between the hydrogens nearly always occurs, especially if these ligands are cis (Scheme 20.4). Scrambling occurs via an $M–H_3$ (trihydrogen) transient (see Section 20.4), where a deuteride (D) ligand differentiates the hydrogens in Scheme 20.4. In most cases an averaged broad singlet results, which may not decoalesce at low temperatures; many complexes are highly fluxional even below $-100\,°C$ (barriers as low as 1 kcal mol^{-1}). The tautomeric equilibria between H_2 and dihydride forms in solution can be observed by NMR in about a dozen complexes, including $W(CO)_3(PR_3)_2(H_2)$ **1**. The 7-coordinate dihydride tautomer **2** exists in about 20% concentration in Scheme 20.9 for P = P^{i}Pr$_3$ and is not isolatable but is detectable by IR and NMR, which shows separate signals for the H_2 and hydride ligands at or below room temperature. Spin saturation transfer experiments using both ^{31}P and ^1H NMR confirm that these signals correspond to equilibrium species [38]. The relaxation time T_1 for the broad H_2 signal is much shorter than for the hydride signals, as will be described below. The inequivalency for both the hydride and phosphine signals in Scheme 20.9 is consistent with the calculated pentagonal bipyramid geometry (Eq. (20.1)) [24], which is also observed crystallographically in the dihydride $MoH_2(CO)(Et_2PC_2H_4PEt_2)_2$ [39].

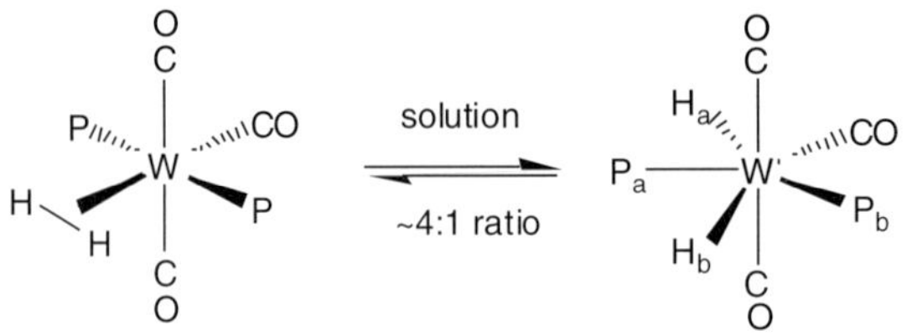

$$J_{HD} = 34 \text{ Hz}$$

$$
\left.
\begin{array}{l}
\delta = -4.3 \ (H_2) \\
\delta = 33.5 \ (P) \\
T_1 = 4 \text{ ms} \ (H_2)
\end{array}
\right\}
\begin{array}{c}
\text{NMR} \\
-92\,°C \\
\text{toluene-}d_8
\end{array}
\left\{
\begin{array}{l}
\delta = -2.15 \ (H_a), \ -4.5 \ (H_b) \\
\delta = \ \ 30.8 \ (P_a), \ 39.5 \ (P_b) \\
T_1 = 1.67 \text{ s} \ (H_a \text{ and } H_b)
\end{array}
\right.
$$

$$
\left.
\begin{array}{l}
\nu_{HH} = 2695 \text{ cm}^{-1} \\
\nu_{CO} = 1969, 1856 \text{ cm}^{-1}
\end{array}
\right\}
\begin{array}{c}
\text{IR} \\
\text{hexane}
\end{array}
\left\{
\begin{array}{l}
\nu_{CO} = 1993, 1913, \\
\ \ \ \ \ \ \ \ 1867, 1828 \text{ cm}^{-1}
\end{array}
\right.
$$

Scheme 20.9

The tautomerization process in Scheme 20.9 has been studied theoretically in regard to the equilibrium isotope effect (EIE), which was studied by calculation and predicts that deuterium should bind more favorably than hydrogen to the nonclassical site at 300 K [24b]. The experimental EIE values for H_2 versus D_2 addition to metal complexes, K_H/K_D, are usually inverse over a large temperature range (260–360 K), showing that counter-intuitively D_2 binds more strongly to metals than H_2 [23].

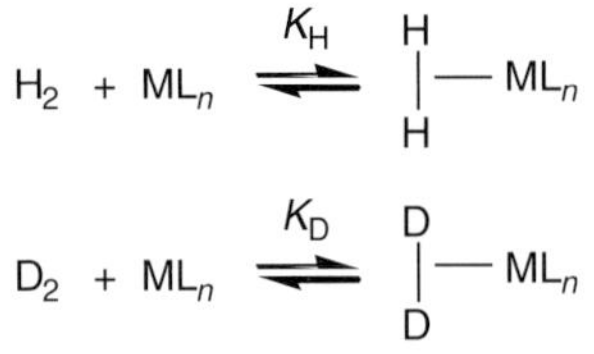

$$
H_2 + ML_n \ \underset{}{\overset{K_H}{\rightleftharpoons}} \ \begin{array}{c} H \\ | \\ | \\ H \end{array} \!\!-ML_n
$$

(20.5)

$$
D_2 + ML_n \ \underset{}{\overset{K_D}{\rightleftharpoons}} \ \begin{array}{c} D \\ | \\ | \\ D \end{array} \!\!-ML_n
$$

This is due mainly to isotopic factors that relate to the large number of vibrational modes for molecularly-bound H_2 or D_2, which offset the weakness of the D–D bond relative to the H–H bond. The values of K_H/K_D observed thus far are 0.36–0.77 for formation of H_2 complexes and 0.47–0.85 for complete splitting of the H–H bond to form hydrides.

The single most important NMR spectroscopic parameter is the scalar coupling constant, $^1J_{HD}$, for the HD isotopomer of an H_2 complex (J_{HD} without the superscript is normally used here to refer to the one-bond coupling). The signal becomes a 1:1:1 triplet (D has spin 1) with much narrower linewidth and is direct proof of the existence of an H_2 ligand, since classical hydrides do not show significant J_{HD} because no residual H–D bond is present. J_{HD} for HD gas is 43 Hz, the maximum possible value (d_{HD} is 0.74 Å). A lower value represents a proportionately shorter d_{HD}. J_{HD} determined in solution correlates well with d_{HH} in the solid state via the empirical relationships developed by both Morris [40] and Heinekey [41]:

$$d_{HH} = 1.42 - 0.0167J_{HD} \text{ Å} \quad \text{[Morris]} \tag{20.6}$$

$$d_{HH} = 1.44 - 0.0168J_{HD} \text{ Å} \quad \text{[Heinekey]} \tag{20.7}$$

The input data for formulating these expressions, which differ only slightly, include d_{HH} from both diffraction and solid-state NMR measurements. Unusual behavior in the temperature dependence of J_{HD} is found in three complexes with elongated d_{HH}, [Cp*Ru(H$\cdots$D)(dppm)]⁺ and *trans*-[OsX(H$\cdots$D)(dppe)₂]⁺ (X = H, Cl), and gave initial indications of the highly delocalized bonding (Scheme 20.6) in these and related species (Cp* = C_5Me_5; dppm = $Ph_2PCH_2PPh_2$; dppe = $Ph_2PC_2H_4PPh_2$) [7, 42–45]. Subsequent NMR studies by Heinekey of the HD, HT, and DT isotopomers of [Cp*Ru(H₂)(dppm)]⁺ show remarkably high isotope and temperature dependence of the bond distance of the coordinated dihydrogen isotopomer (ranging from 1.034 Å for d_{DT} at 204 K to 1.091 Å for d_{HD} at 286 K) as determined by the various NMR J couplings [44]. This is attributed to the extremely flat potential energy surface which defines the H–H and M–H interactions in this complex, which allows the zero point energy differences among the various isotopomers to be directly reflected in d_{HH}. The striking change of d_{HH} with small changes in temperature is due to the thermal population of vibrational excited states that are only slightly higher in energy than the ground state, an unprecedented situation in a readily isolable molecule. The results provide direct experimental verification of the conclusions of DFT (density functional theory) studies by Lledos, Lluch, and coworkers who predicted, for example, that the bond distance for the T–T complex would be 10% shorter than in the corresponding H–H species [46].

A small class of polyhydrides such as [CpIrH₃L]⁺, CpRuH₃L, and Cp₂NbH₃ with hydrides that are separated by approximately 1.7 Å exhibit quantum-mechanical exchange coupling (QEC) with extraordinarily large J_{HH} that can exceed 1000 Hz in some cases [47]. Quantum-mechanical tunneling of two protons through a vibrational potential surface accounts for QEC; covalent bonding between the hydrogens is not needed. Such a phenomenon is extremely rare and has been seen previously for heavy particles (non-electrons) only in studies of ^{3}He at cryogenic temperatures. The current belief is that QEC and related hydride dynamics are simply a manifestation of quantum-mechanical delocalization of light particles in a soft potential energy surface where very slight changes in structure can lead to dramatic changes in dynamic behavior (see Scheme 20.6 and related discussion). An osmium complex best described as a "compressed dihydride" (d_{HH} = 1.46 Å) is a good example here [47e].

Another standard, though more problematic, method for determining d_{HH} in solutions of H_2 complexes involves measuring the minimum value of the relaxation time, T_1^{min}, for the H nuclei of the metal-bound H_2. This provides a reasonably accurate d_{HH} because dipolar relaxation of one H by its close neighbor, the partner H of the H_2 ligand, is dominant and T_1 is proportional to the sixth power of d_{HH}. The broadness of NMR resonances for H_2 ligands was attributed by Crabtree in 1985 to rapid dipole–dipole relaxation, which gives very short values for T_1 (normally <50 ms compared to ≫100 ms for classical hydrides) [48]. This became a useful criterion for η^2-H_2 versus dihydride complexes, although extensive refinements were needed. The theory for dipole–dipole relaxation shows that T_1 varies with temperature and goes through a minimum. Subsequently Crabtree and Hamilton developed a quantitative method for extracting d_{HH} from the experimentally determined T_1^{min} [49]. Because T_1 depends on the sixth power of d_{HH}, it is extremely sensitive to the presence of hydrogens that are close together as in an H_2 complex, and values as low as 3 ms (at 300 MHz) are observed. Furthermore, the dipolar relaxation *usually* dominates the relaxation (e.g. >95% for FeH$_2$(H$_2$)(PEtPh$_2$)$_3$). However, there can be several interfering factors, particularly if the H_2 is bound to certain metals with a high magnetogyric ratio such as Co, Re, and Mn that can make substantial contributions [50]. Also, in polyhydride complexes with cis-hydrogens, all hydride–hydride interactions need to be considered. Corrections for this can be made in some cases, and measurement of T_1^{min} for both the H_2 and HD isotopomers can be used to cancel out relaxation caused by M. For example, using the T_1^{min} values of 15.2 and 116 ms determined, respectively, for [Mn(CO)(dppe)$_2$(H$_2$)]$^+$ and its HD isotopomer, both d_{HH} (0.91(2) Å, compared to 0.89(1) Å by solid state NMR [51]) and d_{MnH} (1.64(3) Å) can be calculated using this subtraction method [52].

The fast internal rotation of η^2-H_2 that causes crystallographic problems can also effect the calculation of d_{HH} from T_1^{min} if the rotational rate is faster than molecular tumbling. Although the term "spinning" H_2 is often used, the H_2 does not really spin like a propeller but undergoes rapid libration combined with less frequent complete rotation. This reorientational motion can effect dipolar relaxation, which Morris addresses by viewing fast H$_2$ rotation much like methyl group rotation studied by Woessner, where d_{HH} calculated from $T_1{}^{min}$ is corrected by a factor of 0.793 [53]. The formulas for calculation are:

$$d(HH) = 5.81[T_1{}^{min} \text{ (slow)}/v]^{1/6} \qquad (20.8)$$

$$d(HH) = 4.61[T_1{}^{min} \text{ (fast)}/v]^{1/6} \qquad (20.9)$$

v = spectrometer frequency in MHz.

This methodology can only set limits on d_{HH}, a longer one when the H$_2$ rotational frequency is lower than the spectrometer frequency and a shorter one when rotation is faster. Structural and spectroscopic data for a large series of cationic complexes, [MH(H$_2$)(diphosphine)$_2$]$^+$ (M= Fe, Ru), strongly reinforce this supposition. These complexes show large values of J_{HD} of 29.5–32.8 Hz, giving d_{HH} = 0.87–0.94 Å from Eqs. (20.6) and (20.7), which is consistent with only the d_{HH} calculated from the fast-rotation formula (0.86–0.90 Å versus 1.09–1.15 Å for slow rotation). The uncorrected neutron diffraction value of d_{HH} is 0.82(3) Å for [RuH(H$_2$)(dppe)$_2$]$^+$ (d_{HH} corrected for H$_2$ rotation is 0.94 Å). While these complexes seem to require the fast time-scale correction factor, Gusev and Caulton [54] point out that a greater number do not, including W(CO)$_3$(P-i-Pr$_3$)$_2$(H$_2$) which clearly has a rapidly reorientating H$_2$. To rationalize this, the character of H$_2$ reorientation must be examined. Inelastic neutron scattering studies generally show a double minimum sinusoidal potential for rotation (Section 20.2.1), but a smaller 4-fold term is often added as a correction for non-sinusoidal behavior such as wobbling off the plane of rotation. However in some cases, e.g. four identical ligands cis to H$_2$ as in [MH(H$_2$)(diphosphine)$_2$]$^+$ species, a 4-fold component to rotation truly exists (90° rotation) giving very low rotational barriers near 0.5 kcal mol^{-1} and a faster spinning H$_2$.

20.3.2
Solid State NMR of H$_2$ Complexes

As established by Zilm and coworkers, solid state ^{1}H NMR is very effective in accurately determining d_{HH} because the measurement is unaffected by rotational or other motion of the bound H$_2$ [55]. Only a small amount of solid in powder form is required to observe the η^2-H$_2$ signals using broad-line techniques (Fig. 20.3). The basic principles are well established: isolated pairs of nuclei in a rigid solid experience a mutual dipolar interaction directly proportional to the average of the inverse cube of the internuclear distance. For a powder sample of an H$_2$ complex, a normal Pake doublet line shape results (Fig. 20.3C), which can be used to calculate internuclear H–H distances within 1%. The patterns are quite sensitive to anisotropic motion, that is hindered rotation or torsion of the side-bound H$_2$ about the M–H$_2$ axis. This will not affect the H$_2$ dipolar splitting when this axis is parallel to the applied magnetic field. One pair of temperature-independent

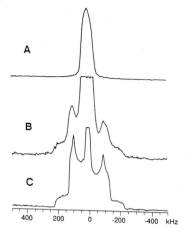

Figure 20.3 A, Solid state ^{1}H NMR spectrum for W(CO)$_3$ (PCy$_3$)$_2$(H$_2$) without suppression of the phosphine ligand signal. B, Same as A except using ligand suppression sequence. C, Same as B except using perdeuterated phosphine ligands.

discontinuities in the Pake pattern is thus unaffected and is used to calculate d_{HH}. The protons in ancillary ligands such as PCy$_3$, where 33 protons are present, can cause problems since they obscure the very wide Pake pattern for the H$_2$. However this can be averted by deuterating the ligands (Fig. 20.3C) or using a hole-burning ligand suppression technique at 77 K to avoid costly deuteration. The latter method relies on the homogeneous line shape of the strongly coupled ligand protons that can be saturated by application of a several millisecond weak pulse. However, the inhomogeneous Pake doublet for η^2-H$_2$ is not saturated, a small hole is burned in the pattern, and the majority of the line shape is unaffected (Fig. 20.3B). The first complex studied, W(CO)$_3$(PCy$_3$)$_2$(H$_2$), showed a d_{HH} of 0.890 ± 0.006 Å [55a], a value that can be considered to be more reliable than the value from neutron diffraction [0.82(1) Å], which must be corrected for the effects of librational motion. As will be discussed below (Section 20.4.1), solid state NMR can also be used to study the dynamics of extremely facile hydrogen exchange processes, such as in IrClH$_2$(H$_2$)(P^{i}Pr$_3$)$_2$ [56].

Limbach and coworkers have carried out extensive solid-state NMR studies of H$_2$ complexes, particularly in regard to hydrogen exchange processes and their quantum mechanical behavior [57]. A unified description of the effects of the coherent and incoherent dihydrogen exchange on the NMR and INS spectra of transition metal hydrides based on the quantum-mechanical density matrix formalism of Alexander-Binsch [58] has been proposed [57a]. The dynamic parameters of the line shape analyses are the exchange couplings or rotational tunnel splittings J of the coherent exchange and the rate constants k of the incoherent exchange. The temperature dependent values J and k were determined for Cp*RuH$_3$(PCy$_3$), including the kinetic HH/HD/DD isotope effects on the incoherent exchange, determined by NMR, and for W(CO)$_3$(PCy$_3$)$_2$(H$_2$), determined by INS. The temperature depen-

dence of J and k was interpreted qualitatively in terms of a simple reaction scheme involving, at each temperature, a ground state and a dominant ro-vibrationally excited state. Using formal kinetics it was shown that a coherent exchange in the excited state contributes to J only if this exchange presents the rate limiting reaction step, i.e., if vibrational deactivation is fast. This is the case for levels located substantially below the top of the barrier. A very fast coherent exchange of levels located close to the top of the barrier contributes only to k. This result reproduces in a simple way the quantum-mechanical results of Szymanski and Scheurer et al. [59]. The results concerning the coherent and incoherent exchange processes in these complexes are discussed in terms of the simplified reaction model.

Solid-state deuterium NMR has also been employed in these types of studies. The ^2H NMR spectra and spin–lattice relaxation rates of $W(CO)_3(PCy_3)_2(D_2)$ have been measured in the temperature regime of 50 K to 300 K [57b]. The spectra have been analyzed employing a model of a combination of homonuclear dipolar D–D interaction and deuterium quadrupolar interaction and a D–D distance of 0.89 ± 0.1 Å. The linewidth of the spectra exhibits a weak temperature dependence at temperatures above 150 K. This temperature dependence is interpreted as a slight decrease in the quadrupolar coupling with increasing temperature, which is an indication of a change in the M–D_2 distance with changing temperatures. The spin–lattice relaxation data of the complex exhibit pronounced deviations from simple Arrhenius behavior at lower temperatures, indicating the presence of a quantum-mechanical tunneling process. This process is analyzed in terms of a simple one-dimensional Bell tunnel model. A comparison with INS data from the H_2 complex reveals a strong isotope effect of 2×10^3 for the exchange rates of the deuterons. Bakhmutov has also utilized ^2H NMR to determine deuterium spin–lattice relaxation times of D_2 ligands and analyzed them in terms of fast internal D_2 motions, including free rotation, librations, and 180° jumps [60]. The results led to a criterion for using the relaxation data to distinguish fast-spinning dihydrogen ligands as discussed in Section 20.3.1.

20.4
Intramolecular Hydrogen Rearrangement and Exchange

Soon after the discovery of H_2 complexes facile intramolecular site exchange of H atoms between H_2 and hydride ligands was found to occur [61]. The ^1H NMR signals of the cis H_2 and hydride in $[Ir(H_2)H(bq)(PPh_3)_2]^+$ coalesce at 240 K because of exchange, and even the hydride *trans* to H_2 in $[Fe(H_2)H(dppe)_2]^+$ exchanges positions with the H atoms of η^2-H_2.

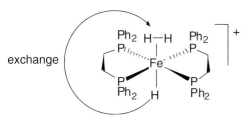

Many new examples include more sophisticated systems with four or more H-donor ligands that display very complex dynamic processes. The complexes encompass early to late metal species such as $[Cp*MoH_4(H_2)(L)]^+$ [62], $[ReH_4(CO)(H_2)(L)_3]^+$ [63], $[RuH(H_2)(CO)_2(L)_2]^+$ [64], and $[TpIrH(H_2)(L)]^+$ [65] (Tp = tris(pyrazolyl)borate; L = PR$_3$). *Ab initio* calculations show that a variety of mechanisms are possible for the site exchange [62, 66]. For example a reductive elimination/oxidative addition pathway through a bis(H_2) intermediate is proposed for the Re complex, but for *cis*-$[FeH(H_2)(L)_4]^+$ and the Cp*Mo species, the preferred pathway is via an M–(H_3) (trihydrogen) transition state (Scheme 20.10). The possible existence of η^3-H_3 ligands has been examined by Burdett in his detailed theoretical studies of polyhydrogen species, H_n ($n = 3$–13) [67]. "Open" (linear H_3^-) or "closed" (triangulo H_3^+) structures are possible. A trihydrogen complex has yet to be isolated, although there is experimental evidence for its intermediacy in facile tautomerization and exchange reactions (see Eq. (20.12) below and Section 20.4.1) [63a]. Polyhydrogen species are known mass spectrometer molecules, and H_3^+ has a closed (triangulo) structure with a calculated d_{HH} of 0.87 Å [68] and can be viewed as an H_2 complex of H^+. However, a trihydrogen ligand is more likely to have an open linear structure best represented as H_3^-, as supported by calculations. The essential features of fluxionality among hydride and H_2 ligands, including mechanistic aspects, are well reviewed by Gusev and others [8, 9, 17]. The intramolecular dynamics will thus only be summarized here and will mainly focus on η^2-H_2 containing systems.

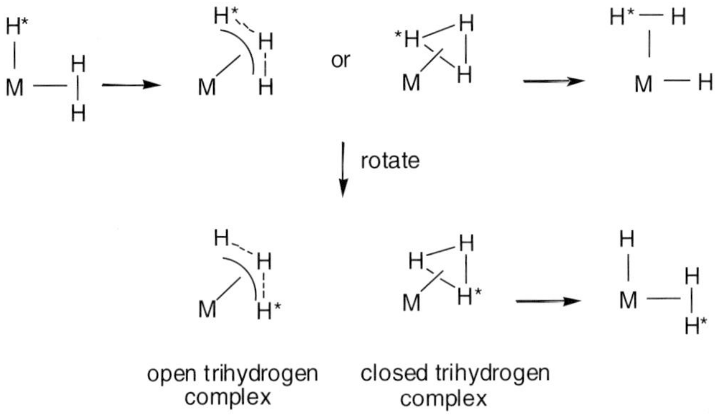

open trihydrogen complex   closed trihydrogen complex

Scheme 20.10

For the simple H_2/hydride situation, two general types of exchange mechanisms can be envisaged. The first is *dissociative* and involves homolysis of the H–H bond to produce a fluxional trihydride intermediate that facilitates intramolecular exchange of H atoms between either adjacent or distal H_2 and H ligands.

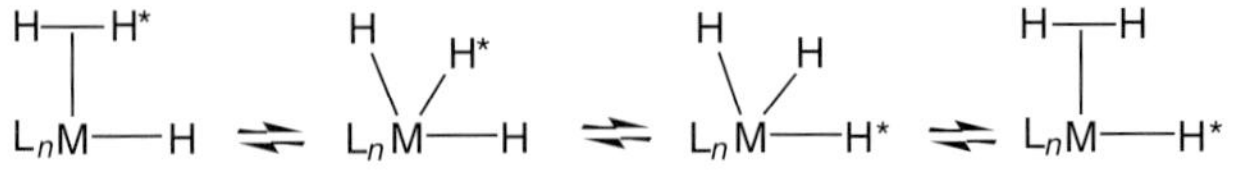

(20.10)

$$
\underset{H}{\overset{H\!\!-\!\!H^*}{L_nM}} \;\rightleftharpoons\; \underset{H}{\overset{H}{L_nM\!-\!H^*}} \;\rightleftharpoons\; \underset{H^*}{\overset{H}{L_nM\!-\!H}} \;\rightleftharpoons\; \underset{H^*}{\overset{H\!\!-\!\!H}{L_nM}} \tag{20.11}
$$

Equations (20.10) and (20.11) are controlled by the same factors that affect the kinetics of homolytic splitting of H_2. The second mechanism is *associative* and implies a trihydrogen intermediate or transition state such as shown in Scheme 20.4. The complex $[Re(H_2)(H)_2(PMe_2Ph)_3(CO)]^+$ provides the first and only good experimental evidence for an associative exchange mechanism that involves such a rotating H_3 intermediate [63]. The latter could occur in other complexes such as $Fe(H)_2(H_2)(PEtPh_2)_3$, which contains a cis-interaction between H_2 and a hydride that can be considered a nascent H_3 ligand [69]. It is not possible to freeze out the J_{HD} for the $M(H)_2(HD)$ isotopomer in such systems containing H_2 plus two or more hydrides, even at the lowest attainable temperature for solution NMR. However even without NMR data, indirect evidence for unstable, fluxional H_2 and/or H_3 intermediates can be obtained merely from isotope exchange reactions using D_2 gas. Before $M\!-\!H_2$ complexes were discovered, Brintzinger proposed that a transient d^0 complex, $Cp^*_2ZrH_2(D_2)$, mediated H/D exchange in $Cp^*_2ZrH_2$ via an associative transition state species $[Cp^*_2Zr(H)(DDH)]^{\ddagger}$ as in Eq. (20.12) [70].

$$
\underset{H}{\overset{\cdots H}{Cp^*\,_2Zr}} \;\underset{-D_2}{\overset{D_2}{\rightleftharpoons}}\; \left[\underset{H}{\overset{H}{Cp^*\,_2Zr\!-\!\underset{D}{\overset{D}{|}}}} \right] \;\rightleftharpoons\; \tag{20.12}
$$

$$
\left[\underset{H}{\overset{H}{Cp^*\,_2Zr\!-\!\underset{D}{\overset{D}{\diagdown}}}} \right]^{\ddagger} \;\underset{HD}{\overset{-HD}{\rightleftharpoons}}\; \underset{D}{\overset{\cdots H}{Cp^*\,_2Zr}}
$$

Here the Zr center could not give a dissociative pathway because Zr^{VI} is an unattainable oxidation state. This was one of the first examples of σ-bond metathesis and postulation of a transient $M\!-\!H_3$ species. In the six-hydrogen system, $[CpM(H)_4(H_2)(PR_3)]^+$ (M = Mo, W), stretching the H_2 toward an adjacent hydride is a low energy process that also leads to a transition state with H_3^- character [62]. The calculated barriers for exchanges are thus only ~4 kcal mol^{-1} (for R = H), in agreement with the inability to decoalesce the hydride NMR signals, even at 133 K. Using ^{13}C solution NMR, Heinekey measured the rate of cis H/H_2 exchange in $[RuH(H_2)(^{13}CO)_2(PCy_3)_2]^+$ to be ~10^3 s^{-1} at 130 K, with $\Delta G^{\ddagger}_{120} = 5.5$ kcal mol^{-1} [64]. Remarkably, decoalescence of the averaged ^{13}CO signal does not occur until 130 K and rapid cleavage of the H–H bond occurs even for this relatively unactivated H_2 complex ($d_{HH} = 0.9$ Å). This is consistent with a highly concerted exchange process with a Ru-trihydrogen-like transition state.

Gusev and Berke highlight two types of principal motion that are often distinguishable in the dynamic behavior of metal hydrides [17]. These are a migratory (M) type and a replacement (R) type (Scheme 20.11). In the M-type exchange one or more ligands migrate from their original inequivalent positions to give inversion of the entire structure. Subsequent replacement of the migrated ligands by each other in their former coordination sites does not occur. The R-type exchange involves a physical rearrangement of identical atoms or ligands that exchange their exact positional coordinates. Hydride H^a replaces H^b at the same time H^b takes the former place of H^a. The simplest example of R exchange is rotation of η^2-H_2 as discussed above. The M and R mechanisms can be distinguished by NMR if they are not simultaneous events on the same time scale.

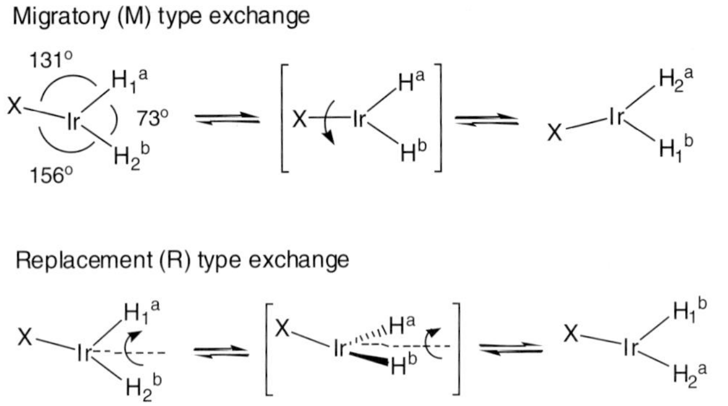

Migratory (M) type exchange

Replacement (R) type exchange

$IrH_2X(P^tBu_2Ph)_2$ (X = Cl, Br, I)

Scheme 20.11

The R mechanism is important because polyhydrides can readily interchange cis-hydrides by transient formation of H_2-like ligands where d_{HH} is shortened to 1.3–1.4 Å (see Scheme 20.3). For example the 16 electron trihydrides OsH_3X $(P^iPr_3)_2$ (X = halide) have a structure with H^a and H^c exerting a strong mutual trans influence resulting in bending toward H^b (Eq. (20.13)).

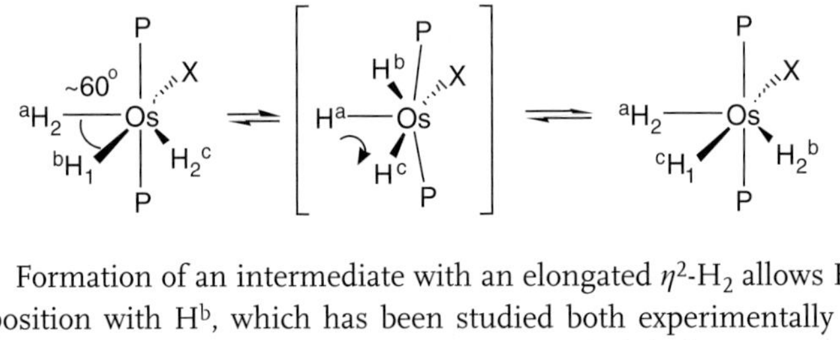

(20.13)

Formation of an intermediate with an elongated η^2-H_2 allows H^c to interchange position with H^b, which has been studied both experimentally and theoretically [71, 72]. The rate of site exchange increases slightly from X = I to X = Cl and the highest barrier, 8.8 kcal mol^{-1} ($\Delta G^{\ddagger}$ at 205 K), is for the former. These complexes display large exchange couplings (QEC) between the hydrides (AB_2 patterns) with $J(H_1$–$H_2)$ values of 920 (Cl), 550 (Br), and 280 Hz (I) at –100 °C (see Section

20.3.1). The observations indicate that exchange couplings can operate between such hydrogens if they are involved in R-type exchange [72]. A complex with an elongated H_2 ligand, $[Cp^*OsH_2(H \cdots H)L]^+$, exchanges hydrogen apparently via a tetrahydride intermediate [22c].

20.4.1
Extremely Facile Hydrogen Transfer in $IrXH_2(H_2)(PR_3)_2$ and Other Systems

Psuedo-octahedral complexes $MXH_2(H_2)L_2$ with H_2 cis to a hydride are extremely fluxional and show M-type exchange. A well-studied case is $IrClH_2(H_2)(P^iPr_3)_2$ where INS studies showed the lowest barrier to H_2 rotation (0.51(2) kcal mol^{-1}) ever measured for a metal complex [73, 74]. Solid-state ^1H NMR studies on a single crystal provided key initial information on the fluxional behavior [56]. A transition state with C_{2v} symmetry is attained in these systems by stretching the H–H bond followed by concerted migration of metal-bound hydrogens. This transient structure inverts with H_a and H_b forming a new H_2 ligand, all of which happens in the equatorial plane of the molecule (Eq. 20.14).

$$(20.14)$$

$$MH_2(H_2)L_3 \quad (M = Fe, Ru)$$
$$MH_2(H_2)(CO)L_2 \quad (M = Ru, Os)$$
$$IrH_2(H_2)XL_2 \quad (X = Cl, Br, I)$$

The NMR data indicate that the hydrogens remain as distinct pairs that do not cross the X–M–L plane, that is H_c does not exchange with the H_a site. Site exchange between H_a and H_b occurs via facile H_2 rotation. Experimental and theoretical studies on $IrXH_2(H_2)(PR_3)_2$ has provided much insight into the mechanism of, and energy barriers to, exchange and attendant rotational dynamics of this system [74]. DFT calculations on model systems for X= Cl, Br, and I and R = H and Me determine the most favorable pathways and corresponding activation parameters for exchange as well as H_2 rotational barriers that are similar in energy. Several mechanisms for exchange are possible in $IrXH_2(H_2)(PR_3)_2$ (including a bis-H_2 intermediate), but calculations strongly support a pathway through a tetrahydride intermediate (Eq. (20.15).

$$(20.15)$$

Exchange barriers of 1.9, 1.8, and 1.7 kcal mol^{-1} were calculated for X = Cl, Br, and I, respectively, (modeled using L= PMe$_3$), in excellent agreement with the experimental value obtained by quasielastic neutron scattering studies for X = Cl (1.5 kcal mol^{-1}) (Section 20.4.2). This low barrier pathway is consistent with the original solid state ^{1}H NMR results that showed a barrier substantially less than 3 kcal mol^{-1} for the hydrogen exchange in IrClH$_2$(H$_2$)(P^{i}Pr$_3$)$_2$ but a much higher barrier for hydrogens crossing the Cl–Ir–P plane [56]. This is a remarkably low barrier for a solid state process, a process involving considerable rearrangement, yet facile enough to persist down to temperatures below 77 K. It is also significant that this is apparently a direct oxidative-addition/reductive-elimination process, one of several possible mechanisms that must be considered in the fluxional behavior of other polyhydride complexes.

The MH$_4$L$_4$ structural type is exceedingly dynamic, displaying both M and R exchanges. The cationic complexes [ReH$_4$(CO)L$_3$]$^+$ (L = PMe$_3$, PMe$_2$Ph) exist in solution as two isomers **A** and **C** in equilibrium, each of which is highly fluxional (Scheme 20.12) [63]. The low-temperature ^{1}H NMR spectra display one exchange-averaged quartet in the hydride region for the tetrahydride **C** plus two decoalesced ReH$_2$ and Re(H$_2$) resonances for the H$_2$ isomer **A**. ^{31}P NMR shows that the phosphine skeleton is rigid in **C** but very fluxional in **A**. Because these species are thermally unstable above –40 °C, X-ray structural data is not available. A dodecahedral structure is probable for **C** from ^{13}C NMR evidence for the CO ligand. However the structure of **A** and the mechanism of exchange is controversial and two possibilities exist [17]. Crabtree proposed the pentagonal bipyramidal structure with H$_2$ in equatorial position shown in Scheme 20.12. A pseudo-octahedral structure [Re(H$_3$)(H)(PMe$_2$Ph)$_3$(CO)]$^+$ with an H$_3$ ligand was proposed as the intermediate or transition state of the exchange reaction within the H$_2$-containing isomer. An H$_3$ intermediate was also proposed earlier by Bianchini in [RuH(H$_2$)(PP$_3$)]$^+$ containing a tripodal phosphine, although the evidence was less clear [75].

The ΔG‡ of 9.9 kcal mol^{-1} for the rate of H-atom exchange is the lowest measured among nearly 30 complexes (including the Ru species) and is consistent with a facile associative process [8]. Thus the H$_3$ intermediate would be no more than 10 kcal mol^{-1} less stable than the H$_2$/dihydride structure, suggesting that isolation of a trihydrogen complex may be attainable.

Recent studies have been carried out on bis(cyclopentadienyl)Mo type complexes, the first complexes with d^2 electronic configurations to have cis hydride-dihydrogen ligands. In contrast to [Cp$_2$MoH$_3$]$^+$, which is a thermally stable *trihydride* complex, the *ansa*-bridged analogs [Me$_2$X (C$_5$R$_4$)$_2$MoH(H$_2$)]$^+$ (X = C, R = H; X = Si, R = Me) have been independently determined by both Heinekey [76] and Parkin [77] to be thermally labile dihydrogen/hydride complexes.

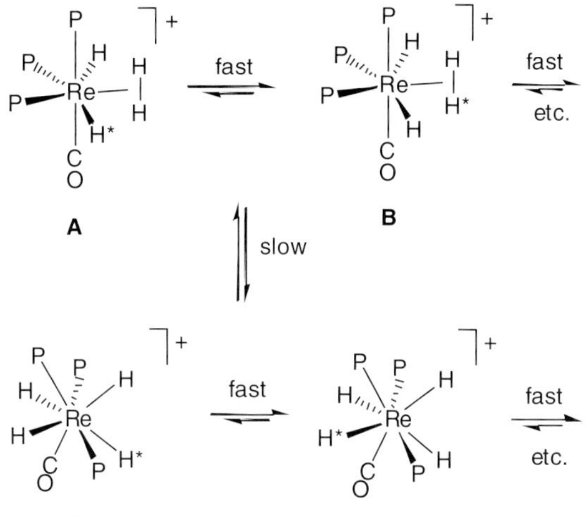

Scheme 20.12

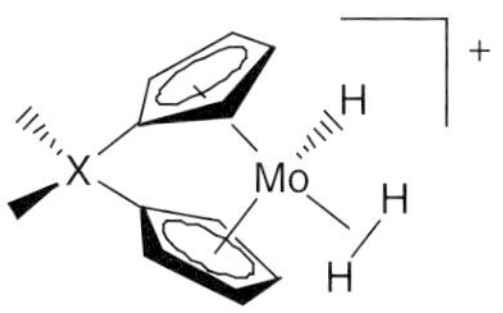

X = C (1), Si (2)

For X = C [76], the presence of the carbon *ansa* bridge decreases the ability of the metallocene fragment to backdonate electrons to the hydrogen antibonding orbitals, thus stabilizing the η^2-H_2 unit. Partial deuteration of the hydride ligands allows observation of J_{H-D} = 11.9 Hz in 1-d_1 and 9.9 Hz in 1-d_2 (245 K), indicative of a dihydrogen/hydride structure. A rapid dynamic process interchanges the hydride and dihydrogen moieties in 1, but it could be determined that the actual (non-averaged) value of J_{H-D} is 30–36 Hz (d_{HH} = 0.84–0.94 Å, compared to a DFT calculated value of 0.98 Å). Low temperature ^{1}H NMR spectra of 1 give a single hydride resonance, which broadens at very low temperature due to rapid dipole–dipole relaxation (T_1 = 23 ms (750 MHz, 175 K). Low temperature ^{1}H NMR spectra of 1-d_2 allow the observation of decoalescence at 180 K into two resonances. The bound H_2 ligand exhibits hindered rotation with $\Delta G^{\ddagger}_{150}$ = 7.4 kcal mol^{-1}, comparable to previously reported observations in d^2 Ta and Nb dihydrogen complexes [34d]. However H-atom exchange is still rapid at temperatures down to 130 K, and Scheme 20.13 depicts the dynamic process envisaged, with the central Mo-trihydrogen structure representing a transition state for atom transfer from one side of the molecule to the other. This process leads to isotopic scrambling of deuterium between hydride and H_2 ligands, where there is a slight preference for deuterium to concentrate in the dihydrogen ligand. This system also apparently has very large exchange coupling (QEC) between the two H atoms of the bound

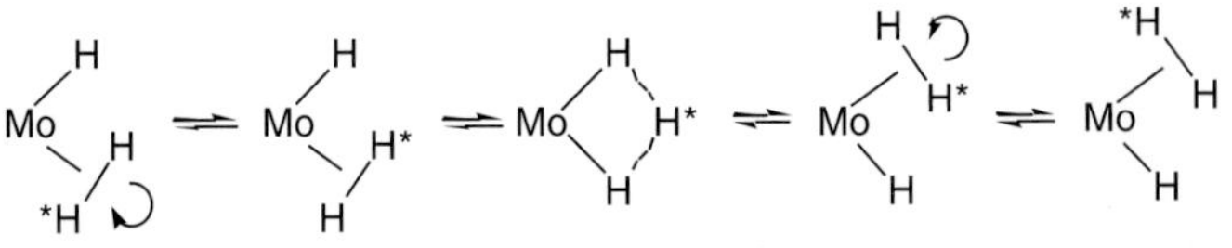

Scheme 20.13

H_2, estimated to be *at least* 24000 Hz, which rationalizes the observation of a single resonance for **1** at all accessible temperatures when no deuterium is incorporated in the hydride ligands. Rapid atom exchange combined with large exchange coupling when two H atoms are adjacent leads to the observed single resonance.

Complex **2** which has an X = Si linker and methyl substituents on the ring carbons [77] offers an excellent comparison to **1**. Low-temperature NMR studies enable direct determination of J_{H-D} for the HD resonances in the Mo(HD)(H) (26.8 Hz) and Mo(HD)(D) (26.4 Hz) isotopomers, from which the H–H distance is estimated to be 0.98 Å by use of the d_{HH}/J_{H-D} correlation [41, 42]. The J_{H-D} values are lower than those in **1**, indicative of a more electron-rich metal center (methyl substituents on cyclopentadienyl carbons favor increased donation of electrons) and longer d_{HH}. Although the proton NMR signal for the three H ligands of **2** is averaged into a singlet resonance at the lowest temperature studied (–95 °C), studies on the MoH_2D and $MoHD_2$ isotopologs provide evidence for hindered rotation of the HD ligand as in **1**. "Side-to-side" motion of the central hydrogen or deuterium atom as in Scheme 20.13, however, remains rapid on the NMR time scale at all temperatures studied. The principal reason why hindered rotation may not be observed for the H_2 ligand is a consequence of a large J_{HH} coupling constant as postulated in **1** and rapid "side-to-side" motion of the central hydrogen, the combination of which causes the highly second-order ABC spectrum to collapse to a singlet. The barrier to rotation of the H_2 ligand is 9.0 kcal mol^{-1} at 25 °C. As in **1**, deuterium exhibits a greater preference than hydrogen to occupy dihydrogen sites in this system. In addition to altering the classical versus non-classical nature of $[Cp^*_2MoH_3]^+$ and $\{[Me_2Si(C_5Me_4)_2]Mo(H_2)(H)\}^+$, the $[Me_2Si]$ *ansa* bridge also influences the stability of the complex with respect to elimination of H_2 and dissociation of H$^+$. Elimination of H_2 from the *ansa* complex is more facile by a factor of ~300 in rate constant and also its acidity is greater than that of $[Cp^*_2MoH_3]^+$, as evidenced by the fact that the *ansa* complex is readily deprotonated by $Cp^*_2MoH_2$ to form $[Cp^*_2MoH_3]^+$. All of the above behavior is rationalized by the *ansa* ligand being overall less electron donating than two unlinked cyclopentadienyl ligands, as proposed for **1**.

There are only a handful of bis-H_2 complexes, which typically additionally have classical hydride ligands and present another example of the very low barriers for exchange of H_2 and hydride ligands situated cis to each other around the equatorial plane of a complex. The complex $[IrH_2(H_2)_2(PCy_3)_2]^+$ is a good example, and separate ^1H NMR resonances for the hydride and H_2 ligands could be observed on cooling of the complex to 188 K [78].

These peaks coalesce at 200 K, and Morris [8] calculates the $\Delta G^{\ddagger}$ at this temperature to be 8.4 kcal mol^{-1}. Chaudret's bis-H_2 complex, $RuH_2(H_2)_2(PCy_3)_2$, is also highly fluxional [79], as is his Tp*RuH(H$_2$)$_2$ complex with the hydride and two η^2-H$_2$ residing on the same side of the complex (Scheme 20.14) [80]. Although crystallographic evidence is unavailable, NMR data is compatible with averaging of the H positions in solution, and cis-interactions between the hydrogen/hydride ligands appear likely here. Calculations indicate that the ground state structure is H(H$_2$)$_2$ rather than a "pentahydrogen ligand," which would have been a marvelous analog of the cyclopentadienyl ligand, which contains delocalized alternating single/double bonds between the 5-membered ring carbons.

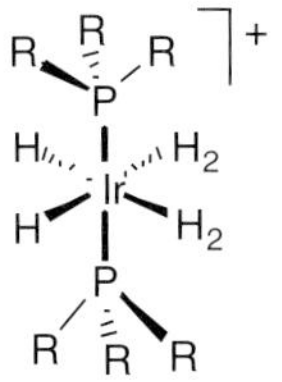

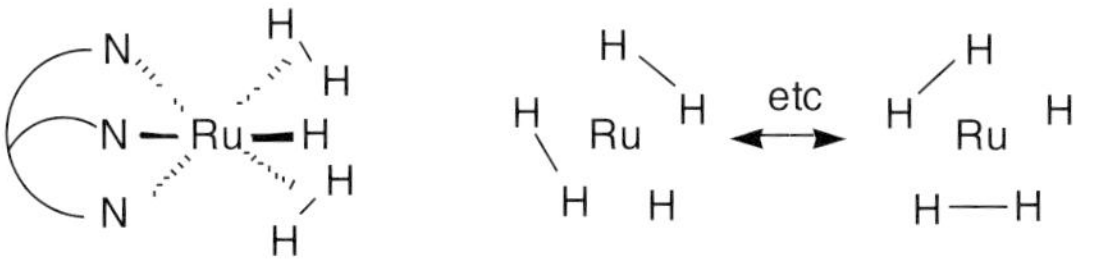

Scheme 20.14

Transfer of H atoms between η^2-H$_2$ and a bridging hydride is seen whenever the two groups are cis to each other (Eq. (20.16)).

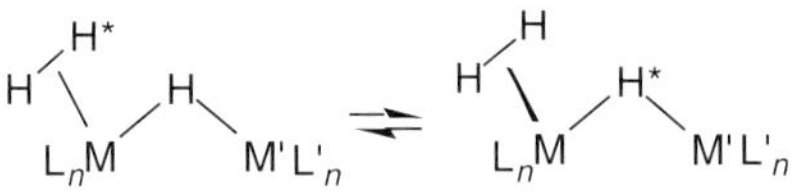

$$\text{(20.16)}$$

Complexes include (L$_2$)(H$_2$)Ru(μ-H)(μ-Cl)$_2$RuH(PPh$_3$)$_2$ (L$_2$ = FeCp(1,2-C$_5$H$_3$-(CHMeNMe$_2$)(P^{i}Pr$_2$)) [81] for which Morris [8] uses rate data to calculate $\Delta G^{\ddagger}$(293 K) = 11.8 kcal mol^{-1}, $\Delta H^{\ddagger}$ = 15.3 kcal mol^{-1} and ΔS = 12 cal mol^{-1} K^{-1}. Although this complex has a cis H$_2$/H interaction that might assist the exchange [82], the $\Delta H^{\ddagger}$ value is still higher than those for mononuclear complexes, apparently because a hydride must shift from its stable bridging position.

Many, if not all, of the above hydrogen transfer/exchange processes can also involve other sigma ligands such as silanes where SISHA (secondary interactions between silicon and hydrogen atoms) [83] reduce the barriers to such processes much as for H/H$_2$ exchanges above. Chaudret has studied σ-ligand substitution mechanisms involving silanes and boranes, primarily for M = Ru [84].

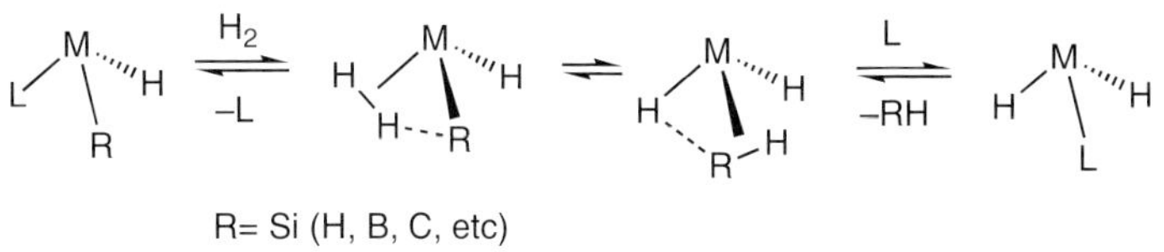

$$\text{(20.17)}$$

R= Si (H, B, C, etc)

The weak SISHA interactions (Si$\cdots$H = 2.0–2.5 Å) play an important role, as their breaking is responsible for the most energetically demanding step. They allow smooth variations leading to the formation of new σ bonds without the necessity of decoordination of a ligand, which can be of considerable importance in catalytic reactions. Related intramolecular hypervalent interactions (IHI) between halosilyl ligands and hydrides are also relevant in this context [85].

20.4.2
Quasielastic Neutron Scattering Studies of H_2 Exchange with cis-Hydrides

Quasielastic neutron scattering (QNS) [86] is valuable for investigating the details of rapid H_2–hydride atom exchange. QNS is actually a form of INS where experiments are carried out at higher temperatures in the regime where quantum-mechanical effects are in transition to classical dynamics (at for example $T >100$ K, but this varies). A complex with hydride(s) cis to H_2 should show increasingly strong interaction between the ligands as a function of T, including possibly exchange. The former affects the rate of rotation of the H_2 ligand as it becomes increasingly "aware" of the neighboring hydride's electrons. This is reflected in non-Arrhenius-like behavior of the quasielastic linewidth, that is broadening of the intense narrow elastic line. QNS data collected at T up to 325 K on [FeH(H_2)(dppe)$_2$]BF$_4$, **3**, FeH$_2$(H_2)(PEtPh$_2$)$_2$, **4**, IrClH$_2$(H_2)(P^{i}Pr$_3$)$_2$, **5**, and RuH$_2$(H_2)$_2$(PCy$_3$)$_2$, **6**, show T dependence of the spectral linewidth that can be fitted to an Arrhenius law which gives an activation energy for the rotation of η^2-H_2 [33, 34, 87]. The latter values for compounds **3**, **4**, and **5** were determined to be 30–50% of the experimentally determined barriers to rotation. This is very surprising, as one would expect that at these T the rotation would be essentially classical, i.e. thermally activated rotational hopping *over* the barrier. It is therefore apparent that even at room T the rotational motion of η^2-H_2 is at least in part quantum-mechanical. Similar effects are known for the translational diffusion of hydrogen in metals, where in many cases experimentally determined activation energies can be substantially lower than potential well depths. Only the data for **3** could be fitted reasonably well to a model for stochastic rotation of a dumbbell molecule in a double-minimum potential, whereas the dynamics of H_2 in **4** and **5** appear to differ substantially from that of **3**. This difference may be attributable to interaction with the cis-hydride in **3** and the very rapid exchange between hydride and η^2-H_2 that is known to occur in **5**. The variable temperature QNS data for **5** represents a breakthrough in showing the first experimental observation by INS of quasielastic scattering attributable to H_2/hydride site exchange and its associated activation energy [74]. As T is increased above 100 K the rotational tunneling transitions for **5** broaden, shift to slightly lower frequencies, and decrease in intensity, while a very broad background appears beneath the peaks. Additionally, the narrow elastic line broadens progressively, indicating that another dynamic process is now fast enough to be observable within the frequency window (i.e. energy resolution) provided by the spectrometer of about 2 cm^{-1} FWHM. Although the intensity of the quasielastic component is quite low, it can be extracted by fitting a Lorentzian convoluted with the measured gaussian resolution function over this part of the spectrum. The extracted Lorentzian linewidths are fitted to an Arrhenius law to effectively provide an activation energy of 1.5(2) kcal mol^{-1} for the exchange. This remarkably low barrier closely matches the DFT calculated activation barrier for site exchange of 1.9 kcal mol^{-1} and is consistent with the mechanistic features in Eq. (20.15) as discussed above.

The bis-H$_2$ complex, RuH$_2$(H$_2$)$_2$(PCy$_3$)$_2$, **6**, is also extremely fluxional, and NMR studies in Freon solvent mixtures at T as low as 143 K still give unresolvable spectra because of rapid exchange of hydrogens and a low H$_2$ rotational barrier [87]. In agreement with this, calculations show that this complex has three isomeric structures within an energy range of only 2 kcal mol^{-1}. The lowest energy structure has all H in the same plane, which is evidence for cis-interaction of H$_2$ and hydride ligands that would promote exchange. Equation (20.18) shows one of many possibilities for exchange that might for example start with the known cis interaction between the η^2-H$_2$ and go through an OA/RE type mechanism as for **5**.

$$(20.18)$$

The INS spectrum of **6** below 50 K consists of the usual pair of bands on either side of the elastic peak corresponding to a low barrier to rotation of 1.1 kcal mol^{-1}, which agrees with calculated barriers [87].

20.5
Summary

The dynamics and transfer of hydrogen within the coordination sphere of metals is astonishingly rich and complex and is much too extensive to fully address here. The metal-catalyzed transfer of hydrogen to organic and inorganic substrates is immensely important in industry, and hydrogenations are the world's largest manmade chemical reactions. This chapter has explored only one facet of this area, that involving dihydrogen ligands, their splitting to hydride, and the exchange/transfer dynamics between these very important moieties that are the initial steps in catalytic hydrogenation. This alone has been a challenge to characterize and has been greatly aided by the extensive synergism between experiment and theory. Because of the diminutive size and extraordinary dynamics of hydrogen, it is often difficult or impossible to distinguish the detailed mechanistic features for transfer of hydrogens, both intramolecularly and intermolecularly. The past 20 years has witnessed surprising revelations such as the perplexing elongated dihydrogen complexes and novel quantum-mechanical phenomena such as rotational tunneling of H_2 and NMR exchange coupling. The hydrogens can be extremely delocalized much as in the superdynamic molecule, CH_5^+ [88]. The future holds further intriguing discoveries as yet more sophisticated techniques are employed to increase our understanding of the marvelous behavior of hydrogen on metals at the molecular level. The surface has only been scratched: virtually everything discussed in this chapter applies also to heteronuclear X–H bonds containing hydrogen such as C–H, Si–H, B–H, and so forth. These all can bind to and transfer hydrogen atoms on metal centers analogously to H–H bonds; the possibilities are nearly endless!

Acknowledgments

We are grateful to the Department of Energy, Office of Basic Energy Sciences, Chemical Sciences Division and Los Alamos National Laboratory for funding.

References

1 (a) Kubas, G. J.; Ryan, R. R.; Swanson, B. I.; Vergamini, P. J.; Wasserman, H. J. *J. Am. Chem. Soc.* **1984**, *106*, 451; (b) Kubas, G. J. *Acc. Chem. Res.* **1988**, *21*, 120.

2 *Metal Dihydrogen and σ-Bond Complexes*, Kubas, G. J., Kluwer Academic/Plenum Publishers: New York, **2001**.

3 Kubas, G.J. *J. Organometal. Chem.* **2001**, *635*, 37.

4 Kubas, G.J. *Catal. Lett.* **2005**, *104*, 79.

5 (a) Maseras, F.; Lledós, A.; Clot, E.; Eisenstein, O., *Chem. Rev.* **2000**, *100*, 601; (b) McGrady, G. S.; Guilera, G. *Chem. Soc. Rev.* **2003**, *32*, 383; (c) Jia, G.; Lin, Z.; Lau, C.P. *Eur. J. Inorg. Chem.* **2003**, 2551; (d) Lin, Z.; Hall, M.B. *Coord. Chem. Rev.* **1994**, *135/136*, 845; (e) Bakhmutov, V. I. *Eur. J. Inorg. Chem.* **2005**, 245.

6 *Recent Advances in Hydride Chemistry*, Peruzzini, M.; Poli, R. (Eds.), Elsevier Science, Amsterdam, **2001**.

7 Heinekey, D.M.; Lledós, A.; Lluch, J.M. *Chem. Soc. Rev.* **2004**, *33*, 175.

8 Jessop, P. G.; Morris, R. H. *Coord. Chem. Rev.* **1992**, *121*, 155.

9 (a) Esteruelas, M. A.; Oro, L. A. *Chem. Rev.* **1998**, *98*, 577; (b) Esteruelas, M. A.; Oro, L. A. *Adv. Organomet. Chem.* **2001**, *47*, 1.

10 Crabtree, R. H. *Angew. Chem. Int. Ed. Engl.* **1993**, *32*, 789.

11 (a) Dewar, M. J. S.; *Bull. Soc. Chim. Fr.* **1951**, *18*, C79; (b) Chatt, J.; Duncanson, L. A. *J. Chem. Soc.* **1953**, 2929.

12 (a) *Transition Metal Hydrides*, E. L. Muetterties, E. L. (Ed.), Marcel Dekker, New York, **1971**; (b) Meakin, P.; Guggenberger, L. J.; Peet, W. G.; Muetterties, E. L.; Jesson, J. P. *J. Am. Chem. Soc.* **1973**, *95*, 1467, and references therein; (c) Jesson, J. P.; Meakin, P. *Acc. Chem. Res.* **1973**, *6*, 269.

13 Hasegawa, T.; Li, Z.; Parkin, S.; Hope, H.; McMullan, R. K.; Koetzle, T. F.; Taube, H. *J. Am. Chem. Soc.* **1994**, *116*, 4352; Malin, J.; Taube, H. *Inorg. Chem.* **1971**, *10*, 2403.

14 Hush, N.S. *J. Am. Chem. Soc.* **1997**, *119*, 1717.

15 Ginsberg, A. P. *Adv. Chem. Ser.*, **1978**, *167*, 201.

16 Jackson, S. A.; Eisenstein, O. *J. Am. Chem. Soc.* **1990**, *112*, 7203.

17 Gusev, D. G.; Berke, H. *Chem. Ber.* **1996**, *129*, 1143.

18 Bergamo, M.; Beringhelli, T.; D'Alfonso, G.; Mercandelli, P.; Sironi, A. *J. Am. Chem. Soc.* **2002**, *124*, 5117.

19 (a) Eckert, J.; Kubas, G. J., Dianoux, A. J. *J. Chem Phys.* **1988**, *88*, 466;

(b) Eckert, J.; Blank, H.; Bautista, M. T.; Morris, R. H. *Inorg. Chem.* **1990**, *29747*; (c) Eckert, J., Kubas, G. J.; Hall, J. H.; Hay, P. J.; Boyle, C. M. *J. Am. Chem. Soc.* **1990**, *112*, 2324.

20 Eckert, J. *Spectrochim. Acta, Part A,* **1992**, *48*, 363.

21 Eckert, J.; Kubas, G. J. *J. Chem Phys.* **1993**, *97*, 2378.

22 (a) Eckert, J. *Trans. Am. Crystallogr. Assoc.* **1997**, *31*, 45; (b) Clot, E.; Eckert, J. *J. Am. Chem. Soc.* **1999**, *121*, 8855; (c) Webster, C. E.; Gross, C. L.; Young, D. M.; Girolami, G. S.; Schultz, A. J.; Hall, M. B.; Eckert, J. *J. Am. Chem. Soc.* **2005**, *127*, 15091.

23 (a) Bender, B. R.; Kubas, G. J.; Jones, L. H.; Swanson, B. I.; Eckert, J.; Capps, K. B.; Hoff, C. D. *J. Am. Chem. Soc.* **1997**, *119*, 9179; (b) Janak, K.E.; Parkin, G. *Organometallics* **2003**, *22*, 4378; (c) Janak, K. E.; Parkin, G. *J. Am. Chem. Soc.* **2003**, *125*, 6889.

24 (a) Tomàs, J.; Lledós, A.; Jean, Y. *Organometallics* **1998**, *17*, 4932; (b) Torres, L.; Moreno, M.; Lluch, J. M. *J. Phys. Chem. A* **2001**, *105*, 4676.

25 Gonzalez, A. A.; Zhang, K.; Nolan, S. P.; de la Vega, R. L.; Mukerjee, S. L.; Hoff, C. D.; Kubas, G. J. *Organometallics* **1988**, *7*, 2429.

26 (a) Wasserman, H. J.; Kubas, G. J.; Ryan, R. R.; *J. Am. Chem. Soc.* **1986**, *108*, 2294; (b) Brookhart, M.; Green, M.L.H.; Wong, L.-L. *Progr. Inorg. Chem.* **1988**, *36*, 1.

27 See for example: (a) *Homogeneous Hydrogenation*, James, B. R., John Wiley and Sons, New York, **1973**; (b) Harmon, R. E.; Gupta, S. K.; Brown, D. J. *Chem. Rev.* **1973**, *73*, 21; (c) *Catalytic Transition Metal Hydrides*, Slocum, D. W.; Moser, W. R. (Eds.), *Ann. N. Y. Acad. Sci.*, **1983**, *415*; (d) *Homogeneous Catalysis: Understanding the Art*, van Leeuwen, P.W.N.M., Kluwer Academic Publishers, Boston, **2004**.

28 Brothers, P.J. *Prog. Inorg. Chem.* **1981**, *28*, 1.

29 Morris, R. H. in *Recent Advances in Hydride Chemistry*, Peruzzini, M; Poli, R. (Eds.), Elsevier Science, Amsterdam, **2001**, pp. 1–38.

30 Kubas, G. J. *Adv. Inorg. Chem.* **2004**, *56*, 127.

31 (a) Morris, R. H. *Can. J. Chem.* **1996**, *74*, 1907; (b) Jia, G.; Lau, C.-P. *Coord. Chem. Rev.* **1999**, *190–192*, 83.

32 (a) Abdur-Rashid, K.; Gusev, D. G.; Landau, S. E.; Lough, A. J.; Morris, R. H. *Organometallics* **2000**, *19*, 1652; (b) Crabtree, R. H.; Siegbahn, P. E. M.; Eisenstein, O.; Rheingold, A. L.; Koetzle, T. F. *Acc. Chem. Res.* **1996**, *29*, 348; (c) Custelcean, R.; Jackson, J. E. *Chem. Rev.* **2001**, *101*, 1963; (d) Epstein, L. M.; Shubina, E. S.*Coord. Chem. Rev.* **2002**, *231*, 165; (e) Cayuela, E.; Jalon, F.A.; Manzano, B.R.; Espino, G.; Weissensteiner, W.; Mereiter, K. *J. Am. Chem. Soc.* **2004**, *126*, 7049; (f) Gruet, K.; Clot, E.; Eisenstein, O.; Lee, D.H.; Patel, B. P.; Macchioni, A.; Crabtree, R. H. *New. J. Chem.* **2003**, *27*, 80; (g) Jimenez-Tenorio, M.; Palacios, M.D.; Puerta, M. C.; Valerga, P. *Organometallics* **2005**, *24*, 3088; (h) Belkova, N. V.; Collange, E.; Dub, P.; Epstein, L. M.; Lemenovskii, D. A.; Lledós, A.; Maresca, O.; Maseras, F.; Poli, R.; Revin, P. O.; Shubina, E. S.; Vorontsov, E. V. *Chem. Eur. J.* **2005**, *11*, 873; (i) Jalon, F. A.; Manzano, B. R.; Caballero, A.; Carrion, M. C.; Santos, L.; Espino, G.; Moreno, M. *J. Am. Chem. Soc.* **2005**, *127*, 15364.

33 Kubas, G. J.; Burns, C. J.; Eckert, J.; Johnson, S.; Larson, A. C.; Vergamini, P. J.; Unkefer, C. J.; Khalsa, G. R. K.; Jackson, S. A.; Eisenstein, O. *J. Am. Chem. Soc.* **1993**, *115*, 569.

34 (a) Antinolo, A.; Carrillo-Hermosilla, F.; Fajardo, M.; Garcia-Yuste, S.; Otero, A.; Camanyes, S.; Maseras, F.; Moreno, M.; Lledos, A.; Lluch, J. M. *J. Am. Chem. Soc.* **1997**, *119*, 6107; (b) Jalon, F. A.; Otero, A.; Manzano, B. R.; Villasenor, E.; Chaudret, B. *J. Am. Chem. Soc.* **1995**, *117*, 10123; (c) Sabo-Etienne, S.; Chaudret, B.; Abou el Makarim, H.; Barthelet, J.-C.; Daudey, J.-C.; Ulrich, S.; Limbach, H.-H.; Moise, C. *J. Am. Chem. Soc.* **1995**, *117*, 11602; (d) Sabo-Etienne, S.; Rodriguez, V.; Donnadieu, B.; Chaudret, B.; el Makarim, H. A.; Barthelat, J.-C.; Ulrich, S.; Limbach,

H.-H.; Moïse, C. *New J. Chem.* **2001**, *25*, 55.

35 Prager, M.; Heidemann, A. *Chem. Rev.* **1997**, *97*, 2933.

36 Beaufils, J. P.; Crowley, T.; Rayment, R. K.; Thomas, R. K.; White, J. W. *Mol. Phys.* **1981**, *44*, 1257.

37 Nicol, J. M; Eckert, J.; Howard, J. *J. Phys. Chem.*, **1988**, *92*, 7117.

38 Khalsa, G. R. K.; Kubas, G. J.; Unkefer, C. J.; Van Der Sluys, L. S.; Kubat-Martin, K. A. *J. Am. Chem. Soc.* **1990**, *112*, 3855.

39 Kubas, G. J.; Ryan, R. R.; Unkefer, C. J. *J. Am. Chem. Soc.* **1987**, *109*, 8113.

40 Maltby, P. A.; Schlaf, M.; Steinbeck, M.; Lough, A. J.; Morris, R.H.; Klooster, W. T.; Koetzle, T. F.; Srivastava, R. C. *J. Am. Chem. Soc.* **1996**, *118*, 5396.

41 Luther, T. A.; Heinekey, D. M. *Inorg. Chem.* **1998**, *37*, 127.

42 Klooster, W. T.; Koetzle, T.F.; Jia, G.; Fong, T.P.; Morris, R.H.; Albinati, A. *J. Am. Chem. Soc.* **1994**, *116*, 7677.

43 Maltby, P. A.; Schlaf, M.; Steinbeck, M.; Lough, A. J.; Morris, R.H.; Klooster, W. T.; Koetzle, T. F.; Srivastava, R. C. *J. Am. Chem. Soc.* **1996**, *118*, 5396.

44 Law, J.K.; Mellows, H.; Heinekey, D. M. *J. Am. Chem. Soc.* **2002**, *124*, 1024.

45 Pons, V.; Heinekey, D. M. *J. Am. Chem. Soc.* **2003**, *125*, 8428.

46 Gelabert, R.; Moreno, M.; Lluch, J. M.; Lledós, A. *J. Am. Chem. Soc.* **1997**, *119*, 9840.

47 (a) Zilm, K. W.; Heinekey, D. M.; Millar, J. M.; Payne, N. G.; Demou, P., *J. Am. Chem. Soc.* **1989**, *111*, 3088; (b) Jones, D. H.; Labinger, J. A.; Weitekamp, D. P. *J. Am. Chem. Soc.* **1989**, *111*, 3087; (c) Sabo-Etienne, S.; Chaudret, B. *Chem. Rev.* **1998**, *98*, 2077; (d) Antinolo, A.; Carillo-Hermosilla, F.; Fajardo, M.; Fernandez-Baeza, J.; Garcia-Yuste, S.; Otero, A. *Coord. Chem. Rev.* **1999**, *193–195*, 43; (e) Schloerer, N.; Pons, V.; Gusev, D. G.; Heinekey, D. M. *Organometallics* **2006**, *25*, 3481.

48 Crabtree, R. H.; Lavin, M. *J. Chem. Soc., Chem. Commun.* **1985**, 1661.

49 Crabtree, R. H. *Acc. Chem. Res.* **1990**, *23*, 95.

50 Desrosiers, P. J.; Cai, L.; Lin, Z.; Richards, R.; Halpern, J. *J. Am. Chem. Soc.* **1991**, *113*, 4173.

51 King, W. A.; Luo, X-L.; Scott, B. L.; Kubas, G. J.; Zilm, K. W. *J. Am. Chem. Soc.* **1996**, *118*, 6782.

52 King, W. A.; Scott, B. L.; Eckert, J.; Kubas, G. J. *Inorg. Chem.* **1999**, *38*, 1069.

53 Morris, R. H.; Wittebort, R. J. *Magn. Reson. Chem.* **1997**, *35*, 243; Woessner, D. E. *J. Chem. Phys.* **1962**, *36*, 1; *37*, 647.

54 Gusev, D. G.; Kuhlman, R. L.; Renkema, K. H.; Eisenstein, O.; Caulton, K. G. *Inorg. Chem.* **1996**, *35*, 6775.

55 (a) Zilm, K. W.; Merrill, R. A.; Kummer, M. W.; Kubas, G. J. *J. Am. Chem. Soc.* **1986**, *108*, 7837; (b) Zilm, K. W.; Millar, J. M. *Adv. Magn. Opt. Reson.* **1990**, *15*, 163.

56 Wisniewski, L. L.; Mediati, M.; Jensen, C. M.; Zilm, K. W. *J. Am. Chem. Soc.* **1993**, *115*, 7533.

57 (a) Limbach, H.-H.; Ulrich, S.; Gruendemann, S.; Buntkowsky, G.; Sabo-Etienne, S.; Chaudret, B.; Kubas, G. J.; Eckert, J. *J. Am. Chem. Soc.* **1998**, *120*, 7929; (b) Wehrmann, F.; Albrecht, J.; Gedat, E.; Kubas, G. J.; Eckert, J.; Limbach, H.-H.; Buntkowsky, G. *J. Phys. Chem. A* **2002**, *106*, 2855.

58 Alexander, S. *J. Chem. Phys.* **1962**, *37*, 971; (b) Binsch, G. *J. Am. Chem. Soc.* **1969**, *91*, 1304; (c) Kleier, D. A.; Binsch, G. *J. Magn. Reson.* **1970**, *3*, 146.

59 (a) Szymanski, S. *J. Chem. Phys.* **1996**, *104*, 8216; (b) Scheurer, C.; Wiedenbruch, R.; Meyer, R.; Ernst, R. R. *J. Chem. Phys.* **1997**, *106*, 1.

60 Bakhmutov, V. I. *Magn. Reson. Chem.* **2004**, *42*, 66.

61 (a) Crabtree, R. H.; Lavin, M. *J. Chem. Soc., Chem. Commun.* **1985**, 794; (b) Morris, R. H.; Sawyer, J. F.; Shiralian, M.; Zubkowski, J. D. *J. Am. Chem. Soc.* **1985**, *107*, 5581.

62 Bayse, C. A.; Hall, M. B.; Pleune, B.; Poli, R. *Organometallics* **1998**, *17*, 4309.

63 (a) Luo, X.-L.; Crabtree, R. H. *J. Am. Chem. Soc.* **1990**, *112*, 6912; (b) Luo, X.-L.; Michos, D.; Crabtree, R. H. *Organometallics* **1992**, *11*, 237; (c) Gusev, D.G., Nietlispach, D.; Eremenko, I. L.; Berke, H. *Inorg. Chem.* **1993**, *32*, 3628.

64 Heinekey, D. M.; Mellows, H.; Pratum, T. *J. Am. Chem. Soc.* **2000**, *122*, 6498.

65 Oldham, W. J., Jr.; Hinkle, A. S.; Heinekey, D. M. *J. Am. Chem. Soc.* **1997**, *119*, 11028.

66 (a) Maseras, F.; Duran, M.; Lledos, A.; Bertran, J. *J. Am. Chem. Soc.* **1992**, *114*, 2922; (b) Lin, Z.; Hall, M. B. *J. Am. Chem. Soc.* **1994**, *116*, 4446.

67 (a) Burdett, J. K.; Phillips, J. R.; Pourian, M. R.; Poliakoff, M.; Turner, J. J.; Upmacis, R. *Inorg. Chem.* **1987**, *26*, 3054; (b) Burdett, J. K.; Pourian, M. R. *Organometallics* **1987**, *6*, 1684; (c) Burdett, J. K.; Pourian, M. R. *Inorg. Chem.* **1988**, *27*, 4445.

68 (a) Pang T. *Chem. Phys. Lett.* **1994**, *228*, 555; (b) Farizon, M.; Farizon-Mazuy, B.; de Castro Faria, N. V.; Chermette, H. *Chem. Phys. Lett.* **1991**, *177*, 451.

69 Van Der Sluys, L. S.; Eckert, J.; Eisenstein, O.; Hall, J. H.; Huffman, J. C.; Jackson, S. A.; Koetzle, T. F.; Kubas, G. J.; Vergamini, P. J.; Caulton, K. G. *J. Am. Chem. Soc.* **1990**, *112*, 4831.

70 Brintzinger, H. H. *J. Organomet. Chem.* **1979**, *171*, 337.

71 Gusev, D. G.; Kuhlman, R. L.; Sini, O.; Eisenstein, O.; Caulton, K. G. *J. Am. Chem. Soc.* **1994**, *116*, 2685.

72 Clot, E.; LeForestier, C.; Eisenstein, O.; Pelissier, M. *J. Am. Chem. Soc.* **1995**, *117*, 1797.

73 (a) Eckert, J.; Jensen, C. M.; Jones, G.; Clot, E.; Eisenstein, O. *J. Am. Chem. Soc.* **1993**, *115*, 11056; (b) Eckert, J.; Jensen, C. M.; Koetzle, T. F.; Le-Husebo, T.; Nicol, J.; Wu, P. *J. Am. Chem. Soc.* **1995**, *117*, 7271.

74 Li, S.; Hall, M. B.; Eckert, J.; Jensen, C. M.; Albinati, A., *J. Am. Chem. Soc.* **2000**, *122*, 2903.

75 Bianchini, C.; Perez, P.J.; Peruzzini, M.; Zanobini, F.; Vacca, A. *Inorg. Chem.* **1991**, *30*, 279.

76 Pons, V.; Conway, S.L.J.; Green, M. L. H.; Green, J.C.; Herbert, B.J.; Heinekey, D.M. *Inorg. Chem.* **2004**, *43*, 3475.

77 Janak, K.E.; Shin, J.H.; Parkin, G. *J. Am. Chem. Soc.* **2004**, *126*, 13054.

78 Lundquist, E. G.; Folting, K.; Streib, W. E.; Huffman, J. C.; Eisenstein, O.; Caulton, K. G. *J. Am. Chem. Soc.* **1990**, *112*, 855.

79 (a) Sabo-Etienne, S.; Chaudret, B. *Coord. Chem. Rev.* **1998**, *178–180*, 381; (b) Grellier, M.; Vendier, L.; Chaudret, B.; Albinati, A.; Rizzato, S.; Mason, S.; Sabo-Etienne, S. *J. Am. Chem. Soc.* **2005**, *127*, 17592.

80 Moreno, B.; Sabo-Etienne, S.; Chaudret, B.; Rodriguez, A.; Jalon, F.; Trofimenko, S. *J. Am. Chem. Soc.* **1995**, *117*, 7441.

81 Hampton, C.; Cullen, W. R.; James, B. R. Charland, J.-P. *J. Am. Chem. Soc.* **1988**, *110*, 6918.

82 Jackson, S. A.; Eisenstein, O. *Inorg. Chem.* **1990**, *29*, 3910.

83 Delpech, F.; Sabo-Etienne, S.; Chaudret, B.; Daran, J.-C. *J. Am. Chem. Soc.* **1997**, *119*, 3167.

84 (a) Atheaux, I.; Delpech, F.; Donnadieu, B.; Sabo-Etienne, S.; Chaudret, B.; Hussein K.; Barthelat, J.-C.; Braun, T.; Duckett, S.B.; Perutz, R.N. *Organometallics* **2002**, *21*, 5347; (b) Lachaize, S.; Essalah, K.; Montiel-Palma, V.; Vendier, L.; Chaudret, B.; Barthelat, J.-C.; Sabo-Etienne, S.; *Organometallics* **2005**, *24*, 2935.

85 (a) Nikonov, G.I. *J. Organometal. Chem.* **2001**, *635*, 24; (b) Nikonov, G.I. *Adv. Organomet. Chem.* **2005**, *53*, 217.

86 Bee, M. *Quasielastic Neutron Scattering* Adam Hilger, Bristol, **1988**.

87 Rodriguez, V.; Sabo-Etienne, S.; Chaudret, B.; Thoburn, J.; Ulrich, S.; Limbach, H.-H. Eckert; J.; Barthelat, J.-C.; Hussein, K.; Marsden, C. J. *Inorg. Chem.* **1998**, *37*, 3475; Borowski, A. F.; Donnadieu, B.; Daran, J.-C.; Sabo-Etienne, S.; Chaudret, B. *Chem. Commun.* **2000**, 543.

88 Thompson, K. C.; Crittenden, D. L.; Jordon, M. J. T. *J. Am. Chem. Soc.* **2005**, *127*, 4954, and references therein.

21

Dihydrogen Transfer and Symmetry: The Role of Symmetry in the Chemistry of Dihydrogen Transfer in the Light of NMR Spectroscopy

Gerd Buntkowsky* and Hans-Heinrich Limbach

21.1
Introduction

The concept of symmetry is one of the basic pillars of modern chemistry and physics. Many fundamental phenomena and laws of nature are related to symmetry or are describable with the help of symmetry arguments. Accordingly, the mathematical description of symmetry, the so-called group theory, is at the heart of both classical mechanics and quantum mechanics. Since the latter field is the basis of modern chemistry, symmetry is also of great importance for chemists. They employ symmetry arguments on a regular basis to help in the understanding of spectra or molecular structures and courses on group theory are a regular part of the chemical education syllabus.

It is less well known that symmetry effects also play an important role in chemical kinetics, in particular when low mass particles like hydrogen or deuterium are involved in the reaction and quantum mechanical tunneling processes are present. Especially when dihydrogen exchange reactions are studied, the exchange of the two hydrogen atoms or more generally hydrons (i.e. ^1H, ^2H or ^3H) is a perfect symmetry operation. This apparently trivial symmetry has far fetching consequences for the reaction dynamics. These consequences stem from two different aspects of symmetry: On the one hand basic quantum mechanics tells us that the wavefunctions of the hydrogen or deuterium atoms have to obey the spin dependent Fermi symmetrization rules: They have to be either symmetric (deuterium, spin 1) or anti-symmetric (hydrogen, spin $1/2$). The result of this symmetrization is the formation of para- and ortho- states, which are the spin isotopomers of dihydrogen and dideuterium [1]. On the other hand group theory tells us that the eigenfunctions of the spatial Hamilton operator are now also eigenfunctions of the operator which exchanges the two hydrogen atoms. This implies that the eigenfunctions are either even or odd functions. As a consequence of this, the whole system behaves quantum mechanically. This is particularly visible at low temperatures when only the rotational ground states of the molecules are occupied.

* Corresponding author

Hydrogen-Transfer Reactions. Edited by J. T. Hynes, J. P. Klinman, H. H. Limbach, and R. L. Schowen
Copyright © 2007 WILEY-VCH Verlag GmbH & Co. KGaA, Weinheim
ISBN: 978-3-527-30777-7

These symmetry related quantum effects are most important in the chemistry of non-classical transition metal hydrides with η^2-bonded dihydrogen ligands [2–24]. Following the pioneering work of Kubas et al., who found the first of these complexes a whole series of transition metal polyhydrides with hydrogen distances varying between 0.8 and 1.7 Å were synthesized [23, 25–29]. Understanding their chemistry has led to a better understanding of catalysis since they may be catalytic precursors or stable models for short lived intermediate steps in catalysis [15, 30–32]. They are of current interest in organometallic chemistry [33–36].

Due to the η^2-binding the two hydrogen atoms forming the dihydrogen part of the system exhibit a much higher mobility than hydrogen atoms in conventional hydride bonds. In particular their exchange is an exact symmetry operation, as discussed above. The mutual exchange of the hydrons is equivalent to a hindered $180°$ rotation around the axis intersecting the M–H_2 angle [37–41]. The rotational barrier is caused mainly by the chemical structure, i.e. the binding between the two hydrogen atoms, the binding of the hydrons to the metal, effects of the ligands and sometimes also by crystal effects from neighboring molecules. For identical hydrogen isotopes, the above discussed quantum mechanical symmetry principles leads to the formation of para-states with anti parallel and ortho-states with parallel nuclear spins.

The various energy levels of the system exhibit a tunnel splitting and the energy eigenfunctions split into two separate manifolds with even and odd symmetry. The height of the barrier determines the energy difference between the lowest even and odd symmetry. This so-called tunnel splitting can be expressed as a tunnel frequency v_t.

The size of the tunnel splitting depends strongly on the hindering potential. It varies from 10^{12} Hz for dihydrogen gas to a few Hz as the depth of the potential is increased. Due to this large range of tunnel frequencies no single spectroscopic technique is able to cover the whole dynamic range. While fast coherent tunneling in the frequency range of GHz to THz were studied by incoherent neutron scattering (INS) [19, 42], relatively slow tunneling processes in the frequency range of Hz to kHz are investigated by ^{1}H liquid state NMR spectroscopy (see for example Refs. [19, 43–49] and many others) or ^{2}H-liquid state NMR [50]. In these ^{1}H liquid state NMR studies the tunnel frequency is usually termed "quantum exchange coupling", due to the fact that the effect of the tunneling on the ^{1}H liquid state NMR spectra is equivalent to the effect of an indirect spin coupling (J-coupling). At higher temperatures incoherent exchange processes are superimposed on the coherent tunneling. They are also visible in the NMR or INS spectra of these hydrides.

An important application of these quantum mechanical symmetry principles is the so-called "Para Hydrogen Induced Polarization (PHIP)" experiment [51, 52]. If hydrogen gas (deuterium gas works similarly) is kept at low temperatures (typically liquid nitrogen or below), it converts after some time into the energetically favorable para-hydrogen, for example by contact with paramagnetic species [1] or adsorption to nuclear spins [53]. These para-hydrogen molecules are in a pure nuclear singlet state, which is associated via the Pauli exclusion principle with the lowest rotational state. They are stable even in liquid solutions [54, 55]. Their high spin polarization can be utilized as an extremely sensitive monitor of the fate of the hydrogen in catalytically induced hydrogenation reactions (see for example Refs. [56–76]).

The rest of the paper is organized as follows: The second chapter gives an introduction into the theoretical description of tunneling phenomena and chemical kinetics. After a short summary about the connection of symmetry and tunneling, the basic properties of coherent and incoherent rotational tunneling and their relation to NMR spectroscopy are discussed and the empiric Bell [77] tunnel model is introduced, which is a powerful semi-empirical tool for the description of chemical kinetics in the transition from the coherent to the incoherent regime. This model has been used for the description of Arrhenius curves of H-transfers as described in more detail in Chapter 6. The next two chapters show applications of these symmetry effects. First the para-hydrogen induced polarization (PHIP) experiments are discussed. There the symmetry induced nuclear spin polarization creates very unconventional NMR lineshape patterns, which are of high diagnostic value for catalytic studies. Then in Section 21.4 symmetry effects on NMR lineshapes and relaxation data of intramolecular hydrogen exchange reactions are discussed and examples from ^{1}H-liquid state and ^{2}H-solid state NMR are presented and compared to INS spectra. The last section gives an outlook on possible future developments in the field.

21.2
Tunneling and Chemical Kinetics

This section gives an introduction to the effects of symmetry and quantum mechanical tunneling on chemical reactions in general and hydrogen transfer in particular.

21.2.1
The Role of Symmetry in Chemical Exchange Reactions

Starting from the early days of quantum mechanics, when the dynamics of the ammonia molecule was analyzed by Hund [78], it is well known that there is a close relationship between the symmetry of a potential and the wavefunction of the system [79]. The eigenfunctions of the system have the symmetry of the irreducible representations of the corresponding symmetry group. This situation gets particularly simple if one considers the motion in a symmetric double well potential, as for example in the case of ammonia, where the three hydrogen atoms can be either at the left or right side of the nitrogen atom.

21.2.1.1 Coherent Tunneling

21.2.1.1.1 Tunneling in a Symmetric Double Minimum Potential

Basic textbook [80] quantum mechanics, already developed by Hund [78], tells us that we can associate two different wavefunctions $|1\rangle$ and $|2\rangle$ with these two states. Both have the same energy E_0. As long as the matrix element $H_{12} = \langle 1|\hat{H}|2\rangle = 0$ between these two states is zero, they are degenerate eigenstates of the system and this is the end of the story. The situation changes dramatically if $H_{12} \neq 0$, since then a level anti-crossing appears and $|1\rangle$ and $|2\rangle$ are no longer the eigenstates of the system. Instead the eigenfunctions are given by the symmetric (**gerade**) and anti-symmetric (**ungerade**) linear combinations of $|1\rangle$ and $|2\rangle$.

$$|g\rangle = \frac{1}{\sqrt{2}}(|1\rangle + |2\rangle) \text{ and } |u\rangle = \frac{1}{\sqrt{2}}(|1\rangle - |2\rangle) \tag{21.1}$$

The corresponding energy levels are symmetrically split

$$E_k = E_0 \pm |H_{12}|. \tag{21.2}$$

The lower energy level belongs to the symmetric and the upper energy level to the anti-symmetric state.

Suppose now that we managed to prepare the system initially in the state $|1\rangle$. Since $|1\rangle$ is no longer an eigenstate of the system there is a periodic motion from $|1\rangle$ to $|2\rangle$ and back, where the probabilities of finding the system in state $|1\rangle$ and $|2\rangle$ change according to

$$p(1) = \cos^2(|H_{12}|t) \text{ and } p(2) = \sin^2(|H_{12}|t) \tag{21.3}$$

The frequency of this oscillation ν_t, the so-called tunnel frequency, is given via

$$\nu_t = \frac{2|H_{12}|}{2\pi} = \frac{|H_{12}|}{\pi} \tag{21.4}$$

This periodic motion is called the coherent tunneling of the system. It simply reflects the fact that the eigenstates of the system are given by Eq. (21.1). The size of the tunnel frequency depends strongly on the hindering potential.

21.2.1.1.2 Tunneling in a Symmetric Double Minimum Potential

In the case of dihydrogen exchange, a linear exchange of the two hydrons is not possible and angular degrees of freedom must be taken into account. Thus the simplest realistic model is a one-dimensional hindered rotation of the two hydrons around their center of mass in a harmonic twofold potential, i.e. a one-dimensional hindered quantum mechanical rotor. In this model it is assumed that the distance between the two hydrons, as well as their distance from the metal, does not change. In this case the angular position, described via an angle φ, is used as the only degree of freedom. The corresponding Schrödinger equation of a rigid rotor in a harmonic twofold potential is:

$$-\frac{\hbar^2}{2\mu r^2}\frac{d^2}{d\varphi^2}|\Psi\rangle - V_0(1 - \cos 2\varphi)|\Psi\rangle = E|\psi\rangle \tag{21.5}$$

where $2V_0$ describes the depth of the hindering potential and μ is the reduced mass of the hydrogen. This differential equation is of the Matthieu type.

For $V_0 = 0$ the dihydrogen is a free one-dimensional rotor with complex eigenfunctions of the type

$$|\Psi_k(\varphi)\rangle = \frac{1}{\sqrt{2\pi}}\exp(ik\varphi) \text{ with } k = 0,\pm1,\pm2,\dots \tag{21.6}$$

Since the corresponding energy eigenvalues

$$E_k = \frac{\hbar^2 k^2}{2mr^2} \tag{21.7}$$

are doubly degenerate $E_{+k} = E_{-k}$, we can use equally well their real and imaginary linear combinations, which are simple sine and cosine functions, respectively a constant ($k = 1, 2, \dots$)

$$|c_0(\varphi)\rangle = \frac{1}{\sqrt{2\pi}}$$

$$|c_k(\varphi)\rangle = \frac{1}{\sqrt{2}}(|\Psi_k(\varphi)\rangle + |\Psi_{-k}(\varphi)\rangle) = \frac{1}{\sqrt{\pi}}\cos(k\varphi) \tag{21.8}$$

$$|s_k(\varphi)\rangle = \frac{1}{\sqrt{2i}}(|\Psi_k(\varphi)\rangle - |\Psi_{-k}(\varphi)\rangle) = \frac{1}{\sqrt{\pi}}\sin(k\varphi)$$

They have even respectively odd symmetry with respect to φ:

$$|c_k(-\varphi)\rangle = |c_k(\varphi)\rangle \text{ and } |s_k(-\varphi)\rangle = -|s_k(\varphi)\rangle. \tag{21.9}$$

For $V_0 > 0$, they are no longer the eigenfunctions of the system. Employing them as base functions of the Hilbert space, the Schrödinger equation (21.5) is converted into a matrix equation, where the kinetic energy operator is diagonal and the only non-vanishing matrix elements are between pairs of even or pairs of odd, base functions which differ in the index k by ±2, i.e. $\langle c_0|\hat{V}|c_{\pm2}\rangle = \frac{1}{\sqrt{2}}V_0$ and

$$\langle c_k|\hat{V}|c_l\rangle = \frac{1}{2}V_0(\delta_{k,l+2} + \delta_{k,l-2})$$

$$\tag{21.10}$$

$$\langle s_k|\hat{V}|s_l\rangle = \frac{1}{2}V_0(\delta_{k,l+2} + \delta_{k,l-2})$$

Thus the matrix is block diagonal and there are two sets of eigenfunctions (Fig. 21.1), namely cosine type functions and sine type functions $|S_n(\varphi)\rangle$, which are linear combinations of the base functions (Eq. (21.9)

$$|C_n(\varphi)\rangle = \sum_{k=0}^{\infty} a_{n,k}|c_k(\varphi)\rangle \text{ and } |S_n(\varphi)\rangle = \sum_{k=1}^{\infty} b_{n,k}|s_k(\varphi)\rangle \tag{21.11}$$

The ground state wavefunction is always a cosine type state with even symmetry and the first excited state is always a sine type function with odd symmetry.

The energy differences between different $C_n(\varphi)$ or $S_n(\varphi)$ depend strongly on the depth of the potential $2V_0$ and vary between zero and the order of typical rotational μ wave or IR transitions (Fig. 21.2). At temperatures around 10 K only the lowest pair of eigenstates is thermally populated.

For identical hydrons, the symmetry postulate of identical particles has to be fulfilled. For protons and tritons this means that the overall wave function must be antisymmetric under particle exchange and for deuterons it must be symmetric under particle exchange. Due to this correlation of spin and spatial state, the energy difference ΔE between the lowest two spatial eigenstates can be treated as a pure spin Hamiltonian, similar to the Dirac exchange interaction of electronic spins.

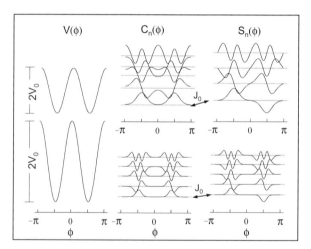

Figure 21.1 Eigenstates and energy eigenvalues of the Schrödinger equation of a rigid D_2 rotor in a harmonic twofold potential for two different depths of the potential barrier (adapted from Ref. [40]). Upper panel: $V_0 = 10^7$ MHz, $J_0 = 6.4 \times 10^3$ MHz, lower panel: $V_0 = 10^8$ MHz, $J_0 = 60$ Hz. Left : potential energy curve $V(\phi)$; middle panel : cosine type eigenfunctions $C_n(\phi)$, right panel: Sine type eigenfunctions $S_n(\phi)$. The energy shift between cosine and sine functions is increased artificially to demonstrate the differences in J_n between energy levels of same n.

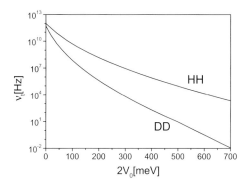

Figure 21.2 Tunnel frequency versus barrier [81]:Tunnel frequency and HH/DD isotope effect as a function of the barrier height $2V_0$ for a proton pair HH and a deuteron pair DD with $R_{DD} = 1\text{Å}$.

Spin $1/2$ case: For a Spin $I = 1/2$ the eigenstates are

$$|S_0\rangle = \frac{1}{\sqrt{2}}(|\alpha\beta\rangle - |\beta\alpha\rangle)$$

$$|T_{+1}\rangle = |\alpha\alpha\rangle$$

$$|T_0\rangle = \frac{1}{\sqrt{2}}(|\alpha\beta\rangle + |\beta\alpha\rangle)$$

$$|T_{+1}\rangle = |\beta\beta\rangle$$

(21.12)

The $|S_0\rangle$ state couples to the symmetric ground para-state to form a singlet manifold and the $|T_k\rangle$ states couple to the odd ortho-state and forms the triplet manifold. The splitting between these states is described by the quantum mechanical exchange interaction, which was given by Dirac in the form

$$\hat{H}_X = X\frac{1}{2}\left(1 + 4\hat{\vec{I}}_1\hat{\vec{I}}_2\right)$$

(21.13)

This Hamilton operator is the product of the energy splitting times the operator

$$\hat{P} = \frac{1}{2}\left(1 + 4\hat{\vec{I}}_1\hat{\vec{I}}_2\right) = \begin{pmatrix} 1 & 0 & 0 & 0 \\ 0 & 0 & 1 & 0 \\ 0 & 1 & 0 & 0 \\ 0 & 0 & 0 & 1 \end{pmatrix}$$

(21.14)

Permutation operator: The operator $\hat{P}$ is an example of a permutation operator in spin space. It exchanges the coordinates of the two spins 1/2. In the product space of the two spins the general definition of such an operator for arbitrary spins is given via

$$\hat{P}\left(\hat{\vec{I}}_1, \hat{\vec{I}}_2\right)|\mu, \nu\rangle = |\nu, \mu\rangle \tag{21.15}$$

An very useful alternate representation of the operator $\hat{P}$ is found if symmetry adapted base functions $|\lambda\rangle$ are employed. In this base the permutation operator is diagonal with diagonal elements $+1$ if the state has even and -1 if the state has odd symmetry.

$$\hat{P}\left(\hat{\vec{I}}_1, \hat{\vec{I}}_2\right)|\lambda\rangle = \pm|\lambda\rangle \tag{21.16}$$

Employing these permutation operators it is easy to write down the exchange Hamiltonian for arbitrary spins.

From the NMR point of view, the eigenfunctions of the spin $1/2$ exchange Hamiltonian are identical to the eigenfunctions of the "normal" homonuclear spin–spin interaction, but the energy eigenvalues are shifted by a constant offset of $X/2$, since

$$\hat{H}_X = \frac{X}{2} + 2X\hat{\vec{I}}_1\hat{\vec{I}}_2$$
$$= \frac{X}{2} + J_X\hat{\vec{I}}_1\hat{\vec{I}}_2 \tag{21.17}$$

Thus for spin $I = 1/2$ nuclei, the quantum mechanical exchange interaction is formally equivalent to an indirect spin–spin interaction. Accordingly the exchange of spin-1/2 particle is usually treated in NMR like a J-coupling (quantum exchange coupling), employing the more simple Hamiltonian

$$\hat{H}'_X = J_X\hat{\vec{I}}_1\hat{\vec{I}}_2 \tag{21.18}$$

These couplings are indeed directly visible in liquid state ^{1}H-NMR spectra [46, 47].

Spin 1 case: For a Spin $I = 1$ the nine eigenstates of a di-deuterium system are

$$|a_1\rangle = |++\rangle \qquad\qquad |a_2\rangle = |00\rangle \qquad\qquad |a_3\rangle = |--\rangle$$

$$|a_4\rangle = \frac{1}{\sqrt{2}}(|+-\rangle + |-+\rangle) \quad |a_5\rangle = \frac{1}{\sqrt{2}}(|+0\rangle + |0+\rangle) \quad |a_6\rangle = \frac{1}{\sqrt{2}}(|-0\rangle + |01\rangle)$$

$$|b_1\rangle = \frac{1}{\sqrt{2}}(|+-\rangle - |-+\rangle) \quad |b_2\rangle = \frac{1}{\sqrt{2}}(|+0\rangle - |0+\rangle) \quad |b_3\rangle = \frac{1}{\sqrt{2}}(|-0\rangle - |01\rangle)$$

$$\tag{21.19}$$

The three b-states have odd and the six a-states have even symmetry. Since the deuterium nucleus is a boson, the overall wavefunction has to be symmetric and the six even a-states couple to the even ground state and the three odd b-states couple to the odd spatial state. The special form of the operator $\hat{P}$ is calculated from Eq. (21.15) or (21.16). For arbitrary base sets, a base independent operator representation of the permutation operator of spin 1 similar to the spin $1/2$ case of Eq. (21.18) is useful. This representation can be found with the help of the following set of normalized single spin operators:

$$B_1 = \frac{1}{\sqrt{2}}S_x \qquad\qquad B_2 = \frac{1}{\sqrt{2}}S_y$$

$$B_3 = \frac{1}{\sqrt{2}}S_z \qquad\qquad B_4 = \frac{1}{\sqrt{2}}E$$

$$B_5 = \frac{1}{2}\sqrt{6}\left(S_z^2 - \frac{2}{3}\right) \qquad\qquad B_6 = \frac{1}{\sqrt{2}}\left(S_y^2 - S_x^2\right) \qquad\qquad (21.20)$$

$$B_7 = \frac{1}{\sqrt{2}}\left(S_x S_y + S_y S_x\right) \qquad\qquad B_8 = \frac{1}{\sqrt{2}}(S_x S_z + S_z S_x)$$

$$B_9 = \frac{1}{\sqrt{2}}\left(S_y S_z + S_z S_y\right)$$

Employing these base operators the permutation operator of a homo nuclear spin 1 pair is given by

$$\hat{P}\left(\hat{I}_1, \hat{I}_2\right) = \sum_k B_k \otimes B_k \qquad\qquad (21.21)$$

where $\otimes$ denotes the tensor or direct product of two vector spaces (see for example Refs. [80, 82]).

As a result we find that in both the hydrogen and the deuterium case the ground state tunneling is describable by a pure spin tunnel Hamiltonian, which describes the tunnel splitting between the spatial pair of states of different symmetry. The implications are discussed in detail in Ref. [83].

Coherent tunneling at higher temperatures: If several pairs of tunnel levels are thermally populated, the thermal average of the different pairs of tunnel levels has to be calculated. As long as only a few levels far below the barrier are contributing, the values of the various tunnel frequencies $J_n = v_{tn}$ will be small compared to the thermal exchange rates between the level pairs and the averaging can be done by summing up the individual values of v_{tn} times their thermal population. This averaging can be approximated using the population of one of the connected levels:

$$v_t = \sum_n v_{tn} \exp\left(-\frac{E_n}{kT}\right) \qquad\qquad (21.22)$$

The situation becomes more difficult if the values of v_t are comparable or greater than the thermal population rates or decay rates. In this regime, a transition from coherent to incoherent exchange will take place (see below), as was shown by density matrix theory [19].

21.2.1.2 The Density Matrix

As soon as large ensembles of particles with statistical populations of the eigenstates and incoherent exchange and relaxation processes between these states are investigated, quantum statistical tools are necessary to describe the system. In this situation the quantum mechanical density operator $\hat{\rho}$ has to be employed. For the coherent evolution of the density operator under the influence of a Hamiltonian $\hat{H}$, the following differential equation is found [80]

$$\frac{d}{dt}\hat{\rho} = -i\left[\hat{H}, \hat{\rho}\right] \tag{21.23}$$

In this equation we have followed the NMR convention and set the constant $\hbar = 1$. This is equivalent to measuring energies in angular frequency units. Employing a suitable set of base functions of the Hilbert space, this equation can be converted into a set of linear differential equations for the matrix elements of $\hat{\rho}$. In the case of a single pair of tunnel levels the Hamiltonian of the two levels with their tunnel splitting can be treated as a two-level system, employing fictitious spin 1/2 operators, describable by the Hamiltonian $\hat{H}$

$$\hat{H} = \begin{pmatrix} E_1 & 0 \\ 0 & E_2 \end{pmatrix} = \frac{E_1 + E_2}{2}\begin{pmatrix} 1 & 0 \\ 0 & 1 \end{pmatrix} + \frac{E_1 - E_2}{2}\begin{pmatrix} 1 & 0 \\ 0 & -1 \end{pmatrix}$$

$$= \frac{E_1 + E_2}{2} + (E_1 - E_2)\hat{S}_z = E + v_t\hat{S}_z \tag{21.24}$$

Here $E = \dfrac{E_1 + E_2}{2}$ is the mean energy of the levels and $v_t = (E_1 - E_2)$ is the tunnel frequency. This set of differential equations is most conveniently written by introducing the so-called Liouville super operator $\hat{\hat{L}}$, which defines the Liouville space, where the density matrix becomes a vector. The equation of motion of the density matrix is

$$\frac{d}{dt}\hat{\rho} = -i\hat{\hat{L}}\hat{\rho} \tag{21.25}$$

The Liouville operator is constructed from the Hamiltonian via

$$\hat{\hat{L}} = \hat{H} \otimes \hat{\hat{E}} - \hat{\hat{E}} \otimes \hat{H} \tag{21.26}$$

Here, $\hat{\hat{E}}$ is the unity matrix of Hilbert space. In the case of the tunnel Hamiltonian it is simply

$$\hat{\hat{L}} = v_t\hat{\hat{S}}_z = v_t\begin{pmatrix} 0 & 0 & 0 & 0 \\ 0 & 1 & 0 & 0 \\ 0 & 0 & -1 & 0 \\ 0 & 0 & 0 & 0 \end{pmatrix} \tag{21.27}$$

21.2.1.3 The Transition from Coherent to Incoherent Tunneling

A rigorous quantum mechanical theory of the transition from coherent to incoherent tunneling was developed by Szymanski [84] and Scheurer [85]. The results of this elaborate theory can be reproduced by a combination of density matrix theory and formal kinetics as was shown in Ref. [19].

The transition from coherent to incoherent tunneling is most easily understood by considering the ground state of the system and the first excited state. In the systems under consideration the typical energy difference between the ground state and the first excited state is in the range 40–50 meV. At temperatures below 50 K practically only the ground state is populated ($n_b < 10^{-5} n_a$) and only a small population is found in the excited state.

As a starting point let us see what happens, if we consider not only the ground state, but also the first excited state of the tunneling pairs. Both states are connected via thermal excitation with rates k_{ab} and k_{ba}. Since the transition between even and odd states is both symmetry and spin forbidden, these rates connect only states of the same symmetry and the spin is conserved in the transition. From the principle of detailed balance we find for the ratio of the rates

$$\frac{k_{ab}}{k_{ba}} = \exp\left(-\frac{E_b - E_a}{k_{\mathrm{B}} T}\right) \ll 1 \tag{21.28}$$

The coherent evolution of the density matrix is described by the corresponding Liouville operators:

$$\hat{L}_a = v_{ta} \begin{pmatrix} 0 & 0 & 0 & 0 \\ 0 & 1 & 0 & 0 \\ 0 & 0 & -1 & 0 \\ 0 & 0 & 0 & 0 \end{pmatrix} \quad \text{and} \quad \hat{L}_b = v_{tb} \begin{pmatrix} 0 & 0 & 0 & 0 \\ 0 & 1 & 0 & 0 \\ 0 & 0 & -1 & 0 \\ 0 & 0 & 0 & 0 \end{pmatrix} \tag{21.29}$$

The exchange between the levels connects the two Liouville spaces

$$\frac{\mathrm{d}}{\mathrm{d}t} \hat{\rho}_a = -(k_{ab} + i L_a) \hat{\rho}_a + k_{ba} \hat{\rho}_b$$

$$\frac{\mathrm{d}}{\mathrm{d}t} \hat{\rho}_b = -(k_{ba} + i L_b) \hat{\rho}_b + k_{ab} \hat{\rho}_a \tag{21.30}$$

In matrix form these coupled differential equations are

$$\frac{\mathrm{d}}{\mathrm{d}t} \begin{pmatrix} \rho_{a1} \\ \rho_{a2} \\ \rho_{a3} \\ \rho_{a4} \end{pmatrix} = \begin{pmatrix} -k_{ab} & 0 & 0 & 0 \\ 0 & -k_{ab} - i v_{ta} & 0 & 0 \\ 0 & 0 & -k_{ab} + i v_{ta} & 0 \\ 0 & 0 & 0 & -k_{ab} \end{pmatrix} \begin{pmatrix} \rho_{a1} \\ \rho_{a2} \\ \rho_{a3} \\ \rho_{a4} \end{pmatrix} + k_{ba} \begin{pmatrix} \rho_{b1} \\ \rho_{b2} \\ \rho_{b3} \\ \rho_{b4} \end{pmatrix}$$

$$\tag{21.31}$$

and

$$\frac{d}{dt}\begin{pmatrix} \rho_{b1} \\ \rho_{b2} \\ \rho_{b3} \\ \rho_{b4} \end{pmatrix} = \begin{pmatrix} -k_{ba} & 0 & 0 & 0 \\ 0 & -k_{ba}-i\upsilon_{tb} & 0 & 0 \\ 0 & 0 & -k_{ba}+i\upsilon_{tb} & 0 \\ 0 & 0 & 0 & -k_{ba} \end{pmatrix}\begin{pmatrix} \rho_{b1} \\ \rho_{b2} \\ \rho_{b3} \\ \rho_{b4} \end{pmatrix} + k_{ab}\begin{pmatrix} \rho_{a1} \\ \rho_{a2} \\ \rho_{a3} \\ \rho_{a4} \end{pmatrix}$$

$$(21.32)$$

This system of equations can be greatly simplified by the fact that the excited state is only weakly populated. This allows us to use a quasi-stationary condition for the excited state by setting the derivative of the population of the excited state to zero. This converts the second differential equation into an algebraic one.

$$\begin{pmatrix} -k_{ba} & 0 & 0 & 0 \\ 0 & -k_{ba}-i\upsilon_{tb} & 0 & 0 \\ 0 & 0 & -k_{ba}+i\upsilon_{tb} & 0 \\ 0 & 0 & 0 & -k_{ba} \end{pmatrix}\begin{pmatrix} \rho_{b1} \\ \rho_{b2} \\ \rho_{b3} \\ \rho_{b4} \end{pmatrix} + k_{ab}\begin{pmatrix} \rho_{a1} \\ \rho_{a2} \\ \rho_{a3} \\ \rho_{a4} \end{pmatrix} = 0 \qquad (21.33)$$

Solving for ρ_b gives:

$$\begin{pmatrix} \rho_{b1} \\ \rho_{b2} \\ \rho_{b3} \\ \rho_{b4} \end{pmatrix} = \begin{pmatrix} \frac{k_{ab}}{k_{ba}}\rho_{a1} \\ \frac{k_{ab}}{k_{ba}+i\upsilon_{tb}}\rho_{a2} \\ \frac{k_{ab}}{k_{ba}-i\upsilon_{tb}}\rho_{a3} \\ \frac{k_{ab}}{k_{ba}}\rho_{a4} \end{pmatrix} \qquad (21.34)$$

Inserting this expression into the equation for ρ_a gives after some simple manipulations:

$$\frac{d}{dt}\begin{pmatrix} \rho_{a1} \\ \rho_{a2} \\ \rho_{a3} \\ \rho_{a4} \end{pmatrix} = \begin{pmatrix} 0 & 0 & 0 & 0 \\ 0 & -k_{ab}-i\upsilon_{ta}+\frac{k_{ab}k_{ba}}{k_{ba}+i\upsilon_{tb}} & 0 & 0 \\ 0 & 0 & -k_{ab}+i\upsilon_{ta}+\frac{k_{ab}k_{ba}}{k_{ba}-i\upsilon_{tb}} & 0 \\ 0 & 0 & 0 & 0 \end{pmatrix}\begin{pmatrix} \rho_{a1} \\ \rho_{a2} \\ \rho_{a3} \\ \rho_{a4} \end{pmatrix}$$

$$(21.35)$$

From the first and the last row it is evident that the density matrix elements ρ_{a1} and ρ_{a4}, which correspond to the populations of the levels, do not change. The center rows become:

$$\frac{d}{dt}\rho_{a2} = \left(-k_{ab}-i\upsilon_{ta}+\frac{k_{ab}k_{ba}}{k_{ba}+i\upsilon_{tb}} \right)\rho_{a2} \qquad (21.36)$$

and

$$\frac{d}{dt}\rho_{a3} = \left(-k_{ab}+i\upsilon_{ta}+\frac{k_{ab}k_{ba}}{k_{ba}-i\upsilon_{tb}} \right)\rho_{a3} \qquad (21.37)$$

They can be rewritten as

$$\frac{d}{dt}\rho_{a2} = \left(-\left(\frac{k_{ab}v_{tb}^2}{k_{ba}^2 + v_{tb}^2}\right) - i\left(v_{ta} + \frac{k_{ab}k_{ba}}{k_{ba}^2 + v_{tb}^2}v_{tb}\right)\right)\rho_{a2} \qquad (21.38)$$

and

$$\frac{d}{dt}\rho_{a3} = \left(-\left(\frac{k_{ab}v_{tb}^2}{k_{ba}^2 + v_{tb}^2}\right) + i\left(v_{ta} + \frac{k_{ab}k_{ba}}{k_{ba}^2 + v_{tb}^2}v_{tb}\right)\right)\rho_{a3} \qquad (21.39)$$

The density matrix elements ρ_{a2} and ρ_{a3} represent coherent superpositions (coherences) between the para and ortho states, which evolve with the tunnelling frequency.

Thus we find that the connection to the higher level causes a shift of the coherent tunnelling frequency to

$$v_{ta}' = v_{ta} + \frac{k_{ab}k_{ba}}{k_{ba}^2 + v_{tb}^2}v_{tb} \qquad (21.40)$$

and in addition a damping of the singlet triplet coherences by a relaxation rate

$$r_{12} = \left(\frac{k_{ab}v_{tb}^2}{k_{ba}^2 + v_{tb}^2}\right) \qquad (21.41)$$

For the interpretation of this relaxation rate it is useful to transform the Liouville operator in the localized base. The transformation matrix in Hilbert space is

$$\hat{S} = \frac{1}{\sqrt{2}}\begin{pmatrix} 1 & 1 \\ 1 & -1 \end{pmatrix} \qquad (21.42)$$

and the corresponding transformation super operator in Liouville space is calculated from the transformation matrix as [82]

$$\hat{\hat{S}} = \hat{S} \otimes \hat{S} = \frac{1}{2}\begin{pmatrix} 1 & 1 & 1 & 1 \\ 1 & -1 & 1 & -1 \\ 1 & 1 & -1 & -1 \\ 1 & -1 & -1 & 1 \end{pmatrix} \qquad (21.43)$$

where we have used the fact that $\hat{S}$ is a real matrix. Employing this operator, the equation in the localized base becomes

$$
\frac{d}{dt}\hat{S}\begin{pmatrix}\rho_{a1}\\\rho_{a2}\\\rho_{a3}\\\rho_{a4}\end{pmatrix}=\hat{S}\begin{pmatrix}0 & 0 & 0 & 0\\0 & -r_{12}-iv_{ta}' & 0 & 0\\0 & 0 & -r_{12}+iv_{ta}' & 0\\0 & 0 & 0 & 0\end{pmatrix}\hat{S}^{-1}\hat{S}\begin{pmatrix}\rho_{a1}\\\rho_{a2}\\\rho_{a3}\\\rho_{a4}\end{pmatrix}
$$

$$
=\left(\begin{pmatrix}-\tfrac{1}{2}r_{12} & 0 & 0 & \tfrac{1}{2}r_{12}\\0 & -\tfrac{1}{2}r_{12} & \tfrac{1}{2}r_{12} & 0\\0 & \tfrac{1}{2}r_{12} & -\tfrac{1}{2}r_{12} & 0\\\tfrac{1}{2}r_{12} & 0 & 0 & -\tfrac{1}{2}r_{12}\end{pmatrix}+i\begin{pmatrix}0 & \tfrac{1}{2}v_{ta}' & -\tfrac{1}{2}v_{ta}' & 0\\\tfrac{1}{2}v_{ta}' & 0 & 0 & -\tfrac{1}{2}v_{ta}'\\-\tfrac{1}{2}v_{ta}' & 0 & 0 & \tfrac{1}{2}v_{ta}'\\0 & -\tfrac{1}{2}v_{ta}' & \tfrac{1}{2}v_{ta}' & 0\end{pmatrix}\right)\hat{S}\begin{pmatrix}\rho_{a1}\\\rho_{a2}\\\rho_{a3}\\\rho_{a4}\end{pmatrix}
$$

(21.44)

The first term has a very simple physical interpretation. It corresponds to an incoherent exchange of the localized particles with an exchange rate

$$
k_{12}=\frac{1}{2}r_{12}=\frac{1}{2}\left(\frac{k_{ab}v_{tb}^2}{k_{ba}^2+v_{tb}^2}\right)
$$

(21.45)

between the localized states, i.e. an incoherent tunneling of the protons or deuterons from the left side to the right side of the potential.

$$
\hat{K}=\begin{pmatrix}-k_{12} & 0 & 0 & k_{12}\\0 & -k_{12} & k_{12} & 0\\0 & k_{12} & -k_{12} & 0\\k_{12} & 0 & 0 & -k_{12}\end{pmatrix}
$$

(21.46)

In the case of $v_{tb}\gg k_{ba}$ this equation simplifies to

$$
k_{12}=\frac{1}{2}k_{ab}
$$

(21.47)

This result has a very intuitive explanation, which is sketched in Fig. 23.3. The rate k_{ab} is the transport rate from the ground state to the excited state. k_{ab} is much smaller then the decay rate k_{ba}. As soon as a molecule is excited to the higher state, the tunneling gets so fast that there is equal probability to find the particle on both sides, when it decays back into the ground state.

Thus, depending on the size of the barrier height, the two hydrons will exhibit strong differences in their dynamic behavior. For a low barrier height, a large tunnel frequency is observed. The dihydrogen pair will be at least partially delocalized and acts more or less like a one-dimensional free quantum mechanical rotor, similar to p-H$_2$ and o-H$_2$, allowing coherent (i.e. strictly periodic) exchange processes of the individual hydrons with the tunnel frequency v_t. For high potential barriers the tunnel splitting goes to zero, no coherent exchange processes take place, each hydron is located in a single potential minimum and the dihydrogen pair is fixed.

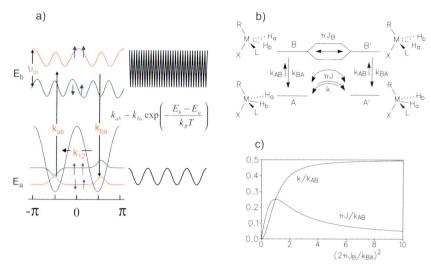

Figure 21.3 Sketch of the transition from coherent to incoherent tunneling:
(a) Quantitative four-level model [86]; (b) corresponding chemisty of coherent
and incoherent dihydrogen exchange [19]; (c) dependence of coherent
and incoherent rates [19].

In this situation, for an exchange of the two hydrons a coupling to external
degrees of freedom is necessary. In this scenario the exchange of the two hydrogen atoms is describable as a thermally activated rate process. Compared to the
previous coherent exchange, the thermally activated rate process corresponds to
an incoherent exchange of the two hydrons, which leads to an exponentially decaying curve for the probability of finding one hydron on its initial position.

21.2.2
Incoherent Tunneling and the Bell Model

As we have seen in the previous section there is a transition from coherent to
incoherent tunneling, caused by the coupling to external bath degrees of freedom.
This second type of tunnel process is the classical forbidden penetration of a barrier (Fig. 21.4), as for example in the Gamow model of α-decay [87, 88] or the field
emission of electrons of Condon [89, 90]. The probability of penetrating the barrier depends on the energy of the incident particle and the width, shape and
height of the potential barrier. For most potentials only approximate solutions, as
for example the well-known Wenzel [91], Kramers [92], Brillouin [93] WKB approximation (see for example the textbook [80]), or numerical calculations of the transition probability are possible. Analytically solvable exceptions include rectangular
potential steps and parabolic potentials. While the former give only very crude
approximations of a real world system, the latter gives reasonably good results,
when compared to experimentally determined rate constants.

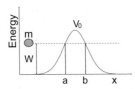

Figure 21.4 Tunneling through an energy barrier: While the particle m with energy $W < V_0$ is classically reflected at a, quantum mechanics allows a tunneling through the barrier from **a** to **b**.

As soon as bound states are considered there are only discrete energy levels. Nevertheless it was shown by Bell [77] that it is possible to employ approximately a continuum of energy levels for the calculations of the tunnel rates, which is adequate for the description of many experimental systems. In the simplest form (see Fig. 21.5) of the Bell model, the potential barrier is an inverted parabola. This allows the use of the known solution of the quantum mechanical harmonic oscillator for the calculation of the transition probability through the barrier. The corresponding Schrödinger equation is

$$\left[\frac{d^2}{dx^2} + \frac{2m}{\hbar^2} \left(E - \frac{1}{2} m\omega_0^2 x^2 \right) \right] |\Psi\rangle = 0 \tag{21.48}$$

The ground state energy level is

$$E_0 = \frac{1}{2} \hbar\omega_0 =: \frac{1}{2} m\omega_0^2 a^2 \tag{21.49}$$

Here $2a$ is the width of the potential barrier at the ground state. Solving for $v_0 = \omega_0/2$ expresses the oscillation frequency via the ground state energy E_0 and the width of the potential at E_0:

$$v_0 = \frac{1}{\pi a} \sqrt{\frac{E_0}{2m}} \tag{21.50}$$

If the oscillator potential is inverted as shown in Fig. 21.5(b)

$$\left[\frac{d^2}{dx^2} + \frac{2m}{\hbar^2} \left(E + \frac{1}{2} m\omega_0^2 x^2 \right) \right] |\Psi\rangle = 0 \tag{21.51}$$

the previous solution can be reused by introducing the imaginary tunnel frequency

$$v_t = \frac{1}{\pi a} \sqrt{\frac{E_0}{2m}} \tag{21.52}$$

From this the probability for transition through the barrier

$$G(W) = \left[1 + \exp\left(\frac{V_0 - W}{h v_t} \right) \right]^{-1} \tag{21.53}$$

is calculated. In typical chemical reactions large numbers of particles N_0 are involved. They are modeled as a stream $J = dN/dt$ of particles hitting the barrier. In thermal equilibrium the number of particles in energy interval $[W, W + dW]$ is given by the Boltzmann distribution

$$
\begin{aligned}
dN &= N_0 \cdot p(W)dW \\
&= N_0 \cdot \frac{1}{kT}\exp\left(-\frac{W}{kT}\right)dW
\end{aligned}
\tag{21.54}
$$

If $T(W)$ is the transition probability at energy W, the number of particles per second, which pass the energy barrier is

$$
J = J_0 \int_0^\infty p(W)T(W)dW
\tag{21.55}
$$

Quantum mechanically the transition rate J_{QM} of the Bell model is calculated by inserting Eq. (21.53) into Eq. (21.55).

$$
\begin{aligned}
J_{QM} &= \frac{J_0}{kT}\int_0^\infty \exp\left(-\frac{W}{kT}\right)G(W)dW \\
&= \frac{J_0}{kT}\int_0^\infty \exp\left(-\frac{W}{kT}\right)\left[1 + \exp\left(\frac{V_0 - W}{h\nu_t}\right)\right]^{-1}dW
\end{aligned}
\tag{21.56}
$$

Comparing this to the classically allowed rate from Arrhenius law

$$
J_c = J_0\exp\left(-\frac{V_0}{kT}\right)
\tag{21.57}
$$

one can define the tunnel correction:

$$
Q_t = \frac{J_{QM}}{J_c} = \frac{\exp(V_0/kT)}{kT}\int_0^\infty \exp\left(-\frac{W}{kT}\right)G(W)dW
\tag{21.58}
$$

For the numerical evaluation, Eq. (21.58) can be approximated by replacing the integration with a discrete sum over a set of energy levels. The result of such an evaluation is displayed in Fig. 21.5(d), which compares the classical Arrhenius rate with the quantum mechanical rates calculated from Eq. (21.56).

21.3
Symmetry Effects on NMR Lineshapes of Hydration Reactions

The symmetry effects associated with the Pauli principle provide an interesting diagnostic tool for the study of hydration and hydrogen transfer reactions. Employing spin-polarized parahydrogen (p-H_2) gas in these reactions, a very

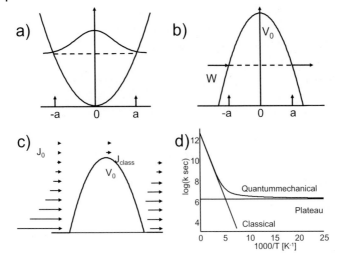

Figure 21.5 The Bell tunnel model: (a) Quantum mechanical harmonic oscillator with its ground state wavefunctions. (b) Inverted harmonic oscillator potential. (c) A stream of particles with a Boltzmann distribution of energies hits the barrier. Classical only those particles with $W > V_0$ can pass the barrier. Quantum mechanically particles with $W < V_0$ may also pass the barrier. (d) Comparison of classical Arrhenius rate and quantum mechanical corrected rate. While classically the rate goes to zero for $T \rightarrow 0$, quantum mechanically a finite plateau is approached (adapted after Bell [77]).

strong signal enhancement and thus selective spectroscopy of the reaction side is possible, as originally proposed by Bowers and Weitekamp [51, 56].

The origins of symmetry induced nuclear polarization can be summarized as follows: as mentioned above molecular dihydrogen is composed of two species, para-H_2, which is characterized by the product of a symmetric rotational wavefunction and an antisymmetric nuclear spin wave function and ortho-H_2, which is characterized by an antisymmetric rotational and one of the symmetric nuclear spin wavefunctions. In thermal equilibrium at room temperature each of the three ortho-states and the single para-state have practically all equal probability. In contrast, at temperatures below liquid nitrogen mainly the energetically lower para-state is populated. Therefore, an enrichment of the para-state and even the separation of the two species can be easily achieved at low temperatures as their interconversion is a rather slow process. Pure para-H_2 is stable even in liquid solutions and para-H_2 enriched hydrogen can be stored and used subsequently for hydrogenation reactions [54].

The transformation of this molecular rotational order into nuclear spin order during the hydrogenation reaction leads to typical polarization patterns in the NMR spectra of the hydrogenation products. Depending on whether the experiment is performed inside or outside of a magnetic field (see Fig. 21.6), these types of experiments have been referred to under the acronyms PASADENA (Parahydrogen and Synthesis Allow Dramatically Enhanced Nuclear Alignment) or

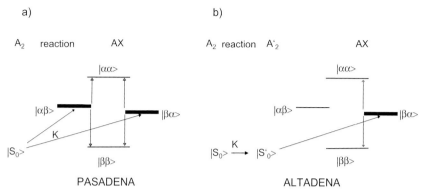

a)

b)

Figure 21.6 Schemes of simple PHIP experiments in an AX-spin system [56]. (a) Ideal PASADENA experiment. With the reaction rate K the population of the para-H_2 $|S_0\rangle$ state is in a sudden change transferred to the $|\alpha\beta\rangle$ and $|\beta\alpha\rangle$ states, which are equally populated. (b) Ideal ALTADENA experiment.

With the reaction rate K the population of the para-H_2 $|S_0\rangle$ state is transferred to the $|S'_0\rangle$ state of the final product. From there the population is adiabatically transferred to the $|\beta\alpha\rangle$ state of the final product, resulting in a selective population of this level (adapted from Bowers et al. [56]).

ALTADENA (Adiabatic Longitudinal Transport after Dissociation Engenders Net Alignment) [56]. All variants are nowadays known under the more generally acronym of PHIP (Parahydrogen Induced Polarization) [52]. The basic theory of the PHIP effect in an AX-spin system was given in the review paper of Bowers and Weitekamp [56].

In the original work only a simple AX-spin system with pure coherent exchange was considered. In practice, however, the situation will be a lot more complicated because there are several coherent and incoherent reaction pathways in the course of a catalyzed hydrogenation reaction, as depicted in Fig. 21.7.

To analyze these situations we first study analytically the PHIP effect for a general two spin system and then the effect of incoherent exchange on the PHIP line shape.

21.3.1
Analytical Solution for the Lineshape of PHIP Spectra Without Exchange

In the case of a simple reaction from the para-H_2 state to the product state it is possible to derive analytical solution of the lineshape of PHIP spectra [65]. In the following an alternative derivation of the lineshape is given. For simplicity the reaction is assumed as a one-way reaction, i.e. no back reaction ($k_{ba}=0$, a, b denote the two different sites).

For the para-H_2 state, the Hamiltonian is given as a pure A_2 spin system.

$$\hat{H}_a = J_a \hat{\vec{I}}_a \hat{\vec{S}}_a \tag{21.59}$$

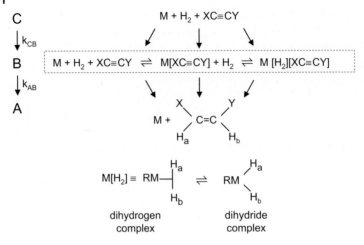

Figure 21.7 Possible pathways of the catalyzed hydrogenation reaction of an unsaturated organic substrate involving various transition metal dihydrogen and/or dihydride intermediates [62]. The initial free dihydrogen is labeled as site C, the intermediates containing the dihydrogen pair as B, and the product containing the dihydrogen pair as A.

For the end product of the reaction a general liquid Hamiltonian is assumed

$$\hat{H}_b = v_S \hat{S}_{zr} + v_I \hat{I}_{zr} + J_b \hat{\vec{I}}_b \hat{\vec{S}}_b,$$
(21.60)

which has the following matrix representation

$$\hat{H}_b = \begin{pmatrix} \frac{v_S+v_I}{2} + \frac{J_b}{4} & 0 & 0 & 0 \\ 0 & \frac{v_S-v_I}{2} + \frac{J_b}{4} & \frac{J_b}{2} & 0 \\ 0 & \frac{J_b}{2} & \frac{-v_S+v_I}{2} - \frac{J_b}{4} & 0 \\ 0 & 0 & 0 & \frac{-v_S-v_I}{2} + \frac{J_b}{4} \end{pmatrix}$$
(21.61)

The initial condition of the problem is that at the beginning all hydrogen atoms are in the singlet S_0 state of the para-H_2.

$$\rho(0) = |S_0 >< S_0|$$

$$= \frac{1}{2}|\alpha\beta - \beta\alpha >< \alpha\beta - \beta\alpha|$$

$$= \frac{1}{4} - \hat{\vec{S}}\hat{\vec{I}}$$
(21.62)

$$= \frac{1}{2}\begin{pmatrix} 0 & 0 & 0 & 0 \\ 0 & 1 & -1 & 0 \\ 0 & -1 & 1 & 0 \\ 0 & 0 & 0 & 0 \end{pmatrix}$$

As a consequence of this the initial dynamics before any pulse is applied occurs only in the subspace spanned by the elements $\rho_a(22,23,32,33)$, $\rho_b(22,23,32,33)$ and the analysis of the dynamics can be restricted to this subspace. In the initial para-H_2 state, the equation of motion is:

$$\frac{d}{dt}\hat{\rho}_a = -i\left[\hat{H}_a, \hat{\rho}_a\right] - k_{ab}\hat{\rho}_a \tag{21.63}$$

Since the Hamiltonian $\hat{H}_a$ commutes with $\hat{\rho}_a(0)$, no oscillations of the coherences are observed and the solution of the equation for $\rho_a(t)$ is simply given by an exponential decay

$$\hat{\rho}_a(t) = \exp(-k_{ab}t)\hat{\rho}_a(0) \tag{21.64}$$

In the next step the differential equation for $\rho_b(t)$ has to be solved:

$$\begin{aligned}\frac{d}{dt}\hat{\rho}_b &= -i\left[\hat{H}_b, \hat{\rho}_b\right] + k_{ab}\hat{\rho}_a - \frac{1}{T_{2b}}\hat{\rho}_b \\ &= -i\left[\hat{H}_b, \hat{\rho}_b\right] + k_{ab}\hat{\rho}_a(0)\exp(-k_{ab}t) - \frac{1}{T_{2b}}\hat{\rho}_b \end{aligned} \tag{21.65}$$

For the solution of this differential equation it is advantageous to transform the matrix equation of Hilbert space into a vector equation in Liouville space (E_b is the identity matrix of the four-dimensional subspace):

$$\hat{\hat{L}}_b = \hat{H}_b \otimes \hat{E}_b - \hat{E}_b \otimes \hat{H}_b \tag{21.66}$$

The matrix representation of $\hat{\hat{L}}_b$ is

$$\hat{\hat{L}}_b = \begin{pmatrix} 0 & -\frac{J_b}{2} & \frac{J_b}{2} & 0 \\ -\frac{J_b}{2} & \Delta v & 0 & \frac{J_b}{2} \\ \frac{J_b}{2} & 0 & -\Delta v & -\frac{J_b}{2} \\ 0 & \frac{J_b}{2} & -\frac{J_b}{2} & 0 \end{pmatrix} \tag{21.67}$$

where we have introduced for abbreviation $\Delta v = v_{bS} - v_{bI}$. The inhomogeneous differential equation (21.65) becomes

$$\frac{d}{dt}\hat{\rho}_b = -2i\pi\hat{\hat{L}}_b\hat{\rho}_b + k_{ab}\hat{\rho}_a(0)\exp(-k_{ab}t) - \frac{1}{T_{2b}}\hat{\rho}_b \tag{21.68}$$

The solution of Eq. (21.68) is

$$\hat{\rho}_b(t) = \left(-2i\pi\hat{\hat{L}}_b + k_{ab} - \frac{1}{T_{2b}}\right)^{-1}k_{ab}$$
$$\times \left(\exp\left(\left(-2i\pi\hat{\hat{L}}_b - \frac{1}{T_{2b}}\right)t\right) - \exp(-k_{ab}t)\right)\hat{\rho}_a(0) \tag{21.69}$$

where the fact has been exploited that

$$\left[-2\pi i \hat{L}_b - \frac{1}{T_{2b}} + k_{ab}, -2\pi i \hat{L}_b - \frac{1}{T_{2b}}\right] = 0 \tag{21.70}$$

However, instead of directly evaluating Eq. (21.68) by Eq. (21.69) it is advantageous to combine the coherent evolution and the relaxation, i.e. the homogeneous part of Eq. (21.68) and transform the resulting matrix into a fictive spin $-1/2$ system, employing the normalized Pauli matrices as base vectors. The corresponding transformation is

$$\hat{S}_1 = \frac{1}{2}\sqrt{2} \begin{pmatrix} 1 & 0 & 0 & 1 \\ 0 & 1 & 1 & 0 \\ 0 & 1 & -1 & 0 \\ 1 & 0 & 0 & -1 \end{pmatrix} \tag{21.71}$$

In this system, the matrix $\hat{L}_b$ becomes the block diagonal

$$\hat{L}_b - \frac{1}{T_{2b}} = \begin{pmatrix} -\frac{1}{T_{2b}} & 0 & 0 & 0 \\ 0 & -\frac{1}{T_{2b}} & 2i\pi\Delta\upsilon & 0 \\ 0 & 2i\pi\Delta\upsilon & -\frac{1}{T_{2b}} & -2i\pi J_b \\ 0 & 0 & -2i\pi J_b & -\frac{1}{T_{2b}} \end{pmatrix} \tag{21.72}$$

It can be diagonalized in a second step by transforming with the matrix

$$\hat{S}_2 = \begin{pmatrix} 1 & 0 & 0 & 0 \\ 0 & s & 0 & c \\ 0 & \frac{c}{\sqrt{2}} & \frac{1}{\sqrt{2}} & \frac{-s}{\sqrt{2}} \\ 0 & \frac{c}{\sqrt{2}} & \frac{-1}{\sqrt{2}} & \frac{-s}{\sqrt{2}} \end{pmatrix} \tag{21.73}$$

with $\quad c = \dfrac{\Delta\upsilon}{\sqrt{\left(\Delta\upsilon^2 + J_b^2\right)}} \quad$ and $\quad s = \dfrac{J_b}{\sqrt{\left(\Delta\upsilon^2 + J_b^2\right)}}$

The resulting matrix of the homogeneous part is

$$\begin{pmatrix} -\frac{1}{T_{2b}} & 0 & 0 & 0 \\ 0 & -\frac{1}{T_{2b}} & 0 & 0 \\ 0 & 0 & -\frac{1}{T_{2b}} + 2i\pi\sqrt{\left(\Delta\upsilon^2 + J_b^2\right)} & 0 \\ 0 & 0 & 0 & -\frac{1}{T_{2b}} - 2i\pi\sqrt{\left(\Delta\upsilon^2 + J_b^2\right)} \end{pmatrix} = \varepsilon_\lambda \delta_{\lambda\mu} \tag{21.74}$$

Here the ε_λ denote the eigenvalues.

Applying the same transformation to the inhomogeneous part of Eq. (21.68) yields ($\hat{\sigma}$ denotes the density matrix in the transformed frame, i.e. $\hat{\sigma} = \hat{S}_2 \hat{S}_1 \hat{\rho}$):

$$\hat{\sigma}_a(t) = \hat{S}_2 \hat{S}_1 \rho_a(t) = \frac{1}{2} \begin{pmatrix} \sqrt{2} \\ \sqrt{2} \\ -c \\ -c \end{pmatrix} exp(-k_{ab}t) \tag{21.75}$$

With this the solution for the density matrix elements becomes ($\lambda = 1.4$, index of eigenvalue ε_k):

$$\sigma_{b\lambda}(t) = \frac{\sigma_{a\lambda} k_{ab}}{\varepsilon_\lambda - k_{ab}} (exp(-k_{ab}t) - exp(\varepsilon_\lambda t)) \tag{21.76}$$

Assuming that the oscillating matrix elements σ_{b3} and σ_{b4} disappear, a quasi-stationary limit of $\sigma_{b\lambda}$ can be calculated. If $T_2 \gg 1/k_{ab}$ and $T_2 \gg 1/(\Delta\nu^2 + J_b^2)^{1/2}$ and $k_{ab} < (\Delta\nu^2 + J_b^2)^{1/2}$ it follows for $t \to \infty$:

$$\sigma_{b1}(\infty) = \frac{1}{2}\sqrt{2}$$

$$\sigma_{b2}(\infty) = -\frac{1}{2}\sqrt{2} \frac{J_b}{\sqrt{(J_b^2 + \Delta\nu^2)}} \tag{21.77}$$

$$\sigma_{b3}(\infty) = 0$$

$$\sigma_{b4}(\infty) = 0$$

Transforming back into the original frame gives:

$$\rho_{b1}(\infty) = \frac{1}{2} \left(1 - \frac{\Delta\nu J_b}{(J_b^2 + \Delta\nu^2)} \right)$$

$$\rho_{b2}(\infty) = -\frac{1}{2} \frac{J_b^2}{(J_b^2 + \Delta\nu^2)}$$

$$\rho_{b3}(\infty) = -\frac{1}{2} \frac{J_b^2}{(J_b^2 + \Delta\nu^2)}$$

$$\rho_{b4}(\infty) = \frac{1}{2} \left(1 + \frac{\Delta\nu J_b}{(J_b^2 + \Delta\nu^2)} \right) \tag{21.78}$$

The resulting spectral line can be calculated numerically from these elements of the density matrix. For the interpretation of the PHIP spectral line shape, the elements ρ_{b1} and ρ_{b4} are of particular interest, since they correspond to the level populations of the $|\alpha\beta\rangle$ and $|\beta\alpha\rangle$ states. Their dependence on J_b is shown in Fig. 21.8.

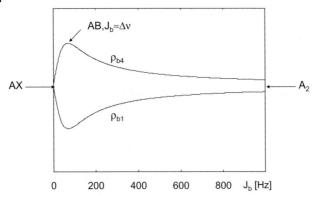

Figure 21.8 Calculation [86] of ρ_{b1} and ρ_{b4}, which are proportional to the populations of the $|\alpha\beta\rangle$ and $|\beta\alpha\rangle$ states in a PASADENA type PHIP experiment as a function of the coupling constant J_b. For a pure AX or A_2 system the populations are equal, while for an AB system deviations exist, which are strongest for $J_b \approx \Delta\nu$. Calculation with $\Delta\nu = 69$ Hz.

21.3.2
Experimental Examples of PHIP Spectra

As mentioned above, PHIP is a versatile tool for the study of catalytic reactions. In the following three different experimental examples of this remarkable power, taken from recent work from the Bargon group followed by a theoretical example of reaction pathway sensitivity from our group are given. The experimental examples show a PHIP experiment performed under ALTADENA conditions, a PHIP experiment performed under PASADENA conditions and a PHIP experiment followed by a heteronuclear polarisation transfer from ^{1}H to ^{13}C.

21.3.2.1 PHIP under ALTADENA Conditions
The first example (see Fig. 21.9) shows a ^{1}H-PHIP spectrum of the hydration of perdeuterated styrene obtained under ALTADENA conditions, i.e. with hydration outside the NMR magnet taken from Ref. [94]. The spectrum of the hydration product ethyl-benzene exhibits the typical strong spin polarized signals at 1 ppm and 3 ppm. Between 5 ppm and 6 ppm there are additional spin polarized signals, which stem from a side reaction. In Ref. [94] it was shown that the ratio of hydration versus geminal exchange is controllable by addition of CO.

21.3.2.2 PHIP Studies of Stereoselective Reactions
PHIP allows also the PASADENA investigation of the stereoselectivity of a reaction. A typical example is given in Ref. [70], where the stereoselective hydrogenation of 3-hexyne-1-ol under the influence of the cationic ruthenium complex [Cp*Ru(η^4-

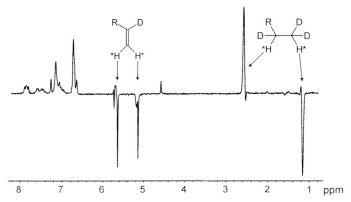

Figure 21.9 ^{1}H *in situ* PHIP NMR spectrum of the hydrogenation of styrene-d$_8$ using H$_2$Ru(PPh$_3$)$_4$ as catalyst measured under ALTADENA conditions (adapted from Ref. [94]).

CH$_3$CH=CHCH=CHCOOH)][CF$_3$SO$_3$] is shown in a ^{1}H-PHIP experiment (see Fig. 21.10). The spectrum reveals the characteristic PHIP polarization pattern of adsorptive and emissive lines. This polarization pattern proves the pair-wise transfer of the para-hydrogen to the substrate. The observed anti-phase coupling constant of 15.5 Hz is a typical value for an olefinic trans coupling constant and identifies the formation of the corresponding (*E*)-alkene by trans-hydrogenation of the substrate.

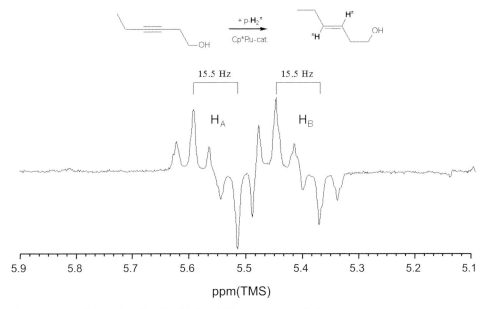

Figure 21.10 Olefinic region of a 200 MHz ^{1}H PHIP spectrum recorded during the hydrogenation of 3-hexyne-1-ol using Cp*Ru catalyst (adapted from Ref. [70]).

21.3.2.3 ^{13}C-PHIP-NMR

The last experimental example shows that PHIP is also a powerful starting point for sensitivity enhancement of NMR spectroscopy of hetero nuclei in catalytic reactions.

The homogeneously catalyzed hydrogenation of an unsaturated substrate with para-hydrogen leads not only to strong signal enhancements in ^{1}H NMR spectra, but can also give rise to strong heteronuclear polarization, in particular in ALTA-DENA type experiments where the hydrogenation is carried out in low magnetic fields. As a typical example taken from Ref. [74], the polarization transfer from protons to ^{13}C nuclei during the hydrogenation of 3,3-dimethylbut-1-yne with para-hydrogen and [Rh(cod)(dppb)]$^+$ as catalyst is shown in Fig. 21.11. In the single shot ^{13}C-NMR spectrum recorded *in situ* all ^{13}C resonances can be observed with good to excellent signal-to-noise ratios. The enhanced SNR is due to the PHIP effect resulting from a transfer of the initial proton polarization of the para-hydrogen to the carbon atoms.

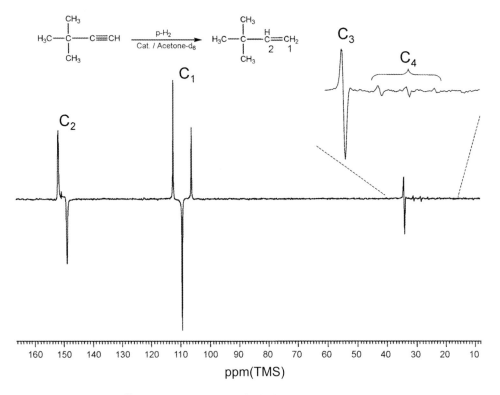

Figure 21.11 ^{13}C *in situ* PHIP spectrum after hydrogenation with para-hydrogen and polarisation transfer from protons to carbons. Note the extremely high signal/noise ratio in the single shot spectrum, which is a measure for the strong signal enhancement obtained by PHIP (adapted from Ref. [74]).

Due to the ALTADENA procedure, nuclear polarization is transferred to all magnetically active nuclei, since all resonance frequencies are virtually the same for all nuclei at the very low magnetic field of the Earth. The corresponding spin systems are of a high order, i.e. the difference in the resonance frequencies between the carbons (^{13}C) and the protons is small, compared with their coupling constants. This is an essential prerequisite for an efficient polarization transfer from protons to a large number of carbons.

21.3.3
Effects of Chemical Exchange on the Lineshape of PHIP Spectra

Equations (21.78) allow the calculation of PHIP spectra in the case of non-exchanging hydrogen atoms, i.e. for $k_{12} = 0$. If $k_{12} > 0$, however, the lineshape of the PHIP spectra does depend on the chemical exchange. Interestingly this allows one to extract information about reaction intermediates, which are not directly visible in the NMR spectra [62] owing to their short lifetime or low concentration.

Employing the quantum mechanical density matrix formalism it is possible to take into account the whole reaction pathway of Fig. 21.12 where both coherent and incoherent reaction pathways are present in the case of a PHIP experiment and convert the PHIP experiment into a diagnostic tool for all stages of a hydrogenation reaction. Such an analysis was performed in Ref. [62]. In these calculations the initial condition was that at the start of the reaction all molecules are in the p-H$_2$ state of site C, i.e., a pure singlet spin state, represented as a circle in Fig. 21.12, which shows the different possible reaction pathways. Moreover, all intermolecular exchange reactions were treated as one-sided reactions, i.e. the rates of the back reactions were set to zero.

Two different scenarios are analyzed, namely where the reaction goes as a two-step process, as depicted in Fig. 21.12(a) or where the reaction goes as a three-step process (Fig. 21.12(b)).

Accordingly in Fig. 21.12(a) only two sites $r = $ C or A are included. For simplicity, only the forward reactions are shown, but in the formalism the backward reactions are also included. In both sites C and A the possibility for incoherent exchange of the two hydrogen atoms H$_a$ and H$_b$, characterized by the rate constants k_{CC} and k_{AA} is included. In addition, the parameter p_{CA} is introduced, which describes the regio-selectivity of the reaction between C and A. It represents the probability of permutation of the two hydrogen atoms during the transfer from C to A and A to C, respectively. The step is completely regio-specific if $p_{CA} = 0$ or 1 and completely non-selective when $p_{CA} = 0.5$ and partially regio-selective for other values of p_{CA}. In the latter two cases the reaction would lead to isotope scrambling if the reaction is performed with an HD pair as substrate. It is found that this regio-selectivity requires that the two protons are labeled prior to the reaction by different Larmor frequencies, otherwise it does not affect the results, as is the case if C corresponds to free dihydrogen; however, the regio-selectivity is important when an additional intermediate B is included in the more general case of Fig. 21.12(b) in which the two former para-H$_2$ protons are chemically different.

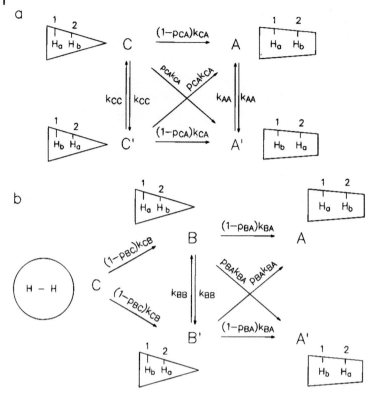

Figure 21.12 Formal two-site (a) and three-site (b) reaction models of a single (a) and two-step (b) hydrogenation reaction [62]. $p_{rs} = p_{sr}$, r, s = A to C represents the parameter characterizing the probability of permutation of the two hydrogen atoms during the interconversion between r and s. The step is called regio-specific if $p_{rs} = 0$ or 1, non-regio-specific if $p_{sr} = 0.5$, and otherwise it is called regio-selective.

In the following, the effects of these exchange processes and isotope scrambling on the level populations and line shapes of the PHIP experiment are shown. Details of the numerical calculations are found in the original paper [62].

Figure 21.13 shows the dependence of the density matrix elements ρ_{22} and ρ_{33} on the mutual exchange rate k_{AA} in a two-step experiment. It is evident that the mutual exchange removes the differences in the populations of the two levels $|\alpha\beta\rangle$ and $|\beta\alpha\rangle$ and thus will change the appearance of the PHIP spectra. Figure 21.14 displays this effect of an incoherent mutual exchange of the two protons in the product site A, characterized by the rate constant k_{AA} on the PHIP spectra. Without mutual exchange there are strong differences in the line intensity pattern and the outer lines are higher than the inner ones. In the case of mutual exchange however, the intensity pattern of the normal NMR spectrum is obtained. These magnetization transfer effects are most pronounced in the case of the AB case where $J_A / \Delta\nu_A \approx 1$, (bottom spectrum). By contrast, AX-type spectra are practically

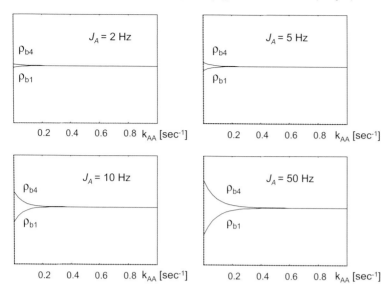

Figure 21.13 Calculation of ρ_{22} and ρ_{33} in a PASADENA type PHIP experiment as a function of the self exchange rate k_a for different spin systems [86]. In all cases the self-exchange leads to an averaging of the populations of ρ_{22} and ρ_{33}.

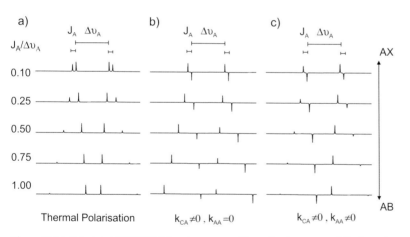

Figure 21.14 Calculated [62] NMR (a) and PHIP NMR (b,c) spectra of a two-proton spin system of compound A as a function of the ratio $J_A/\Delta\nu_A$, produced in a PASADENA experiment with $(\pi/4)$ x pulses by the two-site reaction C → A (reaction time $tr = 10$ s; rate constant $k_{CA} = 1$ s^{-1}). (b) Without self-exchange. The ratios of the absolute outer and inner line intensities differ from those of the normal NMR spectra (not shown). (c) a self-exchange of the two protons during the reaction time tr is introduced with $k_{AA} = 1$ s^{-1} resulting in relative absolute PHIP-signal intensities corresponding to those of the normal NMR spectrum (a) (adapted from Ref. [62]).

not influenced by the exchange, neglecting the minor broadening effect of the lines which is not important for the present discussion.

Calculations of the PASADENA pattern for regio-selecitvities p_{AB} between 0 and 1 showed that the resulting density matrices and, therefore, the calculated PHIP spectra, are independent of this parameter. This is the expected result because in p-H$_2$ the two protons are indistinguishable.

Figure 21.15 shows the dependence of the density matrix elements ρ_{b1} and ρ_{b4} on the mutual exchange rate k_{AA} in a three-step experiment. While the mutual exchange again removes the differences in the populations of the two levels $|\alpha\beta\rangle$ and $|\beta\alpha\rangle$, the strength of the effect now depends on the lifetime of the reaction intermediate B. If this lifetime is short compared to the inverse exchange rate, the density matrix elements are only weakly affected. If the life time is long enough, however, both levels get equally populated (Fig. 21.15(b)). Moreover the region-selectivity of the reaction now also strongly influences the population numbers (Fig. 21.15(c),(d)). Again these changes in the population numbers have a strong influence on the appearance of the PHIP spectra. Figure 21.16 shows the resulting PHIP signal patterns of A formed in the three-site-reaction PASADENA of Fig. 21.12(b). The four sets of spectra illustrate the influence of k_{BB} and of $J_B/\Delta\nu_B$. If B constitutes an AX spin system ($J_B/\Delta\nu_B$), the resulting spectra show no depen-

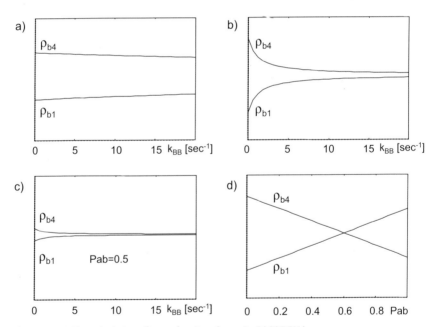

Figure 21.15 The calculation of ρ_{22} and ρ_{33} in a three-site PASADENA experiment as a function of the self-exchange rate k_{BB} for an AB system for different production and decay rates: (a) completely regio-specific, $K_{cb} = 1\ s^{-1}\ K_{ba} = 1\ s^{-1}$; (b) completely region-specific, $K_{cb} = 1\ s^{-1}\ K_{ba} = 100\ s^{-1}$; (c) completely non-regio-specific; (d) dependence on the region selectivity parameter [86].

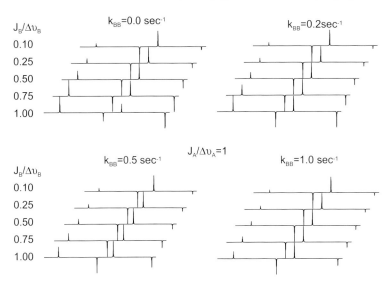

Figure 21.16 Calculated PHIP signal patterns (adapted from Ref. [62]) of a hydrogenation product A resulting in a PASADENA experiment in the presence of the three-site reaction C→B→A as a function of the ratio $J_B/\Delta\nu_B$. Note the differences in the line intensities (inner versus outer lines).

dence on k_{BB} and the ratio between the absolute intensities of the outer and the inner lines corresponds to the normal spectrum. However, if B constitutes an AB spin system, as expected in the case of substantial but not too large exchange couplings, large effects are observed and the ratio between the absolute intensities of the outer and the inner lines has changed. Thus, in principle, the incoherent dihydrogen exchange, the exchange and magnetic couplings, and the chemical shifts of the two protons in the intermediate B, all leave fingerprints which can be deciphered from the PHIP pattern of A.

In conclusion, the numerical simulations of the PHIP spectra show that the PHIP patterns do not only depend on the type of the experiment performed – e.g. ALTADENA in the absence and PASADENA in the presence of a magnetic field – but also on the properties of possible reaction intermediates where the reactants are bound to the employed transition metal catalyst. The important parameters of the intermediate are the chemical shifts and coupling constants of the former p-H$_2$ protons, especially their exchange couplings, as well as the rate of an incoherent dihydrogen exchange. In addition, the regio-selectivity of the hydrogenation step is a factor determining the PHIP-patterns, whereby the individuality of the former p-H$_2$ atoms arises from different chemical shifts in the intermediate.

In summary the calculations presented in Ref. [62] represent the missing theoretical link between the phenomena of incoherent and coherent dihydrogen exchange in transition metal hydrides and the PHIP effect. Moreover, PHIP is identified as a powerful and sensitive tool to study reaction pathway effects via analysis of the polarization patterns of the final hydrogenation products.

21.4
Symmetry Effects on NMR Lineshapes of Intramolecular Dihydrogen Exchange Reactions

Symmetry induced tunneling effects influence not only hydration reactions but also intermolecular hydrogen exchange reactions. In the case of a dihydrogen exchange the spatial Schrödinger equation (Eq. (21.5)) and its solutions (Eq. (21.11)) were discussed in a previous section. Since the eigenfunctions obey the Pauli principle they couple to spin functions in accordance with their symmetry and the spin of the hydrogen isotope (i.e. $1/2$ for ^{1}H, ^{3}H and 1 for ^{2}H) and the whole dynamics of the system can again be described purely in a spin Hilbert space.

The spin Hamiltonian of the system consists of three parts: The chemical shifts and/or quadrupolar interactions define the individual Hamiltonians of the spins $\hat{I}_1$ and $\hat{I}_2$ and the dipolar couplings and exchange interactions define the coupling Hamiltonian $\hat{H}_{1,2}\left(\hat{I}_1,\hat{I}_2\right)$. The mutual exchange of the two nuclei corresponds to a permutation of the two nuclei which exchange their individual chemical shifts and/or quadrupolar couplings with exchange rates $k_{12} = k_{21} = k$.

A relatively formal derivation in Ref. [11], which is based on the NMR lineshape analysis ideas of Alexander and Binsch [95, 96] shows that the whole dynamics is determined by the following Liouville von Neumann equation for the density matrix ρ_g:

$$\frac{d}{dt}|\rho_g) = -(\hat{\hat{W}}_A + \hat{\hat{K}})|\rho_g)$$
(21.79)

Here $\hat{\hat{W}}_A$ is the sum of the Liouville super operator $\hat{\hat{L}}_A$ and the relaxation super operator $\hat{\hat{R}}_A$. $\hat{\hat{K}}$ is the the self-exchange superoperator

$$\hat{\hat{K}} = -k(\hat{\hat{I}}_d - \hat{\hat{P}}_{12})$$
(21.80)

which describes the exchange of the two nuclei in Liouville space. Its elements are the identity operator $\hat{\hat{I}}_d$ and the permutation superoperator $\hat{\hat{P}}_{12}$. The latter is calculated from the permutation operator in Hilbert space $\hat{P}\left(\hat{I}_1,\hat{I}_2\right)$ (Eq. (21.16)) via

$$\hat{\hat{P}} = \hat{P}\left(\hat{I}_1,\hat{I}_2\right) \otimes \hat{P}\left(\hat{I}_1,\hat{I}_2\right)^*$$
(21.81)

where $\hat{P}\left(\hat{I}_1,\hat{I}_2\right)^*$ denotes the complex conjugate of operator of $\hat{P}\left(\hat{I}_1,\hat{I}_2\right)$.

21.4.1
Experimental Examples

The energy differences between the tunnel levels and thus the tunnel frequency depend very strongly on the hindering potential $2V_0$ and vary between zero and

10^{12} Hz (see Fig. 21.2) i.e. over roughly twelve orders of magnitude. As a result of this extremely broad possible dynamic range, no single spectroscopic technique is able to cover the range of possible tunnel splitting. It follows that the experiment must be chosen according to the size of the expected tunnel frequency. While slow tunneling processes can be studied by ^{1}H liquid state NMR spectroscopy, intermediate processes are accessible by ^{2}H solid state NMR spectroscopy and relaxometry and fast processes are accessible by incoherent neutron scattering (INS). Fortunately the dynamic ranges of these techniques overlap partially. From this it follows that, at least in principle, a complete tunneling kinetics can be determined by combining some of these techniques.

In the following three experimental examples of such quantum mechanical exchange processes are discussed. The examples are taken from liquid state NMR spectroscopy, solid state NMR spectroscopy and INS.

21.4.1.1 Slow Tunneling Determined by ^{1}H Liquid State NMR Spectroscopy

As discussed above in ^{1}H liquid state NMR a tunnel splitting, i.e. exchange coupling, and a conventional magnetic J-coupling have the same influence on the ^{1}H liquid state NMR spectra. This theoretical fact is nicely demonstrated in Fig. 21.17). It displays the superimposed experimental and calculated ^{1}H liquid state NMR hydride signals of (C5Me5)RuH3(PCy3) (Cy = cyclohexyl) **1** dissolved in tetrahydrofuran-*d8* [19]. At low temperatures site 2 exhibits a triplet splitting characterized by a temperature dependent exchange coupling constant $J_{12} = J_{23}$. A coupling constant J_{24} with the ^{31}P nucleus in site 4 cannot be resolved. Sites 1 and 3 are equivalent and exhibit the expected doublet splitting with the nucleus in site 2, as $J_{12} = J_{23}$. Furthermore, each line component is split by scalar coupling with the ^{31}P nucleus in site 4 with $J_{14} = J_{34} = 32$ Hz. In contrast to J_{14}, corresponding to a magnetic coupling, J_{12} represents an exchange coupling which increases strongly with temperature, as revealed by the typical AB2X signal pattern. Above 210 K, line broadening and coalescence occurs, eventually leading to a doublet with an average splitting of $J(^1H-^{31}P) = (J_{14} + J_{24} + J_{34})/3 = 22$ Hz. This splitting indicates that the classical exchange process observed is purely intramolecular. By lineshape analysis the exchange coupling constants J_{12} and the rate constants k_{HH} of the classical exchange are obtained.

21.4.1.2 Slow to Intermediate Tunneling Determined by ^{2}H Solid State NMR

Liquid state NMR experiments like above only allow the determination of slow coherent and incoherent tunnel rates, owing to the limited frequency range of the hydrogen chemical shifts. Faster tunneling processes can be studied by ^{2}H solid state NMR spectroscopy [11, 40].

Figure 21.18 compares experimental ^{2}H solid echo NMR spectra and the simulated ^{2}H FID-NMR spectra of the Ru-D$_2$ complex trans-[Ru(D$_2$)Cl(dppe)$_2$]PF$_6$ (Ru-D$_2$). At temperatures below 10 K the singularities of a satellite Pake pattern are visible as a splitting of the spectra at ± 60 kHz. This satellite Pake pattern is the

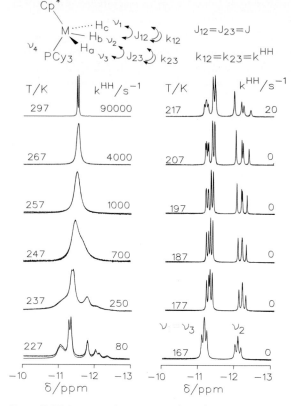

Figure 21.17 Superposed temperature dependent experimental and calculated 500 MHz 1H NMR hydride signals of $(C_5Me_5)RuH_3(PCy_3)$ (Cy = cyclohexyl) **1** dissolved in tetrahydrofuran-d_8 (adapted from Ref. [19]).

result of the coherent tunneling of the two η^2-bound deuterons in the complex. While the satellite transitions are fairly narrow at 5.4 K they start to smear out at higher temperatures. This smearing out is the effect of the incoherent tunneling which starts to dominate the dihydrogen dynamics and thus the spectral line-shape at higher temperatures.

At temperatures above 23 K the ^{2}H NMR line corresponds to a typical ^{2}H NMR quadrupolar Pake pattern with an asymmetry of $\eta = 0.2$. The satellite pattern has completely disappeared. The width of the line decreases slowly with increasing temperature, which is an indication of a weakening of the η^2-bond between the metal and dihydrogen.

Assuming the simple harmonic potential of Eq. (21.5) the height of the rotational barrier can be estimated. Using the value of $R_{HH} = 1$ Å, a rotational barrier of $2V_0 = 270$ meV (6.22 kcal mol^{-1}) is calculated.

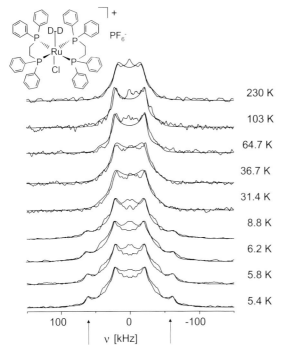

Figure 21.18 Experimental solid echo ^2H NMR spectra of the Ru-D$_2$ complex *trans*-[Ru(D$_2$)Cl(dppe)$_2$]PF$_6$ (Ru-D$_2$), measured in the temperature range 5.4 to 230 K. At temperatures below 8.8 K a splitting in the ^2H NMR lineshape is clearly visible (arrows). This splitting can be explained by a coherent tunneling of the two deuterons in the Ru-D$_2$ sample (simulation as ^2H FID-NMR experiment). The simulations were performed with $q_{zz} = 80\pm3$kHz (i.e. $q_{cc} = 107\pm4$kHz), $\eta = 0$ and a jump angle of $2\beta = 90°$ between the two tensor orientations (adapted from Ref. [40]).

21.4.1.3 Intermediate to Fast Tunneling Determined by ^2H Solid State NMR

Faster incoherent tunnelling processes can be studied by ^2H solid state NMR relaxometry [40, 41]. In these experiments the experimentally determined spin–lattice relaxation rates are converted into incoherent exchange rates. The latter are then evaluated, for example with the Bell tunnelling model described above.

As a first experimental example, Fig. 21.19 displays the result of the T_1 measurements on the same Ru-D$_2$ complex *trans*-[Ru(D$_2$)Cl(dppe)$_2$]PF$_6$ (Ru-D$_2$) as above. Due to the low sensitivity of the sample the spin–lattice relaxation rates were measured only at some selected temperatures. The lowest T_1 value (0.12±0.02 s) was found at 97 K. At low temperatures the T_1 data show strong deviations from simple Arrhenius behavior.

The exchange rates from the relaxation data are obtained for $K^{EFG} = 0.3\pi^2(60\text{ kHz})^2$ and the rate data from the spectra by lineshape analysis.

As a second experimental example, Fig. 21.20 presents the experimental results of the temperature dependence of the ^2H NMR spin–lattice relaxation time mea-

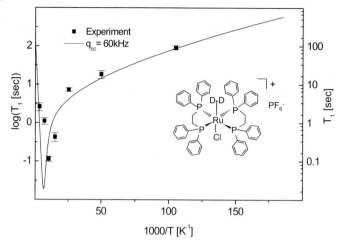

Figure 21.19 T_1 relaxation data of the Ru-D$_2$ complex ((adapted from Ref. [40]). Experimental points from lineshape analysis and relaxation measurements. The solid line is calculated from the exchange rates calculated from the modified Bell model using the value of $K^{EFG} = 0.3\pi^2 (60 \text{ kHz})^2$.

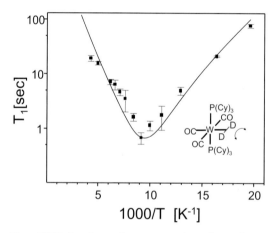

Figure 21.20 Experimental temperature dependence of the ^{2}H spin–lattice relaxation in the W-D$_2$ complex (adapted from Ref. [41]). The data exhibit deviations from Arrhenius behavior at low temperatures. The solid line is calculated from the exchange rates calculated from the Bell model.

surements on the W-D$_2$ complex W(PCy$_3$)$_2$(CO)$_3$(η^2-D$_2$), also known as the Kubas complex, together with a calculation of the relaxation times. The T_1 measurements in the temperature regime from 50 to 230 K show a strong temperature dependence of T_1 with a sharp minimum close to 110 K. At the minimum a T_1

relaxation time of (0.68 ± 0.15) s is found. It is evident that in the low temperature branch of the spin–lattice relaxation curve there are again deviations from a simple Arrhenius behavior, visible in a flattening of the curve.

21.4.1.4 Fast Tunneling Determined by Incoherent Neutron Scattering

Very fast coherent and incoherent tunneling processes can be studied by incoherent neutron scattering (INS). The basic mechanism of interpretation is closely related to the liquid state NMR experiment in the slow tunnel regime, however, now the energy scale of the tunnel splitting is of the order of fractions of meV, i.e. from 10^{10} to 10^{11} Hz. Here the INS lineshape of the energy gain and energy loss transitions are analyzed. They correspond to transitions between the singlet and triplet wavefunctions [19]. From this analysis the coherent tunnel frequency and the incoherent tunnel rates are determined and the spectral parameters J and k are elucidated. J determines the line position, k the increase in line width due to the presence of incoherent exchange.

As an experimental example of such an INS lineshape analysis, the INS spectra of the protonated isotopomer of the same tungsten dihydrogen complex $W(PCy_3)(CO)_3(\eta^2\text{-}H_2)$ are presented.

The superimposed experimental and calculated spectra are depicted in Fig. 21.21. Here the lineshape associated with the two rotational tunnel transitions of the complex is simulated as a function of the parameters J and k. For the sake of clarity, plots of the calculated line shapes of the outer rotational tunnel transitions without the contribution of the quasi-elastic center line are included. J increases only slightly with increasing temperature, in contrast to the Lorentzian line widths W which increase strongly. In other words, the lines broaden with increasing temperature until they disappear. One notes that the relative intensity of the singlet–triplet and the triplet–singlet transitions are almost the same over the whole temperature range covered, in contrast to the case of a thermal equilibrium between the singlet and the triplet states. This indicates that the singlet–triplet conversion rates are very slow in the sample measured, and that the actual relative intensity of the two peaks is arbitrarily dependent on the history of the sample.

21.4.2
Kinetic Data Obtained from the Experiments

The above described experiments allow determination of the coherent and incoherent dihydrogen exchange rates of the two complexes.

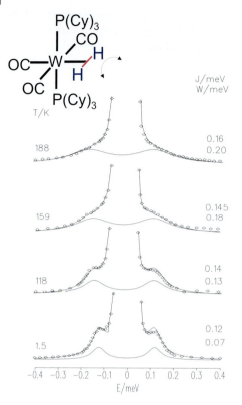

Figure 21.21 Superposed experimental and calculated INS spectra of $W(PCy)_2$ $(CO)_3(\eta\text{-}H_2)$ as function of temperature (adapted from Refs. [19, 42, 97]). W is the total line width in meV; J is the rotational tunnel splitting in meV (1 meV = 2.318×10^{11} Hz = 8.065 cm^{-1}).

21.4.2.1 Ru-D_2 Complex

In the case of the Ru-D_2 complex the data from the lineshape analysis and the T_1-relaxation data are combined in Fig. 21.22 which shows an Arrhenius plot of the temperature dependence of X_{12} and k_{12}.

While the temperature dependence of X_{12} is very weak and nearly linear in the temperature window between 5 and 20 K, the incoherent exchange rate k_{12} exhibits a strong non-Arrhenius behavior and varies from 5×10^3 s^{-1} at 5.4 K to ca. 2.5×10^6 s^{-1} at 103 K and ca. 10^{11} s^{-1} at 300 K. The rate data from both types of experiments overlap between 20 K and 100 K and there is an excellent agreement between the values. This indicates that both rates result from the same motional process.

The simulation of the temperature dependence was performed assuming a thermally activated tunneling process, described by a Bell type of tunneling. The high temperature rate in the tunnel model was chosen as 4×10^{12} s^{-1}, which is expected from the Eyring equation. Since the observed increase in k_{12} at low temperatures is not obtainable by a simple one-dimensional Bell model an effective power law potential was employed:

$$V_{eff}(T^{-1}) = V(T_0^{-1}) + (V(T_1^{-1}) - V(T_0^{-1}))\left(\frac{T^{-1} - T_0^{-1}}{T_1^{-1} - T_0^{-1}}\right)^G \qquad (21.82)$$

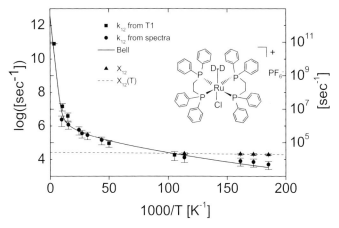

Figure 21.22 Arrhenius plot of the temperature dependence of the coherent tunneling and incoherent exchange rates in the Ru-D_2 sample (adapted from Ref. [40]), extracted from Fig. 21.18 and Fig. 21.19. The solid line is the result of a fit of the temperature dependence of the incoherent rates using a modified Bell tunnel model (see text). The dashed line is a simple linear fit of the coherent tunnel rates.

The best fit of the experimental rates (solid line in Fig. 21.22) was found for an exponent of $G = 0.7$. The effective potential varies between 268 meV(6.18 kcal mol^{-1}) at 5.4 K and 129 meV (2.97 kcal mol^{-1}) at 300 K. This effective potential gives a good reproduction of the experimental data. These rates were used to calculate the whole T_1 dependence (solid line in Fig. 21.19). Moreover there is an excellent agreement between the low temperature value of 268 meV and the value of 270 meV extracted from the ^{2}H NMR lineshape analysis.

This temperature dependent effective potential shows that a complete description of the temperature dependence of the rates needs at least a two-dimensional model, where the average R_{HH} and/or R_{RuH} distances are functions of the temperature.

21.4.2.2 W(PCy)$_3$(CO)$_3$(η-H$_2$) Complex

In the case of the W-H$_2$ complex NMR data from the deuterated complex and INS data from the protonated complex are available. This allows a comparison of the exchange rates and thus a determination of the H/D isotope effect of the reaction rate. For this the T_1 values from Fig. 20.21 are converted to rate constants of the D–D exchange and plotted together with the rate data of the H–H exchange from the INS spectra. The resulting curve (Fig. 21.23) shows a deviation from simple Arrhenius behavior at low temperatures. This deviation is evidence for the presence of a quantum mechanical tunneling process at low temperatures, similar to the tunneling observed in the Ru-D$_2$ sample. Comparison of these rate data with the H–H exchange rates determined from the lineshape analysis of the INS spectra of the protonated species reveals a strong isotope effect, which increases with lower temperatures.

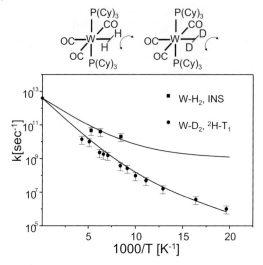

Figure 21.23 Arrhenius plot of the temperature dependence of the incoherent exchange rates in the W-D$_2$ sample, extracted from the ^{2}H T_1 data (adapted from Ref. [41]). The data are compared to data obtained on the W-H$_2$ complex, determined by INS. The solid lines are the results of fits of the temperature dependence of the incoherent rates using a Bell type tunnel model. The fits reveal a strong isotope effect, which is not solely attributable to a simple mass effect. The high temperature limit of the rates was chosen as 4×10^{12} s^{-1}, according to the Eyring equation.

Calculations with the Bell tunnel model reveal that this isotope effect is not solely explainable by the differences of the masses of the two hydrogen isotopes. Thus the activation energy must also have changed. This change in the activation barrier may be caused by isotope effects on the M–D and D–D versus M–H and H–H distances and/or by differences in the zero point energy of the ground or an activated state, which serves as the transition state for the tunneling. At low temperatures the latter is probably the major contribution to the strong isotope effect, since the quadrupolar coupling constant of the low temperature spectra (not shown) is practically constant at low temperatures.

21.5
Summary and Conclusion

This chapter presents some effects symmetry has on the rates and mechanisms of chemical reactions. The reaction kinetics of low mass groups like dihydrogen or dideuterium, in particular at low temperatures, is strongly influenced by quantum mechanical tunneling processes and the Fermi postulate of the symmetry of the wavefunction. These effects are particularly clearly manifested in NMR spectra, where coherent tunnel processes are visible as line splitting and incoherent tunnel processes are visible as line broadenings or relaxation rates. The complex kinetics of a superimposed coherent and incoherent exchange on both INS and NMR lineshapes is describable via two simple, temperature dependent spectroscopic parameters J and k, which are measures of the tunnel splitting and the incoherent exchange rate.

Symmetry effects are also important for the diagnostic application of parahydrogen in para-hydrogen induced polarization (PHIP) effects. While the spinphysics of these isotopomers is well understood there is still a large field of possible applications. Probably the current biggest challenge in this field is the development of biophysical and medical applications of these spin-isotopomers. There the extremely high spin polarization could be employed for sensitivity enhancements in MRI or functional studies of hydrogenase and related enzymes.

Finally, we wish to note that hydrogen is not the only small molecule, where molecular exchange symmetry causes the existence of a para- and an ortho- spin isotopomer. Water is another important example. In the gas phase it exists as para- or ortho-water. They are distinguishable by IR. Their concentration ratio is used in astronomy as a remote temperature sensor. The spin conversion mechanisms of these isotopomers are still an open field for future studies [98, 99].

Acknowledgements

This research has been supported by the Deutsche Forschungsgemeinschaft, Bonn, and the Fonds der Chemischen Industrie (Frankfurt).

References

1 A. Farkas, *Orthohydrogen, Parahydrogen and Heavy Hydrogen*, Cambridge University Press, Cambridge, 1935.

2 G. J. Kubas, R. R. Ryan, B. I. Swanson, P. J. Vergamini, H. J. Wasserman, *J. Am. Chem. Soc.* **116**, 451 (1984).

3 G. J. Kubas, *Acc. Chem. Res.* **21**, 120 (1988).

4 N. Aebischer, U. Frey, A. E. Merbach, *Chem.Comm.*, 2303 (1998).

5 G. Albertin, S. Antoniutti, S. Garciafontan, R. Carballo, F. Padoan, *J. Chem. Soc., Dalton Trans.*, 2071 (1998).

6 I. Alkorta, I. Rozas, J. Elguero, *Chem. Soc. Rev.* **27**, 163 (1998).

7 V. I. Bakhmutov, *Inorg. Chem.* **37**, 279 (1998).

8 T. Y. Bartucz, A. Golombek, A. J. Lough, P. A. Maltby, R. H. Morris, R. Ramachandran, M. Schlaf, *Inorg. Chem.* **37**, 1555 (1998).

9 M. G. Basallote, J. Duran, M. J. Fernandez-Trujillo, M. A. Manez, *J. Chem. Soc., Dalton Trans.*, 2205 (1998).

10 C. Bohanna, B. Callejas, A. J. Edwards, M. A. Esteruelas, F. J. Lahoz, L. A. Oro, N. Ruiz, C. Valero, *Organometallics*, 373 (1998).

11 G. Buntkowsky, H.-H. Limbach, F. Wehrmann, I. Sack, H. M. Vieth,

R. H. Morris, *J. Phys. Chem. A* **101**, 4679 (1997).

12 B. Chaudret, *Coord. Chem. Rev.* **180**, 381 (1998).

13 A. C. Cooper, K. G. Caulton, *Inorg. Chem.* **37**, 5938 (1998).

14 R. H. Crabtree, *J. Organomet. Chem.* **557**, 111 (1998).

15 M. A. Esteruelas, L. A. Oro, *Chem. Rev.* **98**, 577 (1998).

16 R. Gelabert, M. Moreno, J. M. Lluch, A. Lledos, *J. Am. Chem. Soc.* **120**, 8168 (1998).

17 S. Gründemann, H.-H. Limbach, V. Rodriguez, B. Donnadieu, S. Sabo-Etienne, B. Chaudret, *Ber. Bunsenges. Phys. Chem.* **102**, 344 (1998).

18 T. Hasegawa, Z. W. Li, H. Taube, *Chem. Lett.*, 7 (1998).

19 H.-H. Limbach, S. Ulrich, S. Gründemann, G. Buntkowsky, S. Sabo-Etienne, B. Chaudret, G. J. Kubas, J. Eckert, *J. Am. Chem. Soc.* **120**, 7929 (1998).

20 K. S. Macfarlane, I. S. Thorburn, P. W. Cyr, D. Chau, S. J. Rettig, B. R. James, *Inorg. Chim. Acta* **270**, 130 (1998).

21 W. S. Ng, G. C. Jia, M. Y. Huang, C. P. Lau, K. Y. Wong, L. B. Wen, *Organometallics* **17**, 4556 (1998).

22 P. L. A. Popelier, *J. Phys. Chem. A* **102**, 1873 (1998).

23 S. Sabo-Etienne, B. Chaudret, *Chem. Rev.* **98**, 2077 (1998).

24 S. S. Stahl, J. A. Labinger, J. E. Bercaw, *Inorg. Chem.* **37**, 2422 (1998).

25 P. G. Jessop, R. H. Morris, *Coord. Chem. Rev.* **121**, 155 (1992).

26 D. M. Heinekey, W. J. Oldham, *J. Chem. Rev.* **93**, 913 (1993).

27 P. A. Maltby, M. Steinbeck, A. J. Lough, R. H. Morris, W. T. Klooster, T. F. Koetzle, R. C. Srivastava, *J. Am. Chem. Soc.* **118**, 5396 (1996).

28 T. A. Luther, D. M. Heinekey, *Inorg. Chem.* **37**, 127 (1998).

29 A. Toupadakis, G. J. Kubas, W. A. King, L. B. Scott, J. Huhmann-Vincent, *Organometallics* **17**, 5315 (1998).

30 S. Q. Niu, L. M. Thomson, M. B. Hall, *J. Am. Chem. Soc.* **121**, 4000 (1998).

31 M. E. Cucullu, S. P. Nolan, T. R. Belderrain, R. H. Grubbs, *Organometallics* **17**, 1299 (1998).

32 A. J. Lough, R. H. Morris, L. Ricciuto, T. Schleis, *Inorg. Chim. Acta* **270**, 238 (1998).

33 F. Maseras, A. Lledos, E. Clot, O. Eisenstein, *Chem. Rev.* **100**, 601 (2000).

34 J. Matthes, S. Grundemann, A. Toner, Y. Guari, B. Donnadieu, J. Spandl, S. Sabo-Etienne, E. Clot, H. H. Limbach, B. Chaudret, *Organometallics* **23**, 1424 (2004).

35 A. Macchioni, *Chem. Rev.* **105**, 2039 (2005).

36 S. Lachaize, W. Essalah, V. Montiel-Palma, L. Vendier, B. Chaudret, J. C. Barthelat, S. Sabo-Etienne, *Organometallics* **24**, 2935 (2005).

37 H. H. Limbach, G. Scherer, M. Maurer, B. Chaudret, *Angew. Chem., Int. Ed. Engl.* **31**, 1369 (1990).

38 H. H. Limbach, G. Scherer, M. Maurer, *Angew. Chem.* **104**, 1414 (1992).

39 H. H. Limbach, G. Scherer, L. Meschede, F. Aguilar-Parrilla, B. Wehrle, J. Braun, C. Hoelger, H. Benedict, G. Buntkowsky, W. P. Fehlhammer, J. Elguero, J. A. S. Smith, B. Chaudret, in *Ultrafast Reaction Dynamics and Solvent Effects, Experimental and Theoretical Aspects,* Y,.Gauduel, P.J. Rossky (Eds.), American Institute of Physics., New York 1993, p. 225.

40 F. Wehrmann, T. Fong, R. H. Morris, H.-H. Limbach, G. Buntkowsky, *Phys. Chem. Chem. Phys.* **1**, 4033 (1999).

41 F. Wehrmann, J. Albrecht, E. Gedat, G. J. Kubas, H. H. Limbach, G. Buntkowsky, *J. Phys. Chem. A* **106**, 2855 (2002).

42 J. Eckert, G. J. Kubas, *J. Phys. Chem.* **97**, 2378 (1993).

43 T. Arliguie, B. Chaudret, J. Devillers, R. Poilblanc, *C. R. Acad. Sci. Paris, Ser. I* **305**, 1523 (1987).

44 D. M. Heinekey, N. G. Payne, G. K. Schulte, *J. Am. Chem. Soc.* **110**, 2303 (1988).

45 D. M. Heinekey, J. M. Millar, T. F. Koetzle, N. G. Payne, K. W. Zilm, *J. Am. Chem. Soc.* **112**, 909 (1990).

46 K. W. Zilm, D. M. Heinekey, J. M. Millar, N. G. Payne, P. Demou, *J. Am. Chem. Soc.* **111**, 3088 (1989).

47 D. Jones, J. A. Labinger, J. Weitekamp, *J. Am. Chem. Soc.* **111**, 3087 (1989).

48 S. J. Inati, K. W. Zilm, *Phys. Rev. Lett.*, **68**, 3273 (1992).

49 M. T. Bautista, K. A. Earl, P. A. Maltby, R. H. Morris, C. T. Schweitzer, A. Sella, *J. Am. Chem. Soc.* **110**, 7031 (1988).

50 G. A. Facey, T. P. Fong, D. G. Gusev, P. M. MacDonald, R. H. Morris, M. Schlaf, W. Xu, *Can. J. Chem.* **1899–1910** (1999).

51 C. R. Bowers, D. P. Weitekamp, *Phys. Rev. Lett.* **57**, 2645 (1986).

52 T. C. Eisenschmidt, R. U. Kirss, P. P. Deutsch, S. I. Hommeltoft, R. Eisenberg, J. Bargon, *J. Am. Chem. Soc.* **109**, 8089 (1987).

53 G. Buntkowsky, B. Walaszek, A. Adamczyk, Y. Xu, H.-H. Limbach, B. Chaudret, *Phys. Chem. Chem. Phys.* **8**, 1929 (2006).

54 I. F. Silvera, *Rev. Mod. Phys.* **52**, 393 (1980).

55 J. van Kranendonk, *Solid Hydrogen*, Plenum, New York 1983.

56 C. R. Bowers, D. H. Jones, N. D. Kurur, J. A. Labinger, M. G. Pravica, D. P. Weitekamp, *Adv. Magn. Reson.* **15**, 269 (1990).

57 R. Eisenberg, T. C. Eisenschmid, M. S. Chinn, R. U. Kirss, *Adv. Chem. Ser. D*, 47 (1992).

58 S. B. Duckett, C. L. Newell, R. Eisenberg, *J. Am. Chem. Soc.* **116**, 10548 (1994).

59 J. Barkemeyer, M. Haake, J. Bargon, *J. Am. Chem. Soc.* **117**, 2927 (1995).

60 R. Eisenberg, *J. Chin. Chem. Soc.* **42**, 471 (1995).

61 M. Haake, J. Barkemeyer, J. Bargon, *J. Phys. Chem.* **99**, 17539 (1995).

62 G. Buntkowsky, J. Bargon, H.-H. Limbach, *J. Am. Chem. Soc.* **118**, 867 (1996).

63 M. Jang, S. B. Duckett, R. Eisenberg, *Organometallics* **15**, 2863 (1996).

64 S. B. Duckett, C. J. Sleigh, *Progr. NMR. Spectrosc.* **34**, 71 (1999).

65 J. Natterer, O. Schedletzky, J. Barkemeyer, J. Bargon, S. J. Glaser, *J. Magn. Reson.* **133**, 92 (1998).

66 H. G. Niessen, D. Schleyer, S. Wiemann, J. Bargon, S. Steines, B. Driessen-Hoelscher, *Magn. Reson. Chem.* **38**, 747 (2000).

67 S. M. Oldham, J. F. Houlis, C. J. Sleigh, S. B. Duckett, R. Eisenberg, *Organometallics* **19**, 2985 (2000).

68 A. Eichhorn, A. Koch, J. Bargon, *J. Mol. Catal. A* **174**, 293 (2001).

69 A. Koch, J. Bargon, *Inorg. Chem.* **40**, 533 (2001).

70 D. Schleyer, H. G. Niessen, J. Bargon, *New J. Chem.* **25**, 423 (2001).

71 S. Wildschutz, P. Hubler, J. Bargon, *Chem.Phys.Chem* **2**, 328 (2001).

72 D. C. Bregel, S. M. Oldham, R. Eisenberg, *J. Am. Chem. Soc.* **124**, 13827 (2002).

73 A. Permin, R. Eisenberg, *Inorg. Chem.* **41**, 2451 (2002).

74 M. Stephan, O. Kohlmann, H. G. Niessen, A. Eichhorn, J. Bargon, *Magn. Reson. Chem.* **40**, 157 (2002).

75 L. D. Vazquez-Serrano, B. T. Owens, J. M. Buriak, *Chem. Comm.*, 2518 (2002).

76 H. Johannesson, O. Axelsson, M. Karlsson, *Compt. Rend. Phys.* **5**, 315 (2004).

77 R. B. Bell, *The Tunnel Effect in Chemistry*, Chapman &Hall, London & New York 1980.

78 F. Hund, *Z. Phys.* **43**, 805 (1927).

79 F. A. Cotton, *Chemical Applications of Group Theory*, Wiley, New York 1963.

80 C. Cohen-Tannoudji, B. Diu, F. Laloe, *Mechanique Quantique*, Hermann Editeurs, Paris 1977.

81 G. Buntkowsky, *Structural and Dynamical Studies with Dipolar and Quadrupolar Solid State NMR*, Free University, Berlin 1999.

82 R. Ernst, G. Bodenhausen, A. Wokaun, *Principles of NMR in One and Two Dimensions*, Clarendon Press, Oxford 1987.

83 S. Szymanski, P. Bernatowicz, *Annu. Rev. NMR* **54**, 1 (2005).

84 S. Szymanski, *J. Chem. Phys.* **104**, 8216 (1996).

85 C. Scheurer, R. Wiedenbruch, R. Meyer, R. R. Ernst, D. M. Heinekey, *J. Chem. Phys.* **106**, 1 (1997).

86 G. Buntkowsky, unpublished results.

87 G. Gamow, *Z. Phys. Chem.* **51**, 204 (1927).

88 G. Gamow, *Z. Phys. Chem.* **52**, 510 (1928).

89 E. U. Condon, R. W. Gurney, *Nature* **112**, 439 (1928).

90 E. U. Condon, R. W. Gurney, *Phys. Rev.* **33**, 127 (1929).

91 G. Wentzel, *Z. Phys.* **138**, 518 (1926).

92 H. A. Kramers, *Z. Phys.* **39**, 828 (1926).

93 L. Brillouin, *C. R. Acad. Sci.* **153**, 24 (1926).

94 H. G. Niessen, C. Ulrich, J. Bargon, *J. Labelled Compd. Radiopharm.* **43**, 711 (2000).

95 S. Alexander, *J. Chem. Phys.* **37**, 971 (1962).

96 G. Binsch, *J. Am. Chem. Soc.* **91**, 1304 (1969).

97 J. Eckert, G. J. Kubas, A. J. Dianoux, *J. Chem. Phys.* **88**, 466 (1988).

98 V. I. Tikhonov, A. A. Volkov, *Science* **296**, 2363 (2002).

99 H. H. Limbach, G. Buntkowsky, J. Matthes, S. Gründemann, T. Pery, B. Walaszek, B. Chaudret, *Chem. Phys. Chem.* **7**, 551 (2006).

Part VI
Proton Transfer in Solids and Surfaces

In Part VI the environments in which H transfers take place become more complex. In Ch. 22 Sauer reviews the field of proton transfer of positively charged OH groups in zeolites to the carbon atoms of unsaturated organic molecules forming carbenium ions. These processes represent elementary steps of catalytic reactions. Using *ab initio* calculations and Carr Parinello Molecular Dynamic techniques the interactions of substrates with the inner pore surfaces as manifested by vibrational spectroscopy are elucidatedm as well as the different reaction steps and the associated reaction energy reaction profiles. Especially interesting is the role of water which enables the transport of protons via jumps from H_3O^+ to H_2O.

In Ch. 23 Kreuer reviews the mechanisms of proton conduction in solid electrolytes of fuel cells. During operation, a protonic current equivalent to the electronic current passing through the external load is driven through the electrolyte and parts of the heterogeneous electrode structures. It is the proton conducting properties of the diverse electrolytes which are the subject of this chapter. The proton conduction consists of a multitude of consecutive proton transfer reactions in hydrogen bonded chains embedded in channels of solid materials containing water, other hydrogen bonded liquids, or heterocyclic groups such as imidazole derivatives covalently bound to a polymer matrix.

In Ch. 24 Aoki uses FT-IR reflection spectroscopy to monitor the transfer of protons and of water molecules from a layer of H_2O ice to a layer of D_2O ice as a function of time and external pressure. The H/D mutual diffusion coefficient measured at 400 K shows a monotonic decrease by two orders of magnitude as the pressure increases from 8 to 63 GPa. In order to separate molecular from protonic diffusion experiments were also carried out on $H_2{}^{16}O$ /$H_2{}^{18}O$ ice bilayer. Whereas molecular diffusion dominates under normal conditions, it is suppressed at high pressures. The protonic diffusion is assumed to take place via H_3O^+ and OH^- ions which can, however, not be observed.

Ch. 25 by Christmann is devoted to the interaction and reaction dynamics of hydrogen and simple molecules containing a hydroxy group such as water and methanol with transition metal surfaces. In particular, the possibility of H transfer via lateral diffusion or proton tunneling within the adsorbed layers is discussed. It is shown that lateral diffusion and transfer of H atoms does indeed occur via both classical diffusion and tunneling. The growth and structure of the respective layers containing OH groups is largely governed by H bonding effects leading, in practically all of the investigated cases, to a relatively "open" network of water or alcohol molecules. Isotopic scrambling thereby indicates an extraordinarily high mobility of H or D atoms.

Finally, Hempelmann and Skripov review in Ch. 25 hydrogen motions in metals important for the development of new hydrogen storage materials. While the behavior of hydrogen in a number of binary metal–hydrogen systems is well understood, a detailed microscopic picture of H diffusion in more complex compounds of practical importance has not yet evolved. A promising approach to investigation of these compounds is to combine a number of experimental techniques sensitive to different ranges of H jump rates such as NMR, QENS and inelastic relaxation with the neutron diffraction study of hydrogen positional parameters. The relation between the parameters of H motion and the structure of the hydrogen sublattice is emphasized.

22
Proton Transfer in Zeolites

Joachim Sauer

22.1
Introduction – The Active Sites of Acidic Zeolite Catalysts

Catalysis is one of the fundamental principles in chemical reactivity and catalysis by acids is an important subclass common to homogeneous, enzymatic and heterogeneous catalysis. Among the solid acids used in industrial processes, acidic zeolites are most important because they combine the acidic function with selectivity due to their nanoporous crystalline structure. Every drop of gasoline we burn in our car has seen at least one zeolite catalyst on its way through the refinery. Many chemical products, from bulk polymers to fine chemicals, are built up from hydrocarbons in crude oil or natural gas with the help of zeolite catalysts. However, zeolites are of outstanding interest also from the fundamental point of view. Their well-defined crystalline structure makes them very good candidates for studying the role of proton transfer in acidic catalysis.

Zeolites are three-dimensional crystalline networks of corner-sharing SiO_4 and AlO_4 tetrahedra. Because of the large flexibility of the Si–O–Si and Si–O–Al angles, a large variety of different frameworks with channels and cavities is possible, into which external molecules can penetrate. The negative charge of the framework (due to AlO_4 tetrahedra) is compensated by extra-framework cations, and if the charge-compensating cation is a proton the zeolite is a solid acid. The proton attaches to one of the four oxygen atoms of the AlO_4^- tetrahedron, thus forming a bridging hydroxy group, Si–O(H)–Al, which acts as a strong Brønsted site (see formula below).

The concentration of Al in the framework (and its distribution) are additional features by which different acidic zeolites can vary, however within limits. The Löwenstein rule forbids Al–O–Al links between AlO_4^- tetrahedra and the minimum Si/Al ratio is 1 (Löwenstein rule). Moreover, for low Si/Al ratios, not all charge-compensating cations can be protons and typically there are Na^+ ions left. High-silica zeolites are particularly interesting catalysts, because they contain bridging hydroxy groups as perfectly isolated active sites. Two convincing experiments that use the n-hexane cracking activity as test reaction show this. For the H-MFI catalysts (trivial name H-ZSM-5) the activity changes linearly with the

Hydrogen-Transfer Reactions. Edited by J. T. Hynes, J. P. Klinman, H. H. Limbach, and R. L. Schowen
Copyright © 2007 WILEY-VCH Verlag GmbH & Co. KGaA, Weinheim
ISBN: 978-3-527-30777-7

Si   Si   Si   Si   P   Al   P   Al
                                (P) (P)(Al) (Al)

−Si | +H/Al   −P | +H/Si

H   H
| |
Si   Al   Si   Si   P   Al   Si   Al
                                (P) (P)(Al) (Al)

Si/Al ratio between 100 000 and 20 [1]. Two catalysts, H-FAU and H-MFI, with the same Si/Al ratio of 26 have nearly the same specific activities, 11.4 and 8.5 mmol g^{-1} min^{-1}, respectively [2].

There is a third possible variation for acidic zeolite catalysts – the composition of the framework. The active Si–O(H) AlO$_3$ site in a high-silica zeolite is formally created from a nanoporous SiO$_2$ polymorph by replacing Si by Al/H. If we consider an AlPO$_4$ framework instead of an SiO$_2$ framework (as we do when we go from the mineral quartz to berlinite) and then replace P by Si/H (see formula above), we obtain a catalyst with the same Si–O(H)–AlO$_3^-$ active site but a different framework composition.

Strictly, these catalysts are not zeolites (this name is reserved for aluminosilicates), but aluminumphosphates (AlPOs) or silicon-aluminumphosphates (SAPOs). It is indeed possible to synthesize acidic high-silica (H-SSZ-13) [3] and SAPO catalysts (H-SAPO-34) [4] with the same framework structure (CHA) [5].

22.2
Proton Transfer to Substrate Molecules within Zeolite Cavities

It is assumed that, after adsorption into the zeolite, the initial activation of substrate molecules for further conversion is by proton transfer from the zeolite,

$$\{AlO_4\}_ZH + S \rightarrow (\{AlO_4\}_ZH \cdot S \rightleftharpoons \{AlO_4^-\}_Z \cdot HS^+)$$
$$\rightarrow (\{AlO_4^-\}_Z \cdot HP^+ \rightleftharpoons \{AlO_4\}_ZH \cdot P) \rightarrow \{AlO_4\}_ZH + P$$

Inspired by the chemistry in superacidic media, it has been speculated that zeolites may be superacids and able to protonate even saturated hydrocarbon molecules to yield carbonium ions as a first step in catalytic cracking. Later, doubts have been raised as to whether carbenium ions obtained by protonation of unsaturated hydrocarbons are stable intermediates that can be found experimentally or if they are merely transition structures in the catalytic reaction cycle. From the theoretical point of view the question is: Are they minima (stable structures) or saddle points (transition structures) on the potential energy surface? However, even if they are minima, they may be separated by very small barriers from products or their neutral complex counterparts and, hence, transient species that are difficult to detect experimentally.

There may also be additional deactivation channels for protonated species, for example carbenium ions can attach via C–O bonds to the zeolite framework and form alkoxides, as shown in Fig. 22.1 for isobutene.

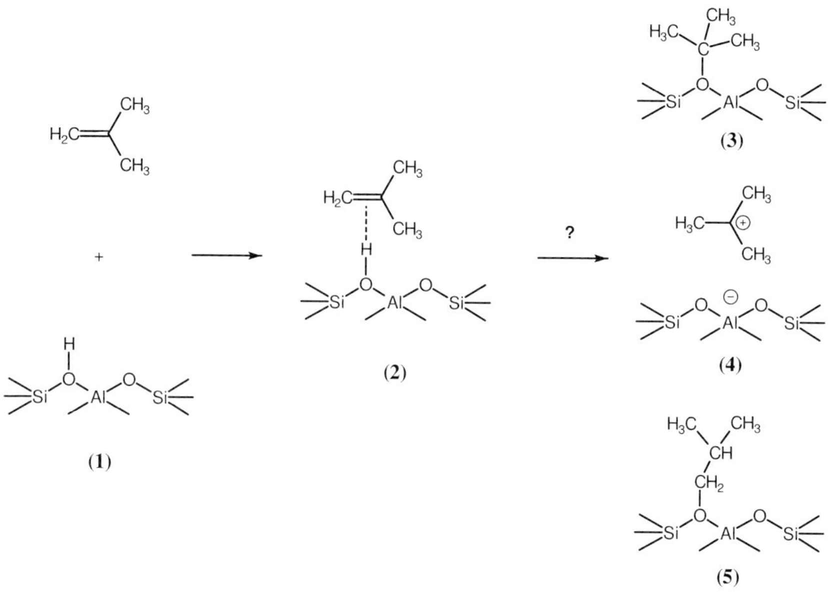

Figure 22.1 Possible products of proton transfer from a zeolitic Brønsted site to isobutene.

Whether or not the neutral adsorption complex or the ion-pair structure is more stable depends on the energy of the proton transfer reaction, ΔE_{PT},

$$\{AlO_4\}_Z H \cdot S \rightarrow \{AlO_4^-\}_Z \cdot HS^+ \tag{22.1}$$

It can be decomposed into the deprotonation energy, $E_{DP}(Z)$ of the zeolite,

$$\{AlO_4\}_Z H \rightarrow \{AlO_4^-\}_Z + H^+ \tag{22.2a}$$

the proton affinity of the substrate, $-E_{PA}(S)$,

$$H^+ + S \rightarrow HS^+ \qquad (22.2b)$$

the binding energy of the substrate on the neutral zeolite surface, $E_{neutral}(S)$,

$$\{AlO_4\}_Z H + S \rightarrow \{AlO_4\}_Z H \cdot S \qquad (22.3a)$$

and the binding energy of its protonated counterpart on the deprotonated zeolite surfaces, $E_{IP}(SH^+)$,

$$\{AlO_4^-\}_Z + HS^+ \rightarrow AlO_4^-\}_Z \cdot HS^+ \qquad (22.3b)$$

$$\Delta E_{PT} = E_{DP}(Z) - E_{PA}(S) - E_{neutral}(S) + E_{IP}(SH^+). \qquad (22.4)$$

Nicholas and Haw concluded that stable carbenium ions in zeolites are observed by NMR if the parent compound (from which the carbenium ion is obtained by protonation) has a proton affinity of 875 kJ mol^{-1} or larger [6]. Simulations by quantum methods showed that this statement is more general and that proton transfer from a H-zeolite to a molecule or molecular cluster occurs if its proton affinity is about that of ammonia (854 kJ mol^{-1}) or larger [7]. In the light of Eq. (22.4) this means that proton transfer occurs ($\Delta E_{PT} \leq 0$)) if $E_{DP}(Z) - E_{neutral}(S) + E_{IP}(SH^+)$ is smaller than 854 kJ mol^{-1}. Table 22.1 shows proton affinities and indicates in which cases and by which method protonated species have been detected.

22.3
Formation of NH$_4^+$ ions on NH$_3$ adsorption

A common experimental means of characterizing the acidity of zeolites is the use of probe molecules. IR spectra leave no doubt that ammonium ions are formed upon adsorption of ammonia in zeolites, the OH band characteristic for bridging Si–O(H)–Al sites disappears and NH$_4^+$ bending bands appear [12]. The energy of ammonia desorption,

$$\{AlO_4^-\}_Z \cdot NH_4^+ \rightarrow \{AlO_4\}_Z H + NH_3 \qquad (22.5)$$

is used to characterize the acid strength of zeolites. Usually it is obtained from temperature-programmed desorption, but true equilibrium values require calorimetric measurements. The adsorption energy defined by the reverse of Eq. (22.5) is composed of the energy of the (hypothetical) desorption of NH$_4^+$ and a subsequent proton transfer from NH$_4^+$ to the AlO$_4^-$ site on the zeolite,

$$E_{ad}(NH_3) = E_{IP}(NH_4^+) + E_{DP}(Z) - E_{PA}(NH_3). \qquad (22.6)$$

Tab. 22.1 Proton affinities, $H_{PA}(298)$ of molecules and clusters (kJ mol^{-1}) and observation of protonated species in zeolites.

Parent compound [a]	Obsd.[a]	MP4[b]	Other[c]	Proof ?
water	691		694	
benzene	750	746		
propene	751	742		
cyclopentene	766	759		
methanol	754		757	
toluene	784			
isobutene	802	805		
water dimer			806	
m-xylene	811			
3-methylphenyl-(2,4-dimethylphenyl)-methane		821[d]		DFT[d]
water trimer			853	DFT[f]
ammonia	854		858	IR[g]
hexamethylbenzene	861			UV–vis[e]
1-methylindene		878		NMR[b]
methanol dimer			887	DFT[h]
water tetramer			895	DFT[f]
1,3-dimethylcyclopentadiene		902		NMR[b]
pyridine	929	917		IR, NMR
1,5,6,6-tetramethyl-3-methylene-cyclohexa-1,4-diene		951		NMR[b]
3,6-dimethylene-cyclohexa-1,4-diene		1031[d]		DFT[d]

a Ref. [8].
b Ref. [6], except where otherwise noted.
c MP2/DZP, unpublished data.
d Ref. [9].
e Ref. [10].
f Ref. [7, 11].
g Ref. [12].
h Ref. [30].

Quantum chemical studies (Table 22.2) confirm the formation of NH_4^+ ions on interaction of ammonia with Brønsted sites in zeolites. The proton transfer energies are around -30 kJ mol^{-1}. No calculation that includes the full periodic structure of the zeolite has found a local minimum for a neutral adsorption complex. A hybrid QM/MM study considered high-silica zeolites with four different zeolite frameworks [13]. The energies of deprotonation indicate that H-FAU releases its proton most easily, yet the heat of NH_3 adsorption (Eq. (22.6)) is largest for H-MOR. The reason is that binding of NH_4^+ onto the zeolite surface, $E_{IP}(NH_4^+)$, is less favorable in the large pore zeolite FAU (12-rings of $SiO_{4/2}$ units) than in the smaller pores of the other zeolites (CHA, MOR, MFI) with 8- and 10-rings. These calculations assumed that the Brønsted site is created at the crystallographic T position at which Al is most stable. However, the energy differences for Al in different crystallographic T positions of a given framework are small, usually a few kJ mol^{-1}. Moreover, the Al distribution can also be controlled by the synthesis process and the preferred positions are not known. For H-MOR DFT calculations have shown that the location of the Brønsted site can have a large effect on the

Tab. 22.2 Proton transfer energy, ΔE_{PT}, deprotonation energy, E_{DP}, hypothetical binding energy of NH_4^+ on the deprotonated zeolite surfaces, $E_{IP}(SH^+)$, and energy of ammonia adsorption, $E_{ad}(NH_3)$, (kJ mol^{-1}) for Brønsted sites in different zeolite frameworks [13].

Zeolite[a]	FAU (47)	CHA (11)	MOR (47)	MFI (95)
ΔE_{PT}	-32	-35	-34	-29
E_{DP}	1252	1271	1277	1283
$E_{IP}(NH_4^+)$	-457 (2)	-476 (3)	-484 (2)	-480 (2)
$E_{ads}(NH_3)^{[b]}$	-109	-109	-116	-106
$E_{ads}(NH_3)$, HF+MP2[c]	-127	-128	-133	-123
$H_{ads}(NH_3)^{[d]}$	-113	-114	-119	-109
$E_{ads}(NH_3)$, DFT		$-123; -133^{[f]}$	$-126 \cdots -128^{[g]}$ $-152^{[h]}$	
$H_{ads}(NH_3)$, obsd.[e]	$-115 \cdots -130$	-155	-160	$-145; -150$

a In parentheses: Si/Al ratio assumed in the calculations.
b Hartree–Fock results.
c Final electronic energy including electron correlation (MP2).
d Calculated heat of adsorption, includes estimates for zero-point
   vibrational energy and thermal corrections (298 K).
e See Ref. [13] and for MOR also Ref. [14] for the original
   references to microcalorimetry.
f B3LYP, Ref. [16].
g PW91, Al in T1 and T2-sites, bidentate, Ref. [14].
h PW91, Al in T3-site, tetradentate, Ref. [14].

heat of NH_3 adsorption. Values of -126 to -128 kJ mol^{-1} (close to the results of Ref. [13]) are found for Al in sites 1 and 2, but for Al in site 3 a much larger value of -152 kJ mol^{-1} is obtained [14]. The reason is that in this case NH_4^+ can bind via four N–H$\cdots$O bonds to the zeolite framework. It coordinates with three N–H bonds to oxygen atoms of an 8-ring and points with one N–H bond into the so-called side pocket of the MOR framework. Only for H-FAU are the calculated heats of adsorption in the range of reported calorimetric results. For H-CHA, H-MFI and H-MOR they are much lower. Even if adsorption in the side pocket is assumed for H-MOR, the predicted heat of adsorption would be about 20 kJ mol^{-1} too low.

For adsorption of a series of structurally related molecules in the same zeolite, Eq. (22.6) implies a correlation between the measured heat of adsorption and the proton affinity of that molecule, provided that the ion-pair binding energy, $E_{IP}(MH^+)$ is constant or changes with the proton affinity. For adsorption of ammonia and methyl-substituted amines on H-MFI and H-MOR such a correlation has indeed been reported [15] with two deviations: The heat of adsorption of trimethylamine is about 30 kJ mol^{-1} smaller than expected from the correlation, and that of n-butylamine is about 25 kJ mol^{-1} larger. Both can be explained by deviations from the prevailing adsorbate structure – interaction via two hydrogen bonds with the zeolite framework. Protonated trimethylamine can form only one hydrogen bond with the zeolite framework, and this explains a lower $E_{IP}(HN(CH_3)_3^+)$. In the case of n-butylamine the van der Waals (dispersion) interaction of the butyl chain contributes to the binding on the zeolite surface in addition to the two hydrogen bonds.

In conclusion, due to its high proton affinity (858 kJ mol^{-1}) ammonia always is protonated in any H-zeolite. The hypothetical neutral adsorption complex, stabilized by hydrogen bonds, would be about 30 kJ mol^{-1} less stable [13] as measured by ΔE_{PT} defined in Eq. (22.4).

22.4
Methanol Molecules and Dimers in Zeolites

The proton affinities (Table 22.1) of methanol (754 kJ mol^{-1}) and water (691 kJ mol^{-1}) are significantly lower than that of ammonia (854 kJ mol^{-1}) and whether or not these molecules are protonated in H-zeolites has created lively debates in the literature.

For H-zeolites with high Si/Al ratios DFT calculations showed that a single methanol molecule in a cavity with one bridging $SiO(H)AlO_3$ site always forms a hydrogen-bonded complex and there is no proton transfer onto methanol. This was first shown for sodalite (SOD) [17] which has a small unit cell and is a hypothetical zeolite catalyst. Its proton form is not known and there is no way that methanol could penetrate into sodalite cages through its six-ring windows. Chabasite (CHA) also has a small unit cell, but is experimentally relevant (H-SSZ-13) [4] and methanol in H-CHA has been studied by Haase et al. [18, 19]. In contrast to the result for sodalite, for chabasite an ion-pair structure of protonated methanol within an 8-ring was found by DFT (energy minimization) [20]. This caused speculations [21] that "a direct correlation between zeolite structure and chemical activation of the adsorbate" might exist which could not be confirmed. Subsequent Car Parrinello molecular dynamics (CPMD) simulations for CHA [18] revealed that the ion-pair complex is a stationary point on the potential energy surface that is reached during MD only 4 times within 2.5 ps for a very short time. The global minimum structure is the neutral complex which is 18 kJ mol^{-1} more stable, see Ref. [22] for a later confirmation. Later studies addressed zeolites with large unit cells, FER and MFI [22] or TON, FER and MFI [19], and all found neutral adsorption structures. The calculated (PW91 functional) heat of adsorption in MFI, 131 kJ mol^{-1}, is close to the experimental value, 115±5 kJ mol^{-1} [23].

Only for methanol adsorption in H-FER [19, 22] did the two DFT studies not lead to a consistent picture. Both simulations find a neutral adsorption complex of methanol in the 10-ring channel. In addition, Stich et al. [22] find a neutral adsorption structure for methanol in the 8-ring channel of FER. However in a dynamics run at 300 K proton transfer occurs and the average structure corresponds to the $CH_3OH_2^+$ ion with two almost equal O_M–H distances [22]. It is presently not clear if this result reflects special properties of the FER framework or more technical difficulties. Use of a unit cell with a very short cell parameter in the c direction could have an effect, but a test with a doubled cell showed that this is probably not the case. The two simulations differed in the location of the bridging hydroxy group in the framework. Stich et al. [22] assume Al in the T4 position and the proton sits on O6, while Haase and Sauer [19] assume Al in the more stable T2 position with the proton on O1. Al in T2 is 38 kJ mol^{-1} more stable than in T4 [24]. For the less stable $Al^{(4)}O^{(6)}$ site, the energy of deprotonation is 29 kJ mol^{-1} lower than for the more stable $Al^{(2)}O^{(1)}$ site [19], and this could explain the tendency to protonate methanol at higher temperatures, cf. Eq. (22.4). Unfortunately, the Al-distribution in the experimental samples is unknown.

All these studies tell us is that it is not enough to look at the stationary points on the PES. Because of the flatness of the potential energy surface and similar energies of neutral and ion-pair adsorption structures, the dynamics of the system at realistic temperatures needs to be considered. The average structures obtained under these conditions may deviate significantly from the equilibrium structures. We will come back to this point in the water adsorption section.

The observed ^{1}H NMR chemical shift for methanol in H-ZSM-5 (9.4 ppm) was much larger than for liquid methanol (4.7 ppm) and in the same range as the shift observed for methanol in the FSO_3H–SbF_5–SO_2 superacid (9.4 ppm) [25]. As it was known that methanol forms stable oxonium species in superacids this was hinting at the formation of methoxonium species in H-zeolites although the authors have been cautious enough not to exclude alternative interpretations. Assignment to a methoxonium species was possible because there was not enough information about its chemical shift when interacting via hydrogen bonds with the zeolite surface. However, quantum calculations could provide such information. Figure 22.2 shows the ^{1}H NMR chemical shifts calculated for methanol and the methoxonium ion, both hydrogen-bonded to the zeolite surface [26]. The

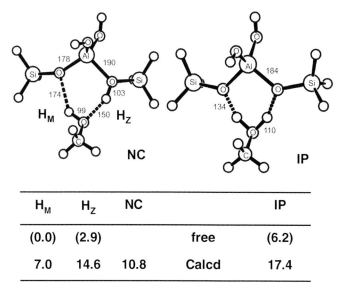

H$_M$	H$_Z$	NC		IP
(0.0)	(2.9)		free	(6.2)
7.0	14.6	10.8	Calcd	17.4

Figure 22.2 ^{1}H NMR chemical shifts (ppm) for methanol (neutral complex – NC) and methoxonium (ion pair complex – IP) interacting with the zeolite surface. H$_M$ – methanol proton, H$_Z$ – zeolite proton in the neutral adsorption complex. Results for the free methanol and methoxonium species are also given.

methoxonium protons undergo a huge downfield shift when interacting with the zeolite surface in the ion-pair complex. The calculated shift of 17.4 ppm is much larger than the observed value 9.4 ppm [25, 27]. In the neutral complex the Brønsted site proton also undergoes a similarly large shift due to the strong hydrogen bond, while the methanol proton extends a weaker hydrogen bond to the zeolite framework and its NMR signal shifts less. Due to a fast exchange of the zeolite and methanol protons an average shift of 10.8 ppm is obtained [26] which is close to the observed value.

Support for a neutral hydrogen-bonded adsorption complex comes also from a measurement of the distance between the methanol and the zeolite protons in H-MOR by wide line ^{1}H NMR at 4 K [28].

The result of 193–200 pm [28] is in good agreement with the calculated equilibrium distances for H-CHA (188 pm) [18], TON (189, 192 pm) and H-FER (190 pm) [19], while the H–H distance calculated for methoxonium adsorbed in CHA is much smaller, 158 pm [18]. The agreement of spectroscopic parameters with predictions for neutral adsorption complexes, of course, does not exclude the possibility that surface methoxonium ions would exist as a minority species in equilibrium. That the methoxonium ion is not a (metastable) local minimum structure but a transition structure is concluded from the quantum calculations mentioned above.

For the adsorption of two methanol molecules per bridging hydroxy groups (2:1 loading) studies on all zeolites, SOD [29], CHA [22, 30], FER [22] agree that a protonated methanol dimer is formed (Fig. 22.3). The obvious reason is the high PA of the methanol dimer that exceeds even the PA of ammonia (Table 22.1).

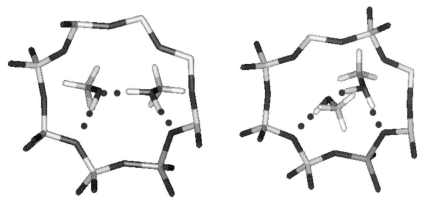

Figure 22.3 Protonated methanol dimer in zeolite chabasite (CHA) as predicted by CPMD simulations [30].

22.5
Water Molecules and Clusters in Zeolites

High-silica zeolites are known to be hydrophobic and it has also been long known that the water uptake at a given pressure is a function of the aluminum content, i.e. of the number of Brønsted sites [31–33]. At standard temperature and modest water pressure (e.g., $p/p_0 = 0.6$) typically four water molecules per Al are adsorbed, suggesting formation of a $H_9O_4^+$ species. Computationally, the interaction of water molecules with H-zeolites was first studied for cluster models of Brønsted sites [34, 35]. These calculations showed that the neutral adsorption complex is a minimum on the PES (stable structure), while the hydroxonium ion corresponds to a transition structure for proton exchange. Infrared spectra obtained for a loading level of a single water molecule per Brønsted site have been interpreted as due either to a neutral hydrogen-bonded molecule or to the formation of a hydroxonium ion. The calculations showed [34] that an ion-pair complex cannot explain the characteristic pair of bands in the hydrogen-bond region (2877 and 2463 cm^{-1}, see Fig. 22.4) while the neutral adsorption complex can, if one assumes that this pair of bands is due to a Fermi-resonance. Due to hydrogen bonding with the adsorbed H$_2$O molecule the zeolitic OH stretching is strongly red-shifted, and resonance with the overtone of the in-plane SiOH bending creates a window at the overtone position (see Ref. [36] for a recent model calculation).

Figure 22.4 shows the observed spectrum [37] and the assignment based on MP2 frequency calculations [34]. The predicted position of the zeolitic OH stretch band (red-shifted and broadened due to hydrogen bonding) falls close to the predicted position of the overtone of the in-plane SiOH bending. The bands at 3698 and 3558 are assigned to OH stretch of the free and hydrogen-bonded protons of the adsorbed H$_2$O molecule, and the bands at 1629 and 1353 to the HOH and SiOH bendings. The crucial experiment was the isotope substitution (^{18}O) of water [37] which showed no effect on the pair of bands at 2877 and 2463 cm^{-1} and thus clearly supported their assignment to vibrations of the zeolitic Brønsted site.

While these computational studies [34] were awaiting publication, a neutron diffraction study on H-SAPO-34 provided evidence for a protonated water molecule [38]. Whereas comments in the more popular press [39] stressed the apparent disagreement with previous calculations ("much of the confusion about how zeolites work stems from quantum calculations") and used the entertaining title "Quantum mechanics proved wrong", a comment to the original paper in the same issue

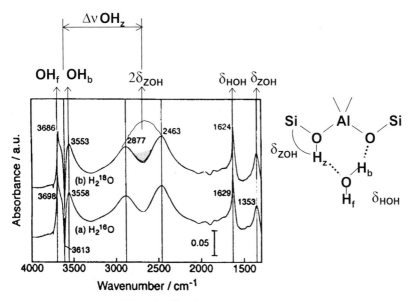

Figure 22.4 IR spectrum of H$_2$^{16}O and H$_2$^{18}O adsorbed on H-ZSM-5 (adapted from Fig. 2 in Ref. [37]). Shown is the assignment based on frequency calculations for models of the neutral adsorption complex [34]. The overtone of the in-plane SiOH bending (δ_{ZOH}) falls onto the red-shifted OH stretching frequency of the bridging hydroxy group of the zeolite, νOH$_z$.

pointed out that a loading higher than one H_2O molecule per Brønsted site may be responsible for protonation of water in the experimentally studied system [40]. Indeed, both cluster studies mentioned [34, 35] and a DFT study applying periodic boundary conditions [29] found that a second water molecule per Brønsted site yields $H_5O_2^+$ attached to the surface as an energy minimum structure. Subsequently, two periodic DFT simulations have been made of the H_2O/H-SAPO-34 system [7, 41] which analyzed the role of an increasing number of water molecules in detail. To match the composition of the experimental sample as closely as possible, a double cell of the CHA structure was chosen with one Brønsted site in one cell and two in the other cell [7] . Even for a loading of two water molecules (per two OH groups in one cell), the neutral water dimer proved to be the energy minimum structure, while $H_5O_2^+$ showed up only a few times during the 2 ps of the MD simulation at room temperature (Fig. 22.5). The lowest energy $H_5O_2^+$ structure is 18 kJ mol^{-1} above the neutral adsorption structure.

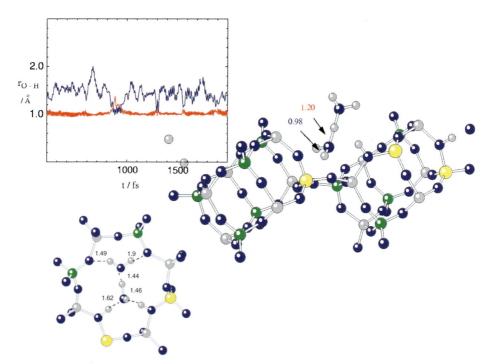

Figure 22.5 CPMD simulation (PW 91 functional) of two H_2O molecules per two Brønsted sites in H-SAPO (the other cell contains only one Brønsted site and one H_2O molecule) [7]. One of the Brønsted protons is residing on the SAPO-framework all the time, the distance of the other Brønsted proton to O of the nearest H_2O molecule shows large variations (upper curve). The lower curve shows the OH distance in this water molecule that is not involved in a H-bond with the second water molecule. Equal OH distances indicate formation of a protonated water dimer ($H_5O_2^+$). Most of the time the distance between the zeolitic proton and the nearest water molecule is much longer – corresponding to an adsorbed H_2O dimer shown at bottom left. Distances are given in Å.

When a simulation with four H_2O molecules per three Brønsted sites was made, something very interesting happened: One water molecule moved from the cage with one OH group into the cage with two OH groups and the three molecules together have a high enough proton affinity to form a $H_7O_3^+$ cluster stabilized by H-bonds with the wall of the SAPO material:

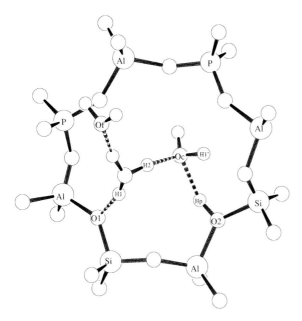

This perfectly fits our picture that the proton transfer depends on the proton affinity of the adsorbed molecule: the proton affinity of the water trimer (853 kJ mol^{-1}, Table 22.1) is about that of ammonia (858 kJ mol^{-1}).

For a single cell of H-SAPO with one Brønsted site, hydrogen-bonded H_2O was found for a loading of one molecule per site by CPMD simulations. For two molecules per site hydrogen-bonded H_2O dimers and protonated dimers, $H_5O_2^+$, were found [41]. Virtually identical energies of adsorption were obtained for both situations (Table 22.3). If the CHA framework is not aluminum phosphate, as in H-SAPO-34, but silica as in H-SSZ-13, for a loading of two to four molecules per site the ion-pair structure is found to be more stable (Table 22.3) [11]. Because of the small energy differences between the neutral complex and the ion-pair structure, the detailed answer depends on the specific density functional applied in the calculations. Table 22.3 shows that the Becke-Lee-Yang-Parr (BLYP) functional yields smaller adsorption energies and gives more weight to neutral adsorption complexes than the Perdew-Wang 91 (PW91) and Perdew-Burke-Ernzerhofer (PBE) functionals. The latter two belong to the same "family" of functionals and are expected to yield very similar results. For a loading of two molecules per site, BLYP predicts the neutral complex to be slightly more stable than the ion-pair structure, but the adsorption energies differ only by about 1 kJ mol^{-1}.

Tab. 22.3 Energy of water adsorption per molecule, $E_{ad}(H_2O)$ (kJ mol^{-1}) on Brønsted sites in zeolites and aluminumphosphates with chabasite (CHA) structure for different H_2O loadings.

		H/Al-SiO$_2$ (H-SSZ-13)				H/Si-AlPO$_4$(H-SAPO-34)	
Loading, n H$_2$O/m H(Al)		BLYP[a]		PBE[a]		PW91[b]	
Average	Cell1;Cell2	NC	IP	NC	IP	NC	IP
1/1	1/1	62	–	76	–	81	
2/1	2/1	46	45	–	61	67	67
2/1	2/1;2/1	48	47	–	63		
2/1	1/1;3/1	49		63			
3/1	3/1	–	45	–	59		
4/1	4/1	–	45	–	60		

a Ref. [11].
b Ref. [41].

For this loading, another interesting result is obtained, if a double cell is used for the simulation. A heterogeneous distribution of one molecule per site in the first cell and three molecules per site in the second cell is energetically slightly more stable than (BLYP) or equally stable as (PBE) the homogeneous distribution. This may have implications for the interpretation of experiments. For example, an IR spectrum obtained for an average loading of 2:1 may be composed of spectra for 1:1 (neutral hydrogen-bonded) and 3:1 complexes (ion-pair structures).

Table 22.4 summarizes the energies of adsorption for the first water molecule on a Brønsted site. The most reliable calculations have been made for H-CHA and

Tab. 22.4 Energy and enthalpy of water adsorption, $E_{ad}(H_2O)$ and $H_{ads}(H_2O)$ (kJ mol^{-1}) on Brønsted sites in zeolites for a loading of 1 H$_2$O/1 H(Al).

Method	Model	$E_{ad}(H_2O)$	$H_{ads}(H_2O)$	Ref.
MP2	cluster	58	45 (0 K)	34
		70	60 (0 K)	35
B3LYP	cluster	72		16
B3LYP/MNDO	cluster/CHA	82		
B3LYP	CHA	84		
PBE	CHA	75		42
MP2/PBE	cluster/CHA	78	73 (298 K)	42
calorimetry	MFI		90 ± 10	23
isotherms	MFI		80 ± 10	33

the most accurate value has been obtained by an MP2/PBE hybrid method and extrapolated to the complete basis set limit [42]. The predicted heat of adsorption at 298 K, 73 kJ mol^{-1}, can be compared with heats of adsorption of 80 ± 10 and 90 ± 10 kJ mol^{-1} obtained from isotherms [33] and calorimetric measurements [23], respectively, for a different zeolite, H-ZSM-5. Comparison of DFT calculations with the most reliable computational result, MP2/PBE, indicates that PBE gives more reliable results than BLYP. For loadings of two to four water molecules, DFT adsorption energies are rather constant and about 15 kJ mol^{-1} lower than for a loading of one molecule per site (Table 22.4). This is in agreement with the observed decrease of the heat of adsorption for H-ZSM-5 from 80 ± 10 ($n = 1$) to 63 ± 10 ($n = 2$–4).

Figure 22.6 (top, right) shows the energy minimum structure of the protonated water trimer in H-CHA [11]. It is an open trimer structure which is protonated on a terminal molecule, $H_3O^+(H_2O)_2$, rather than on the central molecule as in the corresponding gas phase species (Fig. 22.6, top insert bottom right). This is obviously a consequence of the stabilization by a strong hydrogen bond with the negatively charged AlO_4^- framework site nearby. The $H_2O(H_3O^+)H_2O$ structure is

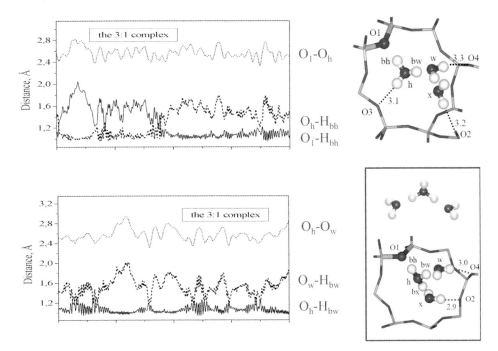

Figure 22.6 CPMD simulation (BLYP functional) of three H_2O molecules per one Brønsted site of H-CHA [11]. Left: characteristic distances along a 4 ps trajectory for the hydrogen bond between the Brønsted site and a terminal H_2O molecule of the trimer and between the terminal and central H_2O molecules of the trimer (bottom). The bottom right insert shows the protonated H_2O trimer in the gas phase with the H_3O^+ in the center, and the corresponding structure in H-CHA, which is a local minimum. Distances are given in Å.

a local minimum in H-SSZ-13 (Fig. 22.6, insert bottom right), 14 kJ mol^{-1} above the global minimum structure. Molecular dynamics simulations on a DFT potential energy surface (CPMD) show frequent proton jumps between the water trimer and the zeolite. Figure 22.6 (left) shows several atomic distances along the MD trajectory. Soon after starting the bh proton leaves the water cluster and makes several attempts to jump back. However, after only about 2 ps proton transfer occurs again and the proton stays for the rest of the simulation time on the water cluster. The consequences for the average structure at 350 K is a shortening of the O1(zeolite)–Oh distance from 258 to 256 pm, and a lengthening of the O_h–H_{bh} distance from 104 to 113 pm, which means a shift of the bridging proton bh from the water cluster (O_h) to the zeolite (O1).

The distances between the "left" (O_h) and the central water molecules (O_w) are given in the left bottom part of Fig. 22.5. There are several attempts by the bw proton to jump to the central water molecule, but they are not successful. The average O_h–O_w distance along the trajectory is 256 nm, while the energy minimum distance is 247 pm. This increase in the intermolecular distance is accompanied by a shortening of the O_h–H_{bw} bond from 111 pm (minimum structure) to 108 pm at 350 K. Hence, even at higher temperature, there is no indication for a proton transfer to the central water molecule. In turn, trajectories started at the $H_2O(H_3O^+)H_2O$ local minimum structure never reached the global minimum structure. This confirms the $H_3O^+(H_2O)_2$ structure for the protonated water trimer in H-SSZ-13 and points to the important role of the environment in the structure of protonated water clusters.

We also conclude that proton transfer between the water trimer and the zeolitic Brønsted site occurs on the picosecond time scale.

22.6
Proton Jumps in Hydrated and Dry Zeolites

In previous sections we have considered proton transfer between the zeolitic Brønsted site and adsorbed proton accepting molecules or clusters. For an unloaded zeolite, there are four oxygen sites around Al to which the proton can be attached. In most zeolites proton affinities are different for these oxygen sites and some will be preferred. For zeolite FAU, two OH frequencies and two ^{1}H NMR shift signals can be experimentally resolved; these are unequivocally assigned to O^{1}H and O^{3}H sites. This is in agreement with quantum calculations for isolated sites (Si/Al = 47) which yield the stability sequence (relative energies, kJ mol^{-1}, in parentheses) O1 (0.0) < O3 (9.5) < O4 (10.0) < O2 (18.3) [43]. For higher temperatures, on-site proton jumps between the different oxygen atoms of an AlO$_4^-$ site may be possible.

Translational proton motion through the zeolite lattice appears to be much less likely because the proton has to leave the Al$^{(-)}$–O–Si site and move to Si–O–Si sites that have a lower proton affinity. Hence, it is expected that the barriers for "intersite" jumps that occur between neighboring TO$_4$ tetrahedra with T = Al$^-$, Si are higher. DFT calculations for a typical Brønsted site in H-MFI (Si/Al = 95) [44] yield an inter-site barrier of 127 kJ mol^{-1} for leaving the AlO$_4$ site. The highest barrier for proton transfer from AlO$_4$ site to AlO$_4$ site along a path of Si–O–Si sites is 202 kJ mol^{-1}. From impedance spectroscopy effective barriers of 100 to 126 kJ mol^{-1} have been inferred for Si/Al ratios between 75 and 500 and temperatures above 423 K. If impedance spectroscopy probes complete translational proton motion between neighboring Brønsted sites the calculated barriers deviate by as much as 70–100 kJ mol^{-1} from the observed ones. This raises the question as to whether defects are responsible for the effective barriers derived from experiments. In contrast to on-site jumps, a "vehicle" mechanism cannot explain this discrepancy because addition of even 3 vol% of water reduces the barrier by not more than 20 kJ mol^{-1} [44].

Quantum calculations for the six different on-site jump paths between the oxygen atoms of the AlO$_4^-$ site (Fig. 22.7) have been made for three different framework structures, CHA, FAU and MFI [43]. For CHA and FAU there is only one crystallographically distinct AlO$_4^-$ site, while for H-MFI there are 12, and one of them (Al7) has been chosen as a representative site for the calculations. For all three structures the barriers vary widely between 70 and 102 kJ mol^{-1} for CHA, between 68 and 106 kJ mol^{-1} for FAU, and between 52 and 98 kJ mol^{-1} for MFI (values include zero-point vibrational energy contributions) [43]. The fact, that the O–Al–O angle substantially narrows in the transition structure (Fig. 22.7(a)) raised speculations that the barrier height may correlate with this angle, either in the transition structure or in the initial structure. However, the calculations for different zeolite frameworks [43] could not confirm that and rather pointed to the importance of (i) the local framework flexibility that allows the Al–O–Al angle to close up to 76–80° without too much energy penalty, together with (ii) the overall flexibility of the zeolite lattice, and (iii) stabilization of the proton in the transition structure by interactions with neighboring oxygen atoms of the framework. All three factors together show up in a plot of the barrier height against the size of the alumosilicate ring that the proton has to pass for a particular jump path

(a)

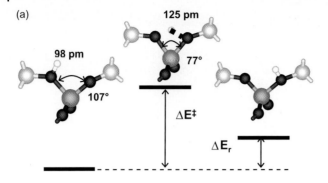

(b)

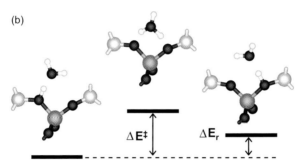

Figure 22.7 Reaction profile for on-site proton jumps on an AlO$_4$H site with inequivalent proton positions for dry zeolites (a) and hydrated zeolites (1:1 loading) (b). $\Delta E^{\ddagger}$ – barrier height, ΔE_r – reaction energy.

(Fig. 22.8). For all zeolites, proton jumps occurring within six-membered rings have the lowest barriers, while higher barriers are found for five- and four-membered rings (due to lower flexibility), but also for eight-membered and larger rings due to fewer oxygen atoms nearby that could stabilize the proton.

For jumps from the most stable proton sites transition state theory yields rates of the order of 1–100 s^{-1} at room temperature and of the order of 10^5–10^6 at 500 K [43]. Experimentally, proton jumps have been studied using various variable temperature ^{1}H NMR techniques [45, 46], but the proton jump barriers reported for H-FAU (61–78 kJ mol^{-1}) and H-MFI (18–45 kJ mol^{-1}) vary widely and are significantly lower than the calculated barriers (68–106 and 52–98 kJ mol^{-1}, respectively). Tunneling cannot be the reason, because the crossover temperature above which tunneling becomes negligible, T_x, is around room temperature [43] and the NMR experiments are carried out above room temperature. The experiments agree in reporting lower barriers for H-MFI than for H-FAU and this is also found in the calculations. The ^{1}H-NMR experiments are based on averaging the dipolar Al–H interaction, which requires that the proton visits all four oxygen atoms of the AlO$_4^-$ site starting from the most stable position. For H-FAU there is no such

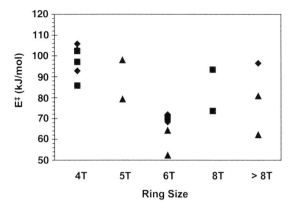

Figure 22.8 Dependence of the proton jump energy barriers on ring size nT for H-CHA (■), H-FAU (◆) and H-MFI (▲) zeolites.

path with a barrier lower than 93 kJ mol^{-1}, while for H-MFI there is a path with the highest barrier of 64 kJ mol^{-1}.

The most likely explanation for the discrepancy between calculated and NMR-derived barrier heights is the presence of residual amounts of small molecules like water or ammonia left over from the preparation process. Such molecules may significantly reduce measured barriers for proton motion by a vehicle mechanism (Fig. 22.7 (b)). A simple calculation shows that already for a coverage of Brønsted sites with water molecules at the ppm level the kinetics is dominated by the very much faster H_2O assisted jumps [47]. A careful computational study (which goes beyond DFT for the H_2O–Brønsted site complex, but includes the full periodic zeolite at the DFT level) yields barriers (including zero-point vibrational contributions) of 65 and 20 kJ mol^{-1} for O^1–O^2 jumps in dry and water-loaded H-CHA (1:1), respectively [42]. At room temperature, this increases the jump rate by eight orders of magnitude from 40 to 30×10^8 per second. The nanosecond time scale at which H_2O-assisted proton jumps can hence be expected is not accessible by CPMD simulations, which typically are run for picoseconds. This explains that during CPMD simulations for the 1:1 H_2O/H-CHA system mentioned above, proton jumps from one framework oxygen to another one via a hydroxonium transition structure (Fig. 22.7(b)) have not been observed.

22.7
Stability of Carbenium Ions in Zeolites

Carbenium ions can be formed by proton transfer from the Brønsted site to an unsaturated hydrocarbon which requires a negative proton transfer energy, Eq. (22.4).

$$\{AlO_4\}_Z H + C_n H_{m-2} \rightarrow \{AlO_4^-\}_Z \cdot C_n H_{m-1}^+ \tag{22.7}$$

Alternatively, carbenium ions can be formed by hydride abstraction from a saturated hydrocarbon,

$$\{AlO_4\}_ZH + C_nH_m \rightarrow \{AlO_4^-\}_Z \cdot C_nH_{m-1}^+ + H_2 \qquad (22.8)$$

The lifetime of the carbenium ion formed will be limited by transferring a proton back to the zeolite, thus completing the dehydrogenation of the hydrocarbon. Hydride abstraction from xylene is assumed to be the initial step in its disproportionation into toluene and trimethylbenzene [9]. The parent compound (7, Fig. 22.9) of the carbenium ion formed (6) has such a high proton affinity (1031 kJ mol^{-1}, Table 22.1) that proton transfer back to the zeolite does not occur at all. However, the lifetime of carbenium ions in zeolites is not only limited by proton transfer, but also formation of a C–O bond between the carbenium ion and a framework oxygen atom, yielding an alkoxide, needs to be considered. In ferrierite (FER) the alkoxide of 6 is found to be 50 to 60 kJ mol^{-1} more stable than the carbenium ion [9].

Table 22.1 lists three examples of cyclic alkenyl carbenium ions that live long enough in zeolites to be detected by NMR [6]. Obviously, alkoxide formation is not favored and the proton affinities of their parent hydrocarbon compounds are so large that they win the competition with the zeolite framework for the proton.

An obvious candidate for a stable noncyclic carbenium ion is the *tert*-butyl cation observed in superacidic media. Even if the proton affinity of isobutene (Table 22.1) does not make it very likely that *tert*-butyl cations will exist in zeolites, several quantum chemical studies have localized stationary points for *tert*-butyl cations in zeolite and found that they are less stable than the adsorption complex, but are similar in stability to surface butoxides. Because of technical limitations vibrational analysis, which could prove that this cation is a local minimum on the potential energy surface, that is a metastable species, have only recently been made. Within a periodic DFT study of isobutene/H-FER a complete vibrational analysis for all atoms in the unit cell was made [48], and as part of a hybrid QM/MNDO study on an embedded cluster model of isobutene/H-MOR a vibrational analysis was made with a limited number of atoms [49]. Both reached the

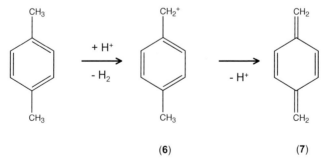

(6) (7)

Figure 22.9 Carbenium ion, **6**, obtained by hydride abstraction from xylene and its deprotonation product, **7**.

conclusion that the *tert*-butyl cation is a local minimum (Fig. 22.10 (b)) and, hence, a possible intermediate with a stability comparable to that of isobutoxide.

However, care has to be taken when applying DFT to hydrocarbon species in zeolites. The currently available functionals do not properly account for dispersion, which is a major stabilizing contribution for hydrocarbon–zeolite interactions. Due to the size of the systems it is difficult to apply wavefunction-based methods such as CCSD(T) or MP2. Thanks to an effective MP2/DFT hybrid approach and an extrapolation scheme energies, including the dispersion contribution, are now available for the different hydrocarbon species of Fig. 22.1 [50]. Fig. 22.10(c) shows the following surprising results: (i) The predicted energy of adsorption (70 kJ mol^{-1} at 0 K) is of the same order of magnitude as estimates based on experiments for related molecules (50–63 kJ mol^{-1}). (ii) With respect to isobutene in the gas phase separated from the zeolite, the *tert*-butyl cation is much less stable (–17 kJ mol^{-1}) than the isobutoxide (–48 kJ mol^{-1}). The reason is that dispersion contributes substantially less to the stabilization of the *tert*-butyl cation than to the stabilization of the adsorption complex or the isobutoxide. As result, the proton transfer energy increases from 24 kJ mol^{-1} (DFT) to 59 kJ mol^{-1} (MP2/DFT) and it seems very unlikely that the *tert*-butyl cation will be detected in zeolites, even as a short-lived species.

(a)

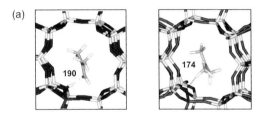

(b)

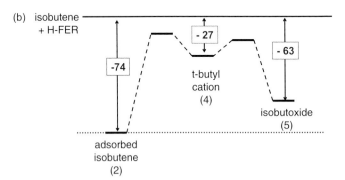

Figure 22.10 Energy profile for possible products of proton transfer from a zeolitic Brønsted site to isobutene, see Fig. 22.1. Standard heats of formation for intermediates 2, 4 and 5 obtained by hybrid MP2/DFT calculations [50], barriers between them are tentative, after Ref. [49]. Structures of (a) adsorbed isobutene (2) and (b) *tert*-butyl cation (4) [48] are also shown.

References

1 W. O. Haag, R. M. Lago, P. B. Weisz, *Nature (London)* **1984**, *309*, 589.

2 J. R. Sohn, S. J. DeCanio, P. O. Fritz, J. H. Lunsford, *J. Phys. Chem.* **1986**, *90*, 4847.

3 S. I. Zones, *J. Chem. Soc., Chem. Commun.* **1995**, 2253.

4 L. J. Smith, L. Marchese, A. K. Cheetham, J. M. Thomas, *Catal. Lett.* **1996**, *41*, 13.

5 W. M. Meier, D. H. Olson, *Atlas of Zeolite Structure Types*, 3rd Revised Edn., Butterworths-Heinemann, London, 1992, http://www.iza-sc.ethz.ch/IZA-SC/.

6 J. B. Nicholas, J. F. Haw, *J. Am. Chem. Soc.* **1998**, *120*, 11804.

7 V. Termath, F. Haase, J. Sauer, J. Hutter, M. Parrinello, *J. Am. Chem. Soc.* **1998**, *120*, 8512.

8 E. P. Hunter, S. G. Lias, in *NIST Chemistry WebBook, NIST Standard Reference Database Number 69*, (http://webbook.nist.gov), P. J. Linstrom, W. G. Mallard (Eds.), National Institute of Standards and Technology, Gaithersburg MD, 20899, 2005.

9 L. A. Clark, M. Sierka, J. Sauer, *J. Am. Chem. Soc.* **2003**, *125*, 2136.

10 M. Bjorgen, F. Bonino, S. Kolboe, K.-P. Lillerud, A. Zecchina, S. Bordiga, *J. Am. Chem. Soc.* **2003**, *125*, 15863.

11 M. V. Vener, X. Rozanska, J. Sauer, *Phys. Chem. Chem. Phys.* **2006**; in preparation.

12 Y. Yin, A. L. Blumenfeld, V. Gruver, J. J. Fripiat, *J. Phys. Chem. B* **1997**, *101*, 1824.

13 M. Brändle, J. Sauer, *J. Am. Chem. Soc.* **1998**, *120*, 1556.

14 T. Bucko, J. Hafner, L. Benco, *J. Chem. Phys.* **2004**, *120*, 10263.

15 C. Lee, D. J. Parrillo, R. J. Gorte, W.E. Farneth, *J. Am. Chem. Soc.* **1996**, *118*, 3262.

16 X. Solans-Monfort, M. Sodupe, V. Branchadell, J. Sauer, R. Orlando, P. Ugliengo, *J. Phys. Chem. B* **2005**, *109*, 3539.

17 E. Nusterer, P. E. Blöchl, K. Schwarz, *Angew. Chem., Int. Ed.* **1996**, *35*, 175; *Angew. Chem.* **1996**, *108*, 187.

18 F. Haase, J. Sauer, J. Hutter, *Chem. Phys. Lett.* **1997**, *266*, 397.

19 F. Haase, J. Sauer, *Microporous Mesoporous Mater.* **2000**, *35–36*, 379.

20 R. Shah, J. D. Gale, M. C. Payne, *J. Phys. Chem.* **1996**, *100*, 11688.

21 R. Shah, M. C. Payne, M.-H. Lee, J. D. Gale, *Science* **1996**, *271*, 1395.

22 I. Stich, J. D. Gale, K. Terakura, M. C. Payne, *J. Am. Chem. Soc.* **1999**, *121*, 3292.

23 C. C. Lee, R. J. Gorte, W. E. Farneth, *J. Phys. Chem. B* **1997**, *101*, 3811.

24 P. Nachtigall, M. Davidova, D. Nachtigallova, *J. Phys. Chem. B* **2001**, *105*, 3510.

25 M. W. Anderson, P. J. Barrie, J. Klinowski, *J. Phys. Chem.* **1991**, *95*, 235.

26 F. Haase, J. Sauer, *J. Am. Chem. Soc.* **1995**, *117*, 3780.

27 M. Hunger, T. Horvath, *J. Am. Chem. Soc.* **1996**, *118*, 12302.

28 L. Heeribout, C. Doremieux-Morin, L. Kubelkova, R. Vincent, J. Fraissard, *Catal. Lett.* **1997**, *43*, 143.

29 E. Nusterer, P. E. Blöchl, K. Schwarz, *Chem. Phys. Lett.* **1996**, *253*, 448.

30 J. Sauer, M. Sierka, F. Haase, in *Transition State Modeling for Catalysis*, D. G. Truhlar, K. Morokuma (Eds.), *ACS Symposium Series 721*, American Chemical Society, Washington, **1999**, p. 358.

31 N. Y. Chen, *J. Phys. Chem.* **1976**, *80*, 60.

32 D. H. Olson, W. O. Haag, R. M. Lago, *J. Catal.* **1980**, *60*, 390.

33 D. H. Olson, W. O. Haag, W. S. Borghard, *Microporous Mesoporous Mater.* **2000**, *35–36*, 435.

34 M. Krossner, J. Sauer, *J. Phys. Chem.* **1996**, *100*, 6199.

35 S. A. Zygmunt, L. A. Curtiss, L. E. Iton, M. K. Erhardt, *J. Phys. Chem.* **1996**, *100*, 6663.

36 V. V. Mihaleva, R. A. van Santen, A. P. J. Jansen, *J. Chem. Phys.* **2004**, *120*, 9212.

37 F. Wakabayashi, J. N. Kondo, K. Domen, C. Hirose, *J. Phys. Chem.* **1996**, *100*, 1442.

38 L. J. Smith, A. K. Cheetham, R. E. Morris, L. Marchese, J. M. Thomas, P. A. Wright, J. Chen, *Science* **1996**, *271*, 799.

39 *Chem. Ind.* **1996**, 117.

40 J. Sauer, *Science* **1996**, *271*, 774.

41 Y. Jeanvoine, J. G. Angyan, G. Kresse, J. Hafner, *J. Phys. Chem. B* **1998**, *102*, 7307.

42 C. Tuma, J. Sauer, *Chem. Phys. Lett.* **2004**, *387*, 388.

43 M. Sierka, J. Sauer, *J. Phys. Chem. B* **2001**, *105*, 1603.

44 M. E. Franke, M. Sierka, U. Simon, J. Sauer, *Phys. Chem. Chem. Phys.* **2002**, *4*, 5207.

45 H. Ernst, D. Freude, T. Mildner, H. Pfeifer, *Proceedings of the 12th International Zeolite Conference*, *Vol. 4*, M. M. J. Treacy, B. K. Marcus, M. E. Bischer, J. B. Higgins (Eds.), **1998**, Baltimore, Maryland, Materials Research Society, Warrendale, PA, **1999**, p. 2955.

46 P. Sarv, T. Tuherm, E. Lippmaa, K. Keskinen, A. Root, *J. Phys. Chem.* **1995**, *99*, 13763.

47 J. A. Ryder, A. K. Chakraborty, A. T. Bell, *J. Phys. Chem. B* **2000**, *104*, 6998.

48 C. Tuma, J. Sauer, *Angew. Chem., Int. Ed.* **2005**, *44*, 4769; *Angew. Chem.* **2005**, *117*, 4847.

49 M. Boronat, P. M. Viruela, A. Corma, *J. Am. Chem. Soc.* **2004**, *126*, 3300.

50 C. Tuma, J. Sauer, *Phys. Chem. Chem. Phys.* **2006**, *8*, 3955.

23
Proton Conduction in Fuel Cells

Klaus-Dieter Kreuer

23.1
Introduction

Fuel cells are devices which electrochemically convert the chemical free energy of gaseous, and sometimes also liquid, reactants into electrical energy. As in a battery the reactants are prevented from reacting chemically by separating them with an electrolyte, which is contacted with electrochemically active porous electrode structures. Apart from effectively separating the anode and cathode gases and/or liquids the electrolyte mediates the electrochemical reactions taking place at the electrodes by conducting a specific ion at very high rates during the operation of the fuel cell. Proton conducting electrolytes are used chiefly as separators for low and intermediate temperature fuel cells such as PEMFCs (polymer electrolyte membrane or proton exchange membrane fuel cells), DMFCs (direct methanol fuel cells), PAFCs (phosphoric acid fuel cells), and AFCs (alkaline fuel cells), but proton conducting oxides [1] and plasic acidic salts of oxo-acids [2] have also been considered recently for fuel cell applications at somewhat higher temperature.

The main features of a fuel cell, including the electrochemical reactions taking place for the most simple case of hydrogen and oxygen as reacting gases, are shown schematically in Fig. 23.1. During operation, a protonic current equivalent to the electronic current passing through the external load is driven through the electrolyte and parts of the heterogeneous electrode structures, and it is the proton conducting properties of the diverse electrolytes which are the subject of this chapter. Since the scope of this Handbook is rather broad, this chapter also gives a more general description of proton conduction phenomena in electrolytes which are currently used in fuel cells or which have the potential to be used for this purpose in the near future. For those readers who are interested in the whole variety of available electrolytes and the specific aspects of their operation in fuel cell environments, appropriate references are given. Of course, this chapter is intended to be complementary to the other chapters in this Handbook with its main focus on the features of proton conduction in fuel cell electrolytes. Many of these electrolytes contain water or other hydrogen bonded liquids, or are hydrogen bonded solid stuctures, which are also discussed in other chapters from different perspectives.

Hydrogen-Transfer Reactions. Edited by J. T. Hynes, J. P. Klinman, H. H. Limbach, and R. L. Schowen
Copyright © 2007 WILEY-VCH Verlag GmbH & Co. KGaA, Weinheim
ISBN: 978-3-527-30777-7

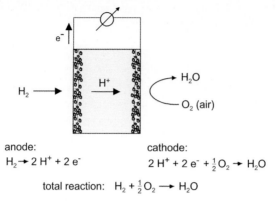

anode:
$$H_2 \rightarrow 2\,H^+ + 2\,e^-$$

cathode:
$$2\,H^+ + 2\,e^- + \tfrac{1}{2}O_2 \rightarrow H_2O$$

total reaction: $H_2 + \tfrac{1}{2}O_2 \rightarrow H_2O$

Figure 23.1 Schematic representation of a hydrogen / oxygen fuel cell, comprising the proton conducting separator (electrolyte) and the heterogeneous gas electrodes. The transport of protons and electrons, the electrode reactions and the total reaction are indicated.

For the non-fuel cell expert, Section 23.2 provides a brief introduction to proto-typical proton conducting fuel cell electrolytes including the rationales for their choice for particular fuel cell systems. Since the key feature of all these electrolytes is their proton conductivity, that is the long range transport of protonic charge carriers, this is first discussed in general in Section 23.3 before describing the proton conduction mechanisms of specific homogeneous media, which have some relevance as the conducting part of heterogeneous separator materials. Indeed, many fuel cell separators exhibit nano-heterogeneities, and the corresponding confinement and interfacial effects sometimes lead to the appearance of qualitatively new features, which are discussed in Section 23.4.

The discussion in all the sections makes use of many results from simulations. For a brief introduction to the underlying models and techniques the interested reader is referred to Ref. [3].

23.2
Proton Conducting Electrolytes and Their Application in Fuel Cells

The fuel cell concept has been known for more than 150 years. It was Christian Friedrich Schönbein who recognized and described the appearance of "inverse electrolysis" [4] shortly before Sir William Grove, the inventor of the platinum/ zinc battery, constructed his first "gas voltaic battery" [5]. Grove used platinum electrodes and dilute sulfuric acid as a proton conducting electrolyte. Sulfuric acid is still used today for the impregnation of porous separators serving as the electrolyte in direct methanol laboratory fuel cells [6], but the most commonly used fuel cell electrolytes today are hydrated acidic ionomers. As opposed to aqueous sulfuric acid, where the dissociated protons and the diverse sulfate anions (conjugated bases) are mobile, such ionomers are polymers containing covalently immobilized sulfonic acid functions. The immobilization reduces anion adsorption on the platinum cathode, which may lead to reduced exchange current densities for oxygen reduction in the case of sulfuric acid as electrolyte. Among the huge number of sulfonic acid bearing polymers [7–12], the most prominent representative of this class of separators is DuPont's Nafion [13, 14]. Such polymers naturally combine, in one macromolecule, the high hydrophobicity of the backbone (green in Fig. 23.2(a)) with the high hydrophilicity of the sulfonic acid functional group, which gives rise to a constrained hydrophobic/hydrophilic nano-separation. The sulfonic acid functional groups aggregate to form a hydrophilic domain that is hydrated upon absorption of water (blue in Fig. 23.2(a)). It is within this continuous domain that ionic conductivity occurs: protons dissociate from their anionic counter ion (SO_3^-) and become solvated and mobilized by the hydration water (red in Fig. 23.2(a)). Water typically has to be supplied to the electrolyte through humidification of the feed gases and is also produced by the electrochemical reduction of oxygen at the cathode. This is the reason for two serious problems relevant to the use of such membranes in fuel cells. Since high proton conductivity is only obtained at high levels of hydration, the maximum operating temperature is limited approximately to the condensation point of water, and any protonic current also leads to transport of water through the membrane (as a result of electroosmotic drag; see also Section 23.4.1) and, if methanol dissolves in the membrane, this is transported at virtually the same rate [12]. The limited operating temperature and the acidity of the electrolyte makes it necessary to use platinum or platinum alloys (the most active but also the most expensive electrocatalysts) to promote the electrochemical reactions in the anode and cathode structures. However, even with platinum, only rather pure hydrogen can be oxidized at sufficient rates. Nevertheless, such acidic polymers are currently very popular as separator materials in PEM fuel cells because they allow very high electrical power densities (up to about 0.5 W cm^{-2}). The smaller conductivities of basic electrolytes (the highest proton conductivities are observed for aqueous KOH solutions) actually limit the power density of AFCs, but the efficiency of such fuel cells is significantly higher than for fuel cells based on acidic electrolytes, the latter showing higher overpotentials for the oxygen reduction reaction. At the operation temperature of state-of-the-art PEM-fuel cells (usually below 90 °C), the rate of direct oxidation of methanol (which is frequently considered an environmentally friendly fuel) is not sufficient for high power applications, and even trace amounts of CO present in any hydrogen-rich reformate (for instance produced by steam reforming of methanol or methane) poison platinum-based catalysts by blocking the reaction sites. The humidification requirements, along with the high electroosmotic drag of water and methanol in solvated acidic ionomers, complicates the water and heat management of the fuel cell and leads to a significant chemical short-circuiting, this is parasitic chemical oxidation of methanol at the cathode. These disadvantages are overcome by using phosphoric acid as the electrolyte in PAFCs. Phosphoric acid keeps its high protonic conductivity even at high temperature (up to about 200 °C) and low humidity. In PAFCs, phosphoric acid is usually adsorbed

(a)

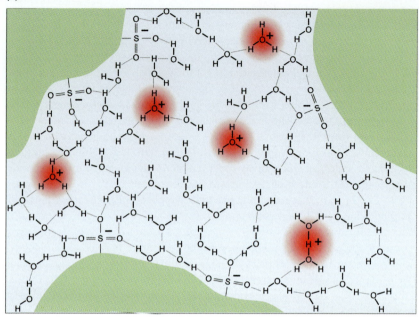

(b)

(c)

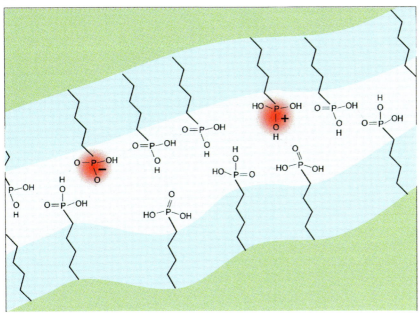

Figure 23.2 Schematic representation of the nanostructures of (a) hydrated acidic ionomers such as Nafion, (b) complexes of an oxo-acid and a basic polymer such as PBI·n H$_3$PO$_4$ and (c) proton solvents fully immobilized via flexible spacers (in this particular case the proton solvent (phosphonic acid) also acts as a protogenic group). Note, that there are different types of interaction between the polymeric matrices (green) and the liquid or liquid-like domains (blue). The protonic charge carriers (red) form within the liquid or liquid-like domain, where proton conduction takes place.

by a porous silicon carbide separator, but more recently adducts of basic polymers (for instance polybenzimidazole) and phosphoric acid have also become a focus of atttention (for reviews see Refs. [15, 16]). The microstructure of such separator materials is illustrated in Fig. 23.2(b), showing the polymer matrix (green) which is protonated by phosporic acid, resulting in the formation of a stable nonconducting complex. It is the excess amount of phosphoric acid absorbed by this complex which leads to the appearance of proton conductivity. Leaching out of phosphoric acid in the presence of water and the poor oxygen reduction reversibility on platinum-based cathodes in the presence of phosphate species are inherent drawbacks of fuel cells using such electrolytes. Therefore, there is currently tremendous effort to develop separator materials which conduct protons in the absence of any low molecular weight solvent such as water or phosphoric acid (for a review see Ref. [17]). One of the recent approaches is illustrated in Fig. 23.2(c): protogenic groups (here phosphonic acid) are immobilized to an inert matrix (green) via flexible spacers. It is within the domain formed by the protogenic groups that proton conductivity occurs. In this type of electrolyte the only mobile species is the proton.

23.3
Long-range Proton Transport of Protonic Charge Carriers in Homogeneous Media

Despite the diversity of proton conducting electrolytes there have been attempts to describe the underlying elementary reactions. Initially the proton was considered to interact chemically with only two electronegative nearest neighbors (mostly oxygen or nitrogen) via hydrogen bonding before the electrostatic interaction with the more distant proton environment was included by considering the environment as a simple dielectric continuum (for a review of the different approaches see Ref. [11]).

Such simple concepts did not include any chemical or structural details other than the chemical nature of the proton donor and acceptor, and the donor / acceptor separation. Also the dynamics of the proton environment was either reduced to the variations of the donor / acceptor separation [19–21] or described qualitatively as being solid-like or liquid-like [22].

However it turned out that the structural, chemical and dynamical details are essential for complex descriptions of long-range proton transport. These parameters appear to be distinctly different for different families of compounds, preventing proton conduction processes from being described by a single model or concept as is the case for electron transfer reactions in solutions (described within Marcus' theory [23]) or hydrogen diffusion in metals (incoherent phonon assisted tunneling [24]).

A common feature of most proton conduction mechanisms is the conflict between high rates of proton transfer and structural reorganization, which are both required to establish long-range proton transport. This has to do with the characteristics of the hydrogen bond interaction which not only provides a path for proton transfer but also has pronounced structure forming properties. Rapid proton transfer is actually favored by short, strong hydrogen bonds, while structural reorganization, requiring the breaking of hydrogen bonds, is hindered by strong hydrogen bonding. This is especially true for small clusters such as the simple proton / acceptor system $H_5O_2^+$. For this the proton transfer barrier equals the energy needed to break the central hydrogen bond at a donor / acceptor separation $Q \approx 300$ pm, where both energies are of the order of 1 eV, which is significantly higher than the activation enthalpy of proton mobility in bulk water (≈ 0.1 eV) [25]. Obviously, it is the mutual interaction of many particles in bulk water, which is essential for the appearance of high rates of proton transfer and structural reorganization, which are both required for fast proton conduction. This is not surprising considering that the regions around protonic defects (excess or defect protons) in condensed matter frequently show pronounced relaxation effects, which in turn suggests a strong coupling of proton conductance to the dynamics of its environment [18]. In the following, details of such complex proton conduction mechanisms are presented for homogeneous media, where effects from confinement and interaction with other phases are not yet considered. These media comprise aqueous solutions, phoshoric (phosphonic) acid and heterocycles such as imidazole, which form the molecular environments in proton conducting electrolytes used or considered for fuel cell applications.

23.3.1
Proton Conduction in Aqueous Environments

The dominant intermolecular interaction in water is hydrogen bonding. The introduction of an excess proton (i.e. the formation of a protonic defect) leads to the contraction of hydrogen bonds in the vicinity of such a defect. This corresponds to the well-known structure forming properties of excess protons in water (see for example Ref. [26]). Thus the isolated dimer $H_5O_2^+$ finds its energetic minimum at an O / O separation of only 240 pm [27, 28] with an almost symmetrical single well potential for the excess proton in the center of the complex.

But due to the presence of additional hydrogen bonds, the central bond of such complexes is weakened to some extent [18] with some small barrier building up in bulk water. In other words, the binding power of a water molecule depends on the number of hydrogen bonds it is already involved in. This also leads to relaxation effects in neighboring hydrogen bonds as a response to hydrogen bond formation or cleavage: when a hydrogen bond is formed, the surrounding bonds are weakened, while the cleavage of a hydrogen bond leads to a strengthening of neighboring bonds. As a consequence, the effective energy for breaking a hydrogen bond in bulk water is significantly lower than the average hydrogen bond energy. This is evidenced, for instance, by the evolution of the fraction of broken hydrogen bonds with temperature for pure bulk water [29]. The apparent activation energy for hydrogen bond cleavage at room temperature only amounts to 50 meV, which is significantly lower than the average hydrogen bond energy ($\approx$ 180 meV). On the other hand the effective activation energy increases to about 100 meV at the critical temperature, although the average hydrogen bond energy decreases with temperature. At such high temperatures the number of intact hydrogen bonds is small, with little interaction remaining between hydrogen bonds. Consequently, the full energy of a hydrogen bond is required to break the bond.

For the well-connected hydrogen-bond network present at low temperature this bond interaction leads to a significant softening of the intermolecular interaction and therefore to strong variations in the hydrogen bond length as well as a rapid breaking and forming of hydrogen bonds.

The above-described features are reproduced in a high level quantum-molecular-dynamics simulation of an excess proton in water [30, 31]. In accordance with results from several other groups, this finds the excess proton either as part of a dimer ($H_5O_2^+$, "Zundel"-ion) or as part of a hydrated hydronium ion ($H_9O_4^+$, "Eigen"-ion) (Fig. 23.3).

Interestingly, the center of the region of excess charge coincides with the center of symmetry of the hydrogen bond pattern [25], i.e. apart from the bonds with the common shared proton, each water molecule of the Zundel-ion acts as a proton donor through two hydrogen bonds, and each of the three outer water molecules of the Eigen-ion acts as a proton donor in two hydrogen bonds and as an acceptor for the hydronium ion and an additional water molecule (Fig. 23.3). Changes to these hydrogen bond patterns through hydrogen bond breaking and forming processes displaces the center of symmetry in space and therefore also the center of

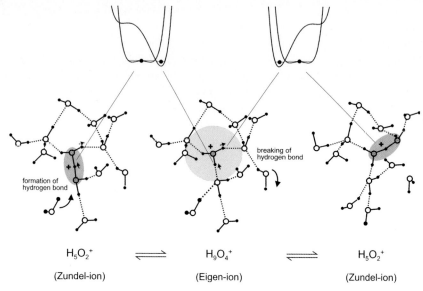

$$H_5O_2^+ \quad \rightleftharpoons \quad H_9O_4^+ \quad \rightleftharpoons \quad H_5O_2^+$$

(Zundel-ion)   (Eigen-ion)   (Zundel-ion)

Figure 23.3 Proton conduction mechanism in water. The protonic defect follows the center of symmetry of the hydrogen-bond pattern, which "diffuses" by hydrogen-bond breaking and forming processes. Therefore, the mechanism is frequently termed "structure diffusion". Note that the hydrogen bonds in the region of protonic excess charge are contracted, and the hydrogen bond breaking and forming processes occur in the outer parts of the complexes (see text). Inserted potentials correspond to nonadiabatic transfer of the central proton in the three configurations (atomic coordinates taken from Ref. [30, 31] with kind permission from *Chemical Reviews*.

the region of excess charge. In this way a Zundel-ion is converted into an Eigen-ion, which then transfers into one of three possible Zundel-ions (Fig. 23.3). This type of mechanism may be termed "structure diffusion" (as suggested by Eigen for a similar mechanism [32, 33]) because the protonic charge follows a propagating hydrogen bond arrangement or structure.

The sum of all proton displacements involved in the hydrogen bond breaking and forming processes and the proton displacements within the hydrogen bonds of the Zundel- and Eigen-ions then corresponds to the net displacement of one unit charge by just a little more than the separation of the two protons in a water molecule (i.e. $\cong 200$ pm). Although there are no individual exceptionally fast protons, even on a short time scale, the fast diffusion of protonic defects leads to a slight increase in the physical diffusion of all protons in the system. This is indeed observed for aqueous solutions of hydrochloric acid, for which mean proton diffusion coefficients were found to be up to 5% higher than the diffusion coefficient of oxygen as measured by ^{1}H- and ^{17}O-PFG-NMR [34] reflecting the slight decoupling of proton and oxygen diffusion in acidic media.

Another interesting feature of this mechanism is that the hydrogen bond breaking and forming (hydrogen bond dynamics) and the translocation of protons within the hydrogen bonds take place in different parts of the hydrogen bond network, albeit in a highly concerted fashion. This is the most thermodynamically favorable transport path, because the hydrogen bonds in the center of the two charged complexes are contracted to such an extent as to allow an almost barrierless proton translocation while the hydrogen bond breaking and forming processes take place in the weakly bonded outer parts of the complexes. This contraction of the center of the complex is probably a direct consequence of the lower coordination of the involved species (3 instead of about 4). The activation enthalpy of the overall transport process is dominated by the hydrogen bond breaking and forming, which also explains the strong correlation of the proton transport rate and the dielectric relaxation [2]. The Zundel- and Eigen-complexes are just limiting configurations, and the simulations indeed produce configurations that can hardly be ascribed to one or the other [36].

The time-averaged potential surfaces for proton transfer in such contracted hydrogen bonds are almost symmetrical (especially for the Zundel-ion) without significant barriers, and the proton is located close to the center of the bond. Whether its location is off-center at any time mainly depends on the surrounding hydrogen bond pattern, and it is the change in this pattern that alters the shape (and asymmetry) of this potential and therefore the position of the proton within the hydrogen bond (see Fig. 23.3 top). In other words, the proton is transferred almost adiabatically with respect to the solvent coordinate [18]. This has important consequences for the mechanism when static asymmetric potential contributions are introduced, for instance by chemical interactions or the presence of ionic charges (see below). The very low barriers are also the reason why the mechanism can be well described classically with respect to the motion of the nuclei (especially the proton), in particular, proton tunneling has only a minor effect on the rate of transfer. Nevertheless, the protonic defect (region of protonic excess charge) may become delocalized through several hydrogen bonds owing to quantum fluctuations [36, 37].

The mechanism also provides insight into the extent to which proton transfer in water is a cooperative phenomenon. In many physical chemistry textbooks one still finds cartoons showing the concerted transfer of protons within extended hydrogen bonded water chains (Grotthuss mechanism) in order to explain the unusually high equivalent conductivity of protons in this environment. However, the creation of the corresponding dipolar moment in an unrelaxed high dielectric constant environment costs far too much energy to be consistent with a very fast process [18, 38]. As anticipated in Ref. [18], the propagation of a protonic defect in a low-dimensional water structure surrounded by a low dielectric environment is obviously between "concerted" and "step-wise" in mechanism [39, 40], but in bulk water the cooperation is restricted to the dynamics of protons in neighboring hydrogen bonds (see also Fig. 23.2).

One should also keep in mind that water is a liquid with a high self-diffusion coefficient ($D_{H2O} = 2.25 \times 10^{-5}$ cm^2 s^{-1} at room temperature) and that the diffusion of protonated water molecules makes some contribution to the total proton conductivity (vehicle mechanism [41]). But, as suggested by Agmon [42], the diffusion of H_3O^+ may be retarded owing to the strong hydrogen bonding in its first hydration shell.

Of course, the relative contributions of "structure diffusion" and "vehicular diffusion" depend on temperature, pressure and the concentrations and kinds of ions present. With increasing temperature, "structure diffusion" is attenuated and with increasing pressure the contribution of "structure-diffusion" increases until it reaches a maximum around 0.6 GPa (6 kbar) [18]. Especially relevant for the later discussion of proton transport in hydrated polymeric fuel cell electrolytes is the observation that structure diffusion strongly decreases with increasing acid concentration [43] which is probably due to changes in the hydrogen bond pattern (there are progressively more proton donors than corresponding proton acceptor "sites") and a consequence of the biasing of the hydrogen bonds in the electrostatic field of the ions suppressing the proton transfer mechanism illustrated in Fig. 23.3.

Since basic aqueous solutions also have some relevance for fuel cell applications (in AFCs, where aqueous KOH solutions are used as electrolyte) the conduction mechanism of defect protons (OH^-) is also summarized here. As opposed to Zundel- and Eigen-complexes, in which the central species are only three-fold coordinated (under-coordinated with respect to water in pure water which is probably the reason for the bond contraction in these complexes, see above) on the average, the hydroxide ion is found to be coordinated by about 4.5 water molecules in an almost planar configuration with the OH proton pointing out of the plane [44]. This is considered to be a true quantum effect and contradicts the common understanding of a threefold coordination [45]. This "hyper-coordination" is suggested to prevent proton transfer from an H_2O to the OH^-, because this would produce an unfavorable H–O–H bond angle of 90°. The proton transfer only occurs when the OH^- coordination is reduced to 3 by breaking one of the 4 hydrogen bonds within the plane and some rearrangement of the remaining bonds, which allows the direct formation of a water molecule with a tetrahedral geometry. Surprisingly, the ground-state coordination of the most favorable configurations around excess protons appears to be close to the coordination of the transition state for the transport of defect protons. However, it should be noted that "hyper-coordination" of the OH^- is still the subject of controversial debate. The statistical mechanical quasichemical theory of solutions suggests that tricoordinated OH^- is the predominant species in the aqeous phase under standard conditions [46, 47]. This finding seems to be in agreement with recent spectroscopic studies on hydroxide water clusters, and is in line with the traditional view of OH^- coordination.

It should also be mentioned that OH^- "hyper-coordination" is not found in concentrated solutions of NaOH and KOH [48]. In contrast to acidic solutions where structure diffusion is suppressed with increasing concentration the transference number of OH^- (for example in aqueous KOH solutions) remains surprisingly high (approximately 0.74) for concentrations up to about 3 M.

In pure water, excess protons (H_3O^+, $H_5O_2^+$) and defect protons (OH^-) are present in identical concentration, but owing to their low concentration (10^{-7} M under ambient conditions) and the high dielectric constant of bulk water the diffusion of these defects is quasi-independent.

The complexity of the above-discussed many-particle conduction mechanisms of excess and defect protons in water reduces the effective activation enthalpy for the long-range transport of protonic defects, but it is also responsible for the relatively low pre-exponential factor of this process, which probably reflects the small statistical probability to form a transition state configuration in this environment.

23.3.2
Phosphoric Acid

The proton conduction mechanism in phosphoric acid has not been investigated to the same extent as is the case for aqueous solutions, but it is evident that the principal features exhibit both similarities and important differences.

Above its melting point, $T_m = 42\,°C$, neat phosphoric acid (H_3PO_4) is a highly viscous liquid with extended intermolecular hydrogen bonding. But in contrast to the situation in water, there are more possible donor than acceptor sites and the amphoteric character is significantly more pronounced: phosphoric acid may act as both a Brønsted acid and base. In terms of equilibrium constants both K_a and K_b are reasonably high (K_a of the conjugate base is low). Consequently, phosphoric acid shows a very high degree of self-dissociation (auto-protolysis) of about 7.4% [49] along with some condensation, $H_2PO_4^-$, $H_4PO_4^+$, H_3O^+ and $H_2P_2O_7^{2-}$ being the main dissociation products. Because of their high concentrations, the separation of the overall conductivity into charge carrier concentration and mobility terms is problematic. Nevertheless, the proton mobility has been calculated from total conductivities by the Nernst–Einstein equation by taking concentrations from Ref. [49]; and the values have been found to be almost two orders of magnitude higher than the values for the diffusion coefficient of the diverse phosphate species obtained directly by ^{31}P PFG-NMR [50] and estimated from viscosity measurements via the Stokes–Einstein relation.

Pure phosphoric acid is a liquid with a low diffusion coefficient of phosphate species but an extremely high proton mobility, which must involve proton transfer between phosphate species and some structural rearrangements. The contribution to the total conductivity is about 98%, in other words, phosphoric acid is an almost ideal proton conductor. The total conductivity at the melting point (42 °C) is 7.7×10^{-2} S cm^{-1} with an estimated proton mobility of 2×10^{-5} cm^2 s^{-1} [50]. Extremely high proton mobilities have also been indirectly determined with ^1H-PFG-NMR and were found to be even higher (by a factor of 1.5–2.3). This has been explained by the correlated motion of the oppositely charged defects ($H_2PO_4^-$, $H_4PO_4^+$) when they are close to one another (this is the case just after their formation (by dissociation of H_3PO_4) and before their neutralization). Correlation effects are actually quite common in proton conductors with high concentrations of charge carriers and they are even more pronounced in other systems with lower dielectric constant [51, 52].

Molecular details of the "structure diffusion" mechanism with the hydrogen bond breaking and forming and the proton transfer between the different phosphate species (essentially $H_2PO_4^-$, H_3PO_4, $H_4PO_4^+$) have not been investigated yet, but the high degree of self-dissociation suggests that the proton transfer events are even less correlated than in water (the system is more tolerant towards protonic charge density fluctuations). The transfer events are probably almost barrierless as indicated by negligible H/D effects of the diffusion coefficients in mixtures of H_3PO_4 and D_3PO_4 [53].

The principal proton transport mechanism seems to be essentially unchanged with the addition of some water with a conductivity increase up to 0.25 S cm^{-1} under ambient conditions. A ^{1}H and ^{31}P-PFG-NMR study also showed an 85 wt% phosphoric acid system to be an almost ideal proton conductor with 98% of the conductivity originating from the structure diffusion of protons [54]. The combination of high intrinsic charge carrier concentration and mobility gives the possiblility of very high conductivities in these systems. In particular, there is no perturbation from extrinsic doping, that is there is no suppression of structure diffusion despite the high concentration of protonic charge carriers. On the other hand, attempts to increase the conductivity of phosphoric acid based systems by doping have expectedly failed [55].

It should be mentioned, that the transport properties of phosphonic acid, which has recently been used as protogenic group in fully polymeric proton conductors (see Section 23.4.2), seem to be similar to these of phosphoric acid

23.3.3
Heterocycles (Imidazole)

Historically, the interest in hydrogen bonding and proton conductivity in heterocycles has its roots in speculation about the participation of hydrogen bonds in energy and charge transfer within biological systems [56, 57]. Even Zundel has worked in the field [58] and it is not surprising that his view of the proton dynamics in imidazole is closely related to that of water. He suggested a high polarizability of the protons within intermolecular hydrogen bonds and, as a consequence, a very strong coupling between hydrogen bonds, as indicated by the intense IR continuum in the NH stretching regime. Surprisingly, he did not suggest the existence of any complex similar to the Zundel complex in water [58] (see Section 23.3.1), whereas Riehl [59] had already suggested "defect protons" or "proton holes" as requirements to maintain a current in solid imidazole. Early conductivity measurements were focused on crystalline monoclinic imidazole, which has a structural hydrogen bond length of 281 pm [60]. The measured conductivities were typically low (approximately 10^{-8} S cm^{-1}) with very poor reproducibility [56, 61, 62]. Later tracer experiments [63] and a ^{15}N-NMR study [64] raised doubts about the existence of proton conductivity in pure crystalline imidazole.

The conductivity of liquid imidazole, however, was found to be several orders of magnitude higher (about 10^{-3} S cm^{-1} at the melting point $T_m = 90\,°C$ [56]) but the conduction mechanism was investigated much later. It was the search for chemical environments different from water in fuel cell membranes that brought heterocycles back into focus. The potential proton donor and acceptor functions (amphoteric character), the low barrier hydrogen bonding between the highly polarizable nitrogen atoms, and the size and shape of the molecule were the reasons why Kreuer et al. started to investigate the usefulness of heterocycles as proton solvents in separator materials for fuel cells [65]. This work also comprises the study of the transport properties of neat and acidified liquid imidazole, pyrazole, and later benzimidazole [66]. An important finding was that the transport coefficients (mobility of protonic charge carriers and molecular diffusion coefficients) are close to those of water at a given temperature relative to the melting point. This is particularly true for their ratio: the proton mobility is about a factor of 4.5 higher than the molecular diffusion coefficient at the melting point of imidazole [65]. This is a direct indication of fast intermolecular proton transfer and the possibility of structure diffusion in this environment. Subsequently, details were revealed by a CP-MD simulation [67]. In contrast to earlier suggestions of concerted proton transfer in extended chains of hydrogen bonds [56, 58] (analagous to the proton conduction mechanism in water presented in most textbooks at that time [38]) a structure diffusion mechanism similar to that for water (Fig. 23.3) was found. The region containing the excess proton (intentionally introduced) is an imidazole with both nitrogens protonated and acting as proton donors towards the two next nearest imidazoles in a configuration Imi – Imi$^+$ – Imi with hydrogen bonds (approximately 273 pm) slightly contracted compared to the average bond length of the system but still longer than the bonds in the isolated complex (in the gas phase) [69]. The position of the protons within these hydrogen bonds depends mainly on the hydrogen bonding between the nearest and next nearest solvating imidazoles (Fig. 23.4). The hydrogen bonded structure in imidazole is found to be chain-like (low-dimensional) with two possible orientations of the hydrogen bond polarization within segments which are separated by imidazoles with their protonated nitrogen directed out of the chain. This may even form a "cross-linking" hydrogen bond with a nonprotonated nitrogen of a neighboring strand of imidazole. The simulation data revealed the existence of imidazole mol-

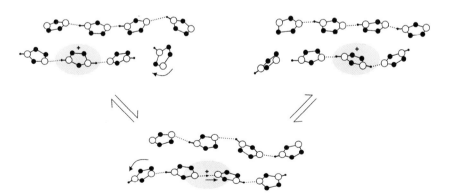

Figure 23.4 Proton conduction mechanism in liquid imidazole as obtained by a CP-MD simulation [67]. As in water, changes in the second solvation shell of the protonic defect (here imidazolium) drive the long-range transport of the defect.

ecules close to the protonic defect in hydrogen bond patterns, which rapidly change by bond breaking and forming processes. Similar to water, this shifts the excess proton within the region and may even lead to complete proton transfer as displayed in Fig. 23.4. There is no indication of the stabilization of a symmetrical complex (Imi – H – Imi)$^+$: there always seems to be some remaining barrier in the hydrogen bonds with the proton being on one side or the other.

As for the CP-MD simulation of water, the simulated configuration is artificial, because there is no counter charge compensating for the charge of the excess proton. This is necessary, methodologically, since self-dissociation is unlikely to occur within the simulation box used (8 imidazole molecules with a single excess proton) and the accessible simulation time (approximately 10 ps). The self-dissociation constants for heterocycles (in particular imidazole) are actually much higher than for water, but degrees of self-dissociation (concentration of protonic charge carriers) around 10^{-3} are still about two orders of magnitude lower than for phosphoric acid (see Section 23.3.2). Site-selective proton diffusion coefficients (obtained by ^{1}H-PFG-NMR of different imidazole-based systems) show surprisingly high diffusion coefficients for the protons involved in hydrogen bonding between the heteroatoms (nitrogen) [70]. Depending on the system, they are significantly higher than calculated from the measured conductivities corresponding to Haven ratios (σ_D/σ) of 3–15. This indicates some correlation in the diffusion of the proton, which may be due to the presence of a counter charge neglected in the simulation.

In pure imidazole, regions containing excess protons must be charge compensated by proton deficient regions with electrostatic attraction between these regions (defects) that depends on their mutual separation distance and the dielectric constant of the medium. Under thermodynamic equilibrium such defects are steadily formed and neutralized. Formally, the creation of a protonic defect pair is initiated by a proton transfer from one imidazole to another with the subsequent separation of the two charged species with a diffusion mechanism as described above (see also Fig. 23.4). However, this transfer is against the electrostatic field of the counter charge, favoring a reversal of the dissociation process. But since the two protons of the positively charged imidazolium (Imi$^+$) are equivalent, there is a 50% chance that another proton will be transferred back, provided that the orientational coherence between the dissociating molecules is completely lost. If the same proton is transferred back, the transient formation and neutralization of an ion pair contributes neither to the proton diffusion nor to the proton conductivity. But if the other proton is transferred back, the protons interchange their positions in the hydrogen bond network, which generates diffusion but no conductivity, since the transient charge separation is completely reversed. The sum of all proton translocation vectors then forms a closed trajectory which is reminiscent of cyclic intermolecular proton transfer reactions known to take place in certain organic pyrazole containing complexes [71] and proton diffusion in hydroxides [51, 52].

23.4
Confinement and Interfacial Effects

As described in Section 23.2, the proton conducting media discussed in Section 23.3 are dispersed within matrices (usually polymers), which not only give the separator material its morphological stability and gas separating properties but also modify the charge carrier distribution and transport properties within the conducting domain as a result of confinement and interaction. Such effects are described in the following for the three type of separator materials illustrated in Fig. 23.2.

23.4.1
Hydrated Acidic Polymers

The hydrophilic domain of hydrated acidic polymers contains only water and excess protons (Fig. 23.2(a)), which is reminiscent of the situation illustrated in Fig. 23.3, but both species interact chemically and electrostatically with the immobile negatively charged sulfonic groups. Traditionally, the distribution of charge carriers within the corresponding space charge layer is described by the Gouy–Chapman theory, which has been developed for semi-infinite geometries, or by numerically solving the Poisson–Boltzmann equation for specific geometries [72]. In either case, one obtains a monotonically decreasing concentration of protonic charge carriers as one moves from the hydrophobic/hydrophilic interface (where the anion charge resides) towards the center of the hydrated hydrophilic domain. This picture, however, is not complete because these continuum theories neglect any structural inhomogeneity in the vicinity of the electrified interface. In the Gouy–Chapman approach, even a homogeneous distribution of the counter charge over the interface is assumed, but the fact that the separation of neighboring sulfonic acid groups (approximately 0.8 nm) and the typical extension of the hydrophilic domain (a few nm) is of similar order does not justify this assumption. Also the assumption of a homogeneous dielectric constant of the aqueous phase breaks down for such dimensions and is indeed not backed up by dielectric measurements as a function of the water content in the microwave (i.e. GHz) range [73, 74]. As known for the near surface region of bulk water or any interface with water on one side, the dielectric constant of the hydrated hydrophilic phase is significantly reduced close to the hydrophobic/hydrophilic interface. In addition, the specific interaction of the sulfonic acid group with water (hydration) also decreases the dielectric constant. Therefore, the spatial distribution of the dielectric constant within hydrated domains depends strongly on the width of the channels (degree of hydration) and the separation of the dissociated sulfonic acid functional groups. This is evidenced by an equilibrium statistical thermodynamic modeling of the dielectric saturation in different types of hydrated polymers [75]. According to these calculations, the dielectric constant reaches the bulk value (81) in the center of the channel (pore) for water contents higher than about 10 water molecules per sulfonic acid group, corresponding to a domain width of about

1 nm, while for lower degrees of hydration even in the center of the channel the dielectric constant is lower than the bulk value. The calculations did not account for specific chemical interactions between the water and the polymer backbone, which are expected to further reduce the dielectric constant at the hydrophilic/hydrophobic interface. The general picture arising from this simulation is that the majority of excess protons are indeed located in the central part of the hydrated hydrophilic nano-channels where the enthalpy of the proton hydration is anticipated to be highest, as a result of the high dielectric constant. It should be mentioned that a multistate empirical valence bond (MS-EVB) simulation finds a marked preference of excess protons for the hydrophilic/hydrophobic interface suggesting an amphiphilic character for the excess proton [76].

However, the experimentally obtained proton conductivities are in favor of a charge carrier stabilization in the center of the channels. In this region the water is bulk-like (for not too low degrees of hydration) with local proton transport properties similar to those described for water in Section 23.3.1. Indeed, the experimentally found activation enthalpies of both proton mobility and water diffusion are close to those of bulk water and only increase slightly with decreasing degree of hydration for intermediate water contents [43, 77–79]. Apart from the slight retardation of the local proton mobility (D_σ) and water diffusion (D_{H2O}) within the hydrophilic domain, the decrease in the macroscopic transport coefficients with decreasing degree of hydration therefore mainly reflects the decreasing percolation within the water-like domain. At the highest degrees of hydration the major proton conduction mechanism is actually structure diffusion ($D_\sigma > D_{H2O}$, Fig. 23.5). With decreasing water content the concentration of excess protons in the aqueous phase increases, which in turn increasingly suppresses intermolecular proton transfer and therefore structural diffusion, an effect which is well known for aqueous solutions [80]. Consequently, proton mobility at intermediate and low degrees of hydration is essentially vehicular in nature. Nonequilibrium statistical mechanics-based calculations of the water and hydronium self-diffusion coefficients in Nafion membranes have addressed this conductivity contribution, and they clearly show that the diffusion of water (vehicle) and hydrated protons (H_3O^+) are retarded for intermediate degrees of hydration as a result of the interaction with the negatively charged sulfonate groups [81].

For very low degrees of hydration (for instance Nafion membranes with less than about 3 water molecules per sulfonic acid group) the decreasing solvent (water) activity leads to a decreasing dissociation of the sulfonic acid group, that is an increasing exclusion of protons from the transport in the aqueous phase. When methanol enters the hydrophilic domain (for example in a direct methanol fuel cell) the proton conductivity may dramatically decrease, even at higher degrees of solvation [12], simply because of an increased ion pairing (decreased dissociation) as a result of the lowering of the dielectric constant.

As discussed above, proton conduction is related to the transport and the local dynamics of water. This water transport shows up not only as water self diffusion, chemical diffusion and permeation [3], but also as electroosmotic drag, which is the transport of water coupled to the drift velocity of protonic defects in an electri-

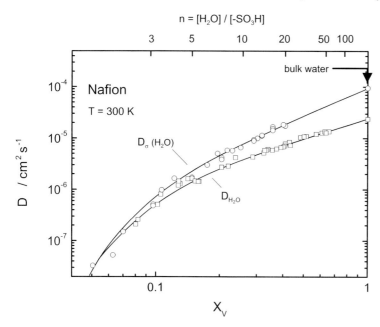

Figure 23.5 Proton conductivity and water diffusion coefficient (D_σ and D_{H2O}) of hydrated Nafion as a function of its water volume fraction. Data are taken from Refs. [12, 43, 77, 78, 108–116] , unpublished data from the author's laboratory are also included.

cal field and is usually expressed as the number of water molecules transported per protonic charge carrier. This is a pronounced effect in hydrated acidic polymers, because the only mobile charge carriers are protonic defects (for example hydronium ions) strongly interacting with the water, while the corresponding hydrated counter ions (sulfonic anion)) are immobilized by covalent bonding to the polymer. The classical mechanistic theory of electroosmosis dates back to the time of Helmholtz [82], Lamb [83], Perrin [84] and Smoluchowski [85] who assumed that transport takes place only close to the wall in electrical double layers of low charge carrier concentration and with extent significantly less than the pore (channel) diameter. The corresponding theories qualitatively describe the electroosmotic drag in wide pore systems, such as clay plugs, but both model assumptions are not valid for typical PEM materials such as Nafion. The width of the hydrated (solvated) channels is orders of magnitude smaller than the Debye length of water and the concentration of charge carriers is very high (typically around 5 M within the hydrophilic domain). For this type of system, Breslau and Miller developed a model for electroosmosis from a hydrodynamic point of view [86]. Recently, electroosmotic drag coefficients became accessible by electrophoretic NMR [3, 12, 87, 88] for a wide range of polymer–solvent volume ratios, and the results clearly confirm the hydrodynamic nature of electroosmosis, particularly at high degrees of solvation. The data presented in Fig. 23.6 [3, 89–95] essentially

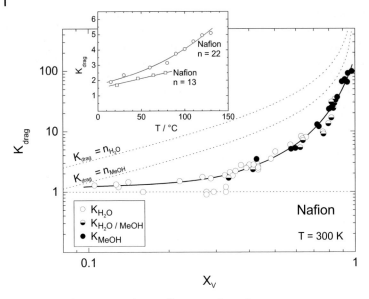

Figure 23.6 Electoosmotic drag coefficient K_{drag} for Nafion as a function of its solvent (water and/or methanol) volume fraction; data from Refs. [12, 87, 88, 117–123] and unpublished data from the author's laboratory. The normalized drag coefficients for water and methanol are plotted together because they are virtually identical. (Reproduced with kind permission from *Chemical Reviews*.)

show two things: (i) at low degrees of hydration the electroosmotic drag coefficient approaches a value of one but does not fall below this value, and (ii) with increasing solvent fraction (increasing channel width) the drag coefficient dramatically increases and reaches about 50% of the maximum possible value (dashed line), which corresponds to an identical drift velocity for all solvent molecules and protonic charge carriers. Considering that at high degrees of hydration about half of the conductivity is carried by structure diffusion, as indicated by the proton mobility (proton self-diffusion coefficient) being about twice the water self diffusion coefficient, almost all water molecules appear to drift at approximately identical velocity (about half of the drift velocity of protonic charge carriers) in extremely swollen samples. This situation corresponds to minor relative motion of water molecules with respect to one another, that is the transport is clearly of a collective nature. The decrease in the drag coefficient with decreasing water content roughly scales with the 4th power of the channel diameter, which is reminiscent of Hagen–Poisseuille type behavior with continously increasing "stripping off" of the water molecules. This stripping comes to an end at low degrees of hydration, where the motion of one water molecule remains strongly coupled to the motion of the excess proton ($K \sim 1$). This is also expected from the high enthalpy of primary hydration (stability of H_3O^+) and the proton conduction mechanism, which is the diffusion of H_3O^+ in a water environment (vehicle mechanism).

There is no quantitative model yet that describes the observed electroosmotic drag coefficients as a function of the degree of hydration and temperature. However, the available data provide strong evidence for a mechanism which is: (i) hydrodynamic in the high solvation limit, with the dimensions of the solvated hydrophilic domain and the solvent–polymer interaction as the major parameters, and (ii) diffusive at low degrees of solvation, where the excess proton essentially drags its primary solvation shell (e.g. H_3O^+).

23.4.2
Adducts of Basic Polymers with Oxo-acids

To date the most relevant materials of this type are adducts (complexes) of poly-benzimidazole (PBI) and phosphoric acid. as illustrated in Fig. 23.2(b). In contrast to water, which exhibits a high mobility for protonic defects but a very low intrinsic concentration of protonic charge carriers, phosphoric acid shows both high mobility and concentration of intrinsic protonic defects (see Section 23.3.2). Phosphoric acid is intrinsically a very good proton conductor with a very small Debye length, and its charge carrier density is hardly affected by the interaction with PBI. Indeed, a strong acid–base reaction occurs between the nonprotonated, basic nitrogen of the PBI repeat unit and the first phosphoric acid absorbed. The transfer of one proton leads to the formation of a benzimidazolium cation and a dihydrogenphosphate anion forming a stable hydrogen bonded complex, as shown by infrared spectroscopy [96, 97]. It is a common observation for all systems of this type that their conductivity strongly increases upon further addition of an oxo-acid, approaching the conductivity of the pure acid for high acid concentrations. In particular, there is no indication of participation of the polymer in the conduction process.

Although no microstructural information is available to date, the macroscopic transport has been investigated in the related system poly (diallyldimethylammonium-dihydrogenphosphate)–phosphoric acid (PAMA$^+$ H$_2$PO$_4^-$) $\cdot$ n H$_3$PO$_4$ [98]. The proton mobility (D_σ) and the self-diffusion coefficient of phosphorus (D_p) as a measure of the hydrodynamic diffusion of the system is shown in Fig. 23.7 for a given temperature as a function of the polymer/acid ratio. Similar to pure phosphoric acid, the mobility of protonic charge carriers is significantly higher than the self-diffusion coefficient of the phosphate species and both transport coefficients decrease with increasing polymer content, virtually in the same manner. Therefore, the main effect is just the decreasing percolation within the liquid-like part of the phosphoric acid domain, which is reminiscent of the situation in hydrated acidic polymers. At very small acid contents, when all the phosphoric acid is immobilized in the 1:1 complex, only very little conductivity is left.

As expected, the confinement of phosphoric acid in the PBI matrix does not give rise to any relevant electroosmotic drag. Of course, the main reason is the fact that proton conductivity is dominated by structure diffusion, that is the transport of protonic charge carriers and phosphoric acid are effectively decoupled. The other reason is that protonic charge carriers are produced by self-dissociation of the proton solvent (phoshoric acid), that is the number of positively and negatively

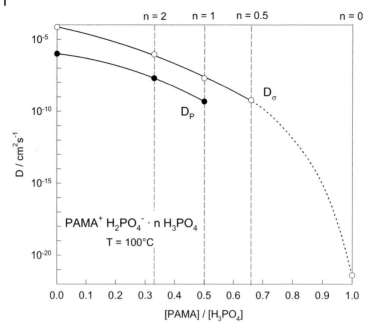

Figure 23.7 Proton conductivity diffusion coefficient (D_σ) and self-diffusion coefficient of phosphorous for poly-(diallyldimethylammonium-dihydrogenphosphate)-phosphoric acid ($(PAMA^+H_2PO_4^-)\cdot n\,H_3PO_4$) as a function of the phosphoric acid content [98]. Note that the ratio D_σ/D_P remains almost constant (see text).

mobile charged defects in the liquid-like domain are virtually identical (see also Section 23.3.2).

As for pure phosphoric acid, the transport properties of PBI and phosphoric acid also depend on the water activity, this is on the degree of condensation (polyphosphate formation) and hydrolysis. There is even indication that these reactions do not necessarily lead to thermodynamic equilibrium, and hydrated orthophosphoric acid may coexist with polyphosphates in heterogeneous gel-like microstructures [99]. There is not much known on the mechanism of proton transport in polymer adducts with polyphosphates and/or low hydrates of orthophosphoric acid. Whether the increased conductivity at high water activities is the result of the plasticizing effect of the water on the phosphate dynamics, thereby assisting proton transfer from one phosphate to the other, or whether the water is directly involved in the conduction mechanism has not been elucidated.

23.4.3
Separated Systems with Covalently Bound Proton Solvents

Both types of heterogeneous systems discussed above comprise a polymeric domain and a low molecular weight liquid-like domain containing the proton solvent (H_2O, H_3PO_4) with weak ionic or hydrogen bond interaction between the two domains. But there are other proton solvents such as heterocycles and phosphonic acid, which allow covalent immobilization. Apart from the proton donor and acceptor sites, such solvents contain sites, which may be used for covalent "grafting" to polymeric structures. If these are hydrophobic (nonpolar), a similar separation as in hydrated acidic polymers may occur, however with covalent bonding bridging the nonpolar/polar "interface" (Fig. 23.2(c))

This approach has been implemented in order to obtain systems with high proton conductivity with structure diffusion as the sole proton conduction mechanism. Of course, the covalent bonding across the nonpolar/polar "interface" mediates a significant influence of the nonpolar part of the structure on the structure and dynamics of the polar proton-conducting domain. If heterocycles are used as proton solvent the two hetero-nitrogens act equally as proton donor and acceptor. Any covalent immobilization must avoid reduction of this symmetry, which is best achieved by using the carbon between the two nitrogens (C2 in imidazole or C4 in pyrazole) for covalent bonding (of course symmetry reduction is not a problem for the covalent bonding of phosphonic groups via C–P bonds). It is interesting to note that the symmetry is broken in the case of histidine, an imidazole-containing amino acid, which is frequently involved in proton translocation processes within biological systems [57]. However, the energetic asymmetry is very small (about 20 meV) in this particular case [100].

The type of bonding appears to be more important, that is only single bonds allow rapid reorientation of the bonded proton solvent, which is a persistent element in the proton conduction mechanism. But even for single bonds, significant barriers appear for the rotation of the proton solvent around this bond of the isolated (non-hydrogen bonded) alkane segment, with higher barriers for the phosphonic acid (~ 0.10 eV) compared to heterocycles (~ 0.04 eV) [101].

In order to minimize the constraints in the dynamical aggregation of the heterocycles, immobilization via flexible spacers, such as alkanes or ethylene oxide (EO) segments, appears to be favoured [17, 102]. The optimum spacer length is then given by the optimum balance between heterocycle aggregation and heterocycle density, on the one hand, and the dynamics of the hydrogen bond network formed by the heterocycles on the other hand. Di-imidazole (a brittle solid with a high melting point), is perfectly aggregated by strong static hydrogen bonding with negligible proton conductivity. Separating the two imidazoles by a soft EO-spacer leads to the appearance of significant proton conductivity and a decrease in the melting point and glass transition temperature with increasing spacer length [103]. The conductivity then displays typical VTF behavior and, for a given concentration of excess protons (dopant), it is very similar for all spacer lengths when plotted versus $1/(T - T_o)$ where T_o is closely related to T_g [17]. For very high spacer lengths, the dilution of the heterocycles by the spacer segments tends to reduce aggregation of the heterocycles and, therefore, once again reduce proton mobility. Corresponding oligomers terminated by phosphonic acid usually show higher melting points, and in the liquid state the conductivity somehow scales with the concentration of the phoshonic fuctional group [104], which resembles the high temperature behavior of heterocycle-based systems.

The aggregation of imidazole leading to a continuous static hydrogen bonded structure in crystalline Imi-2 (two imidazole spaced by two ethylene oxide repeat units) is shown in Fig. 23.8(a) [17]. Upon melting, the situation in most parts of the material is more like that shown in Fig. 23.8(b). This result comes from an NMR study [105] demonstrating that liquid Imi-2 exhibits ordered domains (similar to the crystalline form), a disordered but still aggregated domain with dynamical hydrogen bonding, and a certain fraction of nonbonded molecules. It is only within the disordered domain (Fig. 23.8(b)) that fast proton mobility is observed, again demonstrating the delicate balance of aggregation and dynamics in hydrogen bonded structures with high proton mobility.

Recently, fully polymeric systems with side chain architectures have been developed that still exhibit high proton mobility despite complete long-range immobili-

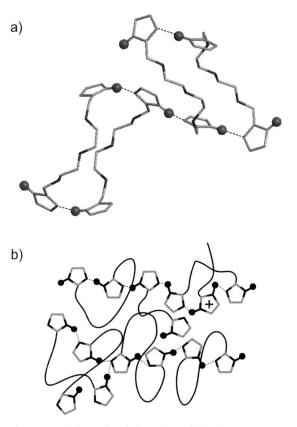

a)

b)

Figure 23.8 Hydrogen-bonded structure of Imi-2 (two imidazoles spaced by two ethylene oxide (EO) repeat units): (a) in the crystalline state as revealed by a X-ray structure alaysis [103] and b) in the liquid state (schematical) as suggested by an NMR study [105]. Note, that the hydrogen bonds in the solid state are long lived, whereas the hydrogen bonding in the molten state is highly dynamic (see text).

zation of imidazole [106] or phosphonic acid SSPC, this is $D_\sigma/D_{solvent} = \infty$. This finding is of paramount importance since it demonstrates that complete decoupling of the long-range transport of protons and the proton solvent is possible. This is directly evidenced by the echo attenuation of the proton resonance in PFG-NMR experiments of phosphonic acid functionalized oligomer [104]. Only the echo of the phosphonic protons is attenuated while the echo of the oligomer protons is only slightly affected by the magnetic field gradient (Fig. 23.9). The reader may recall that complexation of phosphoric acid and a basic polymer does not show any sign of this effect (see Fig. 23.7), which opens the way to the development of true single ion conductors.

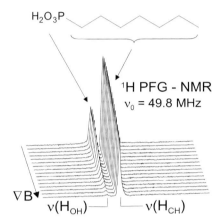

Figure 23.9 Echo attenuation of the proton resonance in PFG-NMR experiments of a phosphonic acid funtionalized oligomer [104]. Only the echo of the phosphonic protons is attenuated while the echo of the oligomer protons is only slightly affected by the magnetic field gradient.

One of the problems associated with the use of heterocycles as proton solvent in fuel cell separators is that the intrinsic concentration of protonic charge carriers can only be moderately increased through acid doping. This is particularly the case when the dynamics within the hydrogen-bonded domain is highly constrained through immobilization (especially in fully polymeric systems), which is probably the direct consequence of the reduced dielectric constant. This also leads to a further increase in the Haven ratio D_H/D_σ as discussed in Section 23.3.3. Similar Haven ratios are also observed for phosphonic acid functionalized oligomers and polymers, but the observed proton conductivities are generally about one order of magnitude higher than for heterocycle-based systems. This is simply the result of the higher amphotericity and therefore higher degree of self-dissociation (in the dry state), that is the higher intrinsic concentration of protonic defects.

23.5
Concluding Remarks

High efficiency and power density of PEM fuel cells are closely related to high proton conductivity and the gas separating property of the used electrolyte (separator). High proton conductivity, that is the long-range diffusion of protonic defects, is preferentially observed in the liquid state of hydrogen-bonded structures, because these provide the proper balance of the structure forming and dynamical properties of the intermolecular hydrogen bond interaction [25]. Aqueous solutions were indeed used as electrolyte in the first fuel cells, and even state of the art separator materials still have a liquid domain (usually water or phosphoric acid) providing the generally heterogeneous structures with their high proton conductivity (Fig. 23.10), while the inert matrix gives the material its separating properties.

However, the vapor pressure of molecular liquids, their miscibility with water and/or methanol and their viscous properties, lead to severe limitations in current fuel cell technology (see Section 23.2). Therefore, the development of non-liquid electrolytes with proton conduction properties close to these of hydrogen-bonded liquids is a key issue of current PEM fuel cell research. However, the fuel cell requirements do not allow much of a compromise with respect to proton conductivity, which should not drop below about $\sigma = 5 \times 10^{-2}$ S cm^{-1}. Such high conduc-

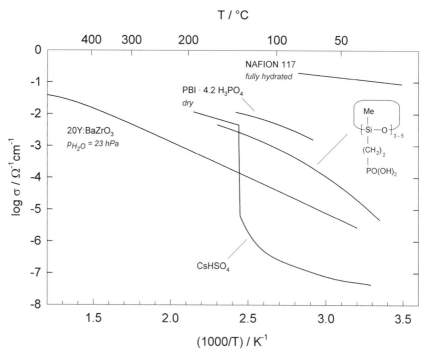

Figure 23.10 Proton conductivity of a few prototypical proton conducting separator materials: Nafion as a representative of hydrated acid ionomers (see also Fig. 23.2(a) [43, 78], a complex of PBI (polybenzimidazole) and phosphoric acid as a representative of adducts of basic polymers and oxo-acids (see also Fig. 23.2(b)) [16], phosphonic acid covalently immobilized via an alkane spacer at a siloxane backbone (see also Fig. 23.2(c)) [127], the acid salt CsHSO$_4$ [125] and an Y-doped BaZrO$_3$ [126].

tivities are several orders of magnitude higher than is known for the proton conductivity of biological systems, for instance transmembrane proteins [18, 57]. The conducting volume increments of such systems contain relatively ordered hydrogen-bonded structures of protein residues and water molecules and exhibit high selectivity for the transport of protonic charge carriers but this is only possible at the expense of high conductivity. The complete decoupling of the transport of protonic charge carriers from their solvating environment in fully polymeric systems with conductivities up to about 10^{-2} S cm^{-1} (Fig. 23.10) is, therefore, a fundamental achievement. However, much is left to be done for the development of competitive fuel cell separators free of any liquid phase. Apart from a further increase in the proton conductivity, stability and the electrochemical reactivity requirements are making this a challenging but also an appealing task [107].

Acknowledgment

The authors thank J. Fleig (Max-Planck-Insitut für Festkörperforschung) and the external reviewers for carefully reading the proofs and for fruitful discussions. We thank A. Fuchs for assisting in producing the figures and the *Deutsche Forschungsgemeinschaft* (KR 794), the *Bundesministerium für Bildung und Forschung* (0329567) and the *Stifung Energie Baden-Württemberg* (A 19603) for financial support.

References

1 K. D. Kreuer, *Annu. Rev. Mater. Res.* **33**, 333 (2003).

2 S. M. Haile, D. A. Boysen, C. R. I. Chisholm, S. M. Merle, *Nature* **410**, 910 (2001).

3 K. D. Kreuer, S. J. Paddison, E. Spohr, M. Schuster, *Chem. Rev.* **104**, 4637 (2004).

4 C. F. Schönbein, *Philos. Mag.* **14**, 43 (1839).

5 W. R. Grove, *Philos. Mag.* **14**, 127 (1839).

6 E. Peled, T. Duvdevani, A. Aharon, A. Melman, *Electrochem. Solid State Lett.* **3**, 525 (2000).

7 O. Savadogo, *J. New Mater. Electrochem. Syst.* **1**, 47 (1998).

8 Q. Li, R. He, J. O. Jensen, N. J. Bjerrum, *Chem. Mater.* **15**, 4896 (2003).

9 M. Rikukawa, K. Sanui, *Progr. Polym. Sci.* **25**, 1463 (2000).

10 J. Rozière, D. J. Jones, *Annu. Rev. Mater. Res.* **33**, 503 (2003).

11 P. Jannasch, *Curr. Opin. Colloid Interface Sci.* **8**, 96 (2003).

12 K. D. Kreuer, in *Handbook of Fuel Cells – Fundamentals, Technology and Applications*, W. Vielstich, A. Lamm, H. A. Gasteiger (Eds.), John Wiley & Sons, Chichester, p. 420, 2003.

13 M. Doyle, G. Rajendran, in Ref. 12, p.351.

14 K. A. Mauritz, R. B. Moore, *Chem. Rev.* **104**, 4535 (2004).

15 R. He, Q. Li, G. Xiao, N. J. Bjerrum, *J. Membr. Sci.*, **226**, 169 (2003).

16 Y. L. Ma, J. S. Wainright, M. H. Litt, R. F. Savinell, *J. Electrochem. Soc. A* **151**, 8 (2004).

17 M. F. H. Schuster, W. H. Meyer, *Annu. Rev. Mater. Res.* **33**, 233 (2003).

18 K. D. Kreuer, *Chem. Mater.* **8**, 610 (1996).

19 K. D. Kreuer, A. Fuchs, J. Maier, *Solid State Ionics* **77**, 157 (1995).

20 W. Münch, K. D. Kreuer, U. Traub, J. Maier, *J. Mol. Struct.* **381**, 1 (1996).

21 W. Münch, G. Seifert, K. D. Kreuer, J. Maier; *Solid State Ionics* **97**, 39 (1997).

22 K. D. Kreuer, *Solid State Ionics* **94**, 55 (1997).

23 R. A. Marcus, *J. Chem. Phys.* **24**, 966 (1956).

24 C. P. Flynn, A. M. Stoneham, *Phys. Rev.* **B1**, 3966 (1970).

25 K. D. Kreuer, *Solid State Ionics* **136/137**, 149 (2000).

26 W. A. P. Luck, *Fortschr. Chem. Forsch.*, **4**, 653 (1964).

27 X. Duan, S. Scheiner, *J. Mol. Struct.* **270**, 173 (1992).

28 R. Janoschek, *J. Mol. Struct.*, **321**, 45 (1994).

29 W. A. P. Luck, *Ber. Bunsenges. Phys. Chem.* **69**, 626 (1965).

30 M. E. Tuckerman, K. Laasonen, M. Sprike, M. Parrinello, *J. Chem. Phys.* **103**, 150 (1995).

31 M. E. Tuckerman, D. Marx, M. L. Klein, M. Parrinello, *Science* **275**, 817 (1997).

32 M. Eigen, L. De Maeyer, *Proc. R. Soc. (London)*, *Ser. A* **247**, 505 (1958).

33 M. Eigen, *Angew. Chem.* **75**, 489 (1963).

34 T. Dippel, K. D. Kreuer, *Solid State Ionics* **46**, 3 (1991).

35 B. Cohen, D. Huppert, *J. Phys. Chem. A* **107**, 3598 (2003).

36 D. Marx, M. E. Tuckerman, J. Hutter, M. Parrinello, *Nature* **397**, 601(1999).

37 R. Vuilleumier, D. Borgis, *Chem. Phys. Lett.* **284**, 71 (1998).

38 K. D. Kreuer, *Solid State Ionics* **94**, 55 (1997).

39 R. R. Sadeghi, H. P. Cheng, *J. Chem. Phys.* **111**, 2086 (1999).

40 Q. Cui, M. Karplus, *J. Phys. Chem. B* **107**, 1071(2003).

41 K. D. Kreuer, A. Rabenau, W. Weppner, *Angew. Chem. Int. Ed. Engl.* **21**, 208 (1982).

42 N. Agmon, *J. Chim. Phys. Phys. – Chim. Bio.* **93**, 1714 (1996).

43 K. D. Kreuer, T. Dippel, W. Meyer, J. Maier, *Mater. Res. Soc. Symp. Proc.* **293**, 273 (1993).

44 M. E. Tuckerman, D. Marx, M. Parrinello, *Nature* **417**, 925 (2002).

45 B. L. Trout, M. Parrinello, *J. Phys. Chem. B* **103**, 7340 (1999).

46 D. Asthagiri, L. R. Pratt, J. D. Kress, M. A. Gomez, *Chem. Phys. Lett.* **380**, 530 (2003).

47 D. Asthagiri, L. R. Pratt, J. D. Kress, M. A. Gomez, *Proc Nat. Acad. Sci (USA)* **101**, 7233 (2004).

48 B. Chen, J. M. Park, I. Ivanov, G. Tabacchi, M. L. Klein, M. Parrinello, *J. Am. Chem. Soc.* **124**, 8534 (2002).

49 R. A. Munson, *J. Phys. Chem.* **68**, 3374 (1964).

50 T. Dippel, K. D. Kreuer, J. C. Lassègues, D. Rodriguez, *Solid State Ionics* **61**, 41 (1993).

51 M. Spaeth, K. D. Kreuer, T. Dippel, J. Maier, *Solid State Ionics* **97**, 291 (1997).

52 M. Spaeth, K. D. Kreuer, J. Maier, *J. Solid State Chem.* **148**, 169 (1999).

53 T. Dippel, N. Hainovsky, K. D. Kreuer, W. Münch, J. Maier, *Ferroelectrics* **167**, 59 (1995).

54 S. H. Chung, S. Bajue, S. G. Greenbaum, *J. Chem. Phys.* **112**, 8515 (2000).

55 A. Schlechter, R. F. Savinell, *Solid State Ionics* **147**, 181 (2002).

56 A. Kawada, A. R. McGhie, M. M. Labes, *J. Chem. Phys.* **52**, 3121 (1970).

57 T. E. Decoursey, *Physiol. Rev.* **83**, 475 (2003).

58 G. Zundel, E. G. Weidemann, *Eur. Biophys. Congr., Proc., 1st* **6**, 43 (1971).

59 N. Riehl, *Trans. N.Y. Acad. Sci.* **27**,772 (1965).

60 G. Will, *Z. Krist.* **129**, 211 (1969).

61 G. P. Brown, S. Aftergut, *J. Chem. Phys.* **38**, 1356 (1963).

62 K. Pigon, H. Chojnacki, *Acta Phys. Pol.* **31**, 1069 (1967).

63 A. R. McGhie, H. Blum, M. M. Labes, *J. Chem. Phys.* **52**, 6141 (1970).

64 B. S. Hickman, M. Mascal, J. J. Titman, I. G. Wood, *J. Am. Chem. Soc.* **121**, 11486 (1999).

65 K. D. Kreuer, A. Fuchs, M. Ise, M. Spaeth, J. Maier, *Electrochim. Acta* **43**, 1281 (1998).

66 K.D. Kreuer, *Solid State Ionics and Technology*, B.V.R. Chowdari et al. (Eds.), World Scientific, Singapore, pp. 263–274, 1998.

67 W. Münch, K. D. Kreuer, W. Silvestri, J. Maier, G. Seifert, *Solid State Ionics* **145**, 437 (2001).

68 J. T. Daycock, G. P. Jones, J. R. N. Evans, J. M. Thomas, *Nature* **218**, 67 (1968).

69 W. Tatara, M. J. Wojcik, J. Lindgren, M. Probst, *J. Phys. Chem. A* **107**, 7827 (2003).

70 Unpublished results from the author's laboratory.

71 F. Toda, K. Tanaka, C. Foces-Foces, A. L. Llamas-Saiz, H. H. Limbach, F. Aguilar-Parrilla, R. M. Claramunt, C. López, J. J. Elguero, *J. Chem. Soc. Chem. Commun.* 1139 (1993).

72 M. Eikerling, A. A. Kornyshev, *J. Electroanal. Chem.* **502**, 1 (2001).

73 S. J. Paddison, G. Bender, K. D. Kreuer, N. Nicoloso, T. A. Zawodzinski, *J. New Mater. Electrochem. Syst.* **3**, 291 (2000).

74 S. J. Paddison, D. W. Reagor, T. A. Zawodzinski, *J. Electroanal. Chem.* **459**, 91 (1998).

75 S. J. Paddison, *Annu. Rev. Mater. Res.* **33**, 289 (2003).

76 M. K. Petersen, S. S. Iyengar, T. J. F. Day, G. A. Voth, *J. Phys. Chem. B* **108**, 14804 (2004).

77 K. D. Kreuer, *Solid State Ionics* **97**, 1 (1997).

78 T. A. Zawodzinski, T. E. Springer, J. Davey, R. Jestel, C. Lopez, J. Valerio, S. Gottesfeld, *J. Electrochem. Soc.* **140**, 1981 (1993).

79 M. Cappadonia, J. W. Erning, S. M. S. Niaki, U. Stimming, *Solid State Ionics* **77**, 65 (1995).

80 T. Dippel, K. D. Kreuer, *Solid State Ionics* **46**, 3 (1991).

81 S. J. Paddison, R. Paul, K. D. Kreuer, *Phys. Chem. Chem. Phys.* **4**, 1151 (2002).

82 H. Helmholtz, *Weid. Ann.* **7**, 337 (1879).

83 H. Lamb, *Philos. Mag.* **5**, 52 (1888).

84 J. Perrin, *J. Chim. Phys.* **2**, 601 (1904).

85 M. Smoluchowski, *Handbuch der Elektrizität und des Magnetismus*, Vol. II, Barth, Leipzig, 1914.

86 B. R. Breslau, I. F. Miller, *Ind. Eng. Chem. Fundam.* **10**, 554 (1971).

87 M. Ise, PhD Thesis, University Stuttgart, 2000.

88 M. Ise, K. D. Kreuer, J. Maier, *Solid State Ionics* **125**, 213 (1999).

89 T. Okada, S. Moller-Holst, O. Gorseth, S. Kjelstrup, *J. Electroanal. Chem.* **442**, 137 (1998).

90 X. Ren, S. Gottesfeld, *J. Electrochem. Soc.* **148**, A87 (2001).

91 T. A. Zawodzinski, J. Davey, J. Valerio, S. Gottesfeld, *Electrochim. Acta* **40**, 297 (1995).

92 T. A. Zawodzinski, C. Derouin, S. Radzinski, R. J. Sherman, V. T. Smith, T. E. Springer, S. Gottesfeld, *J. Electrochem. Soc.* **140**, 1041 (1993).

93 T. A. Zawodzinski, C. Derouin, S. Radzinski, R. J. Sherman, V. T. Smith, T. E. Springer, S. Gottesfeld, *J. Electrochem. Soc.* **140**, 1041 (1993).

94 G. Xie, T. Okada, *J. Electrochem. Soc.* **142**, 3057 (1995).

95 T. F. Fuller, J. Newman, *J. Electrochem. Soc.* **139**, 1332 (1992).

96 X. Glipa, B. Bonnet, B. Mula, D. J. Jones, J. J. Rozière, *Mater. Chem.* **9**, 3045 (1999).

97 R. Bouchet, E. Siebert, *Solid State Ionics* **118**, 287 (1999).

98 A. Bozkurt, M. Ise, K. D. Kreuer, W. H. Meyer, G. Wegner, *Solid State Ionics* **125**, 225 (1999).

99 B. C. Benicewicz, presented as a poster during the *Gordon Research Conference on "Fuel Cells"* 2003.

100 W. Münch, unpublished result.

101 S. J. Paddison, K. D. Kreuer, J. Maier, in preparation.

102 K. D. Kreuer, *J. Membr. Sci.* **185**, 29 (2001).

103 M. F. H. Schuster, W. H. Meyer, M. Schuster, K. D. Kreuer, *Chem. Mater.* **16**, 329 (2004).

104 M. Schuster, T. Rager, A. Noda, K. D. Kreuer, J. Maier, *Fuel Cells*, in press.

105 G. R. Goward, M. F. H. Schuster, D. Sebastiani, I. Schnell, H. W. Spiess, *J. Phys. Chem. B* **106**, 9322 (2002).

106 H. G. Herz, K. D. Kreuer, J. Maier, G. Scharfenberger, M. F. H. Schuster, W. H. Meyer, *Electrochim. Acta* **48**, 2165 (2003).

107 K. D. Kreuer, *ChemPhysChem* **3**, 771 (2002).

108 C. A. Edmondson, P. E. Stallworth, M. E. Chapman, J. J. Fontanella, M. C. Wintersgill, S. H. Chung, S. G. Greenbaum, *Solid State Ionics* **135**, 419 (2000).

109 C. A. Edmondson, P. E. Stallworth, M. C. Wintersgill, J. J. Fontanella, Y. Dai, S. G. Greenbaum, *Electrochim. Acta* **43**, 1295 (1998).

110 M. C. Wintersgill, J. J. Fontanella, *Electrochim. Acta* **43**, 1533 (1998).

111 T. A. Zawodzinski, M. Neeman, L. O. Sillerud, S. Gottesfeld, *J. Phys. Chem.* **95**, 6040 (1991).

112 C. A. Edmondson, J. J. Fontanella, S. H. Chung, S. G. Greenbaum, G. E. Wnek, *Electrochim. Acta* **46**, 1623 (2001).

113 C. A. Edmondson, J. J. Fontanella, *Solid State Ionics* **152–153**, 355 (2002).

114 J. J. Fontanella, M. G. McLin, M. C. Wintersgill, J. P. Calame, S. G. Greenbaum, *Solid State Ionics* **66**, 1 (1993).

115 X. Ren, T. E. Springer, T. A. Zawodzinski, S. Gottesfeld, *J. Electrochem. Soc.* **147**, 466 (2000).

116 S. Hietala, S. L. Maunu, F. Sundholm, *J. Polym. Sci B: Polym. Phys.* **38**, 3277 (2000).

117 T. Okada, S. Moller-Holst, O. Gorseth, S. Kjelstrup, *J. Electroanalytical Chem.* **442**, 137 (1998).

118 X. Ren, S. Gottesfeld, *J. Electrochem. Soc.* **148**, A87 (2001).

119 T. A. Zawodzinski, J. Davey, J. Valerio, S. Gottesfeld, *Electrochim. Acta* **40**, 297 (1995).

120 T. A. Zawodzinski, C. Derouin, S. Radzinski, R. J. Sherman, V. T. Smith, T. E. Springer, S. Gottesfeld, *J. Electrochem. Soc.* **140**, 1041 (1993).

121 X. Ren, W. Henderson, S. Gottesfeld, *J. Electrochem. Soc.* **144**, L267 (1997).

122 G. Xie, T. Okada, *J. Electrochem. Soc.* **142**, 3057 (1995).

123 T. F. Fuller, J. Newman, *J. Electrochem. Soc.* **139**, 1332 (1992).

124 H. Rössler, M. Schuster, K. D. Kreuer, presented at SSPC-12, Uppsala, 2004.

125 A. I. Baranov, L. A. Shuvalov, N. Shchagina, *JETP Lett.* **36**, 459 (1982).

126 K. D. Kreuer, *Solid State Ionics* **125**, 285 (1999).

127 H. Steininger, M. Schuster, K. D. Kreuer, J. Maier, *Solid State Ionics*, in press and electronically available 2006.

24
Proton Diffusion in Ice Bilayers

Katsutoshi Aoki

24.1
Introduction

24.1.1
Phase Diagram and Crystal Structure of Ice

Ice is one of the most familiar substances to human beings and has attracted interest in a wide range of research fields including chemistry, physics, biology and earth or planetary science [1]. The chemical and physical properties specific to ice arise from its bonding nature, that is, hydrogen bonding. Each water molecule is linked to four nearest neighboring molecules tetrahedrally coordinated and the molecules construct a three-dimensional hydrogen-bonded network, geometrically similar to that in diamond. A hydrogen bond has a directional nature like a covalent bond but also has much flexibility in the bond length and bond angle. A free water molecule has an H–O–H angle of 104.52° and an O–H distance of 0.09572 nm. The molecular geometry is modified by a few percent on crystallization into ice [2]. The hydrogen bond is flexible due to its complex bonding nature involving a covalently bonded hydrogen nucleus and electrostatic interactions as a major contribution.

The flexible hydrogen bond results in a rich phase behavior in the pressure–temperature diagram of ice, as illustrated schematically in Fig. 24.1. Each phase has a local structure of tetrahedrally coordinated water molecules connected by hydrogen bonds although a slight change in the bond length or angle leads to modification of the unit cell parameters.

A dramatic change in molecular packing takes place along with the phase transition from VI to VII in the high temperature region and also to VIII in the low temperature region. The high pressure phases VII and VIII appear at pressures above 2 Gpa and have dense structures consisting of interpenetrating diamond-like sublattices [3], while all the low pressure phases, below 2 Gpa, have essentially diamond structures with a large empty space. The interpenetration of sublattices leads to a body centered cubic (bcc) arrangement of oxygen atoms and smears out the empty space (Fig. 24.2). In the phase VII, for instance, each oxygen atom has

Hydrogen-Transfer Reactions. Edited by J. T. Hynes, J. P. Klinman, H. H. Limbach, and R. L. Schowen
Copyright © 2007 WILEY-VCH Verlag GmbH & Co. KGaA, Weinheim
ISBN: 978-3-527-30777-7

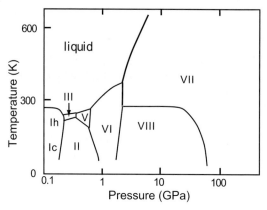

Figure 24.1 The outline of the phase diagram of ice on a logarithmic scale.

eight nearest neighbors at the corners of a cube but is tetrahedrally linked by hydrogen bonds to four of them. Ice VII is present over a wide area of the phase diagram and hence is an appropriate candidate for study of the structural or physical properties modified largely by applying pressure [4–8].

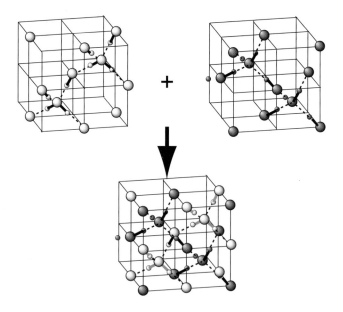

Figure 24.2 Crystal structure of the high pressure phase of ice VII consisting of interpenetrated sublattices (bottom). The free spaces, which can be seen as the empty cubes in the sublattices with a diamond-like geometry (top), are filled with the counterpart sublattice.

24.1.2
Molecular and Protonic Diffusion

Another characteristic feature of ice is the migration of water molecules in it. The diamond structure with the large empty space and the flexible hydrogen bond may allow even whole water molecules to move by either a vacancy or an interstitial mechanism. Molecular diffusion has been a major subject in research on the crystal growth of ice. The diffusion coefficient for the molecular migration has been determined using various techniques such as the isotopically labeled molecule method [9–11]. A tracer labeled with ^2H or ^{18}O is placed on one face of an ice block and then held for a certain fixed time. The analysis of the tracer concentration as a function of the depth of penetration allows us to derive the diffusion coefficient under some assumptions for the boundary conditions. The diffusion coefficients measured at temperatures of 233–273 K range from 10^{-16} to 10^{-14} $m^2 s^{-1}$ with an activation energy of 0.6–0.7 eV. The water molecules can move relatively fast in ice, at a rate of 10–100 nm s^{-1} on average.

In contrast to the well studied molecular diffusion, proton diffusion is less well understood. Protons can move in the hydrogen-bonded network of water molecules by transfer in a hydrogen bond and a successive jump into another hydrogen bond by molecular rotation, as illustrated in Fig. 24.3. This diffusion process is oversimplified but well highlights the dominant proton motions involved. Although the model was proposed a half century ago [12], its process has eluded experimental investigation. The competitive molecular diffusion shades the protonic diffusion. The protonic diffusion coefficient is estimated to be the order of 10^{-20} $m^2 s^{-1}$ for ambient-pressure ice at 263 K, which is four to five orders of magnitude less than the molecular diffusion coefficient. The dielectric property and electrical conductivity, from which we can derive the protonic diffusion coefficient, have been measured for pure and doped ices.

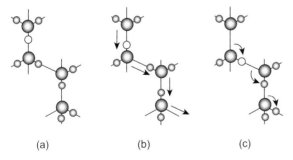

(a)   (b)   (c)

Figure 24.3 Diffusion process model: an excess proton is located at an ionic defect (a). The proton transfers along the hydrogen bond (b) and then jumps to an adjacent hydrogen bond with molecular rotation (c).

24.1.3
Protonic Diffusion at High Pressure

The protonic diffusion is considered to be enhanced significantly at high temperature and high pressure, realized in planets such as jupiter. Theoretical studies
have consistently predicted the presence of a superionic (or superprotonic) phase
characterized by a fast protonic diffusion with a coefficient of ~10^{-8} m^2 s^{-1} at extremely high temperatures and pressures [13,14]. The superionic state would
appear at about 1000 K and 20 GPa and to move to higher temperatures ranging
from 2000 to 4000 K above 100 GPa. At such high temperatures and pressures,
the protons move to occupy the midpoints between the adjacent oxygen atoms
and hence water molecules can no longer be recognized. The protons likely move
fast by jumping successively between their neighboring occupation sites in the
crystal lattice, consisting solely of oxygen atoms (as shown in Fig. 24.2 the oxygen
atoms form a bcc lattice). This implies that the two-step diffusion process proposed for the molecular crystalline ice changes to a single-step process in dense
ices with the bcc lattice of oxygen. Such a superionic phase can be characterized
as a partially melted state and compared with an ionic fluid or a fully melted state
in which neutral or ionized water molecules diffuse freely. The superionic phase
of ice may play a crucial role in the generation of the magnetic fields in giant planets as well as their metallic fluid [14–17]

The experimental fact of the small diffusion rate at ambient pressure and the
theoretical prediction of a superprotonic state under extreme conditions point us
to diffusion measurements for "hot ice" in which protons are thermally activated
to move faster. The high pressure techniques enable us to generate an extreme
condition around 1000 K and 20 GPa where ice is predicted to enter the superprotonic state. As seen in Fig. 24.1, ice VII exists over a wide pressure span above
2 GPa. The interpenetrated dense structure would prevent water molecules from
moving and allow measurement of the protonic diffusion at high temperatures
beyond the melting point of 273 K at ambient pressure [18]. Ice VII thus provides
a great advantage for protonic diffusion measurement.

A diamond-anvil-cell (DAC) is a small high pressure cell most suitable for the
spectroscopic measurement of molecular or atomic diffusion. The DAC is used
for various kinds of spectroscopic investigations on liquids and solids at pressures
up to several tens of GPa [19–22]. The optically transparent nature of diamond
over a wide wavelength span allows *in situ* optical measurements in combination
with conventional equipment such as visible light or infrared spectrometers. The
protonic diffusion in ice is measured by a traditional diffusion-couple method, in
the present case, with an H_2O/D_2O ice bilayer. The mutual diffusion of hydrogen
(H) and deuteron (D) in the ice bilayer is monitored by measuring the infrared
vibrational spectra. The experimental details are described in the following sections.

24.2
Experimental Method

24.2.1
Diffusion Equation

For the bilayer configuration of H_2O/D_2O ice, the equation of diffusion can be described with an analytical form under appropriate boundary conditions. The protons (deuterons) initially contained in the H_2O (D_2O) ice layer diffuse into the D_2O (H_2O) ice layer by H/D exchange reaction. The initial distribution of proton, which is described with a step-function as shown in Fig. 24.4, deforms gradually with time and eventually reaches a homogeneously distributed state. Starting with Fick's second law, we can derive a one-dimensional diffusion equation for the concentration of H at time t and location x under the following boundary and initial conditions [23].

$$\frac{\partial C_H(x, t)}{\partial t} = D\frac{\partial^2 C_H(x, t)}{\partial x^2}$$

$$\frac{\partial C_H(x = 0, t)}{\partial x} = \frac{\partial C_H(x = \ell, t)}{\partial x} = 0$$

$$C_H(x, t = 0) = \begin{cases} 1 & (0 \le x \le l_H) \\ 0 & (l_H \le x \le l\ ) \end{cases}$$

$$C_H(x, t) = \frac{l_H}{l} + \sum_{n=1}^{\infty}\frac{2}{n\pi}\sin\left(\frac{n\pi l_H}{l}\right)\cos\left(\frac{n\pi x}{l}\right)\exp\left\{-\left(\frac{n\pi}{l}\right)^2 Dt\right\} \tag{24.1}$$

Here the mutual diffusion coefficient D is assumed to be the same for the proton and deuteron migrations. Since the lighter element hydrogen can move faster than deuterium which has twice the mass of H, the coefficient D might correspond to the diffusion coefficient for the rather slowly moving deuterium. For the isotope concentrations at the outer back surface of the bilayer, $x = l$, the equation can be deduced to be a simple function described with time t alone. The proton concentration at the back surface of the D_2O ice layer C_H is now presented by

$$C_H(0, t) = \frac{l_H}{l} + \sum_{n=1}^{\infty}\frac{2}{n\pi}\sin\left(\frac{n\pi l_H}{l}\right)\exp\left\{-\left(\frac{n\pi}{l}\right)^2 Dt\right\} \tag{24.2}$$

where l_H and l_D are the initial thicknesses of the H_2O and D_2O ice layers, respectively. Their sum gives the total thickness of the bilayer, l: $l = l_H + l_D$. The exponential term containing time t determines the essential shape of the C_H vs. t curve. The proton concentration C_H rises rapidly in the initial stage of the diffusion process and approaches gradually the steady value given by l_H/l, the first term on the right-hand side in Eq. (24.2). The counterpart equation can readily be derived for the deuteron concentration C_D at the back surface of the H_2O ice layer or $x = 0$.

$c_H(x,t)$

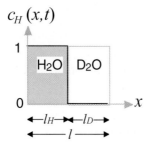

Figure 24.4 The initial state of an H_2O/D_2O ice bilayer used for measuring the mutual diffusion of H and D. C_H: concentration of hydrogen, l_H: thickness of H_2O ice layer, l_D: thickness of D_2O ice layer. The diffusion process along the x-axis can be derived within a one-dimensional approximation as described in the text.

24.2.2
High Pressure Measurement [24]

The optically transparent nature of diamond allows *in situ* measurement of infrared absorption spectra in association with the excitation of vibrational states. The pressure accessible with a DAC reaches a very high pressure of around 100 GPa for a temperature range of 0–1000 K. The size of a sample pressurized with DAC depends on the pressure desired for each experiment, ranging from a few hundred to ten microns in diameter. The DAC is ordinarily used in combination with a microscope system to focus the probe light on a small sample.

Figure 24.5 shows an optical setup for high-pressure diffusion measurement with a DAC. An H_2O/D_2O ice bilayer is already prepared in the sample chamber. The dimension of each ice layer is typically 120 μm in diameter and 20 μm in thickness. The detailed procedure for preparing the ice bilayer has been described in the literature [24]. The surface concentration of proton, C_H, at $x = l$ can be

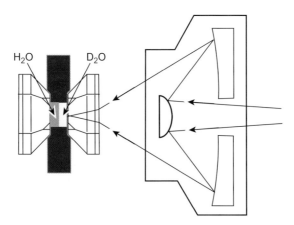

Figure 24.5 Schematic drawing of the optical configuration for measuring the infrared reflection spectra: (left) a diamond anvil cell containing an ice bilayer and (right) a reflection objective for focusing incident infrared lights. (From Ref. [24]).

obtained from infrared reflection spectra measured at the interface between the diamond and the D_2O ice. The protons initially contained in the H_2O ice layer penetrate through the D_2O ice layer by the H/D exchange reaction and eventually reach the back surface. The surface concentration C_H can be related to the peak intensity of the OH stretch vibration, which shows a gradual growth with time until the diffusion process reaches the steady state, as analyzed with a bilayer model in the previous sections. A reflection objective is used to introduce the incident infrared light to the ice–diamond surface and also the reflected light to a detector. It has magnification of 16 and a numerical aperture of 0.6.

24.2.3
Infrared Reflection Spectra

Reflection spectra measured by focusing incident light on the ice/diamond interface involve the extrinsic components arising from the reflection and absorption of the diamond anvil. The incident light is reflected from the air/diamond interface and then absorbed while passing through the diamond anvil. The reflected light from the ice surface undergoes absorption and reflection as well in the back track. Spectral correction is hence an essential procedure for deriving the intrinsic spectral features of the specimen. The spectral features of diamond, which range over a wavenumber region roughly from 1600 to 2600 cm^{-1}, can be eliminated practically and effectively by subtracting an appropriate reference spectrum from each raw spectrum. Spectral correction is well made using a reference spectrum taken for the DAC filled with potassium bromide, KBr (Fig. 24.6).

Peak intensity calibration is another essential procedure required for correct determination of the isotope concentrations from the observed spectra. Here we should note that the peak intensity per one OH bond is not necessary equal to that per one OD bond. Calibration spectra are taken for an ice specimen containing homogeneously distributed H and D isotopes prepared by freezing an

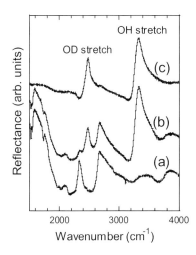

Figure 24.6 Infrared reflection spectra: (a) a reference spectrum measured for the KBr-diamond interface with a DAC, (b) a representative raw spectrum measured for the D_2O back surface of an ice bilayer specimen, (c) a reflection spectrum of ice obtained by dividing the raw spectrum by the reference spectrum. The OD and OH stretch peaks are located approximately at 2400 and 3400 cm^{-1}, respectively. (From Ref. [24].)

$H_2O(50\%)$–$D_2O(50\%)$ solution and are used to determine practically the relative peak intensities between the OH and OD stretch peaks. The reflection spectra measured at pressures from 9.5 to 16.2 GPa reveal that each peak height remains nearly independent of pressure. The ratio of the OH peak height with respect to the OD peak height ranges from 1.32 to 1.48, yielding an average value of 1.40. This value is used for conversion of the peak heights to the isotope concentrations.

Infrared reflection spectra are measured with a microscope FT-IR spectrometer mounted with a reflection objective, with which the incident lights are focused onto the diamond/ice interface and the reflected lights are introduced to a MCT detector cooled by liquid nitrogen. A wavenumber region of 700–5000 cm^{-1} is covered. The reflected lights are focused on the path to the detector to make a magnified real image of the specimen, which is trimmed into a 40×40 μm^2 square in its real scale with an optical mask. Reflection spectra are thus measured for the central sample area; the one side of the masking square 40 μm corresponds to one third of the sample diameter. Equation (24.1) and (24.2) are derived for a bilayer specimen with an infinite radius. The trimming of the measuring area leads to elimination or reduction of undesirable wall effects involved in the actual diffusion process.

24.2.4
Thermal Activation of Diffusion Motion

The high-pressure experiment enables us to investigate the protonic diffusion process thermally enhanced by heating. The diffusion rate of ambient-pressure ice is estimated to be of the order of 10^{-20} m^2 s^{-1} at 258 K [1], indicating that 1000 years is required for the protons to pass through an ice layer of 20 μm in thickness. Thermal annealing is an effective way to increase the diffusion rate, even by several orders in magnitude. At ambient pressure heating ice is limited to the melting point of 273 K; the diffusion coefficient might increase by only one order, insufficient to reduce the astronomical figures. The melting temperature rises rapidly on applying pressure, reaching, for instance, 690 K at 10 GPa. This temperature is 400 K higher than that at ambient pressure and sufficient to accelerate the diffusion rate so that it can be detected within a laboratory scale time.

Thermal annealing of the sample is simply achieved by warming the DAC itself in an electric oven. An appropriate annealing temperature and time are examined with a bilayer sample and chosen to be 400 K and several tens of hours, respectively. Infrared reflection spectra are measured with the DAC taken out of the oven and cooled quickly to room temperature. Thermal annealing and successive spectral measurement are repeated several tens of times until the spectral change is complete. The accumulated annealing time required for one experimental run ranges from several hundreds to a thousand hours. The DAC is warmed up to 400 K and cooled down to 298 K in a few minutes, negligibly short compared with a thermal annealing time of several tens of hours. No correction for the transient time is required in accumulation of the annealing time.

24.3
Spectral Analysis of the Diffusion Process

24.3.1
Protonic Diffusion

The protonic diffusion process in the ice bilayer is clearly monitored by the vibrational spectroscopic measurement. The reflection spectra measured after annealing at 400 K and 10.2 GPa are shown in Fig. 24.7. Panels A and B present the spectral changes with time measured for the back surfaces of D_2O and H_2O ice layers, respectively. The OH stretch peak is located around 3200 cm^{-1}, whereas the OD stretch peak is located around 2500 cm^{-1}, lower by a factor of approximately $1/\sqrt{2}$, as expected from a square root of the mass ratio $\sqrt{m_H}/\sqrt{m_D}$ [5]. The H/D mutual diffusion process is demonstrated as gradual changes in the peak height in the opposite directions with time. For instance, at the back surface of the D_2O ice layer, the OD peak shows a gradual decrease in height with time, while the OH peak shows a gradual increase. The spectra measured for both back surfaces become equivalent after 1287 h, indicating homogeneous distribution of H and D over the ice specimen as a result of the mutual diffusion.

A diffusion coefficient can be derived from the variation of the peak height or the surface concentration of hydrogen (deuteron) with time. The deuteron concentration C_D at the back surface of the H_2O ice layer shows an exponential increase with time, as expected from Eq. (24.2), as displayed in Fig. 24.8. It rises abruptly after a small time lag of several hours (not able to be seen in Fig. 24.8 plotted with a full time-scale of 1200 h) and approaches a steady value of 0.52. The fitting of the experimental results yields a diffusion coefficient of 6.2×10^{-16} m^2 s^{-1}. This

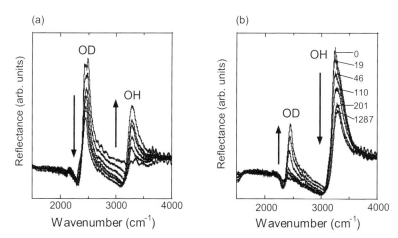

Figure 24.7 Variation of infrared reflection spectra with time measured for the outer surface of D_2O ice layer (A) and for that of H_2O ice layer (B). The numbers attached to the spectra give the accumulated annealing time in hours. (From Ref. [24].)

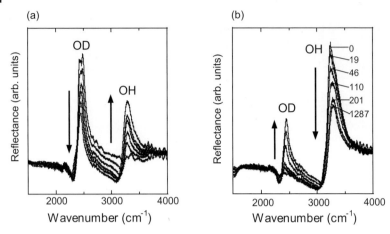

Figure 24.8 Variation of deuteron concentration C_D with time measured for the back surface of an H_2O/D_2O ice bilayer annealed at 400 K at 10.2 GPa. The solid line represents a fit to the diffusion equation. (From Ref. [24].)

value is larger by a factor of 10^4 than that estimated for ambient pressure ice at 258 K. The diffusion motion is significantly enhanced at 400 K by thermal activation. The exact diffusion equation is described in terms of an infinite series of n as given in Eqs. (24.1) and (24.2). The equation with $n = 100$ is capable of reproducing satisfactorily all the features of the concentration variation measured: an abrupt rise in the initial stage of diffusion and a subsequent gradual increase toward the steady value. The C_D–t curve thus reproduced with the coefficient of 6.2×10^{-16} m^2 s^{-1} is represented by a solid line.

24.3.2
Molecular Diffusion

As described in Section 24.1.2, molecular diffusion is dominant in ambient-pressure ice and there still remains the possibility that the spectral change observed is brought about by migration of whole molecules of H_2O and D_2O. Molecular diffusion is therefore carefully examined in an $H_2{}^{16}O$ /$H_2{}^{18}O$ ice bilayer. The substitution of oxygen isotopes does not influence the molecular vibrations, since the difference in atomic mass between ^{16}O and ^{18}O is very small compared with that between ^{1}H and ^{2}D. The resultant frequency difference is estimated to be approximately 20 cm^{-1} for the stretch vibrations.

Molecular diffusion is suppressed in the high pressure phase of ice VII. No signal for molecular diffusion is detected. Reflection spectra measured for the back surfaces of an $H_2{}^{16}O$ /$H_2{}^{18}O$ ice bilayer are shown in Fig. 24.9. The annealing temperature and pressure are 400 K and 10.2 GPa, respectively, the same as those for the protonic diffusion measurement. The abscissa axis is expanded to emphasize the peak positions of the ^{16}OH and ^{18}OH stretch vibrations. Separation by

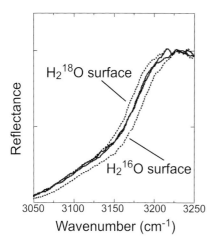

Figure 24.9 Infrared reflection spectra collected from the back surfaces of an $H_2^{16}O/H_2^{18}O$ ice bilayer (dashed lines) annealed at 400 K and 10.2 GPa for 600 h. The spectra measured for an ice specimen once melted at 298 K by releasing pressure and then compressed again to 10.2 GPa showed the peak shifts to the midpoint (solid lines). (From Ref. [24].)

20 cm^{-1} is clearly seen, in agreement with the estimation. Thermal annealing at 400 K results in no spectral change. The stretch peaks remain in the initial positions even after 600 h annealing, indicating that molecular diffusion does not take place. To confirm this, the ice specimen is once melted at 298 K by releasing the pressure carefully (ice melts at 0.9 GPa at room temperature after passing through another high pressure phase of ice VI) and frozen by increasing the pressure quickly to 10.2 GPa. The spectra measured again for the back surfaces show the peak shifts to the midpoint between the original positions. $H_2^{18}O$ and $H_2^{16}O$ molecules are homogeneously mixed by fast molecular diffusion in the liquid state.

24.3.3
Pressure Dependence of Protonic Diffusion Coefficient [25]

The protonic diffusion coefficients measured for ice VII in a pressure range 2.1–63 GPa are shown in Fig. 24.10. Around 60 GPa, the O–H···O bond length decreases to a critical value of 0.24 nm, at which the hydrogen atoms move to occupy the midpoint between the neighboring oxygen atoms, that is, the hydrogen bond becomes symmetric with equal O–H and H–O distances [4–9, 26]. In other words, the molecular crystal is converted to a nonmolecular crystal at this critical pressure. Such a significant change in the hydrogen bond would influence the protonic motions and hence result in some anomalous feature in the diffusion coefficient.

The diffusion coefficients measured at 400 K show a monotonic decrease by two orders of magnitude with increasing pressure up to 63 GPa (Fig. 24.10). The influence of pressure on the diffusion is phenomenologically described by $D = D_0\exp(-\gamma P/kT)$, where γ represents the magnitude of the pressure influence on the activation energy for the protonic diffusion motion. Fitting to the experimental results gives 0.003 eV GPa^{-1} for γ. Although no experimental data are available for the activation energy at ambient pressure, ΔE_0, we may assume it to correspond to that determined from the electric conductivity measurement on pure ice at ambient pressure [27]. The activation energy, $\Delta E = \Delta E_0 + \gamma P$, is finally rewritten as $\Delta E = 0.70 + 0.003P$, where ΔE and ΔE_0 are in eV and P in GPa.

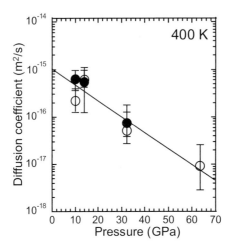

Figure 24.10 Variation of the protonic diffusion coefficient with pressure measured for ice VII. Open and solid circles represent those obtained from reflection spectra measured for the H$_2$O and D$_2$O back surfaces of H$_2$O/D$_2$O bilayer, respectively. (From Ref. [25].)

The variation of the diffusion coefficient with pressure appears contrary to that expected from the distinct change in the hydrogen bonding state in association with the molecular to nonmolecular transition. In the molecular state at low pressures, the protonic diffusion proceeds by two steps: transfer in a hydrogen bond and a successive jump into another hydrogen bond. For ambient-pressure ice, the activation energies are determined experimentally by infrared absorption measurements to be 0.41 eV for the transfer and 0.52 eV for the jump [28]. In the nonmolecular region around 60 GPa, a small energy barrier still exists at the midpoint and hence a double minimum shape of potential remains slightly. The proton, however, can transfer almost freely by tunneling or thermal activation between two occupation sites and form statistically ionized molecular species [29–32]. The diffusion motion would be alternated with a single-step process involving a proton jump alone, and would be expected to be accelerated by an increase in population of the ionized molecular species such as H_3O^+ and OH^-. Nevertheless, the diffusion coefficient still shows a decrease around this pressure region. The potential barrier for the proton jump between the adjacent oxygen sites likely rises with pressure as the molecules approach, and consequently the one-step diffusion motion might be inhibited.

24.4
Summary

Protonic diffusion in ice has been investigated by a spectroscopic method. This method is based on the isotope effect on molecular vibrations. The mass difference between hydrogen and deuteron results in a frequency difference by a factor of $\sqrt{2}$ for the stretch mode. The peak positions are well separated in the spectra and hence their heights are converted to the H(D) concentrations with good accuracy. The diffusion process is monitored by measuring the reflection spectra of an H_2O/D_2O ice bilayer, for which the equation of diffusion is described in analytical form. The H/D mutual diffusion coefficient measured at 400 K shows a monotonic decrease by two orders of magnitude as the pressure increases from 8 to 63 GPa.

The spectroscopic method can be applied to other substances, for instance, solid acids such as $CsHSO_4$ and polymer electrolytes. The range of diffusion rate covered by the present method could be extended by the use of an advanced infrared light-source and detector. The diffusion coefficient of 10^{-16} to 10^{-14} m^2 s^{-1} is currently accessible using a conventional Fourier-transform infrared spectrometer mounted with a ceramics-heater light-source and an MCT detector. A synchrotron radiation facility provides brilliant light in the infrared region; the intensity is higher by several orders of magnitude than that of a conventional light source. Infrared array detectors such as the HgCdTe device provide parallel detection without moving parts and spectral rates much higher than with a FT-IR system.

References

1 V. F. Petrenko, R. W. Whitworth, *Physics of Ice*, Oxford University Press, New York, 1999.

2 E. Whalley, *Hydrogen Bond*, Vol. 3, P. Schuster, G. Zundel, C. Sabdorfy (Eds.), North-Holland, Amsterdam, p. 1425 (1976).

3 W. F. Kuhs, J. L. Finney, C. Vettier, D. V. Bliss, *J. Chem. Phys.* **81**, 3612 (1984).

4 R. J. Hemley, A. P. Jephcoat, H. K. Mao, L. W. Finger, D. E. Cox, *Nature* **330**, 737 (1987).

5 M. Song, H. Yamawaki, M. Sakashita, H. Fujihisa, K. Aoki, *Phys. Rev. B* **60**, 12644 (1999).

6 Ph. Pruzan, E. Wolanin, M. Gaythier, J. C. Chervin, B. Canny, *J. Phys. Chem. B* **101**, 6230 (1997).

7 A. F. Goncharov, V. V. Struzhkin, H. K. Mao, R. J. Hemley, *Phys. Rev. Lett.* **83**, 1998 (1999).

8 M. Song, H. Yamawaki, H. Fujihisa, M. Sakashita, K. Aoki, *Phys. Rev. B* **68**, 014106 (2003).

9 H. Blicks, O. Dengel, N. Riehl, *Phys. Kondens. Materie* **4**, 375 (1966).

10 R. O. Ramseier, *J. Appl. Phys.* **38**, 2553 (1967).

11 K. Itagaki, *J. Phys. Soc. Jpn.* **22**, 427 (1967).

12 N. Bjerrum, *Science* **115**, 385 (1952).

13 P. Demontis, R. LeSar, M. L. Klein, *Phys. Rev. Lett.* **60**, 2284 (1988).

14 C. Cavazzoni, G. L. Chiarotti, S. Scandolo, E. Tosatti, M. Parrinello, *Science* **283**, 44 (1999).

15 D. J. Stevenson, *Rep. Prog. Phys.* **46**, 555 (1983).

16 W. J. Nellis, D. C. Hamilton, N. C. Holmes, H. B. Radousky, F. H. Ree, A. C. Mitchell, M. Nicol, *Science* **240**, 779 (1988).

17 W. J. Nellis, N. C. Holmes, A. C. Mitchell, D. C. Hamilton, *J. Chem. Phys.* **107**, 9096 (1997).

18 F. Datchi, P. Loubeyre, R. LeToullec, *Phys. Rev. B* **61**, 6535 (2000).

19 S. Block, G. J. Piermarini, *Phys. Today* **29**, 44 (1976).

20 A. Jayaraman, *Rev. Mod. Phys.* **55**, 65 (1983).

21 H. K. Mao, R. J. Hemley, A. L. Mao, in *High-Pressure Science and Technology*, S. C. Schmidt, J. W. Shaner, G. A. Samara, M. Ross (Eds.), AIP, New York, 1613 (1993).

22 M. I. Eremets, *High Pressure Experimental Methods*, Oxford University Press, New York, 1996.

23 J. Crank, *The Mathematics of Diffusion*, Clarendon, Oxford, 1975.

24 K. Aoki, Eriko Katoh, H. Yamawaki, H. Fujihisa, and M. Sakashita, *Rev. Sci. Instrum.* **74**, 2472 (2003).

25 E. Katoh, H. Yamawaki, H. Fujihisa, M. Sakashita, K. Aoki, *Science* **295**, 1264 (2002).

26 P. Loubeyre, R. LeToullec, E. Wolanin, M. Hanfland, D. Hausermann, *Nature* **397**, 503 (1999).

27 V. F. Petrenko, R. W. Whitworth, *J. Phys. Chem.* **87**, 4022 (1983).

28 W. B. Collier, G. Ritzhaupt, J. P. Devlin, *J. Phys. Chem.* **88**, 363 (1984).

29 K. S. Schweizer, F. H. Stillinger, *J. Chem. Phys.* **80**, 1230 (1984).

30 P. G. Johannsen, *J. Phys.: Condens. Matter.* **10**, 2241 (1998).

31 M. Benoit, D. Marx, M. Parrinello, *Nature* **392**, 258 (1998); M. Benoit, D. Marx, M. Parrinello, *Solid State Ionics* **125**, 23 (1999).

32 W. B. Holzapfel, *Physica B* **265**, 113 (1999).

25
Hydrogen Transfer on Metal Surfaces
Klaus Christmann

25.1
Introduction

Among the definitions of the term "hydrogen bond" one can find the following explanations: *A weak bond involving the sharing of an electron with a hydrogen atom; hydrogen bonds are important in the specificity of base pairing in nucleic acids and in the determination of protein shape; or: A hydrogen bond is a chemical bond in which a hydrogen atom of one molecule is attracted to an electronegative atom, especially a nitrogen, oxygen, or fluorine atom, usually of another molecule.*

A somewhat closer look into the chemical bonding situation reveals that a hydrogen bond is the consequence of an attractive intermolecular force between two partial electric charges of opposite sign, whereby an H atom participates. The simplest and most common example is perhaps an H atom located between the two oxygen atoms of two neighboring water molecules: This H atom can build up a bond to either oxygen atom, thus forming a *bridge* between these O atoms. This is the reason why, in the German notation, this particular type of bonding is called "Wasserstoff-*Brücken*-Bindung". It is not necessarily an intermolecular bond; considering large molecules such as proteins, it is also possible that H bonds are formed between two parts of the *same* molecule. These intramolecular bonds often decisively influence the shape of the respective molecular entity, they are, for example, responsible for the folding of proteins etc. Accordingly, the significance of H bonding in biochemistry or, more generally, in life sciences cannot be overestimated. The strength of an H bond is usually larger than the common intermolecular (van-der-Waals-like) forces, however, it cannot compete with the strength of typical covalent or ionic bonds. This "intermediate" bond strength is certainly the reason behind the pronounced variability of H bonding effects and their importance in the life sciences.

While one could further expand greatly on the specifics of H bonds, one of their prominent characteristics is that the H atom involved can easily be *transferred* from one electronegative heteroatom (nitrogen, oxygen, chlorine...) to another, and the question arises as to how this transfer process proceeds and where the H atom is actually located or which electronegative center it is associated with.

Hydrogen-Transfer Reactions. Edited by J. T. Hynes, J. P. Klinman, H. H. Limbach, and R. L. Schowen
Copyright © 2007 WILEY-VCH Verlag GmbH & Co. KGaA, Weinheim
ISBN: 978-3-527-30777-7

Indeed, the respective H atom may be considered entirely delocalized between the two negatively polarized atoms or functional groups. Due to its small size it can even tunnel through the potential energy barrier which exists between the two negative centers. Since liquid water is the most common solvent, a wealth of investigations has been carried out to specify hydrogen bonding in water and/or hydrophilic systems (unnecessary to say that just these systems are the dominating ones in biochemistry). The respective studies involve both experimental and theoretical work and often focus on a particularly important aspect, namely, the *transfer* of hydrogen. Detailed information on the energetics and dynamics of this hydrogen transfer is required to understand the mechanisms of the biochemical reactions and processes. Unfortunately, these reactions mostly (if not always) occur in three-dimensional condensed phases (preferably in liquid systems) and usually involve large and complicated molecules. Nevertheless, studies with very simple (in most cases admittedly *too* simple) model systems can be advantageous to disentangle selective reaction steps of hydrogen transfer. Studies on *surfaces*, for example, are often quite helpful to reduce the dimension or the symmetry of a problem, and it is not purely by chance that the "inventors" of the scanning tunneling microscope soon after its invention tried to image a biomolecule (DNA) deposited on a surface [1]. In order to learn something about hydrogen transfer it is therefore a legitimate approach and, in addition, quite helpful, to deposit molecules on solid *surfaces* and scrutinize, e.g., the migration and exchange of H atoms between two molecular entities. We recall that two-dimensional systems involving H bonding and H transfer also play a decisive role in heterogeneous catalysis, in thin solid or liquid films, in micelles and (fuel cell) membranes. At a first glance, these systems may be considered less complex than H bonding in three-dimensional liquid or solid–liquid systems. While this may be true in a few cases, the details of H transfer are nevertheless also complicated in 2D systems, due to the variability of possible routes of interaction. On the other hand, the transition from 2D to 3D systems can be accomplished by continuously increasing the amount of molecules deposited on the surface template until condensation occurs. During the respective multilayer growth process, the change in the system's properties can be followed and attributed to the characteristics of the 3D aggregation.

An approach that has been pursued very successfully in the past is to simplify the system under consideration even further and to consider the interaction of gases with geometrically and electronically well defined two-dimensional *single-crystal* surfaces. *Metal* surfaces are comparatively well understood with respect to both geometrical and electronic structure. This includes simple processes occurring during the interaction of reactive gases (hydrogen, oxygen, carbon monoxide) with these surfaces. Focusing here on dihydrogen interaction with metal surfaces in general, numerous review reports exist [2–4] which provide detailed information on thermodynamic and kinetic properties (hydrogen adsorption and desorption energies, sticking probabilities, frequency factors and activation energies, vibrational frequencies, electronic levels and dipole moments). The experimental studies have been accompanied by an almost equal number of theoretical calculations and simulations [5], often with quite satisfactory agreement between experiment and theory. One might argue that the interaction of a molecule as simple as H_2 with a metal surface should lead to a very clear and distinct view. However, just the interaction of hydrogen bears a variety of complications which are caused, among others, by the small physical size of hydrogen, its quantum-mechanical properties and the ability of many metals to spontaneously dissociate dihydrogen. Instead of a simple one-point adsorption step (which nevertheless occurs in quite a number of instances) the very reactive H atoms formed upon dissociation can (and often will) cause consecutive reactions with the metal surface: Among others, the relatively large heat of hydrogen adsorption can enable metal surface atoms to move and geometrically rearrange themselves to energetically more favorable lattice positions, thus causing a new lateral periodicity of the entire surface (a process referred to as "surface reconstruction"). In a next step, this rearrangement may produce surface channels through which H atoms can more easily enter subsurface or even bulk sites (dissolution and absorption of hydrogen) ending up with metal–hydrogen compounds, i.e., hydridic phases. Metals like palladium, vanadium, titanium, niobium or tantalum are well known to absorb appreciable quantities of H atoms under appropriate thermodynamic conditions [6]. How a dihydrogen molecule can interact with metal surfaces, including absorption and solution, is schematically sketched in the (one-dimensional) potential energy diagram of Fig. 25.1.

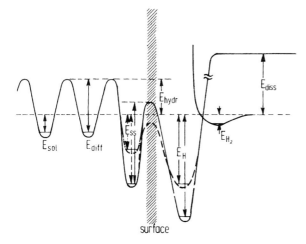

Figure 25.1 One-dimensional potential energy diagram illustrating the changes in the potential energy of a hydrogen molecule which approaches a metal (the location of the surface is indicated by the hatched area). The following processes may occur:
- Trapping of a H_2 molecule in a shallow physisorption minimum of depth E_{H_2}
- Dissociation of the H_2 molecule and formation of a stable chemisorptive bond between each H atom and the surface;

release of adsorption energy E_H. Full line: Sparsely H-covered surface; dashed line: fully H-covered surface (consideration of coverage-dependent interaction potentials, cf. Fig. 25.5)
- Migration of H atoms into subsurface sites, with a (coverage-dependent – full and dashed lines) sorption energy E_{ss}.
- (Possible) absorption of H atoms in interstitial sites with heat of solution E_{sol}. The activation energy of diffusion of the respective H atoms, $E_{diff.}$, is indicated.

A brief outline as to how this chapter is organized may be helpful. In our attempt to consider H bonding and related effects on and at metal surfaces we will largely exclude the aforementioned complications such as surface reconstruction, subsurface-site population or hydrogen sorption effects, since they may obscure the essential H transfer and bonding phenomena.

First, in Section 25.2 we shall familiarize the reader with some general terms and elementary processes that can occur during the interaction of gaseous dihydrogen with metal surfaces and that are crucial in order to understand the details of H transfer on these kinds of surfaces. They include processes like physisorption and chemisorption, activated and nonactivated adsorption and desorption, *a priori* and *a posteriori* energetic heterogeneity, formation of phases with long-range order, and others.

In Section 25.3, we will discuss the elementary steps of hydrogen *transfer* on a metal surface. In the simplest case, this is just the *diffusion* of H atoms or H_2 molecules from one lattice site of a homogeneous periodic surface to another site on the same surface, whereby the H species may travel over distances on the *microscopic* (a few nanometers) or on the *macroscopic* scale (micrometers to millimeters). Especially for the lightest hydrogen isotope $_1^1H$, the 'classical' diffusion may be accompanied (or even replaced) by quantum tunneling processes which can dominate the transport at lower temperatures. Relevant, too, is hydrogen transfer on a *heterogeneous* surface (especially in the area of technical catalysis). Generally, such a surface may consist of different compounds and/or elements arranged in patches of different size and surface geometry and actually represents quite a complex system. Focusing on metallic surfaces (as we do here), we consider alloy or bimetallic surfaces in the first instance (the latter consisting of immiscible components), where one component is *active* with respect to hydrogen adsorption (dissociation) and the other is *not*. There arises an immediate question: Will the H atoms formed on the active part of the bimetallic surface remain trapped on the sites belonging to the active patch, or will they, once formed, be able to migrate also to sites located on the inactive surfaces? This so-called *spillover effect* is believed to be a crucial property in various hydrogenation reactions over metal and supported metal catalyst surfaces [7].

Finally, in Section 25.4 homogeneous surfaces will be considered that are (partially) covered with negatively polarized molecules containing hydroxy (OH) groups. In this context, the adsorption and especially the condensation and network formation of water or alcohol molecules at surfaces deserves interest, because a monitoring of the respective growth processes allows conclusions to be drawn on the two-dimensional $\leftrightarrow$ three-dimensional phase transition and the network formation mainly caused by hydrogen bonding effects.

25.2
The Principles of the Interaction of Hydrogen with Surfaces: Terms and Definitions

A hydrogen molecule arriving from the gas phase first 'feels' the (generally slightly attractive) force fields of the surface and can experience three processes, depending on the strength of the interaction potential: If the H_2 molecule is merely trapped by weak van-der-Waals forces, it undergoes "physical" adsorption; this genuine *physisorption* of hydrogen usually involves interaction energies of merely a few kJ mol^{-1}, and temperatures as low as 15–20 K are already sufficient to make the molecules leave the surface again by thermal desorption. Physisorptive interaction is the rule with chemically inactive surfaces (e.g., alkali metal halide or graphite surfaces) and, accordingly, hydrogen molecules do not adsorb at common temperatures ($T \geq 300$ K) on those materials. Only close to the freezing point of solid molecular hydrogen, i.e., in the temperature interval between, say, 5 and 20 K is it possible to precipitate and condense liquid or solid H_2 layers on these surfaces. In a few special cases the H_2 molecule may interact somewhat more strongly with metal surfaces, provided the respective system offers sites with higher geometrical coordination and a peculiar electronic structure; in this case, interaction energies up to ~ 20 kJ mol^{-1} are involved, leading to markedly elevated hydrogen desorption temperatures between 60 and 90 K. Examples of this weak molecular *chemisorption* have been reported for stepped Ni surfaces [8] and for some fcc(210) surfaces (Pd, Ni) [9, 10]. The molecular nature of the adsorbed hydrogen is clearly proven by vibrational loss (observation of the H–H stretching vibration) and isotope exchange measurements, in that the isotopic scrambling (observation of HD besides H_2 and D_2) does *not* take place. In Fig. 25.2 we present, as an example, thermal desorption spectra of H_2, D_2, and HD after a 'mixed' exposure of a Pd(210) surface to hydrogen gas and deuterium gas at ~40 K. Obviously, there is only a vanishingly small HD contribution, ruling out significant isotopic scrambling [9]. The still low desorption temperatures of the molecular hydrogen thereby reflect the weak interaction forces which are, on the other

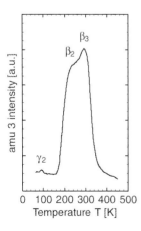

Figure 25.2 Thermal desorption spectrum of hydrogen deuteride (HD, mass 3) from a Pd(210) surface that had received a simultaneous exposure of hydrogen, H_2, and deuterium, D_2, at 40 K. The β states represent atomically adsorbed hydrogen (deuterium), while the γ states are due to the molecular species. Apparently, practically no isotopic scrambling occurs in the γ states (absence of HD), while the exchange is complete in the atomic β states. After Schmidt et al. [9,10].

hand, responsible for the appreciable mobility of the H_2 molecules while being trapped on the surface. In other words, easy hydrogen transfer (H_2) is achieved with these systems.

More common and important, however, is the *chemisorption* of hydrogen, with the interaction energies ranging from ~ 50 to 150 kJ mol^{-1}. In this case, the H_2 molecules undergo homolytic dissociation into two H atoms, either spontaneously (nonactivated adsorption) or, much more slowly, in an activated step (activated adsorption). Which process actually dominates depends on the electronic structure of the metal in question, see below. Activated H adsorption greatly reduces the rate of H uptake at lower temperatures, but this rate increases strongly with temperature T, since then more molecules exist having the required activation energy. In Fig. 25.3, spontaneous and activated adsorption are illustrated by a simple Lennard-Jones potential energy diagram. Note that only in the case of activated

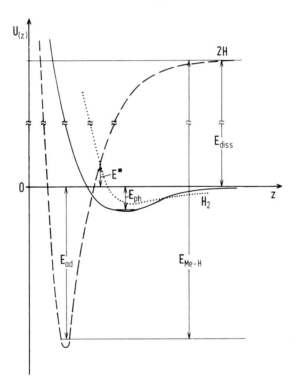

Figure 25.3 Lennard-Jones potential energy diagram of a H_2 molecule interacting with an active (full line) and an inactive metal surface (dotted line) as a schematic one-dimensional description of the activated (non-activated) hydrogen adsorption. The dashed line indicates the potential energy $U(z)$ for a pre-dissociated H_2 molecule (shifted by the dissociation energy E_{diss}, with respect to energy zero) and for the two isolated, reactive, H atoms as they approach the surface. The deep potential energy well (E_{Me-H}) represents the energy of the metal–H bond formed (which is gained twice). In the activated case the intersection between the dotted and dashed curves leads to an activation barrier of height $E*$ that a H_2 molecule getting in contact with the surface has to overcome in order to be chemisorbed.

adsorption does the H_2 molecule have to overcome an activation barrier, E_{ad}^*. Examples are coinage metal surfaces (Cu, Ag, Au), where barriers of up to 50 kJ mol^{-1} have been determined [11] and other 'free electron' metals (alkali, alkaline earth and earth metals) or various elemental semiconductor surfaces (Si, Ge) with barriers of similar height [12, 13]. Even more illustrative is a two-dimensional representation of the homolytic dissociation reaction ('elbow' plot), see Fig. 25.4, where the internuclear H–H distance x is plotted against the distance y of the H_2 molecule with respect to the surface. The resulting elbow-like equipotential lines indicate that only close to the surface are the forces operating between the substrate and the H atoms strong enough to sufficiently stretch the inner-molecular bond and to make it dissociate. We add here that a wealth of theoretical work has been carried out to develop appropriate quantum-chemical models which can explain both the spontaneous and activated hydrogen dissociation [14–16]. Following Harris [16], dissociation generally requires weakening of the H–H bond, and both filling the empty anti-bonding $2\sigma^*$ molecular orbital (MO) of the H_2 molecule with electrons, or emptying its filled 1σ bonding MO can cause the respective bond weakening effect. In addition, since the H_2 molecule is a closed-shell unit, the Pauli repulsion between the filled, delocalized metallic s,p bands of the substrate and the occupied 1σ hydrogen MO is responsible for the appearance of an activation barrier as the H_2 molecule is brought closer to the surface. Only if empty d electron states with similar energy to that of the sp electrons are available, can rehybridization help to circumvent the Pauli repulsion. Harris, who has theoretically modeled this rehybridization, states explicitly that the sp electrons of the metal can 'escape' into the empty d state [16]. From this it is immediately apparent why transition metals (with their high density of empty d states, right at the Fermi level E_F) are *active* and coinage metals (with their filled d bands lying ~ 2–3 eV below E_F) are *inactive* with respect to spontaneous H_2 dissociation.

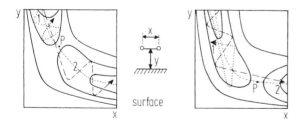

surface

Figure 25.4 Two-dimensional representation ("elbow" plots) of the potential energy situation when a H_2 molecule interacts with an active metal surface. The coordinate x describes the internuclear H–H distance, y is the distance of the H_2 molecule from the surface. Two possible trajectories are indicated: (1) represents a reflection trajectory (unsuccessful event) with no chemisorption, (2) a successful approach that leads to dissociation. The saddle point P can be located either in the 'entrance' channel (relatively far away from the surface, left-hand side) or in the 'exit' channel (closer to the surface, right-hand side). In the first case, the H_2 molecule needs mainly translational energy for a successful passage across the barrier, while vibrational excitation is advantageous in the second case.

Once the H atoms have reached the bottom of the deep potential energy well (cf., Fig. 25.1 and Fig. 25.3) two metal–H bonds, each with energy E_{Me-H}, are formed and, in addition, the excess energy is released as adsorption energy, E_{ad}, or heat of hydrogen adsorption. This quantity can easily be measured experimentally as will be pointed out further below. From the overall energy balance

$$E_{Me-H} = \frac{1}{2}(E_{ad} + E_{diss})$$

(25.1)

(E_{diss} = heat of H_2 dissociation = 432 kJ mol^{-1}), the energy of a metal–H bond, E_{Me-H} can immediately be deduced. Generally, E_{ad} is a crucial property of any H–metal interaction system, and numerous experimental and theoretical studies have been, and still are being, devoted to the determination of this quantity. Concerning the experimental methods, E_{ad} can either be determined by means of equilibrium measurements, for example by taking hydrogen adsorption isotherms [17], or by analyzing thermal desorption data [9]. Many different theoretical methods are used to obtain hydrogen–metal binding and adsorption energies, such as tight-binding, cluster and slab (supercell) approaches. Particularly powerful are calculations based on density functional (DFT) methods plus generalized gradient approximations (GGA); for more details we refer to reviews and monographs [3, 5].

In view of H transport across surfaces, it is worth mentioning that E_{ad} also decisively governs this transport. The reason is that quite generally the heat of adsorption is internally correlated with the activation energy for (classical) lateral diffusion: An empirical relation states that for any adsorbate system the activation energy for diffusion, i.e. for hopping events from one surface site to another, is about one tenth of the adsorption energy, i.e., the depth of the adsorbate–surface interaction potential. Therefore one can get at least a crude estimate of the magnitude of the lateral diffusion energies (see Section 3.1) from known E_{ad} values. In Table 25.1 [3] we present some experimental values for E_{ad} along with values for E_{Me-H} calculated according to Eq. (25.1). Interestingly, these latter numbers are relatively similar for different systems, because the fairly large heat of dissociation of the H_2 molecule is always involved. The E_{ad} data thereby refer to vanishing hydrogen coverages, since elevated H surface concentrations can and will induce lateral H–H interactions between adjacent adsorbed H atoms. Usually, these interactions are repulsive (since the adsorbed H atoms share the charge density of the metal atom(s) underneath), in a few cases, however, they may also be attractive and, hence, support H chain or H island formation, the H-on-Ni(110) system being a good example [17]. The respective lateral interactions have quantum-chemical origin [19–21] and can significantly alter the potential energy situation across the surface as illustrated schematically in Fig. 25.5. For a single adsorbed H atom, see Fig. 25.5(a), there is a regular sinusoidal potential energy situation along the *x,y*-direction leading to entirely equivalent adsorption sites; for a pair of adjacent H atoms, repulsive lateral interactions of size ω *lower* the adsorption energy E_{ad} (Fig. 25.5(b)), while attractive interactions lead to an *increase* in E_{ad} in the

Tab. 25.1 Adsorption energies of hydrogen on selected metal single crystal surfaces [3].

Surface orientation [hkl]	Fe(110)	Ni(100)	Cu(110)	Mo(100)	Ru(0001)	Rh(100)
Adsorption energy [kJ mol^{-1}]	109 ± 5[a]	96.3[b]	77.1 ± 1[c]	101.3[d]	125[e]	79.9 ± 2[f]
Metal–hydrogen binding energy [kJ mol^{-1}]	270.5	264.1	254.6	266.7	278.5	255.9
Surface orientation [hkl]	Pd(100)	Ag(111)	W(110)	Re(0001)	Pt(111)	
Adsorption energy [kJ mol^{-1}]	102.6[g]	43.6[h]	146.4[i]	134[j]	71[k]	
Metal–hydrogen binding energy [kJ mol^{-1}]	267.3	237.8	289.2	283.0	251.5	

a Bozso, F., Ertl., G., Grunze, M., Weiss, M., *Appl. Surf. Sci.* 1 (1977) 103;
b Christmann, K., Schober, O., Ertl, G., Neumann, M., *J. Chem. Phys.* 60 (1974) 4528;
c Goerge, J., Zeppenfeld, P., David, R., Büchel, M., Comsa, G., *Surf. Sci.* 289 (1993) 201;
d Zaera, F., Kollin, E. B., Gland, J. L., *Surf. Sci.* 166 (1986) L149;
e Feulner, P., Menzel, D., *Surf. Sci.* 154 (1985) 465;
f Kim, Y., Peebles, H. C., White, J. M., *Surf. Sci.* 114 (1982) 363;
g Behm, R. J., Christmann, K., Ertl, G., *Surf. Sci.* 99 (1980) 320;
h Parker, D. H., Jones, M. E., Koel, B. E., *Surf. Sci.* 233 (1990) 65;
i Nahm, T.-U., Gomer, R., *Surf. Sci.* 380 (1997) 434;
j He, J.-W., Goodman, D. W., *J. Phys. Chem.* 94 (1990) 1502;
k Poelsema, B., Mechtersheimer, G., Comsa, G., *Proc. IVth Int. Conf. Solid Surfaces and IIIrd ECOSS*, eds. Degras, D. A., Costa, M., (Cannes 1980), p. 834.

local area of the H–H pair (Fig. 25.5(c)). This modification of ω also has, of course, an effect on the activation energy for diffusion; i.e., the lateral periodic potential is modulated by the H_{ad}–H_{ad} pair potential, an energetic heterogeneity is *induced* (*a posteriori* heterogeneity). For comparison, there exist surfaces (e.g. crystallographically 'open' (high-Miller-index) surfaces such as Pd(210) [9, 10]) which possess inherently energetically different adsorption sites. Likewise, alloys with chemically different constituents can exhibit energetically heterogeneous surfaces, see Section 3.2. In the course of adsorption, the respective energetically inequivalent sites become successively filled by hydrogen. The overall phenomenon is called *a priori heterogeneity* and sensitively influences the lateral mobility of adsorbed H atoms or molecules.

The aforementioned lateral interactions are also responsible for the formation of H (H$_2$) phases with two-dimensional long-range order, since the repulsive or attractive forces between the adsorbed H atoms or H$_2$ molecules can force the spe-

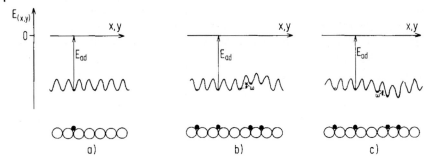

Figure 25.5 Variation of the potential energy of an adsorbed H atom parallel to the surface, $E(x,y)$. Three different cases are shown: (a) single particle adsorption with no lateral interactions; equivalence of all adsorption sites; (b) repulsive interactions, ω, between neighboring adsorbed H atoms with the consequence of energetically inhomogeneous adsorption sites; (c) attractive interactions, ω, between neighboring H atoms leading to energetically more favorable adsorption sites in the direct vicinity of an already adsorbed H atoms.

cies into periodic lattice sites at sufficiently low temperatures. Examples exist for both H atoms on metal and H_2 molecules on graphite surfaces, whereby the latter systems often exhibit a wealth of complicated structures [22]. Ordered phases of H *atoms* on metal surfaces and their thermal stability are interesting subjects to study, since they allow (via statistical mechanics) conclusions on the sign and magnitude of the mutual H–H interaction forces [23, 24]. An example for ordered H phases is given in Fig. 25.6 which displays the phase diagram of the three ordered H structures reported for the Ru(0001) surface below a temperature of ~ 75 K [25]. Raising the temperature to values above the critical value, T_c, destroys

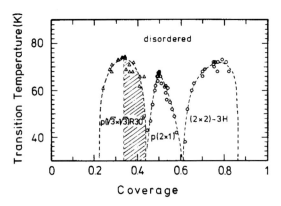

Figure 25.6 Phase diagram (T–Θ diagram) for the H-on-Ru(0001) adsorption system. Three different ordered H phases are formed: A $p(\sqrt{3}\times\sqrt{3})R30°$ phase around $\Theta = 0.33$; a $p(2\times1)$ phase at $\Theta = 0.50$, and a (2×2)-3H phase at $\Theta = 0.75$ with critical temperatures of 74 K, 68 K, and 72 K, respectively. After Sokolowski et al. [25].

the long-range order and leads to two-dimensional lattice gas behavior. This can conveniently be followed by means of temperature-dependent LEED experiments (LEED = low-energy electron diffraction) and evaluated (e.g., by using Monte Carlo calculations) with respect to order parameters, critical exponents, and lateral interaction energies [26]. Another frequently studied example of an ordered H phase is the $c(2\times2)$-2H structure on the Ni(111) surface with a critical temperature of $T_C = 273$ K and a characteristic asymmetric phase diagram [27, 28].

It is worth noting that for systems with chemisorption energies at the upper end (120–150 kJ mol^{-1}) the H chemisorption process is frequently accompanied by the aforementioned structural changes (reconstruction) and the consecutive processes of H sorption and hydride formation [29]. In addition, for systems with 'normal' chemisorption energies (80–120 kJ mol^{-1}] elevated hydrogen pressures and temperatures may also favor these more vigorous interactions and increasing chemical attack towards hydrogenation. Surfaces which can form volatile hydrides (Li, Al etc.) can therefore easily be chemically eroded by exposure to hydrogen, especially at elevated temperatures, despite the large activation barrier for spontaneous H_2 dissociation [30]. Similar effects have been reported for semiconductor surfaces (Si, Ge, GaAs), which can undergo successive hydrogenation and form volatile hydrides, i.e., SiH_4 (silane), GeH_4 (germane), GaH_3 (gallane) and AsH_3 (arsine), as soon as reactive H atoms are available at the surface [31, 32].

25.3
The Transfer of Hydrogen on Metal Surfaces

25.3.1
Hydrogen Surface Diffusion on Homogeneous Metal Surfaces

The most efficient process for transferring hydrogen at surfaces is diffusion – the only requirement is that the H atoms or H_2 molecules are trapped in the respective chemisorption or physisorption potentials and possess a sufficient residence time, τ, in this state – this can be achieved by choosing the appropriate surface temperature, T. Weakly bonded hydrogen species require low, sometimes very low, temperatures, whereas strongly chemisorbed H atoms remain trapped, even at elevated temperatures. Slowing down the lateral diffusion by lowering T leads to 'immobile' adsorption, where the H atoms or H_2 molecules remain in their local sites, whilst higher temperatures (below the desorption temperatures though) favor the mobility of the H adlayer.

Generally, surface diffusion of adsorbed particles is crucial for many processes occurring on and at surfaces; among others, it influences the rate of adsorption and desorption, the formation of phases with long-range order and, finally, the turn-over numbers of catalytic processes. For the sake of brevity, *bulk* diffusion phenomena will be excluded here, although they also involve transfer of particles from the surface region to the bulk of a crystal and often play an important role just in hydrogen–metal interaction systems: note that certain metals (e.g., V, Ti,

Zr, Nb, Pd and Ta) can absorb large quantities of hydrogen and, therefore, act as hydrogen storage materials. We note that Pd surfaces especially show a rich diffusion scenario involving overlayer–underlayer (surface–subsurface) transitions and interstitial diffusion leading to absorption and hydride formation processes. For more details here we recommend the respective literature [6]. Concerning genuine *surface* diffusion, articles by Morris et al. [33] and Naumovets and Vedula [34] reviewed the state of the art until the mid-eighties (including a description of experimental methods) and gave, in addition, a useful description of the general laws and relationships as well as mechanistic details of the diffusion process(es).

For a brief introduction, we refer again to Fig. 25.5(a), which immediately reveals that it requires an activation energy of E^*_{diff} for an atom (or a molecule) to be transferred from a site "A" to another, geometrically identical site "B" on the same periodic surface. In the classical view, this two-dimensional diffusion process can be thought of as a sequence of individual and statistical 'hopping' events of frequency v, each activated with an energy E^*_{diff}, as pointed out, for example, by Roberts and McKee [35]. The inverse of this frequency then yields the residence time, τ', of the particle in the respective site. For thermally equilibrated particles, the temperature dependence of the classical surface diffusion is described by the well-known Arrhenius relation

$$D(T) = D_0 \exp\left(-\frac{E_{\mathrm{diff}}}{kT}\right) \qquad (25.2)$$

with D_0 being the pre-exponential factor [cm^2 s^{-1}], and $D(T)$ the temperature-dependent diffusion coefficient. The pre-exponential factor may be associated with an attempt frequency to overcome the activation barrier. According to Fick's first law which assumes stationary diffusion, i.e., a constant concentration gradient $\left(\frac{\partial c}{\partial x}\right)_t$, $D(T)$ can be expressed as the ratio of the particle flux through the concentration front and the actual concentration gradient at time t. Alternatively, the diffusion progress with time t can be monitored by the mean square displacement $\langle x^2 \rangle$ of a given particle on a surface, based on Einstein's equation:

$$\sqrt{\langle x^2 \rangle} = \sqrt{2Dt} \qquad (25.3)$$

Another frequently used expression is based on random walk events between fixed sites and relates the pre-exponential factor D_0, to the jump length a and the vibrational frequency parallel to the surface, v. For a surface with fourfold symmetry one has

$$D_0 = \frac{1}{4} a^2 v \qquad (25.4)$$

The factor $1/4$ arises from the four possible diffusion probabilities on this surface. Depending on the surface structure and corrugation, there may exist 'easy' and 'difficult' pathways for diffusion with low and high activation barriers, respectively; hence, on a crystallographically well defined single crystal surface, the diffusion coefficient is usually strongly direction-dependent. Most of the experiments focus on a determination of activation energies for diffusion and diffusion coefficients, whereby, as mentioned above, the 'classical' regime must be distinguished from the (low-temperature) regime, in which quantum tunneling dominates, as will be shown further below.

Confining our considerations to *hydrogen* surface diffusion, the 'mobility' of a hydrogen 'diffusion front' was studied already 50 years ago in Gomer's laboratory on Pt and W field emitters [36, 37]. At 4.2 K, a sharp front of adsorbed hydrogen (presumably consisting of H_2 (D_2) molecules) was produced by partially exposing a Pt field emitter tip to hydrogen (deuterium) [37]. As the tip temperature was raised to 20 K, the front became very mobile, in that the hydrogen molecules spread over the entire tip area. Pre-dissociating the low-temperature hydrogen layer led to a more strongly bound species (very likely chemisorbed H atoms), whose front became mobile only around 105 K tip temperature. Using the relations (25.3) and (25.4) and assuming $x \approx 600$ Å, $a = 6$ Å and 10^{12} attempts s^{-1} for the jump frequency v the authors concluded on an activation energy for H diffusion of ~ 4.5 kcal mol^{-1} (= 19 kJ mol^{-1}). From the known H_2 adsorption energy of 16 kcal mol^{-1} (= 67 kJ mol^{-1}), a value of 58 kcal mol^{-1} (≈ 243 kJ mol^{-1}) was deduced for the depth of the Pt–H potential-energy well – thus basically confirming the aforementioned 10:1 relation between the adsorption and diffusion energy. Some years later, the field emission fluctuation method was developed in Gomer's laboratory, a powerful technique to determine diffusion coefficients and activation energies, especially for H on W surfaces. For more details about the sophisticated technique we refer to the original work by DiFoggio and Gomer [38].

In 1972, Ertl and Neumann introduced the laser-induced thermal desorption technique to determine the rate of diffusion [39]; later, this method was further refined by Seebauer and Schmidt [40] and Mak and collaborators [41–43]: In principle, the technique is based on the 'hole refilling' phenomenon and is relatively straightforward: A laser beam of known cross section is incident on an adsorbate-covered, well-defined patch on the surface, whereby the power of the laser beam is just sufficient to thermally desorb all the particles in the illuminated area. After the laser shot, refilling of the hole from the cold, H-rich surrounding sets in, which can be followed as a function of time by subsequently fired laser pulses into the same spot. The refilling signal, which is monitored with a fast and sensitive mass spectrometer, is then expressed in terms of Fick's second law of diffusion. Despite its simplicity, this technique bears some problems, among which are the appropriate adjustment of the beam power and the determination of the surface temperature within the burnt hole; furthermore, it is difficult to consider directional and coverage dependences of the diffusion fluxes [44]. Mak and George have published a simplified method to determine the coverage dependence of surface diffusion coefficients [42]. In Table 25.2, we have compiled some diffusion coefficients and activation energies for this 'classical' atomic H diffusion on metal surfaces.

So far we have not considered the most prominent property of an H atom, viz., its peculiar *quantum character*. This is based on the light mass of the proton (or

Tab. 25.2 Activation energies of hydrogen diffusion on selected metal single crystal surfaces [3].

Surface orientation [hkl]	Ni(100)	Cu(100)	Ru(0001)	Rh(111)	W(100) at T>220	W(100) at T=140	Pt(111) FEM tip	Pt(111)
Diffusion energy [kJ mol^{-1}]	16.7 ± 2[a]	19.0[b]	16.7[c]	15.5[d]	> 16.7[e]	> 4.2[e]	18.8[f]	> 29.3[g]

a George, S. M., DeSantolo, A. M., Hall, R. B., *Surf. Sci.* 159 (1985) L425;
b Lauhon, L. J., Ho, W., *Phys. Rev. Lett.* 85 (2000) 4566;
c Mak, C. H., Brand, J. L., Deckert, A. A., George, S. M., *J. Chem. Phys.* 85 (1986) 1676;
d Seebauer, E. G., Kong, A. C. F., Schmidt, L. D., *J. Chem. Phys.* 88 (1988) 6597;
e Daniels, E. A., Gomer, R., *Surf. Sci.* 336 (1995) 245;
f Lewis, R., Gomer, R., *Surf. Sci.* 17 (1969) 333;
g Seebauer, E. G., Schmidt, L. D., *Chem. Phys. Lett.* 123 (1986) 129.

the neutral H atom), i.e., $m_H = 1.67 \times 10^{-27}$ kg, only heavier than an electron by a factor of 1836. Just this property, compared to the other elements, is specific for hydrogen. In addition, the availability of two significantly heavier hydrogen isotopes deuterium $_1^2H$, and tritium $_1^3H$, allows one to perform comparison diffusion experiments and to judge directly the influence of tunnel processes. Particularly revealing is, of course, a measurement of the T dependence of the diffusion rate: Note that for genuine tunnel processes the temperature coefficient should tend to zero. On the other hand a distinct dependence of the tunnel probability on the mass of the hydrogen isotope used is to be expected. Isotope experiments of this kind have been frequently performed in the past, especially with W surfaces; again, Gomer's work has to be mentioned in the first instance: Careful studies of atomic hydrogen (deuterium, tritium) adsorbed on W single crystal surfaces at different temperatures and coverages revealed detailed insight into the dynamics of adsorbed H atoms on these surfaces. On W(110), DiFoggio and Gomer [38] and Auerbach et al. [45] found an anomalous isotope dependence of the hydrogen (deuterium) diffusion rates in the low-coverage regime: Low temperature tunneling diffusion showed an isotope effect several orders of magnitude smaller than predicted by simple rigid-lattice models, and high-temperature (activated) diffusion exhibited an *inverse* isotope effect several orders of magnitude larger than expected from the rigid lattice model. These effects were explained by peculiar W lattice–hydrogen interactions, i.e., by a large difference in the time scales of the motions of hydrogen and the tungsten lattice. Activated diffusion was described as a many-phonon process in which the vibron is thermally excited as a result of phonon–vibron coupling. The interesting data are reproduced in Fig. 25.7, taken from Auerbach's et al. work [45].

Returning to the possibility of quantum delocalization, one could expect that the light H atoms adsorbed on a surface would behave similar to the electrons in free-electron metal surfaces, provided the energy barrier for the H atoms to move across the surface, i.e., the diffusion barrier E^*_{diff}, is very small. This barrier has to

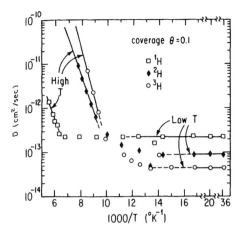

Figure 25.7 Arrhenius-type plot of the diffusion coefficient D [cm^2 s^{-1}] versus the inverse temperature [1/K] for the three hydrogen isotopes (H, D, and T) adsorbed on the tungsten (110) surface (low coverage regime ($\Theta = 0.1$)). Clearly evident is that (T-independent) tunneling dominates in the low temperature range, whereas classical diffusion takes over at higher temperatures. After Auerbach et al. [45].

be compared with the vibrational energy associated with a H atom adsorbed in a given surface site. For many H-on-metal systems, these vibrational ground state energies are of the order of 50–150 meV and, hence, compete with the diffusion barriers, E^*_{diff}, which amount to 200–300 meV (~20–30 kJ mol^{-1}), cf., Table 25.2. It is therefore by no means surprising that experimental evidence of hydrogen's quantum character has been searched for by LEED and, especially, vibrational loss measurements. On certain (preferentially densely packed) metallic surfaces the lateral H–metal potential is often sufficiently flat as to suggest the possibility of efficient tunneling between neighboring adsorption sites. In an extreme view, this is equivalent to the existence of protonic bands, thus underlining the delocalized character of H atoms adsorbed in the respective surface sites. This behavior was first concluded for the Ni(111) + H$_{\mathrm{ad}}$ system in conjunction with LEED experiments to explain the observed structural disorder [27]; if diffusion barriers are neglected, the de Broglie wavelength resulting from the thermal energy of H atoms moving parallel to the surface is of the order of 1 Å only. This motion parallel to the surface was proposed to occur in a band-like fashion with band gaps caused by diffraction of H atoms from the two-dimensional periodic potential. The principal idea can be taken from Fig. 25.8 which shows the potential energy situation both parallel (Fig. 25.8(a)) and perpendicular to the surface (Fig. 25.8(b)) with the delocalized protonic bands and the band gaps indicated [27]. A few years later, the idea was picked up again by Puska et al. and subjected to a detailed theoretical consideration using the effective medium theory [46, 47]. In Fig. 25.9 we present the respective results for H adsorbed on a Ni(100) surface:

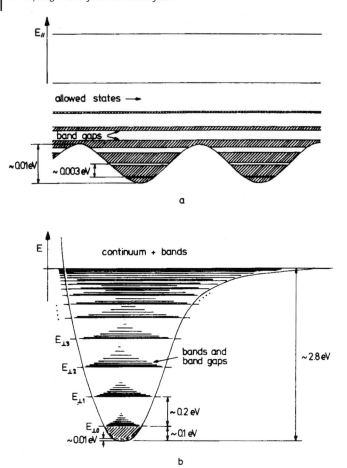

Figure 25.8 Schematic representation of the atomic band structure
for H atoms chemisorbed on a metal surface (assumed adsorption
energy ~ 2.8 eV = 270 kJ mol⁻¹) with very low diffusion barriers between
the adsorption sites. Top: Atomic band structure parallel to the surface;
bottom: Total atomic band structure with the motion perpendicular
to the surface. Bands and band gaps are sketched. After Ref. [27].

Shown is the band structure (i.e., the $E(k)$ relation) for hydrogen chemisorbed on
the Ni(100) surface along the high-symmetry directions of the surface Brillouin
zone (shown in the upper right). Apparently, bands formed by delocalized H wavefunctions can be distinguished, which are separated by gaps of several tens of
meV. Particularly the higher excited bands show noticeable dispersion. The calculations revealed that the bandwidth of the ground state is a few meV only, whereas
those of the excited states can reach several tens of meV. The authors also pointed
out that the motion of H perpendicular to the surface should significantly couple
to the parallel motion because of the anharmonicity of the combined perpendicu-

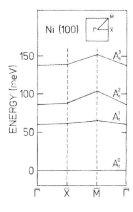

Figure 25.9 The band structure of hydrogen atoms adsorbed on a Ni(100) surface ($E(\vec{k})$ relation) along the high-symmetry directions of the surface Brillouin zone, $\bar{\Gamma}$–$\bar{X}$ and $\bar{\Gamma}$–$\bar{M}$. Only the states belonging to the A_1 representation of the C_{4v} point group are shown. After Puska et al. [46].

lar and parallel potentials in conjunction with the delocalized nature of the H adsorption. Somewhat more recent theoretical considerations on the quantum diffusion of hydrogen on metal surfaces have been published by Whaley et al. [48]. The authors present a quantum mechanical theory for the low-temperature diffusion of hydrogen atoms on metal surfaces based upon a band model for the hydrogen motion. At low coverages the hydrogen band motion is restricted by collisions between the adsorbate particles causing a lowering of the diffusion coefficient with increasing concentration. Additional features of H–H interactions have to be introduced to explain the coverage dependence at higher surface concentrations. The model satisfactorily reproduces the experimental observations for the coverage dependence of H, D, and T diffusion on W(110) surfaces by Gomer et al. [38, 49].

In principle, vibrational loss measurements (mostly and conveniently performed by high-resolution electron-energy loss spectroscopy (HREELS)) should be capable of detecting this band-like behavior, but it took several years until appropriate (electron energy loss) spectrometers with sufficient resolution and sensitivity were available to prove the respective excitations. The first experimental evidence came from a study by Mate and Somorjai focusing on the H(D)-on-Rh(111) system [50]. At H coverages of ~ 0.4 monolayers, the authors observed a prominent loss peak at 450 cm^{-1} which they attributed to transitions from the ground-state band to the first excited-state band for the motion of the H atoms on the Rh(111) surface. This conclusion was mainly motivated by a comparison with the aforementioned theoretical report by Puska et al. [46, 47] who calculated for the H-on-Ni(111) system (a surface with the same geometry as Rh(111)) that the first excited band for H motion parallel to the surface has E symmetry and is located ~ 320 cm^{-1} above the ground-state band, close to the 450 cm^{-1} observed for Rh(111)/H. Furthermore, the absence of dipole scattering contributions in this band (as expected for excitations of E-type symmetry) was also taken as evidence for the aforementioned assignment, as well as the absence of this band in the deuterium loss spectra. Besides the 450 cm^{-1} band, there appeared also (weak) loss features at higher energies which were, according to the model, interpreted as

Figure 25.10 Calculated eigenvalues and eigenstates lead to a series of vibrational states with quantum numbers (n, k); for the $k = 0$ state the Bloch orbitals of each branch were examined. At low energies, the respective orbitals are mainly confined around the fcc, top, or hcp sites of the Pt(111) surface. The figure shows for selected bands the localization/delocalization of the orbitals $n = 1$ (Fig. 25.9(a, b)); $n = 3$ (Fig. 25.9 (c, d)); $n = 4$ (Fig. 25.9 (e, f)); $n = 26$ (Fig. 25.9 (g, h)); $n = 15$ (Fig. 25.9 (k, l)), and $n = 16$ (Fig. 25.9 (m, n)), whereby the right sequence of graphs displays the probability density $\rho(r)$ in transversal sections. Ten equidistant contour lines are used in each graph. After Badescu et al. [55].

reflecting excitations from the ground-state band to the higher excited states. In full agreement with the protonic band model, no ordered H overlayer was found in LEED at 80 K adsorption temperature, which led the authors to call the delocalized adsorbed H layer a 'hydrogen fog'. In addition, the extended line widths of the vibrational bands could be taken as a hint to nonlocal properties of the adsorbed H atoms.

The next study to be mentioned here was concerned with time-of-flight scattering and recoiling spectroscopy of hydrogen adsorbed on a W(211) surface and included also an effective-medium-theoretical treatment [51]. This shows a shallow H–W potential, with an activation barrier to motion along the [1-1-1] troughs of only 100 meV. The lowest excited states correspond to vibrations parallel to the surface with large amplitudes, fill a large portion of the trough and can be populated thermally. At 450 K, the calculations reveal that the H atoms are delocalized to a greater extent than expected from the shadow-cone radius of a W atom arising from the ion scattering. In 1992, Astaldi et al. took coverage-dependent vibrational loss spectra of H and D adsorbed at 110 K on a Cu(110) surface and concluded that there was a protonic band structure at low coverages ($\Theta \leq 0.15$), while the H atoms were more localized at higher coverages [52]. The HREELS measurements were supported by parallel LEED observations: Below $\Theta = 0.15$ there was no H-induced superstructure visible, pointing to a disperse lattice-gas H phase, whilst for $\Theta > 0.15$ 'extra' spots of a (1×3) phase appeared, indicating rather more localized H atoms in the troughs of the Cu(110) surface. Particularly revealing were the isotope effects: spectra taken with deuterium showed a much lower coverage dependence. About four years later Takagi et al. published vibrational loss data for H and D adsorbed (likewise in low concentrations) at 90 K on a Pd(110) surface [53] and performed parallel model calculations from which they also concluded that the delocalized protonic band model was valid. The latest work dealing with that issue appeared in 2001, when Badescu et al. studied the H-on-Pt(111) system by means of high-resolution electron-energy loss spectroscopy (HREELS) at 85 K and again found vibrational bands and systematic excitations in the respective energy range which were attributed to the existence of protonic bands and the respective excitation. From their most recent experimental and theoretical work [54, 55] we reproduce in Fig. 25.10 their orbital model with the probability density of the 'protonic orbitals' for low excited ($n = 1$, 3 and 4) and high excited states

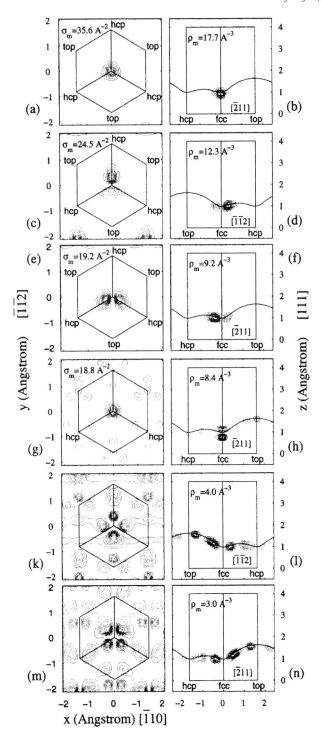

($n = 15$, 16, and 26), the H coverage being 1 monolayer. Other more recent experimental [56] and theoretical work [57] also favored at least a partial delocalization of adsorbed H atoms on transition metal surfaces. Quite recently, an exhaustive review article appeared on the issue of quantum delocalization of hydrogen on metal surfaces [58] which is recommended for further reading.

It should be mentioned here, however, that the concept of highly delocalized hydrogen has also been questioned several times, and there exist various reports for a variety of metal surfaces, in which it is denied that quantum delocalization plays a dominant role in hydrogen adsorption [59, 60]. As a good example, we refer to a recent study by Kostov et al. [61] on the H-on-Ru(0001) system, where very careful measurements of H-induced vibrational bands were performed (including angle dependences and linewidths) and a thorough discussion conducted concerning the evidence for or necessity to assume delocalization. Actually, the (0001) surface of the hexagonal close-packed (hcp) system with its shallow corrugation represents an ideal 'candidate' for delocalized hydrogen. However, Kostov et al. pointed out explicitly that all their observed vibrational phenomena and isotope effects could be consistently explained by 'classical' adsorption of H atoms in distinct sites of the Ru(0001) surface. In concluding this paragraph, one can state that more (and more careful) vibrational measurements are required to shed more light on the (undoubtedly interesting) issue of quantum motion of hydrogen on metallic surfaces.

Turning to classical diffusion again, significant progress in understanding the *mechanism* of this diffusion has been achieved since the late eighties and early nineties by performing scanning tunneling microscopy (STM) experiments and watching the adsorbed particles during their hopping and site exchange events [62–64]. A direct counting and subsequent statistical analysis of the number of migrating N (O) atoms on a Ru surface as a function of time and coverage revealed much insight into the principal surface hopping, diffusion, and lateral ordering phenomena of O and N atoms at and around room temperature. By performing rapid scans it was even possible to make an STM movie of the dynamical surface scenario during oxygen atom migration [64]. However, in order to watch diffusing *hydrogen* atoms with their appreciably larger diffusion rate, the surfaces have to be kept at significantly lower temperatures; a possible solution is provided by performing STM observations in combination with inelastic electron tunneling spectroscopy (IETS) in a 4 K-STM [65, 66]. Recently, there arose much interest in single atom diffusion and its direct observation. Details of this novel technique which is being developed in the laboratory of W. Ho with emphasis on hydrogen adsorbed layers can be taken from the internet site http://www.physics.uci.edu/~wilsonho/stm-iets.html. The space limitations do not allow us to expand further on both experimental and theoretical investigations on hydrogen diffusion. There exist numerous theoretical articles dealing with diffusive H motion on surfaces, many of them focusing on the interesting nonthermally activated quantum tunneling processes [67, 68].

25.3.2
Hydrogen Surface Diffusion and Transfer on Heterogeneous Metal Surfaces

While the foregoing section was concerned only with homogeneous, elemental metal surfaces (consisting of a single kind of atoms only), the admixture of a second (guest) metal can change the situation significantly, depending on the chemical differences between the host and guest metal. As far as the interaction of hydrogen with the respective alloy or 'bimetallic' surface is concerned, these changes include both the *dissociation probability* of a H_2 molecule and the *adsorption energy* of the H atoms in a given adsorption site. Certainly also affected will be, of course, the shape and height of the activation barriers for H diffusion (transfer) parallel to the surface, especially in the vicinity of the admixed guest atoms and, hence, the overall H transfer. Alloy or bimetallic systems (often in conjunction with oxidic support materials) are of utmost importance in heterogeneous catalysis, because they can supply hydrogen atoms for reactions with co-adsorbed molecules which either simply add the H atoms to unsaturated bonds (hydrogenation) or use the H atoms to cleave internal bonds, preferentially C–C bonds (hydrogenolysis), see below. In recent years, however, other technical applications have also led to an enormous interest in physisorption and chemisorption of hydrogen on a variety of mixed metal surfaces, either in dispersed form on a support or in a more compact fashion. We recall the development of gas sensors, where often Pd (or Pd alloys) is deposited on tin oxide or other semiconducting surfaces and used to activate molecular hydrogen which can then, in conjunction with the surface oxide, dehydrogenate hydrides and hydrocarbons [69].

Even more important is the application of Pt group metals and their alloys in fuel cell technology. Fuel cells serve as very clean electricity sources in that they electrochemically oxidize a fuel, typically hydrogen, and, at the same time, reduce oxygen to give water. In this respect, they can be assigned as 'proton pumps'. The basic ingredients of a fuel cell are (i) the electrolyte, usually a proton conducting membrane (NAFION®), which is embedded in a sandwich-like fashion between (ii) two gas-porous electrodes which contain the redox catalysts [70]. In low-temperature fuel cells, very often Pt is the essential part of the catalyst material, however, it is mostly doped with other metals (Cr, Co, Ni) to enhance the catalyst's activity for both oxidation and reduction reactions. Therefore, the interaction of hydrogen with Pt-based alloys is frequently being investigated with respect to H_2 dissociation, H binding states, and H-induced surface restructuring phenomena. Since, however, both Pt and Ni readily dissociate hydrogen and the H atoms formed are chemisorbed with similar adsorption energies it is not at all easy to distinguish chemically between the two constituents of the alloy and their specific interaction with both H_2 molecules and H atoms. In some cases the vibrations of the respective different H–metal chemisorption complexes, i.e., H–Pt and H–Ni, are different, an analysis of the local H–metal vibrations can yield some insight into the local distribution and filling of the respective adsorption sites. For surfaces with larger areas, inelastic neutron scattering (INS) is a unique analytical tool to study the vibrational dynamics of hydrogen-containing materials including dihydrogen. Therefore, this technique can be applied for dispersed, polycrystalline catalyst materials. More information can be taken from work by Mitchell et al. who have, among others, used Pt and PtRu alloys deposited on carbon supports and studied the adsorbed states of dihydrogen and H spillover to the C surface by means of INS [71, 72].

However, there exists one particularly interesting class of materials which we shall expand on in the following section, namely binary alloys that consist of a 'hydrogen-active' and a 'hydrogen-inactive' metal: As pointed out in Section 2, the chemical affinity of hydrogen with respect to a 'noble' metal (NM) containing only filled, deep-lying, d-bands (Cu, Ag, Au) differs largely from that of a transition metal (TM) with its high density of d electron states right at the Fermi level (Ni, Pd, Pt etc.). Within this class of materials, it is reasonable to further distinguish (i) homogeneous alloys with completely miscible components forming a continuous series of solid solutions (Cu+Ni, Ag+Pd), (ii) miscible components which form stoichiometric intermetallic compounds and phases with long-range order (Cu+Pd, Cu+Pt), and (iii) constituents that are immiscible in the bulk (Cu+Ru, Ag+Ru, Au+Ru). In all three cases both the electronic structure of the alloy surface (density of electronic states) and its topography can, and often will, be greatly modified by the mutual concentrations of the constituents. In addition, surface enrichment may occur, whereby the more volatile constituent usually segregates at the surface [73]. Taking an example from sub-category (i), we consider an alloy between Ni and Cu, where Ni is the 'active' and Cu the 'inactive' component. Depending on the composition and heat treatment, a certain statistical distribution of Cu atoms in a Ni matrix will arise, and ensembles containing few or many Cu atoms are formed in the surface. In the simplest case, a given Cu atom will arithmetically block an H adsorption site of the Ni. More often, however, it induces a long-range effect in that it 'spoils' (i.e., deactivates) fairly large ensembles of active Ni atoms. The hydrogen adsorption capacity of this ensemble is then greatly reduced or even completely extinguished by a single Cu atom ('ensemble effect'). On the other hand, a Cu atom in the direct vicinity of a Ni atom can also be electronically activated by a local charge transfer and then, despite its inherent inactivity, be activated to bind H atom(s) ('ligand effect'). Both ensemble and ligand effect are decisive and well-known issues in heterogeneous catalysis [74, 75].

Similar scenarios are encountered with (ii) alloys which form ordered superlattices and/or intermetallic compounds with a defined stoichiometry, for instance Cu/Pd or Cu/Pt alloys. Depending on the copper concentration, either Cu-rich Cu_3Pd (Cu_3Pt) or Pd(Pt)-rich alloys (Pd_3Cu (Pt_3Cu) can be adjusted. These materials often exhibit a regular surface distribution of the constituents and, hence, form geometrically well-defined ensembles which enables the researcher to relate the ensemble size and geometry with the adsorption property of a given bimetallic surface [76–78]: For a regular $Cu_3Pt(111)$ surface, a given Pt atom is actually surrounded by 6 Cu atoms and, hence, geometrically isolated from its next Pt neighbors. In a LEED experiment, 'extra' diffraction spots indicate a defined surface composition and geometry, and, in recent years, scanning tunneling microscopy

(STM) with atomic resolution could sometimes directly image the distribution of the elements in the surface alloy, provided there is sufficient 'chemical contrast' between the two elements [79]. For CO adsorption on Pt-Co alloys, direct STM investigations made it possible to pinpoint the ligand effect for the first time [80]. In technical catalysis, *bimetallic* systems (category (iii)) often play a more important role than *homogeneous* alloys. With these systems, a limited thermodynamic miscibility controls the lateral dispersion of the NM deposit on the TM substrate and often leads to a very inhomogeneous two-dimensional distribution in that extended islands of the NM are formed on top of the TM surface, with additional adatom, edge and kink sites which may provide particular centers of chemical activity. Typical experiments aiming at the determination of ensemble effects and hydrogen spill-over (see below) date back to the sixties and seventies: Sinfelt from Exxon laboratories was one of the first who developed the so-called bimetallic cluster catalysts and pointed out their peculiar catalytic activity concerning hydrogenation and hydrogenolysis reactions [81–83]. By appropriate co-precipitation and calcination, a cluster of a catalytically active transition metal (Ni or Ru) was partially covered with an inactive coinage metal (Cu, Ag, or Au), whereby the coinage metal existed in the form of a flat, 'raft-like' surface array [84]. Therefore, surface scientists became interested in modeling these bimetallic cluster catalysts and prepared, especially, Cu deposits on a flat Ru(0001) surface [85–95]. A survey of the scientific activities focusing on bimetallic surface chemistry has been given by Campbell [96]. In this context it is also of vital interest to scrutinize the influence of nonmetallic, often strongly polarized, additives to metal surfaces, such as oxygen, chlorine, sulfur, or phosphorus (electronegative species) or alkali metals (Na, K, Cs etc., electropositive species) on the adsorption properties of a metal single crystal surface. Here we refer to Goodman's review [97] in which the influence of electronegative, neutral, and electropositive impurity atoms is discussed in terms of promoting or inhibiting effects for catalytic reactions involving hydrogen and/or carbon monoxide. Mostly, electronic effects (ligand, ensemble effects) are invoked to explain the observed poisoning which is, for a Ni(100) surface, stronger with sulfur than with chlorine or phosphorus; mainly, the hydrogen uptake is suppressed, likely due to a reduction of the hydrogen sticking probability.

Returning to the interaction of an alloy or bimetallic surface with hydrogen, both the hydrogen adsorption/desorption *kinetics* and the *energetics* of adsorption and diffusion can be affected. A typical kinetic effect occurs, if the H_2 *sticking coefficient* is modified by the alloy's surface topography (which is frequently the case). Since theories of the H_2–surface interaction dynamics [98] predict a strong influence of hydrogen dissociation on the local geometry of an adsorption site, defined *ensembles* of adjacent TM atoms are believed to be required for nonactivated H_2 dissociation. This effect has been proven many times by recording the H uptake as a function of the alloy composition $(TM)_x(NM)_y$. Increasing the concentration of the NM resulted in an over-proportional decrease in the H uptake [86, 87, 94], and it was concluded that ensembles of up to four adjacent TM atoms are required to dissociate the H_2 molecule and to appropriately adsorb the formed H atoms. Very recently, this expectation was directly confirmed by observing the H_2 dissociation reaction on a Pd(111) surface with atomic resolution by means of low-temperature STM. Three adjacent Pd atoms were found to form an 'active' ensemble for H_2 dissociation [99].

At this point the important question arises as to whether H atoms formed by dissociation on the active part of a surface will be able to 'spill over' to empty sites or patches of the inactive deposit. Many researchers assume that hydrogen transfer of this kind does indeed occur. On heterogeneous catalyst materials consisting of small Ni, Ru, Pd, Ir or Pt clusters dispersed on a silica, alumina, titania, zirconia or carbon support, H spillover has frequently been reported [72, 100–103]. It should be noted here that the literature available for the issue "spill-over of hydrogen" certainly fills several book shelves [100], and there are very important novel aspects of energy technology where the transfer of hydrogen, e.g., from transition metal nanoparticles to carbon nanotubes is considered in order to develop new materials (carbon–metal composites) for hydrogen storage and fuel cell applications [72, 73, 103]. For alloys or bimetallic systems, fewer reports exist in favor of H spill-over. Cruq et al. measured hydrogen adsorption isotherms on pure Ni and (polycrystalline) Cu–Ni alloys and observed hydrogen adsorption on mixed ensembles of Cu and Ni atoms via a spill-over mechanism; H_2 molecules were assumed to dissociate only over Ni ensembles [104]. Shimizu et al. studied the adsorption of hydrogen on Cu+Ru(0001) surfaces and could not detect significant transfer of hydrogen atoms from Ru sites to Cu sites at 100 K adsorption temperature [86]. A couple of years later, the same system was re-investigated by Goodman and Peden, and these authors reported non-negligible H spill-over to Cu by raising the adsorption temperature to 230 K; apparently, H transfer from the active Ru sites to the less active Cu sites required a slightly thermally activated surface diffusion step [88]. We reproduce, from their work, the experimental evidence of spill-over, cf. Fig. 25.11. Shown are three H_2 thermal desorption traces obtained from a Ru(0001) surface covered with 0.7 monolayers of Cu: curve (a) refers to a H_2 saturation exposure at 100 K, (b) to a saturation exposure at 230 K, and (c) is the difference curve (b) – (a) and clearly shows the amount of spillover hydrogen at 230 K. Apparently, it is possible for H_2 molecules to dissociate on Ru sites into H atoms which can then migrate to Cu sites where they chemisorb with a binding energy close to that found with genuine Cu(111) surfaces, i.e., ~65 kJ mol^{-1} [105–107]. Note that this binding (adsorption) energy does not really deviate from the respective values measured for typical transition metals such as Ni or Pd – this suggests that the crucial elementary step is the *dissociation* (and not the adsorptive binding) which is activated on the noble metal (Cu, Ag, Au...) and nonactivated on the transition metal. From the (slight) thermal activation energy for H migration from Ru to Cu sites suggested by the data of Goodman and Peden [88] it turns out, however, that a H atom chemisorbed on a bimetallic surface can definitely distinguish between TM and NM sites which are apparently separated from each other by slight diffusion barriers.

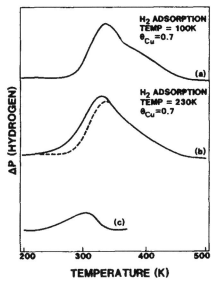

Figure 25.11 Selected hydrogen thermal desorption traces obtained from a bimetallic Cu–Ru surface (Cu coverage = 0.7 monolayers on a Ru(0001) surface) as a function of adsorption temperature: The top curve (a) was obtained after the system had received a saturation exposure at 100 K; curve (b) H_2 desorption trace after a saturation exposure at 230 K. The dashed line indicates the direct superposition of (a) onto (b). The bottom curve (c) represents the difference (b) – (a) and, hence, is equal to the amount of hydrogen spilled over from Ru to Cu sites at 230 K. After Goodman and Peden [88].

25.4
Alcohol and Water on Metal Surfaces: Evidence of H Bond Formation and H Transfer

25.4.1
Alcohols on Metal Surfaces

In this section some features of hydrogen transfer will be discussed in conjunction with the interaction (adsorption and partial dissociation) of water and aliphatic alcohols (methanol CH_3OH, in particular) with selected metal surfaces. Common to these molecules is that they contain a hydroxy group which is (formally) coupled to either a H atom (water) or to an organic rest, for example a methyl group CH_3. In the hydroxy group we realize the basic principle of a hydrogen bond mentioned in the introductory section: the close vicinity of an H atom to a strongly electronegative atom, here the oxygen atom (atom "A"). As another electronegative atom (in the simplest case the oxygen atom "B" of an adjacent water or alcohol molecule) is brought close to the H atom in question, the latter can be split off and transferred to the atom "B", depending on the mutual distance H–B. The situation "A–H–B" is typical for H bonding with its characteristic double-potential well. Of course, nitrogen, sulfur, or phosphorus atoms can also play the part of the second electronegative atom "B" with the consequence that the double potential well is no longer symmetric.

If a hydroxy group-containing molecule interacts with a metal surface, several effects can occur, depending on temperature, on the nature of the adsorbing metal and the adsorbate's surface concentration. Besides mere adsorption of the molecular entity in various bonding configurations and without or with long-range order, dissociation reaction(s) may readily take place. In quite a number of cases, the hydroxy hydrogen atom is actually split off leaving behind either a nude hydroxy group (in the case of water) or an aliphatic oxide (for example, methoxide in the case of methanol). Thermodynamically decisive here is the affinity of the metal surface with respect to adsorbing the split-off hydrogen. The overall energy gain (including the release of the hydrogen adsorption energy) determines whether or not the dissociation process is thermodynamically favored. However, even if the free enthalpy gain of the system suggests the dissociation path, possible kinetic barriers may still prevent a spontaneous dissociation or at least greatly slow down the rate of H transfer to the surface. Looking at Table 25.1, spontaneous O–H dissociation processes of this kind are principally possible with most transition metal surfaces, among others with Ni, Ru, Rh, Pd, Ir or Pt , and have indeed been reported in quite a number of cases. Relatively easy to survey is the adsorption of alcohols on surfaces, because the progress of the O–H dissociation can be followed conveniently by monitoring the molecular orbitals of the fragments by means of UV photoemission or by following their characteristic vibrations by means of electron-energy loss spectroscopy. To some extent, this holds also for the dissociation of water on surfaces, where X-ray photoelectron spectroscopy (XPS), thermal desorption and vibrational loss measurements have been used to prove the existence of the dissociation products – although the respective features are somewhat more difficult to unravel.

In 1977 Rubloff and Demuth investigated the interaction of methanol with a Ni(111) surface by means of UV photoemission (UPS) and reported on the formation of a methoxide species, CH_3O, plus adsorbed hydrogen [108]; this conclusion was somewhat later confirmed by HREELS measurements [109]. Christmann and Demuth studied the adsorption and decomposition of methanol on a Pd(100) surface [110,111] using UPS, HREELS, work function ($\Delta\varphi$) and thermal desorption spectroscopy (TDS) measurements. At 120 K, a small fraction of methanol underwent dissociation into CH_3O and H, while the major part adsorbed molecularly in a disordered fashion. During the formation of multilayers, however, TD spectra revealed hints of hydrogen bonding effects, and the formation of a chain-like network made up by H-bonded CH_3OH units as displayed in Fig. 25.12 was concluded. Ehlers et al. used infrared reflection–absorption spectroscopy (IRAS) and UPS in conjunction with TDS to follow the interaction of methanol with Pt(111)

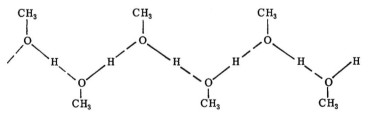

Figure 25.12 Schematic sketch of the formation of a H-bonded chain of methanol molecules on a Pd(100) surface as deduced from the energetic and kinetic behavior of the CH_3OH thermal desorption spectra. After ref. [110].

and reported on the formation of a first undissociated CH_3OH layer with the molecules being strongly chemisorbed at their oxygen end, while the IR data suggested strong hydrogen bonding within the second and third monolayer phase [112].

In view of the increasing use of methanol in fuel cells (DMFC = direct methanol fuel cell) the interaction of CH_3OH with metal surfaces, especially Pd and Pt surfaces, has been extensively studied in recent years [113–119], based on the idea that the electrooxidation of methanol on the Pt or PtRu anode is one of the decisive reactions in the DMFC. The first step will certainly be the adsorption of methanol on the metal surface, followed by a subsequent dehydrogenation of the hydroxo and/or the methyl group. Stuve et al. [113] have reviewed some of the recent gas phase adsorption and decomposition studies of methanol interaction with metal surfaces and concluded that either a hydroxymethyl (CH_2OH) or a methoxide (CH_3O) intermediate can be formed on the Pt (PtRu) surface. Both species will then further react with oxygen via other short-living intermediates. After the removal of water a transient CO species is obtained which is then oxidized to the final stable product CO_2. An important point in this reaction scenario is whether or not the C–O bond of the methanol molecule is actually cleaved. Here, some discrepancy exists with respect to the electro-oxidation reaction in the condensed phase (DMFC application) and the reaction sequence reported for methanol adsorbed under ultra-high vacuum (UHV), conditions, see below.

Chen et al. [118] showed for CH_3OH vapor adsorbed on a Pd(111) surface that the dissociation of the C–O bond requires methanol coverages close to one monolayer: Upon heating, most of the methanol (75%) desorbs, but the remaining part (25%) becomes partially dehydrogenated, while some other fraction of CH_3OH molecules undergo a bimolecular reaction via

$$2\ CH_3OH = CH_3O + CH_3 + H_2O,$$

whereby the C–O bond is opened. A hydrogen bridge between the two methanol molecules was considered essential for this C–O bond scission reaction. The same system was recently re-investigated by Schennach et al. (combining experimental (TDS) and theoretical (DFT) work) [119]; these authors, too, pointed out that a hydrogen bond between two neighboring methanol molecules adsorbed on a Pd(111) surface is *necessary* to break the C–O bond.

In continuation of the work of Ehlers et al. [112] H bonding effects for methanol monolayers adsorbed on a Pt(111) surface have been deduced from combined $\Delta\varphi$ and IRAS studies by Villegas and Weaver [117] who emphasized that adsorbed CH_3OH layers provided a particularly suitable 'solvent' to model a double layer network on a surface, since both the O–H stretching (ν_{OH}) and the C–OH stretching vibrations (ν_{C-OH}) are sensitive to the local coordination environment. At low coverages, CH_3OH adsorbs on Pt(111) in the form of monomers, while at higher coverages clusters stabilized by H bonding are formed. In a recent DFT study Desai et al. [116] have calculated the routes of interaction of methanol with a Pt(111) surface and considered the two reaction channels mentioned above, viz., the dissociation of chemisorbed methanol into either methoxide CH_3O plus H or into hydroxyl methyl CH_2OH plus H fragments. Intact methanol adsorbs on Pt(111) in an atop site at 25% surface coverage with an adsorption energy of ~ 43.2 kJ mol^{-1}; this relatively weak van-der-Waals-like Pt–CH_3OH bonding is concluded also from the rather extended (calculated) Pt–O bond length of 2.59 Å. The methoxy intermediate, on the other hand, favorably adsorbs in a three-fold hollow site at low coverages Θ (~ 10% of total monolayer) and switches to the Pt–Pt bridge site as soon as Θ reaches 25%. The adsorption energy then yields the rather considerable value of 161 kJ mol^{-1}. In this configuration, the CH_3O intermediate interacts with the Pt surface through the oxygen atom, forming two Pt–O bonds with 2.51 Å bond length. The internal C–O bond is tilted by an angle of 65° with respect to the surface plane. In another theoretical contribution by Greeley and Mavrikakis [120] who performed a periodic, self-consistent, DFT calculation, likewise the gas-phase decomposition of methanol on Pt(111) was explored. The reaction starts with O–H dissociation (the rate-limiting step) and proceeds via sequential hydrogen abstraction from the methoxy intermediate towards the final products CO and hydrogen. For several decomposition pathways, the authors present a potential energy diagram as a function of the reaction coordinate.

The long-standing controversy in methanol electro-oxidation over Pt and Pd surfaces as to whether the scission of the hydroxy group or the cleavage of a C–H bond of a methyl group (leading to CH_2OH) is the primary step has already been touched upon. Davis and Barteau point out [121] that thermodynamic arguments favor the latter mechanism, since the energy required to break a C–H bond is 393 kJ mol^{-1}, whereas 435 kJ mol^{-1} is the dissociation energy of an O–H bond. However, and this underlines the important role of hydrogen bonding effects, there is – at least for reactions in a condensed aqueous environment – a tendency of the OH group to be solvated by as many as three water molecules [122] which keeps it away from the Pt electrode surface and thus increases the probability that the reaction proceeds via the methyl end of the molecule. Without going further into any details, we simply refer the reader to Franaszczuk et al.'s exhaustive comparison of the electrochemical and gas-phase decomposition pathways of methanol on Pt surfaces [122].

25.4.2
Water on Metal Surfaces

While these considerations should just illustrate the importance of H bonding effects in electro-catalysis of methanol (which is, as pointed out before, essential for mimicking the oxidation of methanol on the Pt-containing catalyst-electrodes of the DMFC), we move on now to the interaction of *water* with metal surfaces which is largely dominated by H bonding and H transfer processes. Based on the importance of water in all kinds of biochemical systems a voluminous literature has been accumulated in the past concerning 'soft' biological systems which contain water in all kinds of aggregations. Because of space limitations we are unable to enter this (admittedly interesting) field; instead we focus on some of the UHV work on the interaction of H_2O with metal surfaces. The state of the art in this field until 1987 has been reviewed by Thiel and Madey [123]; in 2002 Henderson [124] revisited the same topic in an exhaustive review article the reader is referred to for further details. Again, we examine some typical transition metal surfaces, especially the 4d and 5d TMs (Ru, Rh, Pd, Ir, Pt), and pay particular attention to evidence of H bonding and H transfer. As with the alcohol adsorption studies, the experimental techniques of UV photoelectron spectroscopy and especially vibrational loss spectroscopy (HREELS) have proven to be very sensitive tools to identify adsorbed water and its fragments and to monitor the structure of the H_2O surface phases formed. As will be shown below these are dominated by H bonding effects. Pioneering HREELS reports about water adsorption on platinum surfaces appeared already in 1980 when Ibach and Lehwald studied the H_2O/Pt(100) [125] and Sexton [126] the H_2O/Pt(111) system. The observation of three different O–H stretching vibrations at 2850 cm^{-1}, 3380 cm^{-1} and 3670 cm^{-1} for water adsorbed on the Pt(100) surface in submonolayer quantities was interpreted as indicating H bonding to Pt, H bonding between oxygen atoms, and O–H bonding in 'free' OH groups, respectively [125]. For the Pt(111) surface, Sexton likewise inferred evidence for H bonding from both the presence of a broad librational region between 100 and 1000 cm^{-1} in the HREEL spectra of water and the exposure (coverage) dependences of the water bending and stretching modes. While the bending (δ) modes were almost invariant with coverage (since their frequency is not strongly affected by the O–H bond strength), there was a marked frequency downshift of the stretching (ν) modes, caused by a continuous weakening of the O–H bonds due to H bonding [126]. As an example of the water clustering induced by H bonding, we present Ibach's and Lehwald's structure model in Fig. 25.13. Structures of this kind are typical for H-bonded water networks and ice formed by continuous exposure of cold metal surfaces to water vapor. Griffiths et al. [127] followed, by means of infrared absorption spectroscopy, the development of the H bonded O–H stretch as a function of water coverage on three different metal surfaces, Ni(110); Pt(100) and Al(100). Only on Ni(110), could H_2O monomers be detected, whereas on Pt(100) water clustering very readily took place at 130 K and even at low coverages. On Al(100), a broad distribution of H_2O cluster sizes was found which was taken as evidence for a restricted mobility of the adsorbing water molecules after sticking to the surface. In a sense, these observations are symptomatic for the water interaction with metal surfaces, whereupon clustering and network formation far below the monolayer coverage take place, often in the form of bilayers representing a buckled hexagonal arrangement of water molecules interconnected by H bonds.

An important observation was communicated by Sexton [126] who found, by means of isotopic labeling (co-adsorption of H_2O and D_2O), that complete isotopic exchange of hydrogen occurred, indicating a very high mobility of the H and D atoms in the adsorbed water layers. Although isotope exchange is not unusual for ice at room temperature [128], it is nonetheless surprising that such rapid isotopic scrambling takes place within a network of H bonded water molecules on a Pt surface kept at a temperature as low as 100 K. This observation underlines the

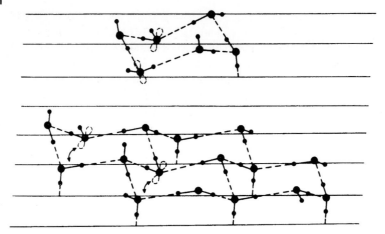

Figure 25.13 Sketch of the structure of water adsorbed at 150 K on a Pt(100) surface stabilized by H bonding. Top: Formation of a (tilted) cyclic water hexamer that accommodates to the Pt surface by forming a "chair"-like arrangement known from the cyclohexane molecule. Bottom: Structure of the network formed by a monolayer of water on the Pt surface: The oxygen atoms are held at distance to the Pt surface atoms by H bonds, thereby forming an open bilayer structure. The lone pair orbitals of the O atoms are indicated. After Ibach and Lehwald [125].

possibility of rapid H transfer even within water clusters or water networks on metal surfaces despite their rigid structure. A more recent HREELS study of the H_2O/Pt(111) system over a wide exposure range and with improved vibrational resolution was reported by Jacobi et al. [129]. They followed the development of networks of H-bonded water molecules from monomers to three-dimensional ice. On Pt(111), a H_2O bilayer is formed, and two perpendicular vibrational modes could be distinguished for the top H_2O molecules (266 cm^{-1}) and the bottom molecules in contact with the Pt surface (133 cm^{-1}).

Very extensive and revealing studies of water interaction with the Ru(0001) surface were performed by Thiel et al. [130, 131] using a variety of techniques (TDS, AES, HREELS, UPS, and ESDIAD = electron-stimulated desorption ion angular distribution). The authors observed several binding states in their TD spectra which develop simultaneously with H_2O exposure, again indicating that a first and a second water layer grow at the same time, whereby the second layer is H-bonded to the first layer. In other words, H_2O clusters are formed even at low exposures, in agreement with the observations for the Pt surfaces discussed above. The respective bilayer of H_2O on Ru(0001) even gave rise to the formation of an ordered $p(\sqrt{3} \times \sqrt{3})R30°$ phase with a well-developed diffraction pattern. Accordingly, a careful structure determination of adsorbed D_2O on Ru(0001) was performed by Held and Menzel [132] by analyzing the LEED intensities of the $p(\sqrt{3} \times \sqrt{3})R30°$ structure. The main result was that the D_2O bilayer is almost flat, caused by a periodic displacement of surface Ru atoms in anti-correlation to the water molecules. The construction principle of the bilayer was confirmed to contain chemical bonds between the Ru surface atoms and the O atoms of the closer water molecules, and H bondings from the latter to the other D_2O molecules. Apparently, the strength of the Ru–O bonding compares somewhat with the Ru–Ru bond strength leading to the situation that the underlying metal surface must no longer be regarded as a rigid lattice but represents a system that responds dynamically to the presence of adsorbing D_2O molecules.

The adsorption of water (D_2O) on the Rh(111) surface at 20 K has been studied by Yamamoto et al. [133] by means of IRAS, and again, the development of amorphous ice layers beginning with the formation of water monomers, dimers and larger clusters $(D_2O)_n$ with $3 < n < 6$, via subsequent two-dimensional islands and three-dimensional clusters and ending with bulk amorphous ice layers. The vibrational data were collected with high resolution and sensitivity and carefully assigned to the various stretching and bending frequencies for D_2O clusters of different aggregations, n. Interesting conclusions could be drawn with respect to the mobility of the individual water clusters on the Rh surface in comparison to Pt(111) and Ni(111) surfaces: On Rh(111), the migration of water molecules is claimed to be especially hindered, leading to comparatively small 2D islands.

This brings us to another peculiar property of adsorbed water molecules and clusters in conjunction with the formation of H bonds, namely, their mobility and dynamical behavior in the adsorbed state. The respective studies are closely tied to the possibility to directly image adsorbed water molecules and clusters by means of scanning tunneling microscopy (STM). The application of this technique to adsorbed water requires low-temperature studies and an appropriately designed low-T-scanning tunneling microscope. Pioneering studies here were communicated by Morgenstern who examined in a sequence of papers the principles of imaging water molecules on Ag(111) surfaces by means of STM [134–136]. In agreement with the studies mentioned above she found that the H_2O molecules form simultaneously H-bonded networks of different dimensionalities, in addition to an apparently zero-dimensional structure consisting of a cyclic hexamer with the molecules being adsorbed in on-top positions on the Ag surface [135]. The adsorption, clustering and diffusion of water molecules adsorbed at 40 K on a Pd(111) surface were then followed by means of low-T-STM by Mitsui et al. [137]. At low exposures, water adsorbs in the form of isolated molecules which laterally diffuse and collide with adjacent monomers. H bonding then leads to dimers, trimers, tetramers etc. The most interesting observation was, however, that the mobility of these aggregates on the Pd(111) surface increased by several orders of magnitude for the tetramers, but decreased again for larger aggregates (pentamers etc.). Cyclic hexamers were found to be particularly stable, in agreement with Morgenstern's observations for H_2O on Ag(111). With increasing exposure these hexamers combine to a hexagonal honeycomb network that is commensurate to the underlying Pd(111) surface, forming the same $p(\sqrt{3}\times\sqrt{3})R30°$ structure described above for the Ru(0001) surface. The large mobility of small water clusters is explained by a combination of strong H bonding between molecules and the misfit between the O–O distance in the dimer, which is 2.96 Å in the gas phase, but only 2.75 Å (= the nearest Pd–Pd distance) for the O atoms of an adsorbed H_2O dimer (if these O atoms are assumed to be bonded in atop positions on adjacent Pd atoms). The respective observations underline the change in relative strength between intermolecular hydrogen bonds and molecule–substrate bonds as a function of water cluster size and are related to the (macroscopic) wetting properties of a metal surface.

A final point remains to be discussed, namely, the ability of the various metal surfaces to dissociate the adsorbing water molecules into OH and H, whereby the split-off H atoms could be transferred to other (co-adsorbed) species – certainly a catalytically very relevant issue, because the respective H transfer is necessary for hydrogenation reactions. Various previous studies into the dissociation properties of water adsorbed on metal surfaces agree insofar as spontaneous dissociation is a relatively unlikely event on the 'classical' transition metal surfaces, whereby crystallographically 'open' surfaces are expected to be more active with respect to dissociation than flat low-index surfaces. For more reactive materials, e.g., the Al(100) surface, thermally activated water dissociation has been reported [127]. Wittrig et al. investigated the interaction of water with the (1×2)-reconstructed Ir(110) surface [138] and found that at most ~ 6% of the adsorbed H_2O molecules dissociated at 130 K. Similar results were obtained for other surfaces, i.e., spontaneous water dissociation is apparently not a favored route on TM surfaces, regardless of their crystallographic orientation. However, the situation changes considerably, if oxygen atoms are pre-adsorbed. In this case, O–HOH complexes may easily be formed, interconnected by H bonds. From HREELS data, the existence of these complexes was deduced by Thiel et al. [135] for the Ru(0001) surface, from the appearance of certain vibrational losses. Furthermore (and more importantly), a dramatic increase in the water *dissociation probability* into O and OH was found in the presence of preadsorbed oxygen, for example for the Pt(111)+O system by Fisher and Gland [139, 140] or for the Ir(110)(1×2) surface by Wittrig et al. [138] using X-ray photoelectron spectroscopy (XPS). The respective distinction was based on the position of the oxygen 1s orbital energy level: For the species $(OH_2)_{ad}$, $(OH)_{ad}$, and O_{ad} on Pt(111) Fisher and Gland reported orbital energies at 532.2, 530.5, and 528.8 eV, respectively [140].

Another possibility to dissociate water into H and OH is provided by illumination with UV light: As was demonstrated by Gilarowski et al. [141], H_2O dissociation could be achieved on a Pt(111) surface by shining the light of a xenon arc lamp (wave length cut-off at 190 nm) on the water-covered surface and watching the development of various vibrational frequencies within the adsorbed layer. A two-route reaction was emphasized. The irradiation led to molecular desorption of water on the one hand, but also to increasing dissociation into OH and H fragments, as monitored by the gradual development of new vibrational losses characteristic of OH bending and the hindered translation of OH. Furthermore, the appearance of a new peak in the thermal desorption spectra could be associated with the OH + H recombination reaction. Energetic arguments suggest that the mechanism of the photo-dissociation with a threshold of 5.2 eV occurs via a substrate excitation.

25.5
Conclusion

The interaction of hydrogen and simple molecules containing a hydroxy group (water, methanol) with transition metal surfaces has been surveyed with respect to the possibility of H transfer via lateral diffusion or proton tunneling within the respective adsorbed layers. It was shown that lateral diffusion and transfer of H atoms does indeed occur via both classical diffusion and tunneling. The growth and structure of the respective layers containing OH groups is largely governed by H bonding effects leading, in practically all of the investigated cases, to a relatively 'open' network of water or alcohol molecules. Isotopic scrambling thereby indicates an extraordinarily high mobility of H or D atoms.

Acknowledgements

The author gratefully acknowledges financial support of the Deutsche Forschungs-gemeinschaft (through SPP 1091) and helpful discussions with Chr. Pauls.

References

1 Baró, A. M., Miranda, R., Alamán, J., Garcia, N., Binnig, G., Rohrer, H., Gerber, Ch., Carrascosa, J. L., *Nature* 315 (1985) 253.

2 Knor, Z., in *Catalysis* Vol.3, J. R. Anderson, M. Boudart (eds.), Springer, Berlin 1982, p.231.

3 Christmann, K., *Surf. Sci. Rep.* 9 (1988) 1.

4 Davenport, J.W., Estrup, P.J., in *The Chemical Physics of Solid Surfaces and Heterogeneous Catalysis*, D.A. King, D.P. Woodruff, (eds.), Vol. 3A, Elsevier, Amsterdam 1990, p.1.

5 Groß, A., *Theoretical Surface Science*, Springer-Verlag, Berlin 2003.

6 Alefeld, G., Völkl, J. (eds.), *Hydrogen in Metals*, Springer-Verlag, Berlin 1978, Vol. 1 and 2.

7 Conner, Jr., W. C., in *Hydrogen Effects in Catalysis*, Z. Páal, P. G. Menon, (eds.), Marcel Dekker, New York 1988, Ch. 12, p.311 ff.

8 Mårtensson, A.S., Nyberg, C., Andersson, S., *Phys. Rev. Lett.* 57 (1986) 2045.

9 Schmidt, P.K., Christmann, K., Kresse, G., Hafner, J., Lischka, M., Groß, A., *Phys. Rev. Lett.* 87 (2001) 096103.

10 Schmidt, P. K., Ph.D. Thesis, FU Berlin 2002.

11 Balooch, M., Cardillo, M. J., Miller, D. R., Stickney, R. E., *Surf. Sci.* 46 (1974) 358.

12 Kolasinski, K. W., Nessler, W., Bornscheuer, K.-H., Hasselbrink, E., *Surf. Sci.* 331-333 (1995) 485.

13 Bratu, P., Höfer, U., *Phys. Rev. Lett.* 74 (1995) 1625.

14 Groß, A., Wilke, S., Scheffler, M., *Phys. Rev. Lett.* 75 (1995) 2718.

15 Eichler, A., Hafner, J., Groß, A., Scheffler, M., *Phys. Rev.* B59 (1999) 13297.

16 Harris, J., *Appl. Phys. A* 47 (1988) 63.

17 Christmann, K., Schober, O., Ertl, G., Neumann, M., *J. Chem. Phys.* 60 (1974) 4528.

18 Alemozafar, A. R., Madix, R. J., *Surf. Sci.* 557 (2004) 231.

19 Einstein, T. L., Schrieffer, J. R., *Phys. Rev. B* 7 (1973) 3629.

20 Grimley, T. B., *Proc. Phys. Soc. (London)* 90 (1967) 751.

21 Grimley, T. B., *Proc. Phys. Soc. (London)* 92 (1967) 776.

22 Cui, J. H., Fain, S. C., *Phys. Rev. B* 39 (1989) 8628.

23 Roelofs, L. D., Hu, G. Y., Ying, S. C., *Phys. Rev. B* 28 (1983) 6369.

24 Roelofs, L. D., Einstein, T. L., Bartelt, N. C., Shore, J. D., *Surf. Sci.* 176 (1986) 295.

25 Sokolowski, M., Koch, T., Pfnür, H., *Surf. Sci.* 243 (1991) 261.

26 Sandhoff, M., Pfnür, H., Everts, H.-U., *Surf. Sci.* 280 (1993) 185.

27 Christmann, K., Behm, R. J., Ertl, G., van Hove, M. A., Weinberg, W. H., *J. Chem. Phys.* 70 (1979) 4168.

28 Behm, R. J., Christmann, K., Ertl, G., *Solid State Commun.* 25 (1978) 763.

29 Christmann, K., *Z. Phys. Chem. (NF)* 154 (1987) 145.

30 Hara, M., Domen, K., Onishi, T., Nozoye, H., *J. Phys. Chem.* 95 (1991) 6.

31 Wagner, H., Butz, R., Backes, U., Bruchmann, D., *Solid State Commun.* 38 (1981) 1155.

32 Butz, R., Memeo, R., Wagner, H., *Phys. Rev. B* 25 (1982) 4327.

33 Morris, M. A., Bowker, M., King, D. A., Simple Processes at the Gas–Solid Interface, in *Chemical Kinetics*, Vol. 19, Bamford, C.H., Tipper, C.F.H., Compton, R.G., (eds.), Elsevier, Amsterdam 1984, Ch. I.

34 Naumovets, A. G., Vedula, Y. S., *Surf. Sci. Rep.* 4 (1984) 365.

35 Roberts, M. W., McKee, C. S., *Chemistry of the Metal–Gas Interface*, Clarendon Press, Oxford 1979, p.92.

36 (a) Wortman, R., Gomer, R., Lundy, R., *J. Chem. Phys.* 27 (1957) 1099; (b) Gomer, R., *Field Emission and Field ionization*, Harvard University Press, Cambridge 1961.

37 Lewis, R., Gomer, R., *Surf. Sci.* 17 (1969) 333.

38 DiFoggio, R., Gomer, R., *Phys. Rev. B* 25 (1982) 3490.

39 Ertl, G., Neumann, M., *Z. Naturforsch.a* 27 (1972) 1607.

40 Seebauer, E. G., Schmidt, L. D., *Chem. Phys. Lett.* 123 (1986) 129.

41 Mak, C. H., Brand, J. L., Deckert, A. A., George, S. M., *J. Chem. Phys.* 85 (1986) 1676.

42 Mak, C. H., George, S. M., *Surf. Sci.* 172 (1986) 509.

43 Mak, C. H., Brand, J. L., Koehler, B. G., George, S. M., *Surf. Sci.* 191 (1987) 108.

44 Mann, S. S., Seto, T., Barnes, C. J., King, D. A., *Surf. Sci.* 261 (1992) 155.

45 Auerbach, A., Freed, K. F., Gomer, R., *J. Chem. Phys.* 86 (1987) 2356.

46 Puska, M. J., Nieminen, R. M., Chakraborty, B., Holloway, S., Nørskov, J. K., *Phys. Rev. Lett.* 51 (1983) 1083.

47 Puska, M. J., Nieminen, R. M., *Surf. Sci.* 157 (1985) 413.

48 Whaley, K. B., Nitzan, A., Gerber, R. B., *J. Chem. Phys.* 84 (1986) 5181.

49 Wang, S. C., Gomer, R., *J. Chem. Phys.* 83 (1985) 4193.

50 Mate, C. M., Somorjai, G. A., *Phys. Rev. B* 34 (1986) 7417.

51 Grizzi, O., Shi, M., Rabalais, J. W., Rye, R. R., Nordlander, P., *Phys. Rev. Lett.* 63 (1989) 1408.

52 Astaldi, C., Bianco, A., Modesti, S., Tosatti, E., *Phys. Rev. Lett.* 68 (1992) 90.

53 Takagi, N., Yasui, Y., Takaoka, T., Sawada, M., Yanagita, H., Aruga, T., Nishijima, M., *Phys. Rev. B* 53 (1996) 13767.

54 Badescu, S. C., Salo, P., Ala-Nissila, T., Ying, S. C., Jacobi, K., Wang, Y., Bedürftig, K., Ertl, G., *Phys. Rev. Lett.* 88 (2002) 136101.

55 Badescu, S. C., Jacobi, K, Wang, Y., Bedürftig, K., Ertl, G., Salo, P., Ala-Nissila, T, Ying, S. C., *Phys. Rev. B* 68 (2003) 205401; 209903.

56 Okuyama, H., Ueda, T., Aruga, T., Nishijima, M., *Phys. Rev. B* 63 (2003) 233403; 233404

57 Nobuhara, K., Kasai, H., Nakanishi, H., Okiji, A., *Surf. Sci.* 507-510 (2002) 82.

58 Nishijima, M., Okuyama, H., Takagi N., Aruga, T., Brenig, W., *Surf. Sci. Rep.* 57 (2005) 113.

59 Lopinski, G. P., Prybyla, J. A., Estrup, P. J., *Surf. Sci.* 296 (1993) 9.

60 Sprunger, P. T., Plummer, E. W., *Phys. Rev. B* 48 (1993) 14436.

61 Kostov, K. L., Widdra, W., Menzel, D. Surf. Sci. 560 (2004) 130.

62 Zambelli, T., Trost, J., Wintterlin, J., Ertl, G., *Phys. Rev. Lett.* 76 (1996) 795.

63 Trost, J., Zambelli, T., Wintterlin, J., Ertl, G., *Phys. Rev. B* 54 (1996) 17850.

64 Wintterlin, J., Trost, J., Renisch, S., Schuster, R., Zambelli, T., Ertl, G., *Surf. Sci.* 394 (1997) 159.

65 Stipe, B. C., Rezaei, M. A., Ho, W., *J. Chem. Phys.* 107 (1997) 6443.

66 Stipe, B. C., Rezaei, M. A., Ho, W., *Science* 280 (1998) 1732.

67 Freed, K. F., *J. Chem. Phys.* 82 (1985) 5264.

68 Baer, R., Zeiri, Y., Kosloff, R., *Surf. Sci.* 411 (1998) L783.

69 Kohl, D., *J. Phys. D: Appl. Phys.* 34 (2001) R125.

70 Carette, L., Friedrich, K. A., Stimming, U., *Fuel Cells* 1 (2001) 5.

71 Mitchell, P. C. H., Parker, S. F., Tomkinson, J., Thompsett, D., *J. Chem. Soc., Faraday Trans.* 94 (1998) 1489.

72 Mitchell, P. C. H., Ramirez-Cuesta, A. J., Parker, S. F., Tomkinson, J., Thompsett, D., *J. Phys. Chem. B* 107 (2003) 6838.

73 Watanabe, K., Hashiba, M., Yamashina, T., *Surf. Sci.* 61 (1976) 483.

74 Sachtler, W. M. H., *Le Vide* 164 (1973) 67.

75 Ertl, G., in *The Nature of the Surface Chemical Bond*, Rhodin, T. N., Ertl, G. (eds.), North-Holland, Amsterdam 1979, p. 335.

76 Schneider, U., Castro, G. R., Wandelt, K., *Surf. Sci.* 287/288 (1993) 146.

77 Linke, R., Schneider, U., Busse, H., Becker, C., Schröder, U., Castro, G. R., Wandelt, K., *Surf. Sci.* 307–309 (1994) 407.

78 Holzwarth, A., Loboda-Cackowic, J., Block, J. H., Christmann, K., *Z. Phys. Chem.* 196 (1996) 55.

79 Hebenstreit, E. L. D., Hebenstreit, W., Schmid, M., Varga, P., *Surf. Sci.* 441 (1999) 441.

80 Gauthier, Y., Schmid, M., Padovani, S., Lundgren, E., Buš, V., Kresse, G., Redinger, J., Varga, P., *Phys. Rev. Lett.* 87 (2001) 036103.

81 Sinfelt, J. H., *Bimetallic Catalysis*, Wiley & Sons, New York 1983.

82 Sinfelt, J. H., *Acc. Chem. Res.* 10 (1977) 15.

83 Sinfelt, J. H., Lam, Y. L., Cusamano, J. A., Barnett, A. E., *J. Catal.* 42 (1976) 227.

84 Prestridge, E. B., Via, G. H., Sinfelt, J. H., *J. Catal.* 50 (1977) 115.

85 Christmann, K., Ertl, G., Shimizu, H., *J. Catal.* 61 (1980) 397 {Cu on Ru}.

86 Shimizu, H., Christmann, K., Ertl, G., *J. Catal.* 61 (1980) 412 {Cu on Ru}.

87 Vickerman, J. C., Christmann, K., *Surf. Sci.* 120 (1982) 1 {Cu on Ru}.

88 Goodman, D. W., Peden, C. H. F., *J. Catal.* 95 (1985) 321 {Cu on Ru}.

89 Yates Jr., J. T., Peden, C. H. F., Goodman, D. W., *J. Catal.* 94 (1985) 576 {Cu on Ru}.

90 Houston, J. E., Peden, C. H. F., Blair, D. S., Goodman, D. W., *Surf. Sci.* 167 (1986) 427 {Cu on Ru}.

91 Houston, J. E., *Phys. Rev. Lett.* 56 (1986) 375.

92 Harendt, C., Christmann, K., Hirschwald, W., Vickerman, J. C., *Surf. Sci.* 165 (1986) 413 {Au on Ru}.

93 Goodman, D. W., Yates, Jr., J. T., Peden, C. H. F., *Surf. Sci.* 164 (1985) 417 {Cu on Ru}.

94 Lenz, P., Christmann, K., *J. Catal.* 139 (1993) 611 {Ag on Ru}.

95 Lenz-Solomun, P., Christmann, K., *Surf. Sci.* 345 (1996) 41 {Ag on Ru}.

96 Campbell, C. T., *Annu. Rev. Phys. Chem.* 41 (1990) 775 {review}.

97 Goodman, D. W., *Annu. Rev. Phys. Chem.* 37 (1986) 425 {review}.

98 Groß, A., *Theoretical Surface Science – A Microscopic Perspective*, Springer, Berlin 2003, p. 148 ff., and references given therein.

99 Mitsui, T., Rose, M. K., Fomin, E., Ogletree, D. F., Salmeron, M., *Nature* 422 (2003) 705.

100 *Spillover and Migration of Surface Species on Catalysis*, Proc. 4th Intern. Conf. on Spillover, Dalian, China, Can Li, Quin Xin, (eds.), Elsevier, Amsterdam 1997.

101 Baumgarten, E., Denecke, E., *J. Catal.* 95 (1985) 296.

102 Jung, K.-W., Bell, A. T., *J. Catal.* 193 (2000) 207.

103 Yang, F. H., Lachawiec, Jr., J., Yang, R. T., *J. Phys. Chem. B* (2006) 6236.

104 Cruq, A., Degols, L., Lienard, G., Frennet, A., in *Proceedings of the First International Symposium on the Spillover of Adsorbed Species*, G. M. Pajonk, S. J. Teichner, J. E. Germain, (eds.), Elsevier, Amsterdam 1983, p. 137.

105 Greuter, F., Plummer, E. W., Solid State Commun. 48 (1983) 37.

106 Anger, G., Winkler, A., and Rendulic, K. D., *Surf. Sci.* 220 (1989) 1.

107 Luo, M. F., MacLaren, D. A., Allison, W., *Surf. Sci.* 586 (2005) 109.

108 Rubloff, G. W., Demuth, J. E., *J. Vac. Sci. Technol.* 14 (1977) 419.

109 Demuth, J. E., Ibach, H., *Chem. Phys. Lett.* 60 (1979) 395.

110 Christmann, K., Demuth, J. E., *J. Chem. Phys.* 76 (1982) 6308.

111 Christmann, K., Demuth, J. E., *J. Chem. Phys.* 76 (1982) 6318.

112 Ehlers, D. H., Spitzer, A., Lüth, H., *Surf. Sci.* 160 (1985) 57.

113 Jarvi, T. D., Stuve, E. M., in *Electrocatalysis*, Lipkowski, J., Ross, P. N., (eds.), Wiley-VCH, Weinheim 1998, p. 75.

114 Hartmann, N., Esch, F., Imbihl, R., *Surf. Sci.* 297 (1993) 175.

115 Solymosi, F., Berkó, A., Tóth, Z., *Surf. Sci.* 285 (1993) 197.

116 Desai, S. K., Neurock, M., Kourtakis, K., *J. Phys. Chem. B* 106 (2002) 2559.

117 Villegas, I., Weaver, M. J., *J. Chem. Phys.* 103 (1995) 2295.

118 Chen, J.-J., Jiang, Z.-C., Zhou, Y., Chakraborty, B. R., Winograd, N., *Surf. Sci.* 328 (1995) 248.

119 Schennach, R., Eichler, A., Rendulic, K. D., *J. Phys. Chem. B* 107 (2003) 2552.

120 Greeley, J., Mavrikakis, M., *J. Am. Chem. Soc.* 124 (2002) 7193.

121 Davis, J. L., Barteau, M. A., *Surf. Sci.* 187 (1987) 387.

122 Franaszczuk, K., Herrero, E., Zelenay, P., Wieckowski, A., Wang, J., Masel, R. I., *J. Phys. Chem.* 96 (1992) 8509.

123 Thiel, P. A., Madey, T. E., *Surf. Sci. Rep.* 7 (1987) 211–365.

124 Henderson, M. A., *Surf. Sci. Rep.* 46 (2002) 1–308.

125 Ibach, H., Lehwald, S., *Surf. Sci.* 91 (1980) 187.

126 Sexton, B. A., *Surf. Sci.* 94 (1980) 435.

127 Griffiths, K., Kasza, R. V., Esposto, F. J., Callen, B. W., Bushby, S. J., Norton, P. R., *Surf. Sci.* 307–309 (1994) 60.

128 *Water*, Franks, F., (ed.), Vol. 2, Plenum Press New York 1973.

129 Jacobi, K., Bedürftig, K., Wang, Y., Ertl, G., *Surf. Sci.* 472 (2001) 9.

130 Thiel, P. A., Hoffmann, F. M., Weinberg, W. H., *J. Chem. Phys.* 75 (1981) 5556.

131 Thiel, P. A., Hoffmann, F. M., Weinberg, W. H., *Phys. Rev. Lett.* 49 (1982) 501.

132 Held, G., Menzel, D., *Surf. Sci.* 316 (1994) 92.

133 Yamamoto, S., Beniya, A., Mukai, K., Yamashita, Y., Yoshinobu, J., *J. Phys. Chem. B*109 (2005) 5816.

134 Morgenstern, K., Rieder, K.-H., *Chem. Phys. Lett.* 358 (2002) 250.

135 Morgenstern, K., *Surf. Sci.* 504 (2002) 293.

136 Morgenstern, K., Nieminen, J., *J. Chem. Phys.* 120 (2004) 10786.

137 Mitsui, T., Rose, M. K., Fomin, E., Ogletree, D. F., Salmeron, M., *Science* 297 (2002) 1850.

138 Wittrig, T. S., Ibbotson, D. E., Weinberg, W. H., *Surf. Sci.* 102 (1981) 506.

139 Fisher, G. B., Gland, J. L., *Surf. Sci.* 94 (1980) 446.

140 Fisher, G. B., Sexton, B. A., *Phys. Rev. Lett.* 44 (1980) 683.

141 Gilarowski, G., Erley, W., Ibach, H., *Surf. Sci.* 351 (1996) 156.

26
Hydrogen Motion in Metals

Rolf Hempelmann and Alexander Skripov

26.1
Survey

Many metals dissociatively dissolve hydrogen [1–6]. At low H content the metal host lattice is unchanged (apart from a slight lattice expansion), and the hydrogen atoms occupy random sites in the interstitial lattice (e.g. octahedral interstices in Pd or tetrahedral interstitial sites in Nb). In the metal/H phase diagram this regime is called the solid-solution (*a*) phase. At higher H concentration stoichiometric hydride phases appear, in which the hydrogen atoms form an interstitial lattice with long-range order and in which the host lattice structure may differ from the 'empty' host lattice.

The hydrogen atom on its interstitial site in the metal (or metal alloy or intermetallic compound) generally may perform motional processes on very different time scales. At very short times the H atom vibrates against its metallic neighbors which, due to their much heavier masses, do not participate in these high frequency vibrations. Depending on H concentration and H–H interaction they can be considered either as local or optical modes. On the time scale of acoustic vibrations of the host lattice, the H atoms move more or less adiabatically according to the distortion pattern imposed by the host phonons and mirror the host density of states. This type of motion is also called a band mode.

At much longer times the hydrogen is able to leave its interstitial site and to perform 'jumps' to other sites. The quantum mechanical nature of these 'jumps' and their spatial/temporal evolution in the lattice are denoted as hydrogen diffusion in metals, and the present chapter is focused on this type of hydrogen motion. Experimentally, hydrogen diffusion can be studied either by measuring appropriate concentration dependent macroscopic physical properties while a metal hydrogen system, starting from a nonequilibrium situation approaches the new equilibrium, or by methods which are sensitive to single jumps. From the latter methods the most relevant ones – anelastic relaxation, nuclear magnetic resonance and quasielastic neutron scattering, are explained in Section 26.2.

Experimental results for hydrogen diffusion coefficients and their relevance for the kinetics of hydrogen absorption are presented in Section 26.3. Section 26.4

Hydrogen-Transfer Reactions. Edited by J. T. Hynes, J. P. Klinman, H. H. Limbach, and R. L. Schowen
Copyright © 2007 WILEY-VCH Verlag GmbH & Co. KGaA, Weinheim
ISBN: 978-3-527-30777-7

deals with the diffusion mechanism, i.e. the evolution of the jump processes on atomistic scales of time and space, for hydrogen in systems of increasing complexity and disorder: pure metals, alloys, intermetallic compounds, and amorphous metals.

Section 26.5 is devoted to the single hydrogen site change event ('jump') which due to the light mass of hydrogen at all temperatures is governed by quantum mechanical tunneling (quantum diffusion) instead of thermally activated over-barrier jumps (classical diffusion). Some highlights are the hydrogen tunneling experiments on Nb doped with impurities and on a-MnH_x as well as the measurements of rapid low-temperature hopping of hydrogen in a-$ScH_x(D_x)$ and in $TaV_2H_x(D_x)$. In the concluding remarks of Section 26.6 the essential features of hydrogen motion, particularly in metals, are briefly summarized.

26.2
Experimental Methods

In this section we present a brief overview of experimental methods used to study hydrogen motion in metals. The methods giving microscopic information on the hydrogen jump motion are emphasized. We restrict ourselves to a discussion of the basic principles of these methods only. More detailed consideration of the application of different methods to studies of the hydrogen diffusion in metals can be found in the reviews [7–14].

26.2.1
Anelastic Relaxation

When a hydrogen atom occupies an interstitial site, it causes an expansion of the host–metal lattice. If the site symmetry is lower than cubic, such a local lattice distortion can be characterized by its orientation. In the case of uniaxial distortion the strain field is described in terms of the strain tensor with the principal components λ_1, λ_2 and $\lambda_3 = \lambda_2$. Such defects can interact with a shear stress to produce reorientational relaxation (the Snoek effect) [7, 15]. Before application of the stress, equivalent sites with different orientations have the same energy, so that hydrogen atoms are equally distributed among them. Depending on orientation, the applied stress splits the site energies and initiates a relaxation to a new equilibrium hydrogen configuration with preferential occupation of sites with lower energy. Thus, the reorientational relaxation involves local hydrogen jumps between the sites with different orientations of the strain field. If a constant stress is applied, a metal–hydrogen system responds with an immediate elastic strain ε_e followed by the anelastic strain ε_a which grows with time to the limiting value ε_a^0. The ratio $\varepsilon_a^0/\varepsilon_e$ is a convenient measure of the magnitude of the relaxation; it is called the relaxation strength Δ_R. For the Snoek effect, Δ_R is proportional to $(\lambda_1-\lambda_2)^2$. The Snoek effect is usually probed in the dynamical regime, for example, by deforming the sample at a certain frequency and measuring the mechanical damping (internal friction) as a function of temperature. The maximum damping is expected to occur at the temperature at which the reorientation jump rate τ_R^{-1} becomes equal to the circular frequency ω of the external excitation. In particular, the condition $\omega\tau_R = 1$ determines the position of the sound attenuation peak in ultrasonic experiments. In the simple case of a reorientation relaxation with a single type of reorientation jumps, the ultrasonic loss $1/Q_u$ is given by the Debye expression

$$\frac{1}{Q_u} = \Delta_R \frac{\omega\tau_R}{1 + \omega^2 \tau_R^2} \qquad (26.1)$$

Using Eq. (26.1) it is possible to obtain the temperature dependence of the hydrogen jump rate from the experimental data on ultrasonic loss. The Snoek relaxation measurements are especially informative if they are performed at a number of excitation frequencies. It should be noted that the Snoek effect can be observed only for sufficient elastic anisotropy, $\lambda_1 - \lambda_2$, of hydrogen sites. For hydrogen in pure b.c.c. metals, the Snoek effect has not been found [16], in spite of the uniaxial symmetry of tetrahedral sites occupied by hydrogen in these materials. It is believed that the absence of the observable Snoek effect is due to the small value of $\lambda_1 - \lambda_2$ for hydrogen in the tetrahedral sites of b.c.c. metals.

Another method using mechanical relaxation for studies of hydrogen motion in metals is based on the dilatational relaxation caused by the *long-range* hydrogen diffusion (the Gorsky effect) [17, 18]. The Gorsky relaxation is usually measured quasi-statically as an elastic after-effect. The relaxation is initiated by bending a bulk sample of suitable shape (wire, ribbon, disk, etc.). On application of the bending stress, the instantaneous elastic deformation is followed by an additional deformation which is caused by a redistribution of hydrogen atoms over the sample volume. In fact, the stress gives rise to the flow of hydrogen atoms from the compressed side to the stretched side of the sample. Since H atoms cause the lattice expansion, such a process of 'uphill' diffusion results in the additional (delayed) deformation. The relaxation is complete when the flow of H atoms caused by the stress is compensated by the diffusion flow due to the H concentration gradient. The chemical diffusion coefficient D_{ch} of hydrogen can be obtained from the relaxation time τ_G of the elastic after-effect. For example, for a strip of thickness d, the relation between τ_G and D_{ch} is given by

$$\tau_G = d^2/\pi^2 D_{ch} \qquad (26.2)$$

In contrast to the Snoek effect, the Gorsky effect does not depend on the elastic anisotropy. The relaxation strength determined from the Gorsky effect measurements is proportional to the square of the trace of the strain tensor, i.e. to $(\lambda_1 + 2\lambda_2)^2$. The measured relaxation strength also contains important information on the thermodynamic factor $\partial\mu/\partial c$ [19], where μ is the chemical potential of hydrogen and c is the hydrogen concentration. The relation between the chemical diffusion coefficient D_{ch} (as determined by macroscopic methods employing

H concentration gradients) and the tracer diffusion coefficient D (as determined by microscopic methods at equilibrium conditions) is given by [20]

$$D_{ch} = \frac{cD}{f_c k_B T} \frac{\partial \mu}{\partial c} \tag{26.3}$$

where f_c is the correlation factor (or Haven's ratio). For small c the value of f_c is 1; with increasing H concentration, f_c decreases down to values typically between 0.5 and 0.8. Since both D_{ch} and $\partial \mu / \partial c$ can be determined from the same Gorsky relaxation experiment, the Gorsky effect measurements allow one to obtain the tracer diffusion coefficient D. Therefore, these measurements provide a bridge between the macroscopic and microscopic methods of investigation of hydrogen motion.

26.2.2
Nuclear Magnetic Resonance

Nuclear magnetic resonance (NMR) is widely used to evaluate the parameters of hydrogen motion in metals. All three hydrogen isotopes as well as many host-metal nuclei can be employed as natural probes of hydrogen motion. The only serious limitation of applicability of NMR to studies of hydrogen motion is that the samples cannot be magnetically ordered or strongly paramagnetic. There are basically two types of NMR experiments probing the motional behavior of hydrogen: (i) measurements of the NMR linewidths and the nuclear spin relaxation rates giving information on the jump rate of hydrogen atoms, and (ii) measurements of the spin-echo attenuation in applied magnetic field gradients allowing one to determine the tracer diffusion coefficient D.

The former type of experiments are based on the sensitivity of NMR parameters to fluctuations of the local magnetic and electric fields. These fluctuations originate from the jump motion of hydrogen atoms. For protons (^1H) and tritium nuclei (^3T) having nuclear spin $I = 1/2$, the motional contributions to the spin relaxation rates are determined by the modulation of the magnetic dipole–dipole interaction between nuclear spins. Since the deuterium (^2D, $I = 1$) and host-metal nuclei with $I > 1/2$ have nonzero electric quadrupole moments, they can also interact with fluctuating electric-field gradients. Therefore, for these nuclei the fluctuating electric quadrupole interaction gives an important contribution to the nuclear spin relaxation rates.

Measurements of the nuclear spin–lattice (or longitudinal) relaxation rate R_1 are most commonly used to obtain information on the hydrogen jump rates. In favorable cases such measurements allow one to trace the changes in the hydrogen jump rate over the range of four decades. However, the measured values of R_1 normally also contain additive contributions not related to hydrogen motion, for example, the contribution due to the hyperfine interaction between nuclear spins and conduction electrons. The motional contribution to the spin–lattice relaxation rate, R_{1m}, can be extracted using the difference in the temperature and frequency dependences of different contributions to R_1 [12]. Such a procedure may be effective if the R_1 measurements are performed over wide ranges of temperature and resonance frequency. The motional contribution is described by a sum of several terms, each of the general form [12]

$$R_{1m} = <M^2>J(\omega_I, \omega_S, \tau_c) \tag{26.4}$$

where $<M^2>$ is the part of the interaction (dipole–dipole or quadrupole) of the nuclear spin with its environment that fluctuates due to the motion, and $J(\omega_I, \omega_S, \tau_c)$ is the spectral density function that describes the dependence of the fluctuations in M on the resonance frequencies ω_I and ω_S of the resonant and nonresonant nuclei, respectively, and on the correlation time τ_c of the fluctuations. For hydrogen diffusion in metals, τ_c is approximately equal to the mean residence time τ_d of a hydrogen atom in an interstitial site. The characteristic feature of the temperature dependence of R_{1m} is the maximum that occurs when $\omega_I \tau_d \approx 1$. In other words, the R_{1m} maximum is observed at the temperature at which the hydrogen jump rate τ_d^{-1} becomes nearly equal to the (circular) resonance frequency ω_I. In the limit of slow motion ($\omega_I \tau_d \gg 1$) R_{1m} is proportional to $\omega_I^{-2} \tau_d^{-1}$, and in the limit of fast motion ($\omega_I \tau_d \ll 1$) R_{1m} is proportional to τ_d, being frequency-independent.

Since, for long-range diffusion, the appropriate spectral densities $J(\omega_I, \omega_S, \tau_c)$ cannot be calculated in the analytic form, in order to extract $\tau_d^{-1}(T)$ from the R_{1m} data, one has to rely either on the results of numerical (Monte Carlo) calculations [21–23] or on the model spectral densities [24–26]. The most widely used model is that based on the Lorentzian form of the spectral densities, as introduced by Bloembergen, Purcell and Pound (BPP) [24]. The BPP model correctly describes the main features of the R_{1m} data (including the asymptotic behavior in the limits of fast and slow motion) and provides rather simple expressions relating R_{1m} and τ_d. Comparison of the BPP results with those of more sophisticated calculations shows that although the absolute τ_d values derived from the BPP model analysis may be in error by as much as a factor of 2, their relative temperature-dependent behavior is usually far more reliable, resulting in diffusion activation energies that agree to within about 10% with those obtained by other techniques [8]. The range of the jump rates τ_d^{-1} that can be effectively probed by the spin–lattice relaxation measurements is 10^6–10^{10} s^{-1}.

Measurements of the rotating-frame spin relaxation rate $R_{1\rho}$ are more demanding from the experimental point of view than the R_1 measurements. In metal–hydrogen systems, $R_{1\rho}$ can usually be measured only for protons (^{1}H). However, measurements of $R_{1\rho}$ are sensitive to much lower jump rates than those of R_1; the typical range of τ_d^{-1} that can be probed by $R_{1\rho}$ is 10^3–10^7 s^{-1}. Qualitatively, the behavior of $R_{1\rho}$ in metal–hydrogen systems is similar to that of R_1. The maximum of $R_{1\rho}(T)$ is determined by the condition $\omega_1 \tau_d \approx 1$, where ω_1 is the circular frequency of nuclear spin precession in a (weak) rf magnetic field. Since the typical values of ω_1 are three orders of magnitude smaller than those of the NMR frequency ω_I, the maximum of $R_{1\rho}$ is observed at much lower temperature than that of R_1. The maximum value of $R_{1\rho}$ is much higher than that of R_1; therefore, the contributions not related to hydrogen motion appear to be less important for $R_{1\rho}$.

NMR methods of measuring the tracer diffusion coefficient are based on registration of an additional attenuation of the spin-echo signal due to displacement of nuclear spins in an external magnetic field gradient. The application of the field gradient serves to set a spatial scale in a certain direction. In most of the modern experiments, the magnetic field gradient is not static; it is applied in the form of two or more gradient pulses. The fundamental aspects of this pulsed field gradient (PFG) technique are discussed in [8]. The echo attenuation is usually measured as a function of the strength or the length of the gradient pulses. This attenuation is simply related to the displacement of nuclear spins along the direction of the gradient during the time interval between the gradient pulses. Therefore, the PFG technique provides a *direct* way of measuring the tracer diffusion coefficient D. In metal–hydrogen systems, the range of D values that can be measured by the PFG technique is typically 10^{-8}–10^{-4} cm^2 s^{-1}. The lower limit is determined mainly by the maximum values of the gradient available from the pulse circuitry and by the spin–spin (transverse) relaxation limiting the characteristic time of the echo decay. Modern NMR-PFG spectrometers can generate field gradients up to 250 mT kG cm^{-1} [27]. The most serious difficulty in applying the PFG technique to metal–hydrogen systems is the presence of the 'background' field gradients associated with nonuniform sample magnetization in a static magnetic field [8]. These 'background' gradients may be considerable for powder samples with high magnetic susceptibility.

The tracer diffusion coefficient and the jump rate are related by the expression

$$D = f_t L^2 / 6\tau_d \tag{26.5}$$

where L is the jump length of the migrating atoms and f_t is the tracer correlation factor. The value of f_t is very close to 1 for dilute metal–hydrogen systems, while for concentrated hydrides it may drop to about 0.7. Using the PFG measurements of D and the nuclear spin relaxation measurements of τ_d for the same sample, it is possible to evaluate the jump length L from Eq. (26.5). Thus, the two types of NMR experiments can be combined to obtain information on the diffusion path.

26.2.3
Quasielastic Neutron Scattering

One of the advantages of quasielastic neutron scattering (QENS) experiments over the relaxation and resonance measurements described above is that they provide information on spatial as well as temporal aspects of elementary processes of diffusion [14, 28, 29]. This is due to the fact that the wavelength of thermal and cold neutrons is comparable to interatomic distances, and at the same time their energy is of the order of typical solid state excitations. Furthermore, the neutron–nucleus interaction is weak. As a consequence neutrons penetrate deeply into matter and are sensitive to bulk properties; multiple scattering processes which, for example, dominate electron scattering because of the strong Coulomb interaction, are only second-order contributions. The proton exhibits a very large incoherent neutron scattering cross section ($\sigma_{inc} = 79.9$ barn) which is more than one

order of magnitude larger than the incoherent or coherent scattering cross sections of all other nuclei. Incoherent QENS contains information on self-diffusion.
Collective hydrogen diffusion, which can be elucidated from coherent QENS on
the D isotope, is beyond the scope of the present chapter.

For H in metals the measured QENS intensity, after the necessary raw data
corrections (background subtraction, detector efficiency, etc.) is proportional to
the incoherent scattering function $S_{inc}(\mathbf{Q}, \omega)$ which can be written as the two-fold
Fourier transform (in space and time) of the single-particle, space-time van Hove
correlation function, $G_s(\mathbf{r},t)$,

$$G_s(\mathbf{r},t) - \text{FT in space} \rightarrow I_s(\mathbf{Q},t) - \text{FT in time} \rightarrow S_{inc}(\mathbf{Q},\omega) \tag{26.6}$$

where $I_s(\mathbf{Q},t)$ denotes the self-part of the intermediate scattering function. $G_s(\mathbf{r},t)$
means the probability of finding a particle at $\mathbf{r}$ after time t has elapsed since it (the
same particle!) started from the origin ($\mathbf{r} = 0$).

For translational long-range jump diffusion of a lattice gas the stochastic theory
(random walk, Markov process and master equation) [30] eventually yields the
result that $G_s(r,t)$ can be identified with the solution (for a point-like source) of the
macroscopic diffusion equation, which is identical to Fick's second law of diffusion but with the tracer (self-diffusion) coefficient D instead of the chemical or
Fick's diffusion coefficient,

$$G_s(r,t) = (4\pi D\ |t|\ )^{-3/2} \exp\left(- r^2/4D\ |t|\right). \tag{26.7}$$

Spatial Fourier transformation of the self-correlation function (Eq. (26.7)) yields
the intermediate scattering function,

$$I_s(Q,t) = \exp\left(-DQ^2 t\right) \tag{26.8}$$

which is a measure of the probability of finding a H atom, which has been at a
certain site at $t = 0$, still at that site after time t has elapsed; after sufficiently long
time this probability vanishes. Subsequent temporal Fourier transformation of
the self-part of the intermediate scattering function (Eq. (26.8)) yields a Lorentzian
shape for the incoherent scattering function

$$S_{inc}(Q, \omega) = \frac{1}{\pi} \frac{\Lambda}{\Lambda^2 + (\hbar\omega)^2} \tag{26.9}$$

with the width (half width at half maximum, HWHM)

$$\Lambda = \hbar D Q^2 \tag{26.10}$$

Thus a plot of Λ vs. Q^2 yields a straight line, and from its slope the self-diffusion
coefficient is derived. This so-called Q^2 law (valid for sufficiently small Q) is indicative of translational diffusion.

In some intermetallic compounds clusters of isoenergetical sites occur which are well separated from other sites, see for instance the case of TaV_2H_x in Fig. 26.4, and which form closed loops; on those loops the H atom performs a spatially restricted jump motion (jump rotation) for some time until, by thermal fluctuations, eventually it is able to jump into the neighboring loop. For rotational diffusion over a loop of N sites after sufficiently long time the H atom can be found on any of the N sites with equal finite probability. The time-independent (i.e. long-time) contribution to $I_s(\mathbf{Q},t)$ yields an elastic contribution ($\omega = 0$) after temporal Fourier transformation:

$$I_s(\mathbf{Q}, \infty) = \frac{1}{N^2} \left| \sum_{j_{\text{sites}}} \exp(i\mathbf{Q} \cdot \mathbf{R}_j) \right|^2 = S_{\text{inc}}(\mathbf{Q}, 0) \qquad (26.11)$$

this quantity is called the elastic incoherent structure factor, EISF. Although incoherent neutron scattering is single-particle scattering, in this case it bears structural information, namely information about the spatial arrangement of the sites which the H atom visits in the course of its rotational motion.

Experimentally, the EISF is the fraction of the total 'quasielastic' intensity contained in the purely elastic peak. For an 'exact' experimental determination of the EISF the neutron spectra have to be fitted with the correct scattering function which consists of the elastic and a series of quasielastic terms (see [14, 28, 29]). Approximately, however, the EISF can be determined in a kind of model-independent data evaluation by fitting the neutron spectrum with a single Lorentzian plus an elastic term. Although in this way the line shape is not correct (except for $N = 2$ and 3), the EISF thus obtained allows statements to be made about the geometry for the localized motion. For the present chapter two EISFs, both spatially averaged (for powder samples), are of particular relevance:

$N = 2$ (dumb-bell)

$$\langle EISF_2 \rangle = \frac{1}{2} \left(1 + \frac{\sin(Qs)}{Qs} \right) = \frac{1}{2} (1 + j_0(Qs)) \qquad (26.12)$$

where s denotes the distance between the two sites;

$N = 6$ (hexagon)

$$\langle EISF_6 \rangle = \frac{1}{6} \left[1 + 2j_0(Qr) + 2j_0 \left(Qr\sqrt{3} \right) + j_0(2Qr) \right] \qquad (26.13)$$

where six sites are equally spaced on a circle of radius r.

The complete scattering functions for different situations of rotational diffusion can be found in Ref. [29].

QENS is a particularly powerful technique for the investigation of highly ordered metal hydrogen systems. A precondition is a single crystal as a sample, which for intermetallic compounds and their hydrides in most cases is not available up to now. The theoretical description is outlined in Section 26.4.1. QENS is also very useful for the study of diffusion in disordered systems. A very general and widely used approach is the so-called two-state model; this model was originally developed by Singwi and Sjölander [31] in order to describe the diffusion of water molecules in liquid water and later adapted by Richter and Springer [32] for the diffusion of hydrogen in metals in the presence of impurities. The resulting incoherent scattering function consists of two Lorentzians and bears information about the following spatial and temporal quantities:

τ^{-1} jump rate over undisturbed parts of the lattice
τ_0^{-1} escape rate from the impurities acting as traps
τ_1^{-1} trapping rate
ℓ jump length in the undisturbed past of the lattice
s distance between two traps

This model has been applied by Richter and Springer [32] for H diffusion in Nb doped with nitrogen impurities, by Hempelmann [33] for H diffusion in some intermetallic compounds and by Richter et al. [34] for H diffusion in amorphous metals.

26.2.4
Other Methods

The determination of diffusion coefficients D_{ch} from the evaluation of concentration profiles (mostly after a high temperature diffusion period and subsequent quenching to room temperature), as is common in the investigation of metal atom diffusion in metals or ion diffusion in ionic crystals, is hardly feasible for hydrogen in metals. A very special exemption is the *in situ* resistance measurement, where a unidirectional flow of hydrogen is detected by measuring the temporal variation of the electrical resistance of a number of sections of a long specimen [35]. Instead, two main types of experiments are possible which enable the macroscopic measurements of hydrogen diffusion in metals, namely relaxation methods and permeation methods.

In relaxation experiments, starting from the metal hydride sample in an "old" equilibrium, at $t = 0$ suddenly a "new" thermodynamic state is created, and it is observed how the sample approaches the "new" equilibrium. For this purpose, the time dependence of any physical property which is concentration dependent can be measured. A special example is the Gorsky effect (elastic after-effect) described in Section 26.2.2 . In an analogous way magnetic after-effect measurements are possible if the metal hydride is ferromagnetic [36]. These two techniques have the advantage that they are not influenced by surface dependent effects. In other cases the change in the chemical potential is produced by a change in the H concentration, initiated by a step-change of the hydrogen gas pressure. Then the time dependence of the hydrogen absorption process can be followed gravimetrically or volumetrically. In order to be able to extract the hydrogen diffusion coefficient from these data, bulk diffusion (and not, e.g., surface penetration) must be the rate determining step of the complex hydrogen absorption and desorption process. Gravimetric or volumetric measurements are useful for brittle systems which disintegrate into powder upon hydrogenation. If bulk diffusion is the slow step, the hydrogen content n_t at time t of spherical powder particles with radius r is obtained as a solution of Fick's second law [37]:

$$\frac{n_t}{n_\infty} = 1 - \frac{6}{\pi^2} \sum_{m=1}^{\infty} \frac{1}{m^2} \exp\left(-\frac{m^2\pi^2 D_{ch}t}{r^2}\right) \qquad (26.14)$$

If t is not too small, the higher-order terms can be neglected and the reaction rate constant is given by

$$k^{diff} = \pi^2 D_{ch}/r^2 \qquad (26.15)$$

In permeation methods, the temporal variation of a diffusion flow into or out of a specimen is measured after a sudden application or removal of an external driving force – the hydrogen pressure or the hydrogen electrochemical potential [38, 39]. A diffusion foil separates two vessels representing the "entrance" and "detection" side. A change in hydrogen concentration or in hydrogen influx rate at the entrance side – starting from a uniform hydrogen distribution within the foil – appears with a certain time lag at the detection side. For instance for a step-change of the surface H concentration at the "entrance" side and an electrochemical detection of the surface H concentration at the "detection" side, the time-lag or breakthrough time t_b is related to the diffusion coefficient by

$$t_b = 0.755 d^2/\pi^2 D_{ch} \qquad (26.16)$$

where d is the thickness of the foil. A common difficulty of permeation methods is that, in many cases, the surface of the sample acts as a barrier to the flow of hydrogen, and the overall flow rate depends on the surface conditions. This difficulty can be removed, to some extent, by sputter-deposition of a Pd film in ultrahigh vacuum [40]. Usually, possible surface effects are checked by performing experiments on samples of different thicknesses [41].

26.3
Experimental Results on Diffusion Coefficients

Extensive compilations of experimental data on hydrogen diffusion coefficients in binary metal–hydrogen systems have been published by Völkl and Alefeld [9] and Wipf [42]. These reviews can be referred to as sources of information on H diffusivities in different M–H systems. In this section we shall discuss some general features of H diffusivity in metals resulting from numerous experimental studies.

For most of the studied metal–hydrogen systems, the temperature dependence of the measured tracer diffusion coefficient of hydrogen follows the Arrhenius law,

$$D = D_0 \exp(-E_a/k_B T) \qquad (26.17)$$

over wide ranges of T. Here E_a is the activation energy for hydrogen diffusion. Usually Eq. (26.17) is associated with the classical over-barrier jump mechanism of diffusion, E_a being the height of the energy barrier between the potential wells at the nearest-neighbor interstitial sites. However, in many metal–hydrogen systems the experimental values of E_a appear to be considerably lower than the actual depth of the corresponding potential wells. This indicates that tunneling through the barriers should play an important role in the processes of hydrogen diffusion in metals, even at temperatures of the order of room temperature. In fact, the quantum-mechanical theories of diffusion of light particles [10, 43] predict a temperature dependence of the form of Eq. (26.17) for phonon-assisted tunneling at temperatures above about half of the Debye temperature. However, the nature of E_a in these quantum-mechanical approaches differs from that in the classical theory; moreover, the apparent value of E_a in the regime of so-called nonadiabatic transitions (lower temperatures) is expected to be lower than that in the regime of adiabatic transitions (higher temperatures) [10, 43].

We shall emphasize here the following features of hydrogen diffusivity in metals: (i) the strong dependence of D on the host-metal structure, (ii) deviations from the Arrhenius-type temperature dependence of D, and (iii) isotope effects on hydrogen diffusivity. Comparison of the available experimental data on the hydrogen diffusivity (at low H concentrations) in b.c.c. and f.c.c. metals shows that the values of D in b.c.c. metals are generally much higher than in f.c.c. metals, and the corresponding activation energies for H diffusion in b.c.c. metals are lower than in f.c.c. metals. For example, the measured value of $D(300\text{ K})$ in b.c.c. V is 5.4×10^{-5} cm^2 s^{-1} and $E_a = 0.045$ eV (143–667 K), whereas in f.c.c. Pd the corresponding values are $D(300\text{ K}) = 4.0 \times 10^{-7}$ cm^2 s^{-1} and $E_a = 0.23$ eV (230–900 K) [42]. The same trend has also been found for metals which can exist in both b.c.c. and f.c.c. modifications. In fact, the transition from the high-temperature γ (f.c.c.) phase to the low-temperature α (b.c.c.) phase of Fe leads to considerable increase in the H diffusivity and to an order of magnitude decrease in E_a [9]. Even more spectacular results have been reported [9, 44] for hydrogen diffusion in the Pd$_{0.47}$Cu$_{0.63}$ alloy that can have either an ordered b.c.c. structure or a disordered f.c.c. structure in the same range of T including room temperature. The measured value of $D(300\text{ K})$ in the b.c.c. modification appears to be 4 orders of magnitude higher than that in the f.c.c. modification, and the value of E_a in the b.c.c. phase is an order of magnitude lower than that in the f.c.c. phase. The strong difference between the H diffusion parameters in b.c.c. and f.c.c. metals is believed to originate mainly from the geometry of hydrogen sublattices. In b.c.c. metals hydrogen usually occupies tetrahedral interstitial sites which form the sublattice with short nearest-neighbor distances (1.0–1.2 Å). Therefore, the energy barriers between such sites are expected to be low, and the matrix elements for hydrogen tunneling may be relatively large. On the other hand, in f.c.c. metals hydrogen usually occupies octahedral interstitial sites; the nearest-neighbor distances between these sites are at least a factor of two greater than those for the tetrahedral sites in b.c.c. metals.

High H diffusivities and low activation energies have also been found for hydrogen dissolved in many Laves-phase intermetallic compounds [45]. The sublattice of interstitial sites partially occupied by hydrogen in Laves phases is characterized by short nearest-neighbor distances (1.0–1.2 Å). Thus, short intersite distances are definitely favorable for the occurrence of fast hydrogen diffusion. The geometry of the hydrogen sublattice in Laves phases will be discussed in Section 26.4.2.

For a number of metal–hydrogen systems the temperature dependence of D has been found to deviate from the Arrhenius law at low T. As an example of the data, Fig. 26.1 shows the temperature dependences of the diffusion coefficients of hydrogen isotopes H, D and T in b.c.c. metals V, Nb and Ta, as determined from the Gorsky effect measurements [46]. All the data correspond to very low hydrogen concentrations. It can be seen that there is a marked break in the slopes of the Arrhenius plots for H diffusion in Nb and Ta near 200–250 K. A somewhat less pronounced break in the slope of the Arrhenius plot is also observed for H diffusion in V near 300 K. However, for the diffusion of heavier hydrogen isotopes (D and T) in these metals such breaks in the slope have not been found. It is believed that the break in the slope of the Arrhenius plots for H diffusion reflects the transition from the adiabatic regime of tunneling at high temperatures to nonadiabatic tunneling at lower temperatures. As mentioned above, such a transition is expected to result in a decrease in the apparent activation energy. For heavier hydrogen isotopes this transition should occur at considerably lower temperatures (presumably, out of the actual experimental temperature range).

The break in the slope of the Arrhenius plots has also been found for H diffusion in a number of Laves-phase intermetallic compounds where the hydrogen diffusivity is very high. As an example of the data, Fig. 26.2 shows the behavior of D for hydrogen diffusion in the cubic (C15-type) Laves-phase compounds $ZrCr_2H_{0.2}$ and $ZrCr_2H_{0.5}$, as determined from the pulsed-field-gradient NMR measurements [47]. It can be seen that the change in the slope of the Arrhenius plots for these compounds occurs below ~ 200 K. The solid lines in Fig. 26.2 represent the fits of the sum of two Arrhenius-like terms (with different activation energies and pre-exponential factors) to the data. The origin of this effect is assumed to be the same as in the case of H diffusion in b.c.c. metals.

We now turn to the discussion of isotope effects in hydrogen diffusivity. The classical diffusion theory predicts that the pre-exponential factor D_0 in Eq. (26.17) is inversely proportional to the square root of the mass of a diffusing particle, while the activation energy does not depend on this mass. According to these predictions, the diffusion coefficient of D atoms should be $\sqrt{2}$ times lower than that of H atoms over the entire temperature range. However, the isotope dependence of hydrogen diffusivity in all metal–hydrogen systems studied so far shows deviations from the predictions of the classical theory. In particular, the measured effec-

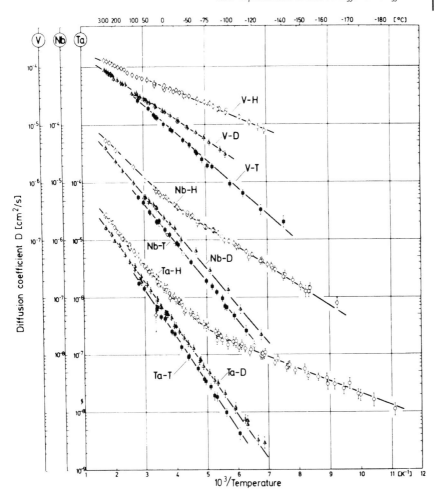

Figure 26.1 The diffusion coefficients of hydrogen isotopes H, D and T in b.c.c. metals V, Nb and Ta as functions of the inverse temperature (from Ref. [46]). Note that the vertical scales for the three metals are shifted with respect to each other.

tive activation energy appears to depend on the mass of a diffusing particle. This effect is clearly seen for b.c.c. metals V, Nb and Ta (Fig. 26.1) where $E_a^H < E_a^{D,T}$. Because of the difference between the activation energies for H and D, the ratio of the diffusion coefficients for H and D in these metals increases with decreasing temperature. These results are consistent with the concept of phonon-assisted hydrogen tunneling. Moreover, the numerical calculations based on this concept [43, 48, 49] show a surprisingly good agreement with the experimental data on hydrogen diffusion coefficients in Nb and Ta.

The isotope dependence of the hydrogen diffusivity in f.c.c. metals differs from that in the case of b.c.c. metals. While the ratio of the pre-exponential factors

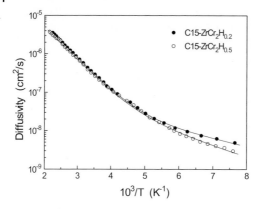

Figure 26.2 The hydrogen diffusivities in C15-type $ZrCr_2H_{0.2}$ and $ZrCr_2H_{0.5}$ as functions of the inverse temperature (from Ref. [47]). The solid lines represent the fits of the sum of two Arrhenius terms to the data.

D_0^H/D_0^D for f.c.c. metals is usually close to $\sqrt{2}$ (as predicted by the classical theory), the activation energy for deuterium in a number of f.c.c. metals appears to be *lower* than that for H. For example, the activation energy for deuterium in Pd derived from the Gorsky effect measurements [50] is 0.206 eV (~220–350 K). This value should be compared to that for H in Pd, $E_a^H = 0.23$ eV. Because of the inequality $E_a^D < E_a^H$, the diffusion coefficient of the heavier isotope becomes *higher* than that of the lighter isotope below a certain temperature (~ 600 K for Pd). The diffusivity of tritium in Pd near room temperature has also been found to exceed that of H [51]. The inequality $E_a^D < E_a^H$ has also been reported for hydrogen diffusion in other f.c.c. metals (Cu, Pt, Ni) [42, 52]. A plausible qualitative explanation of the inverse isotope effect on E_a is based on the idea that some excited vibrational states of hydrogen may play an important role in the process of transitions between the interstitial sites in f.c.c. metals [53]; the excitation energy for D is expected to be lower than that for H.

In contrast to binary metal–hydrogen systems, little is known about isotope effects on the hydrogen diffusivity in intermetallic compounds. In most of the studied Laves-phase hydrides [54–58] the long-range H(D) diffusion is characterized by the inequality $E_a^D > E_a^H$; in this case the diffusive motion of D atoms is slower than that of H atoms (normal isotope effect). The inverse isotope effect corresponding to the inequality $E_a^D < E_a^H$ has been found only for concentrated Laves-phase hydrides $HfV_2H_x(D_x)$ and $ZrV_2H_x(D_x)$ with $x > 3.5$ [55]. However, the inverse isotope effect in these systems is likely to result from the difference between the occupancies of inequivalent sites for the hydrides and deuterides [55]. Thus, the origin of this inverse isotope effect may differ from that in f.c.c. metals.

26.4
Experimental Results on Hydrogen Jump Diffusion Mechanisms

Detailed atomistic information of the elementary process of jump diffusion can be obtained from QENS at Q values comparable with the inverse jump distances, i.e. at 'large' Q [14, 28, 29]. Hydrogen in the a-phases of the cubic f.c.c. and b.c.c. metals Pd, Nb, Ta and V was investigated very early, and the quite detailed results which are now available will be outlined first in Section 26.4.1. One of the rare cases where a hydride single crystal was available is β-V$_2$H, and the corresponding QENS study will be presented in some more detail in the second part of Section 26.4.1.

The theoretical basis of all these QENS studies is the Chudley–Elliott model [59], which in its basic form involves the following postulates.

1. All sites involved in the diffusion process are crystallographically and thus also energetically equivalent (Bravais sublattice).
2. All jumps (to nearest neighbor sites only) have the same jump length ℓ and are characterized by jump vectors s_i, $i = 1$, .., z where z is the coordination number and $|s_i| = \ell$ for all i.
3. The particle stays at a site for a mean residence time τ; then instantaneously it jumps to a neighboring site; i.e. the jump time is negligibly small compared to the residence time; $\Gamma = (z\tau)^{-1}$ denotes the jump rate.
4. Successive jumps are uncorrelated, i.e. the jump direction is completely random.

This physical picture has been cast into a mathematical model in the form of a so-called master equation

$$\frac{\partial}{\partial \tau} P(\mathbf{r}, t) = -\frac{1}{z\tau} \sum_{j=1}^{z} \left\{ P(\mathbf{r}, t) - P\left(\mathbf{r} + \mathbf{s}_j, t\right) \right\} \tag{26.18}$$

which says that the temporal change of the occupation probability P of site $\mathbf{r}$ (left-hand side of Eq. (26.18)) is due to jumps away from that site (first term on the right-hand side of Eq. (26.18)) and jumps into that site from the neighboring sites $\mathbf{r} + \mathbf{s}_j$ which all exhibit the same occupancy P. Solution of Eq. (26.18) in Fourier space and subsequent spatial Fourier transformation yields a single Lorentzian for the incoherent scattering function (Eq. (26.9)) with the HWHM

$$\Lambda(\mathbf{Q}) = \frac{\hbar}{z\tau} \sum_{j=1}^{z} \left(1 - \exp(-i\mathbf{Q}\mathbf{s}_j)\right) \tag{26.19}$$

In this basic form the Chudley–Elliott model describes the H diffusion in f.c.c. Pd.

A first generalization of the Chudley–Elliott model has been developed in order to deal with crystallographically different hydrogen sites. Crystallographically different sites occur, e.g., for H diffusion over the tetrahedral interstices in b.c.c. metals like Nb. In this case the hydrogen sublattice consists of six superimposed b.c.c.

lattices, and the single master equation (Eq. (26.18)) becomes a six-fold differential equation system, which also can be solved in Fourier space yielding a 6×6 Hermitean jump matrix and eventually an incoherent scattering function consisting of a superposition of six Lorentzians; the negative eigenvalues of the jump matrix are the linewidths whereas the respective weights are given by the eigenvectors. It is practically impossible to fit an individual experimental QENS spectrum with six Lorentzians and to determine six half-widths in this way, but one has to record many QENS spectra (at many $|Q|$ values) in a number of different crystallographic directions and then perform one simultaneous fit of the Q and ω dependences in order to extract the atomistic diffusional information.

A further generalization of the Chudley–Eliott model allows the sites to be not only crystallographically but also energetically different. In this case the forth and back jump rates Γ_{mn} and Γ_{nm} between two sites m and n are not equal; they are, however, related by the detailed balance condition to the different site energies,

$$\Gamma_{nm}\exp\left(\frac{-E_n}{k_B T}\right) = \Gamma_{mn}\exp\left(\frac{-E_m}{k_B T}\right) \tag{26.20}$$

where E_m and E_n are the potential energies of sites m and n, respectively. The system of differential equations (master equation) can again be transformed into a jump matrix which in this case, however, is not Hermitean; but anyway eigenvalues and eigenvectors can be determined and transformed into the line widths and weights of the set of Lorentzians which represents the incoherent scattering function. Experimentally, again a single crystal as a sample and QENS spectra recorded in different crystallographic directions are necessary to extract atomistic diffusional information which in this case also involves the energetical differences of the sites involved.

26.4.1
Binary Metal–Hydrogen Systems

The very first quasielastic neutron scattering experiments to study the diffusive motion of hydrogen in metals were performed by Sköld and Nelin [60] on polycrystalline Pd in the diluted disordered α-phase where the hydrogen atoms occupy octahedral sites. Later Rowe et al. [61] performed a more detailed study in two symmetry directions (<100> and <110>) using single crystals. The resulting line widths are in excellent agreement with the predictions of the Chudley –Elliott model. For reasons of comparison with other methods the resulting mean residence times τ taken from a number of quasielastic neutron scattering experiments and from permeation Gorsky effect and NMR measurements are extrapolated to a hydrogen concentration $c = 0$. The comparison shows that there is fairly good agreement and consistency of these data.

In contrast to f.c.c. Pd, striking discrepancies were found for $A(Q)$ for hydrogen in the α-phases of b.c.c. metals, in particular in Nb where hydrogen occupies tetrahedral sites. For hydrogen in the α-phases of b.c.c. metals it was observed by Rowe et al. [62, 63], and Lottner et al. [64, 65] that the experimental linewidths $\Lambda(Q)$ at large Q are considerably lower than expected according to the Chudley–Elliott model. This could formally be attributed to effective jump distances considerably larger than expected for jumps into adjacent sites. Lottner et al. [64] have proposed a two-state model: the hydrogen atom alternates between a state of high mobility, the so-called free state, and an immobilized state, the trapped state; in the free state it stays for a time τ_1 and performs during this time a sequence of 'transport' jumps to the respective nearest neighbour sites with 'transport jump rate' τ^{-1}; in the trapped state the hydrogen stays well localized on a site for a relatively long time τ_0. The corresponding change-over rates are the trapping rate τ_1^{-1} and the escape rate τ_0^{-1}. The physical reason for this seemingly strange behavior is the metal lattice relaxation around an occupied hydrogen site (local lattice expansion) which lowers the local site energy but takes some time. The relaxed sites represent the traps; once the hydrogen has managed (by thermal fluctuations) to escape from such a site, it jumps quickly over unrelaxed sites until, after time τ_1, somewhere apart again lattice relaxation takes place and the hydrogen is trapped.

At high hydrogen concentration metal–hydrogen systems form ordered hydrides, particularly at lower temperatures. β-V$_2$H is an example, and its structure is shown in Fig. 26.3. The hydrogen atoms occupy sets of octahedral sites with nearly tetragonal point symmetry which form sheets in (110) directions in pseudocubic notation. The actual symmetry is monoclinic [66]. In this way occupied sheets alternate with empty layers, see Fig. 26.3. A vacancy diffusion mechanism on the hydrogen sublattice, as mentioned above, is ineffective in this case. Actually, as will be explained below, hydrogen diffusion in β-V$_2$H proceeds via an interstitial mechanism on two interstitial sublattices: some hydrogen atoms

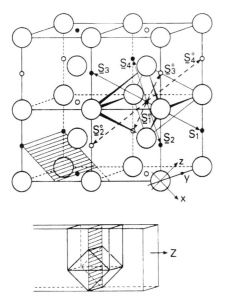

Figure 26.3 Structure of β-V$_2$H, explanation in the text.

occupy sites in the forbidden layer, i.e. anti-structure sites or, in other words, they form Frenkel defects. These are the hydrogen atoms which are significant for the diffusion process; usually, however, they cannot be investigated by means of QENS since their number is too small. In β-V$_2$H, however, the situation is fortunate: with increasing temperature more and more of the sites in the 'forbidden' layers are occupied, and at 448 K an order–disorder transition to the ε-phase occurs in which the hydrogen atoms are randomly distributed over both types of layers. At 390 K $\leq T \leq$ 440 K, the temperatures of the QENS study of Richter et al. [67], the number of H atoms on anti-structure sites turned out to be sufficient to be detectable by means of QENS. The master equation system for H diffusion in β-V$_2$H is given by

$$\frac{\partial P(\mathbf{r},t)}{\partial t} = -\left(\frac{1}{\tau}+\frac{1}{\tau_0}\right)P(\mathbf{r},t) + \frac{1}{4\tau}\sum_{i=1}^{4} P(\mathbf{r}+\mathbf{S}_i,t) + \frac{1}{4\tau_1}\sum_{i=1}^{4} U\left(\mathbf{r}+\mathbf{S}_i^0,t\right) \quad (26.21)$$

$$\frac{\partial U(\mathbf{r},t)}{\partial t} = -\left(\frac{1}{\tau_u}+\frac{1}{\tau_1}\right)U(\mathbf{r},t) + \frac{1}{4\tau_u}\sum_{i=1}^{4} U(\mathbf{r}+\mathbf{S}_i,t) + \frac{1}{4\tau_0}\sum_{i=1}^{4} P\left(\mathbf{r}+\mathbf{S}_i^0,t\right) \quad (26.22)$$

Here $P(\mathbf{r},t)$ and $U(\mathbf{r},t)$ denote the probabilities of finding a hydrogen in the occupied and unoccupied layers, respectively; τ^{-1} and τ_u^{-1} are the jump rates within the occupied and unoccupied layers, respectively (both along the jump vectors $\mathbf{S}_1$ to $\mathbf{S}_4$), and τ_0^{-1} and τ_1^{-1} are the change-over rates from the occupied to the empty layer and vice versa, respectively (both along the jump vectors $\mathbf{S}_1^0$ to $\mathbf{S}_4^0$). This jump model yields a 2×2 jump matrix and subsequently an incoherent scattering function consisting of two Lorentzians.

The quantitative data evaluation in terms of this model allowed the derivation of the four jump rates and led to the following picture: Let us consider a hydrogen atom starting in the filled layer. Then, depending on temperature (in the temperature range 390 K $\leq T \leq$ 440 K), the jump probability to change into the empty layer is 6 to 3 times higher than to move among the occupied sites. If the hydrogen has changed into the empty layer, then on average it performs 5 to 2 jumps before it drops back into a vacancy in one of the adjacent filled layers. Thus the diffusion parallel to the sheets consists of repeated trapping and release processes between the filled and empty layers. In particular, at lower temperatures, where the ratio between the fast jump rate in the empty layer and all other rates is the largest, hydrogen diffusion is strongly anisotropic.

26.4.2
Hydrides of Alloys and Intermetallic Compounds

In alloys and intermetallic compounds the sublattices of interstitial sites usually have rather complex structures. Furthermore, hydrogen atoms dissolved in alloys and intermetallics can occupy a number of inequivalent types of interstitial sites.

These features may give rise to a coexistence of several types of H motion with different characteristic jump rates. Here we shall discuss the experimental results on hydrogen jump diffusion mechanisms in a number of representative interme- tallic compounds. A comprehensive review of the experimental studies in this field before 1992 can be found in Ref. [11].

Among intermetallics, hydrogen diffusion has been most extensively studied in Laves phases AB_2 which can have either the cubic (C15-type) or the hexagonal (C14-type or C36-type) structures. Many of the Laves-phase compounds are known to absorb considerable amounts of hydrogen forming solid solutions AB_2H_x with wide homogeneity ranges [68, 69]. NMR experiments on $TiCr_{1.8}H_x$ (C15) and $TiCr_{1.9}H_x$ (C14) [70] and on C15-type ZrV_2H_x and HfV_2H_x [71] have revealed sig- nificant deviations of the measured nuclear spin relaxation rates from the behav- ior expected for a single frequency scale of H motion. The unambiguous evidence for the coexistence of two hydrogen jump processes with different characteristic frequencies in a cubic Laves phase was found in the series of NMR measurements on the C15-type TaV_2 – H(D) system [54, 72, 73]. These measurements have revealed the temperature dependence of the nuclear spin–lattice relaxation rate R_1 with two well-separated peaks.

The two frequency scales of H jump motion may be related to the structural features of the sublattice of interstitial sites. In cubic Laves phases, H atoms usually occupy only 96g sites (coordinated by $[A_2B_2]$ tetrahedra) at low and inter- mediate hydrogen concentrations (up to $x \approx 2.5$) [68, 69]. In particular, only g sites are occupied by H(D) atoms in $TaV_2H_x(D_x)$ over the entire range of attainable H(D) concentrations ($x \leq 1.7$) [74]. For cubic Laves phases absorbing greater amounts of hydrogen, 32e sites (coordinated by $[AB_3]$ tetrahedra) start to be filled at $x > 2.5$. The spatial arrangement of interstitial g and e sites in the C15-type lat- tice is shown in Fig. 26.4. The sublattice of g sites consists of regular hexagons lying in the planes perpendicular to the <111> directions. Each g site has three nearest neighbors: two g sites (on the same hexagon) at a distance r_1 and one g site (on the adjacent hexagon) at a distance r_2. The ratio r_2/r_1 is determined by the positional parameters (X_g and Z_g) of hydrogen atoms at g sites. Examination of the available neutron diffraction data for cubic Laves-phase deuterides reveals strong changes in the ratio r_2/r_1 from one compound to another. For example, the value of r_2/r_1 is 1.45 for TaV_2D_x [76], 1.07 for $ZrCr_2D_x$ [77] and 0.78 for YMn_2D_x [78]. In the case of $TaV_2D_x(H_x)$, the g-site hexagons are well separated from each other. Therefore, a hydrogen atom is expected to perform many jumps within a hexagon before jumping to another hexagon. In this case, the faster jump rate τ_l^{-1} can be attributed to the localized hydrogen motion within g-site hexagons, and the slower jump rate τ_d^{-1} can be associated with hydrogen jumps from one g-site hexagon to another.

The results of QENS measurements for TaV_2H_x [76] are consistent with this mi- croscopic picture of H motion. First, on the frequency scale of τ_l^{-1} the measured QENS spectra $S(Q, \omega)$ are well described by the sum of a narrow elastic line and a broader quasielastic line having Q-dependent intensity, but Q-independent width. These features are typical of the case of spatially-confined (localized) motion [14].

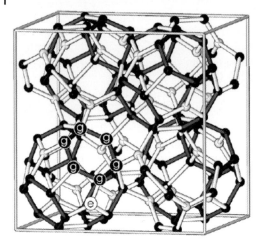

Figure 26.4 The spatial arrangement of interstitial
g sites (dark spheres) and *e* sites (light spheres)
in the C15-type lattice [75].

Second, the Q-dependence of the measured elastic incoherent structure factor
(EISF) appears to be in excellent agreement with the predictions of the model of
localized atomic motion over a hexagon (Eq. (26.13)) with the distance between
the nearest-neighbor sites equal to the experimental r_1 value. As an example of
these results, Fig. 26.5 shows the behavior of the EISF for $TaV_2H_{1.1}$ as a function
of Q at several temperatures. The solid curves represent the fits of the six-site
model to the data. In these fits the distance between the nearest-neighbor sites
has been fixed to its value resulting from the structure, $r_1 = 0.99$ Å, so that the

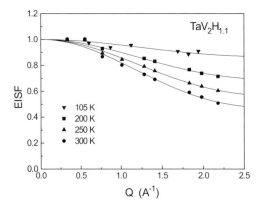

Figure 26.5 The elastic incoherent structure factor for
$TaV_2H_{1.1}$ as a function of Q at $T = 105, 200, 250$ and 300 K [76].
The solid lines represent the fits of the six-site model with the
fixed $r_1 = 0.99$ Å to the data.

only fit parameter is the fraction p of H atoms participating in the fast localized motion. The results presented in Fig. 26.5 indicate that the fraction p increases with increasing temperature.

A similar microscopic picture of hydrogen motion has been observed for other cubic Laves-phase hydrides with exclusive g-site occupation and $r_2/r_1 > 1$: $ZrCr_2H_x$ [77], $ZrMo_2H_x$ [79], $HfMo_2H_x$ [80] and ZrV_2H_x [81]. However, in these systems the difference between the two frequency scales of H motion appears to be smaller than in TaV_2H_x. For example, at 300 K the value of τ_d/τ_l is 5.2×10^3 for TaV_2H_x, 240 for ZrV_2H_x and 20 for $ZrCr_2H_x$ [45]. This decrease in τ_d/τ_l correlates with the decrease in r_2/r_1 [45]. Furthermore, the observed variations of r_2/r_1 caused by changes in the positional parameters of hydrogen atoms at g sites can be rationalized in terms of the metallic radii R_A and R_B of elements A and B forming the AB_2 intermetallic. In fact, since g sites are coordinated by two A and two B atoms, one may expect that the positional parameters X_g and Z_g (and hence, r_2/r_1) are related to R_A/R_B. The experimental values of r_2/r_1 for paramagnetic C15-type hydrides $AB_2H_x(D_x)$ where both A and B are transition metals exhibit nearly a linear decrease with increasing R_A/R_B [45]. Thus, the ratio R_A/R_B gives a key to understanding the systematics of the two frequency scales of H motion in cubic Laves-phase hydrides. In particular, the highest value of τ_d/τ_l for TaV_2H_x can be related to the anomalously low R_A/R_B ratio (= 1.090) for TaV_2.

For cubic Laves-phase compounds with $R_A/R_B > 1.35$, the r_2/r_1 ratio becomes less than 1. In this case, each g site has only one nearest neighbor lying at the adjacent hexagon. Such a transformation of the g-site sublattice may lead to a qualitative change in the microscopic picture of H jump motion: the faster jump process is expected to be transformed into the back-and-forth jumps within *pairs* of g sites belonging to adjacent hexagons. The results of recent QENS experiments [78] on YMn_2H_x ($R_A/R_B = 1.425$, $x = 0.4$, 0.65 and 1.26) are consistent with these expectations.

At high H content ($x > 2.5$) hydrogen atoms start to occupy e sites in C15-type hydrides, the relative occupancy of e sites increasing with x. Each e site has three nearest-neighbor g sites (see Fig. 26.4) at a distance r_3 comparable to the g–g distances r_1 and r_2; the exact value of r_3 depends on the positional parameters of H atoms at e sites (X_e) and g sites (X_g, Z_g). The partial e-site filling makes the microscopic picture of H motion less tractable. However, one may generally expect that the partial occupation of e sites leads to an increase in the long-range H mobility due to the opening of new diffusion paths. This effect is well documented for ZrV_2H_x [55, 82].

The coexistence of at least two frequency scales of hydrogen jump motion has also been found in a number of hexagonal (C14-type) Laves-phase hydrides [77, 83, 84]. As an example of the data, Fig. 26.6 shows the temperature dependences of τ_l^{-1} and τ_d^{-1} obtained from QENS measurements for C14-type $HfCr_2H_{0.74}$ [84]. It can be seen that in the studied temperature range the behavior of both $\tau_l^{-1}(T)$ and $\tau_d^{-1}(T)$ is satisfactorily described by the Arrhenius relation; the values of the activation energies derived from the Arrhenius fits are 122 meV for τ_l^{-1} and 148 meV for τ_d^{-1}. The microscopic picture of H motion in C14-type compounds

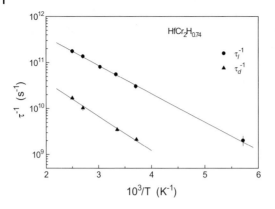

Figure 26.6 The hydrogen jump rates τ_l^{-1} and τ_d^{-1} in C14-type HfCr$_2$H$_{0.74}$ as functions of the inverse temperature. The jump rates are obtained from QENS experiments [84]. The solid lines are the Arrhenius fits to the data.

has been addressed in the QENS study of ZrCr$_2$H$_x$ [77] (note that ZrCr$_2$ may exist in the form of either the hexagonal or the cubic C15 modification). At low hydrogen concentrations H atoms occupy the tetrahedral sites with [A$_2$B$_2$] coordination. In contrast to the C15 structure where all [A$_2$B$_2$] sites are equivalent (g sites), in the C14 structure there are four inequivalent types of [A$_2$B$_2$] site (h_1, h_2, k and l). The sublattice of [A$_2$B$_2$] sites in the C14 structure [85] also consists of hexagons; however, these hexagons are formed by inequivalent sites: $h_1 - h_2 - h_1 - h_2 - h_1 - h_2$ (type I hexagons) and $k - l - l - k - l - l$ (type II hexagons). All the distances between the nearest sites within the hexagons appear to be shorter than the distances between the nearest sites on different hexagons [77]. Therefore, the general features of the microscopic picture of H motion in C14-ZrCr$_2$H$_x$ are expected to be similar to those of H motion in C15-type compounds with g-site occupation and $r_2/r_1 > 1$. The experimental QENS results for C14-ZrCr$_2$H$_{0.5}$ [77] have been interpreted in terms of the model neglecting the small difference between type I and type II hexagons and the difference between the l–l and k–l distances in type II hexagons. The observed Q-dependence of the EISF for C14-ZrCr$_2$H$_{0.5}$ is well described by the model of localized H motion over hexagons with the intersite distance $<r> = 1.16$ Å, where $<r>$ is the weighted average of the intersite distances for type I and type II hexagons. Since the sublattice of [A$_2$B$_2$] sites in C14-type compounds is more complex than that in C15-type compounds, the detailed microscopic picture of H motion in hexagonal Laves phases may imply more than two frequency scales. In order to clarify the systematics of H jump processes in hexagonal Laves-phase hydrides, further QENS experiments (combined with neutron diffraction studies of hydrogen positions) are required.

Another important class of hydrogen-absorbing intermetallics is represented by LaNi$_5$ and the related materials (the hexagonal CaCu$_5$-type structure). The detailed microscopic picture of H motion in this lattice has been investigated by QENS on single-crystalline samples of a-LaNi$_5$H$_x$ [86]. Figure 26.7 shows the structure of a-LaNi$_5$H$_x$. Hydrogen atoms can occupy two types of sites: $3f$ (corresponding to lower site energy) and $6m$ (higher site energy). While the sublattice of $3f$ sites forms infinite layers in the basal planes, the sublattice of $6m$ sites consists of regular hexagons well separated from each other (Fig. 26.7). The sublattices of f and m sites are interconnected by jump vectors such that a diffusing H atom can reach eight m sites from each f site and four f sites from each m site. Since the studied a-LaNi$_5$H$_x$ samples were single-crystalline, the analysis of the QENS data [86] benefited from the possibility to direct the neutron momentum-transfer vector $\mathbf{Q}$ either parallel or perpendicular to the c axis. It has been found that the faster jump process in this system corresponds to localized H motion within the hexagons formed by m sites. The slower frequency scales are associated with H jumps between the sublattices of f and m sites and with H jumps in the f-site layers.

Another example of an interesting H jump diffusion mechanism has been reported for hydrogen dissolved in the cubic A15-type compound Nb$_3$Al [87]. In this compound H atoms occupy the tetrahedral $6d$ sites coordinated by four Nb atoms. The $6d$ sites form three sets of nonintersecting chains in the <100>, <010> and <001> directions. The distance between the nearest-neighbor d sites in the

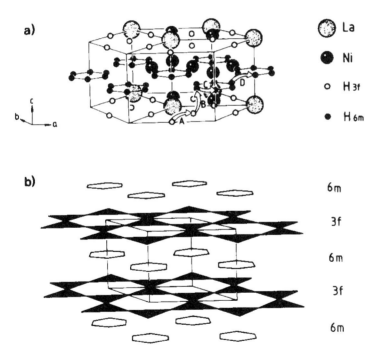

Figure 26.7 (a) The structure of a-LaNi$_5$H$_x$. A–D denote four possible jumps between neighboring interstitial sites. Host metal atoms are partially omitted for clarity. (b) The geometrical representation of the hydrogen sublattice. The $3f$ sites form infinite layers in the basal plane. The $6m$ sites are grouped to form regular hexagons in the $z = 1/2$ plane (from Ref. [86]).

chains is 22% shorter than the shortest distance between d sites on different chains. In this case the faster jump process corresponds to one-dimensional H diffusion along the chain, while the slower process implies H jumps from one chain to another. Other well-documented examples of two coexisting frequency scales of H jump motion include the high-temperature cubic phase of Mg_2NiH_4 [88], the cubic Ti_2Ni-type compounds Ti_2CoH_x [89] and the rhombohedral Th_2Zn_{17}-type compound $Pr_2Fe_{17}H_5$ [90]. The motion of hydrogen dissolved in disordered alloys is usually described in terms of broad distributions of H jump rates [91, 92].

26.4.3
Hydrogen in Amorphous Metals

Hydrides of amorphous metals have been investigated to some extent in the last twenty years, mainly for their potential use in hydrogen storage technology; see, for instance Ref. [93] and references therein. Pressure–composition isotherms, for example of Zr–Ni alloys, deviate strongly from Sievert's law at higher H concentration. These positive deviations have been attributed to a distribution of site energies in the amorphous structure, and H atoms entering successively higher-energy states. This behavior is reviewed by Kirchheim [93] and brought into correlation with diffusion. Generally, interstitial H diffusion is more or less strongly dependent on the H concentration, where this dependence is due to site energy disorder giving rise to a saturation effect of low-energy sites. In the simplest approach, a Gaussian site-energy distribution is assumed [39]. Additionally it is assumed that the energy levels of the potential barriers (in the picture of classical over-barrier jumps) have the same energy value throughout the sample, i.e. do not exhibit an energetic distribution (constant saddle-point energy). In the literature this model is known as the Gaussian model.

Kondratyev et al. [94] have analyzed the available experimental data of hydrogen diffusion in amorphous alloys and presented a review of the existing theoretical approaches noting that the influence of short range order on the hydrogen diffusion was not properly taken into account in previous studies. They propose a model with specific features of the respective amorphous structure and derive general expressions for the diffusion coefficient of hydrogen in amorphous metals and binary alloys with f.c.c.-like short-range crystalline order.

QENS studies of H in amorphous metal hosts give contradictory results. While Schirmacher et al. [95] interpret their results on H diffusion in amorphous (a) $Zr_{76}Ni_{24}$ in terms of a broad continuous distribution of activation energies, Richter et al. [34] find the existence of energetically well-separated interstitial sites in a-$Pd_{85}Si_{15}H_{7.5}$. So-called anomalous diffusion [96] means that in the time scale of interest the mean square distance walked by the diffusing particle increases sublinearly instead of linearly with time. Also the conjecture that the network of diffusion paths exhibits fractal character leads to such a sublinear time dependence: $\langle r^2(t)\rangle \propto t^{2/(2+\Theta)}$, where Θ describes the range dependence of the diffusion coefficient on the fractal network $D \propto r^{-\Theta}$ [97]. For the Q dependence of the quasielastic linewidth Λ this implies $\Lambda \propto Q^{2+\Theta}$, i. e. Λ is expected to grow faster than Q^2. The experimental observation of $\Lambda \propto Q^{1.54}$ does not support the assumption of anomalous diffusion or of a dominating fractal structure for the diffusion paths. The QENS data on amorphous $Pd_{85}Si_{15}H_{7.5}$ could successfully be evaluated in terms of a diffusion and trapping model. Figure 26.8 shows the Q dependences of the weight of the narrow component and of both line widths. The temperature dependence of the weight shows that the hydrogen atoms are activated from energetically more stable sites, which could be called traps, into a state of high mobility.

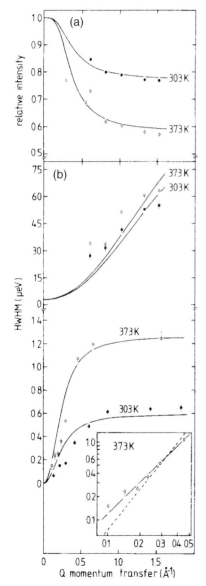

Figure 26.8 (a) Weight of narrow QENS component of H in a-$Pd_{85}Si_{15}H_{7.5}$ as a function of temperature and momentum transfer. (b) T and Q dependences of the linewidths. The solid lines represent a fit of the two-state model to these data. Inset: magnification of the small Q behaviour of the linewidths at 373 K (dashed line: HWHM $\propto Q^2$) (from Ref. [34]).

From the Q dependence a mobility range of about 10 Å is obtained. For the mobile state the isotropic Chudley–Elliott model is assumed. The result of a fit with this two-state model is shown by the solid line in Fig. 26.8; obviously a reasonable agreement between model and fit is achieved. In particular, the diffusion process in between traps is considerably faster than in crystalline Pd, whereas, as a consequence of trapping, the long-range diffusion coefficient is of the same order. At small Q the character of the narrow mode already crosses over from diffusive ($\Lambda \propto Q^2$) to localized behavior ($\Lambda \propto 1/\tau_0$); this explains a Q exponent smaller than 2, as found experimentally. The observation of two well-separated regimes of jump rates is evidence for the existence of two different kinds of interstices. Actually, by means of neutron vibrational spectroscopy [98], vibrational modes typical for octahedral and tetrahedral sites have been detected. Schirmacher [99] and Wagner and Schirmacher [100] interpret their QENS data on a-$Zr_{76}Ni_{24}H_8$ in terms of an effective medium description of anomalous diffusion in disordered systems with a broad distribution of activation energies. On the other hand, Apih et al. [101] in their NMR study of the hydrogen tracer diffusion coefficient in a $Zr_{69.5}Cu_{12}Ni_{11}Al_{7.5}$ metallic glass have observed an Arrhenius law for the temperature dependence, which suggests a well-defined hydrogen site energy, and they have not found any appreciable H concentration dependence of the diffusion coefficient.

26.5
Quantum Motion of Hydrogen

While a rigorous description of elementary H jumps in metals should imply tunneling even at room temperature [6, 43], for most of the studied systems the experimental data on hydrogen jump rates can still be approximated by the Arrhenius-like temperature dependences over extended T ranges. In this section we shall discuss the behavior of hydrogen in a number of systems where a quantum-mechanical description is essential for understanding the basic features of hydrogen motion. Hydrogen dynamics in these systems is characterized either by unusual temperature dependences of the jump rates or by quantum delocalization of hydrogen. For particles moving in a periodic crystal field, one may expect the formation of Bloch states at very low temperatures. However, such a long-range coherent tunneling dynamics has not been observed for hydrogen in metals, most probably because of lattice distortions and interactions between interstitial atoms that prevent the formation of band-like states for relatively heavy particles (as compared to electrons). On the other hand, the effects of quantum delocalization of hydrogen over *small groups* of closely-spaced interstitial sites have been found experimentally in some metal–hydrogen systems to be discussed below.

The simplest model describing general features of hydrogen tunneling is that of a particle in a double-well potential (Fig. 26.9). If the positions of the two minima are close to each other, and the barrier between them is not too high, one may expect an overlap of the ground-state wavefunctions of a particle oscillating in

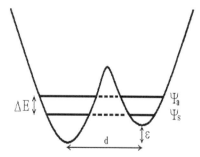

Figure 26.9 Schematic representation of a double-well potential with asymmetry ε. The tunnel-split ground state levels of hydrogen are shown by solid horizontal lines.

either of the two wells. In this case, the wavefunctions of the system are given by linear combinations of the wavefunctions of a particle oscillating in the left (Ψ_{L0}) and right (Ψ_{R0}) wells:

$$\Psi_s = a\Psi_{L0} + \beta\Psi_{R0} \tag{26.23}$$

$$\Psi_a = a\Psi_{R0} - \beta\Psi_{L0} . \tag{26.24}$$

For a symmetric double-well potential, $a^2 = \beta^2 = 1/2$, so that the particle is delocalized and equally distributed over both potential wells. In analogy to the binding and antibinding orbitals of the H_2 molecule, the symmetric and antisymmetric wavefunctions exhibit slightly different energy eigenvalues, and this small energy difference is called the tunnel splitting J. The value of J is determined by the matrix element of the Hamiltonian between the states Ψ_{L0} and Ψ_{R0} in the localized representation; it can be calculated for some specific shapes of the double-well potential [53, 102, 103]. All these results imply that J decreases exponentially with the increasing barrier height, particle mass, and well separation, although precise numerical factors change from one model to another.

If the double-well potential is characterized by the asymmetry ε (see Fig. 26.9), the splitting of the ground state is given by

$$\Delta E = (J^2 + \varepsilon^2)^{1/2} \tag{26.25}$$

and the normalization factors, $a^2 + \beta^2 = 1$, are related as

$$\left(\frac{a}{\beta}\right)^2 = \frac{J^2 + \varepsilon^2}{J^2} \tag{26.26}$$

Thus, the asymmetry of the potential leads to the increase in the splitting and to the partial localization of the particle in the minimum with lower energy.

The tunnel splitting of the ground-state energy level of hydrogen can be directly probed by inelastic neutron scattering (INS). If $\Delta E \gg k_B T$, the energy exchange between neutrons and the tunneling system corresponds to the transitions between the two well-defined states. In this case, the INS spectrum is expected to show two peaks at the energy transfer of $\pm\Delta E$ (in addition to the elastic peak at zero energy transfer). The intensities of these two peaks are related by the condition of detailed balance, and the width of the peaks is determined by the coupling of the tunneling system to its environment (phonons and conduction electrons). The expressions for the double differential neutron scattering cross-section in the case of a tunneling system interacting with its environment were derived and discussed in Refs. [104, 105].

26.5.1
Hydrogen Tunneling in Nb Doped with Impurities

Hydrogen trapped by interstitial impurities (O, N or C) in niobium represents a particularly interesting case of localized motion governed by quantum-mechanical effects. The impurity atoms occupy the octahedral interstitial sites in the b.c.c. lattice of Nb, and the trapped hydrogen can jump between a small number of nearby tetrahedral sites. Most probably, the localized H motion occurs between two equivalent tetrahedral sites (at a distance of ≈ 1.17 Å from each other) located at a distance of ≈ 3.40 Å from the trapping atom [42]. It should be noted that in impurity-free b.c.c. Nb hydrogen diffusion can hardly be studied below 100 K, since in this temperature range all hydrogen atoms are in the precipitated hydride phase. The trapping of hydrogen by interstitial impurities provides an alternative to the phase separation preventing the precipitation of the hydride. Certainly, this scenario is operative only for small hydrogen/impurity concentrations (up to approximately 1 at.%).

The existence of internal excitations in these systems involving very small energies was first revealed in heat capacity measurements at low temperatures [106, 107]. These measurements have demonstrated that the heat capacity anomalies below 2 K previously observed in V, Nb and Ta are, in fact, due to the presence of small amounts of dissolved hydrogen. Moreover, the strong effect of isotope (H$\leftrightarrow$D) substitution on the heat capacity anomaly suggests the possibility of some tunneling motion of hydrogen. Subsequent measurements of the heat capacity of Nb samples doped with O(N) and H(D) [108–110] have clarified the nature of the low-temperature excitations. It has been shown that the low-temperature heat capacity results are consistent with the model of H(D) tunneling in a double-well potential produced by a trapping atom. The quantitative analysis taking into account a distribution of the energy shifts ε for the hydrogen ground-state levels due to lattice distortions yields the following values of the tunnel splitting [110]: $J = 230\pm10$ µeV for H trapped by O and $J = 170\pm10$ µeV for H trapped by N. For deuterium the values of the tunnel splitting derived from the heat capacity data appear to be an order of magnitude smaller than those for H (20 ± 1 µeV for D trapped by O and 14 ± 1 µeV for D trapped by N). The strong isotope dependence of the tunnel splitting is also consistent with the ultrasonic attenuation data for Nb with N traps [111]; however, the absolute value of J for the trapped hydrogen obtained from the ultrasonic experiments is about 20% smaller than that found from the heat capacity measurements.

Direct evidence for the existence of the tunnel-split hydrogen states has been obtained from the high-resolution neutron spectroscopic experiments [104, 105, 112, 113]. These experiments have revealed well-defined peaks at the energy transfer of ~ 0.2 meV in the low-temperature neutron spectra of $Nb(OH)_x$, $Nb(NH)_x$ and $Nb(CH)_x$. As an example of the data, Fig. 26.10 shows the neutron spectra of $Nb(OH)_{0.0002}$ at 0.2 K and 4.3 K [112]. The peak at the energy transfer of ~ 0.2 meV (which at $T = 4.3$ K is clearly seen both for the neutron energy loss and the neutron energy gain) originates from excitations between the tunnel-split ground states of hydrogen. The values of the tunnel splitting obtained from the neutron scattering measurements are in good agreement with those derived from the analysis of the heat capacity data. In particular, the tunnel splitting for hydrogen trapped by N or C is found to be about 15% smaller than that for hydrogen trapped by O. Figure 26.10 also demonstrates that the width of the 0.2 meV peaks strongly depends on the state of conduction-electron system. In fact, since the superconducting transition temperature for niobium is 9.2 K, in zero magnetic field the spectra at both 0.2 K and 4.3 K correspond to the superconducting state. However, a magnetic field of 0.7 T suppresses the superconductivity, so that the sample is in the normal-conducting state at both 0.2 K and 4.3 K. As can be seen from Fig. 26.10, the 0.2 meV peaks in the normal state are considerably broader than those in the superconducting state at the same temperature; this effect is more pronounced at 4.3 K. These results show that, in $Nb(OH)_x$ samples with very small x, conduction electrons play a dominant role in the damping of the tunnel-split H states at low tempera-

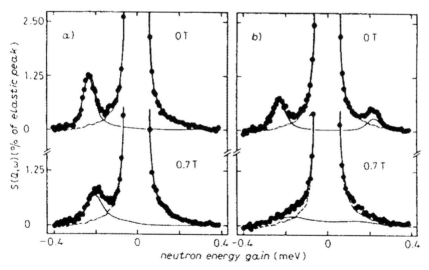

Figure 26.10 Neutron spectra of a $Nb(OH)_{0.0002}$ sample at 0.2 K (a) and 4.3 K (b). For both temperatures, the spectra are taken in the superconducting (0 T) and normal-conducting (0.7 T) electronic state. The thick and thin solid lines represent the fit curves for the total and inelastic scattering intensity, respectively. The broken lines are for the elastic intensity (from Ref. [112]).

tures. The width of the 0.2 meV peaks is also found to increase with increasing defect concentration x [104, 105]. This is believed to result from distributions both in ε and J which are introduced by lattice distortions.

As the temperature increases above ~ 5 K, the well-defined tunnel-split H states are progressively destroyed, and hydrogen starts to perform diffusive jumps between the two nearest-neighbor tetrahedral sites (incoherent tunneling). This is reflected in the following transformation of the neutron spectra: the inelastic peaks (corresponding to excitations between the tunnel-split states) broaden and merge into the quasielastic line. In the regime of local jump motion, the width of the quasielastic line is proportional to the jump rate. Figure 26.11 shows the temperature dependence of the H jump rate derived from the measured width of the quasielastic line in $Nb(OH)_x$ and $Nb(NH)_x$ [114, 115]. The most striking feature of the data is that in the temperature range 10–70 K the hydrogen jump rate *increases* with decreasing temperature both for $Nb(OH)_x$ and $Nb(NH)_x$. Such behavior has been attributed to nonadiabatic effects in the interaction between hydrogen and conduction electrons [43, 116, 117]. The strength of the nonadiabatic coupling between hydrogen and conduction electrons is described by the dimensionless Kondo parameter K; in the case of weak coupling ($K \ll 1$) the theory [43, 118, 119] gives the jump rate τ_l^{-1} as

$$\tau_l^{-1} = \frac{1}{2} \frac{\Gamma(K)}{\Gamma(1-K)} \frac{J}{\hbar} \left(\frac{2\pi k_B T}{J}\right)^{2K-1} \tag{26.27}$$

where $\Gamma(x)$ denotes the gamma function and J is the tunnel splitting in the normal-conducting state for $T \to 0$. As can be seen from Fig. 26.11, the observed temperature dependence of τ_l^{-1} below 60 K is consistent with the T^{2K-1} power law; the value of K derived from the experimental data is 0.055. At $T < 60$ K the hydrogen jump rates in the case of N traps are lower than those in the case of O traps. This is also consistent with Eq. (26.27), since the measured values of J for H trapped by N are smaller than those for H trapped by O. The dashed and solid lines in Fig. 26.11 show the results of calculations based on Eq. (26.27) with the values of J and K determined from the neutron scattering measurements for $Nb(OH)_x$ and $Nb(NH)_x$ at $T < 10$ K (without any adjustable parameters). It can be seen that the values of J and K found from experiments in the regime of H tunneling also provide a good description of the data in the regime of H jump motion, at least up to 60 K. Thus, for hydrogen trapped by interstitial impurities in Nb, the scenario of H tunneling in a double-well potential with nonadiabatic coupling between hydrogen and conduction electrons appears to be self-consistent.

Above 70 K the hydrogen jump rate starts to increase with increasing temperature (Fig. 26.11). In this temperature range the interaction between hydrogen and phonons is believed to become the dominant one, so that Eq. (26.27) no longer describes the data. The appropriate theoretical description of H motion in this range is given in terms of phonon-assisted incoherent tunneling [43]. As the temperature increases above ~ 150 K, there is an increasing probability that a hydrogen atom can leave its trap site. The detrapped H atom is expected to perform a

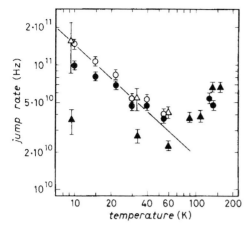

Figure 26.11 The jump rates of trapped hydrogen in Nb(OH)$_x$ (open symbols, circles: $x = 0.002$, triangles: $x = 0.011$) and Nb(NH)$_x$ (full symbols, triangles: $x = 0.004$, polycrystalline sample, circles: $x = 0.004$, single-crystalline sample) as functions of the temperature. The broken (O traps) and solid (N traps) lines are the theoretical predictions based on Eq. (26.27) (from Ref. [115]).

long-range diffusive motion before it is trapped again by another impurity atom. In this regime the hydrogen jump motion is characterized by two different jump rates corresponding to the localized motion in the trapped state and to the long-range diffusion in the free state. These two jump rates, as well as the ratio of the mean lifetimes of a hydrogen atom in the free and trapped states, have been determined from QENS measurements on Nb(OH)$_{0.011}$ in the range 150–300 K [118]. It has been found that at 150 K the hydrogen jump rate for the localized motion is considerably higher than that for the long-range diffusion; however, above 250 K these two jump rates are nearly the same.

It should be noted that low-temperature dynamical processes were also observed for hydrogen trapped by some substitutional impurities (Ti, Zr) in niobium using the internal friction [119, 120] and heat capacity [121] measurements. In particular, the heat capacity of a NbTi$_{0.05}$ alloy doped with H(D) shows considerable hydrogen- or deuterium-induced contributions between 0.05 and 2 K [121]. However, the available experimental information is not sufficient to elucidate the microscopic picture of hydrogen motion in these systems. The same also refers to the low-temperature dynamical processes of hydrogen in b.c.c. V and Ta [106, 107, 122].

26.5.2
Hydrogen Tunneling in α-MnH$_x$

Neutron scattering studies of hydrogen dynamics in solid solutions of hydrogen in α-manganese [123, 124] have revealed a remarkable tunneling effect. This is the first well-documented case of intrinsic (that is, not related to impurities) *coherent* H tunneling in a metal–hydrogen system.

The maximum solubility of hydrogen in a-Mn increases from about 0.3 at.% at ambient pressure to a few at% at hydrogen pressures of 0.6–0.9 GPa [125]. The solid solutions a-MnH$_x$ ($x \leq 0.073$) obtained by high-pressure quenching retain long-term stability at room temperature after the pressure release [123]. The neutron diffraction study of a-MnH$_{0.07}$ [123] has shown that H atoms randomly occupy the interstitial sites $12e$ of the space group $I\bar{4}3m$ in the complex cubic unit cell of a-Mn composed of 58 manganese atoms. The sublattice of $12e$ sites consists of pairs of closely-spaced sites, the distance between the sites in a pair being only 0.68 Å (see inset in Fig. 26.12). Because of such a small intersite distance, each pair can accommodate only one hydrogen atom. On the other hand, these pairs are well-separated from each other, the distance between the centers of the pairs being about 4.5 Å. It is evident that this configuration of $12e$ sites is favorable for hydrogen tunneling.

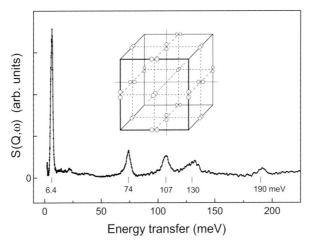

Figure 26.12 The inelastic neutron scattering spectrum of a-MnH$_{0.07}$ measured at 23 K on the TFXA spectrometer (Rutherford-Appleton Laboratory). Inset: the arrangement of closely spaced pairs of $12e$ sites in the unit cell of a-Mn [126].

Inelastic neutron scattering (INS) studies of a-MnH$_{0.07}$ [123, 124] have revealed that, in addition to the band of hydrogen optical vibrations (peaks at 74, 107 and 130 meV), there is a strong peak at the energy transfer of 6.4 meV (Fig. 26.12). The intensity of the 6.4 meV peak corresponding to the neutron energy loss is found to decrease with increasing temperature. At temperatures near 200 K this peak shows relaxation behavior merging into the quasielastic line. The 6.4 meV peak has been attributed [123, 124] to the excitation between the tunnel-split vibrational ground-state levels of hydrogen. This conclusion is supported by the following observations [124].

1. The position of the peak is found to depend strongly on the isotope (H↔D) substitution. In fact, for a-MnD$_{0.05}$ the low-energy peak appears at 1.6 meV. Because of a certain H contamination of the deuterated sample in experiments [124] and the large incoherent neutron scattering cross-section of H, the peaks at both 1.6 meV and 6.4 meV are present in the INS spectra of a-MnD$_{0.05}$. As an example of the data, Fig. 26.13 shows the INS spectra (neutron-energy gain side) of a-MnD$_{0.05}$ at a number of temperatures. The observed isotope effect on the splitting J of the vibrational ground states is consistent with the estimates (J_H = 5 meV, J_D = 1.5 meV) based on the theoretical model [127], if the actual parameters for a-MnH$_x$(D$_x$) are used [124].

2. The temperature dependences of the integrated intensities of the 6.4 meV and 1.6 meV peaks are well described in terms of the Boltzmann thermal population factors of the split ground-state levels, both for the neutron-energy gain and neutron-energy loss. On the other hand, the observed temperature dependences of the peak intensities differ qualitatively from those expected for phonons or harmonic oscillators [124].

3. The behavior of the intensity of the 6.4 meV peak in a-MnH$_{0.07}$ as a function of momentum transfer, Q, is consistent with that expected in the case of tunneling between two sites [105],

$$S(Q, \omega_{\text{tun}}) \propto \left[\frac{1}{2} - \frac{\sin Qd}{2Qd} \right] \exp(-Q^2 u^2) \qquad (26.28)$$

where d is the distance between two sites and u^2 is the effective mean-square displacement due to vibrations of H atoms. Figure 26.14 shows the Q-dependence of the 6.4 meV peak intensity measured at 20 K using the MARI neutron spectrometer at the Rutherford-Appleton Laboratory [124]. As can be seen from Fig. 26.14, Eq. (26.28) with the fixed experimental value d = 0.68 Å and a realistic mean-square displacement (u^2 = 0.0256 Å^2) gives a satisfactory description of the data.

It should be noted that the splitting due to H tunneling in a-MnH$_x$ is anomalously large (about 30 times larger than for H trapped by impurities in b.c.c. metals). Because of its large magnitude, the splitting is observed up to temperatures exceeding 100 K, which is very unusual for such quantum effects. Moreover, the a-MnD$_x$ system gives the unique possibility to observe the level splitting due to tunneling of deuterium.

The tunneling modes in a-MnH$_x$ are found to be suppressed by elastic stresses [128]. In fact, the INS peak due to tunneling disappears at a pressure of 0.8 GPa in a quasi-hydrostatic regime of a sapphire anvil cell. On the other hand, the appli-

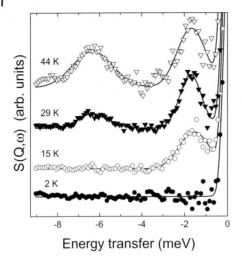

Figure 26.13 The difference between the INS spectra (neutron-energy gain side) of a-MnD$_{0.05}$ and a-Mn measured on the IN6 spectrometer (Institute Laue-Langevin). The spectra corresponding to different temperatures are shifted along the vertical axis. The a-MnD$_{0.05}$ sample is contaminated with about 0.5 at.% H which manifests itself by the peak at 6.4 meV [126].

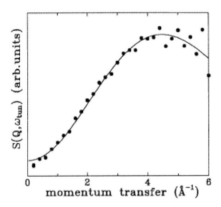

Figure 26.14 The H tunneling peak intensity in the INS spectrum of a-MnH$_{0.07}$ measured at 20 K as a function of Q [124]. The solid line shows the result of a fit based on Eq. (26.28) with the fixed $d = 0.68$ Å.

cation of the same pressure under purely hydrostatic conditions does not cause any changes in the parameters of the 6.4 meV peak [128]. The suppression of the hydrogen tunneling modes by applied inhomogeneous elastic stresses is believed to result from a shift of the energy levels in the adjacent potential wells caused by static displacements of atoms.

26.5.3
Rapid Low-temperature Hopping of Hydrogen in a-ScH$_x$(D$_x$) and TaV$_2$H$_x$(D$_x$)

The behavior of hydrogen in Sc and the related h.c.p. metals Y and Lu shows a number of unusual features. In contrast to most binary metal–hydrogen systems, hydrogen in ScH$_x$, YH$_x$, and LuH$_x$ is not precipitated into a hydride phase at low temperatures; it remains in the solid-solution (a) phase up to $x = 0.2$–0.3. It is believed that the a-phase stabilization is related to the peculiar short-ranged ordering of hydrogen [129, 130], which develops with decreasing temperature. Neutron diffraction [131] and diffuse scattering [129, 130] measurements have shown that hydrogen occupies only the tetrahedral interstitial sites in the h.c.p. Sc lattice and tends to form next-nearest-neighbor pairs with a bridging metal atom in the c axis direction. As the temperature is lowered, these pairs arrange themselves into a longer-range structure, predominantly along the c direction. However, truly long-range order is not achieved in these systems.

The dynamical properties of hydrogen in Sc are also quite remarkable. NMR measurements of the proton spin–lattice relaxation rate R_1 in a-ScH$_x$ [132] have revealed a localized H motion with the characteristic jump rate τ_i^{-1} of about 10^8 s^{-1} at 50 K. This localized motion is evident from an additional frequency-dependent R_1 peak at low temperatures (35–80 K). The structure of the sublattice of tetrahedral interstitial sites in a h.c.p. metal suggests that the localized H motion corresponds to jumps between two nearest-neighbor sites separated by about 1.0 Å in the c direction. QENS measurements on a-ScH$_x$ [133, 134] have revealed the existence of a still faster localized motion with the jump rate passing through a *minimum* of approximately 7×10^{10} s^{-1} near 100 K and increasing to 10^{12} s^{-1} at 10 K. This result is illustrated by Fig. 26.15 showing the width of the quasielastic component of the QENS spectrum for ScH$_{0.16}$ as a function of temperature. Since this width is proportional to the H jump rate, it can be seen that the temperature dependence of the H jump rate for a-ScH$_x$ is similar to that for Nb(OH)$_x$ and

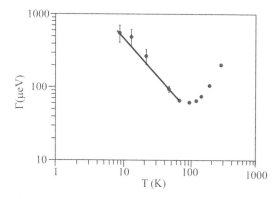

Figure 26.15 The full width at half-maximum of the Lorentzian quasielastic line for ScH$_{0.16}$ as a function of temperature. The solid line shows the fit based on of Eq. (26.27) to the data below 100 K.

$Nb(NH)_x$ (see Fig. 26.11). In particular, the power law (26.27) also appears to describe the temperature dependence of the H jump rate for $ScH_{0.16}$ below 70 K (the solid line in Fig. 26.15); the corresponding fit parameters are $K = 0.039$ and $J = 0.32$ meV [133]. The Q-dependence of the elastic incoherent structure factor for $ScH_{0.16}$ is consistent with H jumps between two sites separated by ~ 1.0 Å. The temperature dependence of the quasielastic line intensity indicates that the fraction p of H atoms participating in the fast localized motion (on the frequency scale determined by the energy resolution of the QENS experiments, ~ 70 μeV) decreases with decreasing temperature; for $ScH_{0.16}$ this fraction is less than 5% below 50 K. It has been suggested [133, 134] that the immobile fraction, $1-p$, corresponds to H atoms forming pairs with a bridging Sc atom, while the mobile fraction p is associated with the unpaired H atoms.

It should be noted that, because of the limited energy resolution (70 μeV), the QENS experiments [133, 134] could not detect the slower H motion found by NMR [132]. This slower motion cannot be attributed to H atoms involved in the pairing, since the NMR results show that only a small fraction of protons participate in this motion, while most of the H atoms are paired at low temperatures. Therefore, it is difficult to elucidate the nature of this slower motion. Note that the long-range H diffusion in a-ScH_x is completely 'frozen out' on the NMR frequency scale at $T < 300$ K. Subsequent NMR measurements of the ^{45}Sc spin-lattice relaxation rates in a-$ScH_x(D_x)$ [135, 136] have revealed strong isotope effects on the localized hydrogen motion. For ^{45}Sc the main motional contribution to R_1 originates from the electric quadrupole interaction modulated by H(D) hopping. Since, for the quadrupole interaction, only *charge* fluctuations are important, H and D atoms are expected to give the same contributions to the host-metal relaxation rate, if their motional parameters are the same. However, the measured amplitude of the low-temperature R_1^{Sc} peak for a-ScD_x samples appears to be much higher than for a-ScH_x samples with comparable hydrogen content [135, 136]. This unusual isotope effect indicates that the fraction of D atoms participating in the localized motion in a-ScD_x on the frequency scale of 10^8–10^9 s^{-1} is considerably larger than the corresponding fraction of H atoms in a-ScH_x.

The whole set of the ^1H, ^2D and ^{45}Sc spin–lattice relaxation measurements at different resonance frequencies [135, 136] and the high-resolution (~ 1.2 μeV) QENS measurements [136] is consistent with the following picture of localized hydrogen motion in a-$ScH_x(D_x)$. The motion of D atoms in a-ScD_x is characterized by a distribution of the jump rates τ_l^{-1}; the most probable value of τ_l^{-1} decreases with decreasing temperature (following an Arrhenius-like law with an activation energy of ~ 50 meV) and passes through the NMR frequency ($\sim 10^8$ s^{-1}) near 100 K. It should be noted that the temperature of the ultrasonic attenuation peak observed for a-$ScD_{0.18}$ (50 K in the frequency range of 1 MHz) [137] is in agreement with the NMR data for a-ScD_x. For H atoms in a-ScH_x, the distribution of the jump rates appears to be shifted to much higher frequencies, so that only a tail of this distribution can be probed by NMR measurements. A large difference between the characteristic jump rates of H and D atoms is consistent with a quantum origin of the localized hydrogen motion in scandium.

Another example of fast low-temperature hydrogen hopping is the localized H(D) motion over g-site hexagons (see Section 26.4.2) in cubic Laves phases. This motion has been most extensively investigated for the TaV_2–H(D) system (where it is the fastest among the studied Laves phases). Therefore, the discussion in this section will be based on the experimental results obtained for $TaV_2H_x(D_x)$. Three interesting features of the localized hydrogen motion will be emphasized: (i) the strong dependence of the jump rate τ_l^{-1} on hydrogen concentration, (ii) the isotope effects on the parameters of the localized motion, and (iii) the non-Arrhenius temperature dependence of the jump rate at low T.

The hydrogen jump rate τ_l^{-1} in $TaV_2H_x(D_x)$ has been found to increase strongly with decreasing H(D) content [54, 72, 73]. In NMR measurements, this is reflected in the marked shift of the position of the low-T maximum of R_1 to lower temperatures. For example, at the frequency of 90 MHz the low-T maximum of the proton R_1 is observed at 187 K for $TaV_2H_{1.33}$, at 125 K for $TaV_2H_{0.87}$ [54] and at 45 K for $TaV_2H_{0.06}$ [73]. The strong dependence of τ_l^{-1} on x has also been found in the QENS measurements on TaV_2H_x [76] and in the ultrasonic experiments on TaV_2D_x [138]. Qualitatively, such a dependence is consistent with the fact that the distance between the nearest-neighbor sites in the hexagon, r_1, becomes shorter with decreasing x due to the decrease of the lattice parameter. In this case, the hydrogen jump rate τ_l^{-1} should be extremely sensitive to changes in r_1, which suggests that the localized H motion is governed by tunneling transitions.

As in the case of a-$ScH_x(D_x)$, NMR measurements of the host-metal ([51]V) spin–lattice relaxation rate in $TaV_2H_x(D_x)$ [54] have revealed a strong isotope effect: the amplitude of the low-temperature R_1^V peak for TaV_2D_x is nearly three times higher than that for TaV_2H_x with the same x. These results also suggest that the fraction of D atoms participating in the fast localized motion is considerably larger than that of H atoms. Measurements of the ultrasonic attenuation in $TaV_2H_x(D_x)$ in the frequency range of 1 MHz [58, 138, 139] have found even more dramatic isotope effects. For example, for $TaV_2D_{0.17}$ the ultrasonic attenuation shows a distinct peak near 20 K which can be attributed to localized D motion; however, for $TaV_2H_{0.18}$ the low-T attenuation peak is observed near 1 K, and its amplitude is about 8 times lower than that for $TaV_2D_{0.17}$ [138]. These results demonstrate that, at low temperatures, the jump rate of H atoms is at least an order of magnitude faster than that of D atoms for similar concentrations. The ultrasonic data [138] are also consistent with the temperature-dependent fraction of 'mobile' atoms, the value of p decreasing with decreasing temperature.

The temperature dependence of the jump rate τ_l^{-1} for both H and D in $TaV_2H_x(D_x)$ is found to be strongly non-Arrhenius. The behavior of $\tau_l^{-1}(T)$ obtained from the proton R_1 measurements in TaV_2H_x [54] is reasonably described by the exponential function,

$$\tau_l^{-1} = \tau_0^{-1} \exp(T/T_0) \tag{26.29}$$

over the temperature range 30–200 K. At low temperatures this dependence is much weaker than the Arrhenius one. Although the relation given by Eq. (26.29) should be considered as empirical, it is worth noting that a similar term in the jump rate has been found in the framework of the quantum diffusion theory taking into account the effects of barrier fluctuations [140]. While for protons the behavior of the measured spin–lattice relaxation rate at low T is affected by the temperature dependence of p [76], the ^2D spin–lattice relaxation rates can be measured separately for the 'mobile' and 'static' fractions [72]. Therefore, the temperature dependence of τ_l^{-1} derived from the ^2D relaxation data for the 'mobile' fraction of D atoms should be more reliable. Figure 26.16 shows the behavior of $\tau_l^{-1}(T)$ obtained from the ^2D relaxation rate data for $TaV_2D_{0.5}$ and $TaV_2D_{1.3}$ [72]. It can be seen that $\tau_l^{-1}(T)$ for D atoms is satisfactorily described by Eq. (26.29). The corresponding fit parameters are T_0 = 38 and 33 K, τ_0^{-1} = 2.0×10^7 and 1.5×10^6 s^{-1} for $TaV_2D_{0.5}$ and $TaV_2D_{1.3}$, respectively. For comparison, the low-temperature behavior of τ_l^{-1} for H in $TaV_2H_{0.56}$ (the dashed line in Fig. 26.16) is described by T_0 = 50.3 K, τ_0^{-1} = 1.1×10^8 s^{-1} [54]. It should be noted that Eq. (26.29) with the values of T_0 and τ_0^{-1} derived from the ^2D NMR data for $TaV_2D_{0.5}$ also gives a reasonable description of the ultrasonic attenuation results for $TaV_2D_{0.5}$ [138]. The ultrasonic experiments on TaV_2H_x with low x [138] show that the hydrogen jump rate τ_l^{-1} remains finite down to the lowest temperature of the measurements (0.3 K), being well above the ultrasound frequency of $\sim$ 1 MHz.

Inspection of the experimental data for the systems with fast localized hydrogen motion suggests that the applicability of Eq. (26.29) is not restricted to hydrogen in Laves phases. In particular, this equation appears to describe the behavior of $\tau_l^{-1}(T)$ for the trapped hydrogen in $Nb(OH)_{0.011}$ [118] in the range 100–300 K and for the 'mobile' H atoms in a-ScH$_{0.16}$ [133] in the range 125–300 K. The values of T_0 estimated from these data are 130 K for $Nb(OH)_{0.011}$ and 146 K for a-ScH$_{0.16}$.

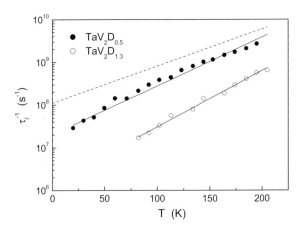

Figure 26.16 The temperature dependence of the jump rates τ_l^{-1} for deuterium in $TaV_2D_{0.5}$ and $TaV_2D_{1.3}$, as determined from the ^2D spin lattice relaxation data [72]. The solid lines show the fits of Eq. (26.29) to the data. The dashed line represents the behavior of $\tau_l^{-1}(T)$ for H in $TaV_2H_{0.56}$, as derived from the fit of Eq. (26.29) to the proton spin–lattice relaxation data (Ref. [54]).

26.6
Concluding Remarks

Information on hydrogen diffusion rates is of crucial importance for the applicability of materials for reversible hydrogen storage. The understanding of hydrogen diffusion processes in metals is expected to contribute to the development of new hydrogen storage materials with fast kinetics of hydrogen release/uptake at moderate temperatures. Therefore, the theoretical and experimental investigation of H motion in metals remains a very active field of research. The present chapter gives a brief review of experimental studies of the mechanisms and parameters of hydrogen diffusion in metals. We emphasize the relation between the parameters of H motion and the structure of the hydrogen sublattice. The general experimental trend is to investigate systems of increasing structural complexity. While the behavior of hydrogen in a number of binary metal–hydrogen systems is well understood, a detailed microscopic picture of H diffusion in complex compounds of practical importance has not yet evolved. A promising approach to investigation of these compounds is to combine a number of experimental techniques sensitive to different ranges of H jump rates (such as NMR, QENS and anelastic relaxation) with the neutron diffraction study of hydrogen positional parameters. A new challenge is to elucidate the mechanisms of H diffusion in nanocrystalline materials which are used for hydrogen storage.

Further studies are also necessary to understand the processes of H tunneling and low-temperature hopping. Recent work [141] has demonstrated the potential of the first-principles density functional theory accurately predicting the rates of activated hopping and quantum tunneling of H in b.c.c. Nb and Ta. Progress in numerical methods may lead to application of the first-principles calculations to H motion in more complex systems. From the experimental side, in the near future one may anticipate the discovery of new compounds showing high H mobility down to low temperatures.

Acknowledgement

We are grateful to the Alexander von Humboldt Foundation for awarding a Research Fellowship to A.S. The present chapter results from a number of research visits of A.S. (supported by the Humboldt Foundation) to the University of Saarbrücken.

References

1 G. Alefeld, J. Völkl (Eds.), *Hydrogen in Metals I*, Springer, Berlin, **1978**.
2 G. Alefeld, J. Völkl (Eds.), *Hydrogen in Metals II*, Springer, Berlin, **1978**.
3 H. Wipf (Ed.), *Hydrogen in Metals III*, Springer, Berlin, **1997**.
4 L. Schlapbach (Ed.), *Hydrogen in Intermetallic Compounds I*, Springer, Berlin, **1988**.
5 L. Schlapbach (Ed.), *Hydrogen in Intermetallic Compounds II*, Springer, Berlin, **1992**.
6 Y. Fukai, *The Metal-Hydrogen System*, Springer, Berlin, **1993**.
7 A. S. Nowick, B. S. Berry, *Anelastic Relaxation in Crystalline Solids*, Academic Press, London, **1972**.
8 R. M. Cotts, in *Hydrogen in Metals I*, G. Alefeld, J. Völkl (Eds.), Springer, Berlin, **1978**, p. 227.
9 J. Völkl, G. Alefeld, in *Hydrogen in Metals I*, G. Alefeld, J. Völkl (Eds.), Springer, Berlin, **1978**, p. 321.
10 Y. Fukai, H. Sugimoto, *Adv. Phys.* **1985**, *34*, 263–326.
11 D. Richter, R. Hempelmann, R. C. Bowman, in *Hydrogen in Intermetallic Compounds II*, L. Schlapbach (Ed.), Springer, Berlin, **1992**, p. 97.
12 R. G. Barnes, in *Hydrogen in Metals III*, H. Wipf (Ed.), Springer, Berlin, **1997**, p. 93.
13 D. K. Ross, in *Hydrogen in Metals III*, H. Wipf (Ed.), Springer, Berlin, **1997**, p. 153.
14 R. Hempelmann, *Quasielastic Neutron Scattering and Solid State Diffusion*, Clarendon Press, Oxford, **2000**.
15 J. L. Snoek, *Physica* **1939**, *6*, 591.
16 J. Buchholz, J. Völkl, G. Alefeld, *Phys. Rev. Lett.* **1973**, *30*, 318–321.
17 G. Schaumann, J. Völkl, G. Alefeld, *Phys. Rev. Lett.* **1968**, *21*, 891–893.
18 J. Völkl, *Ber. Bunsenges. Phys. Chem.* **1972**, *76*, 797.
19 H. Wagner, in *Hydrogen in Metals I*, G. Alefeld , J. Völkl (Eds.), Springer, Berlin, **1978**, p. 5.
20 G. E. Murch, *Philos. Mag. A* **1982**, *45*, 685–692.

21 L. D. Bustard, *Phys. Rev. B* **1980**, *22*, 1–11.
22 D. A. Faux, D. K. Ross, C. A. Sholl, *J. Phys. C* **1986**, *19*, 4115–4133.
23 D. A. Faux, C. K. Hall, *Z. Phys. Chem. N. F.* **1989**, *164*, 859–864.
24 N. Bloembergen, E. M. Purcell, R. M. Pound, *Phys. Rev.* **1948**, *73*, 679–712.
25 H. C. Torrey, *Phys. Rev.* **1953**, *92*, 962–969.
26 C. A. Sholl, *J. Phys. C* **1988**, *21*, 319–324.
27 P. Galvosas, F. Stallmach, G. Seiffert, J. Kärger, U. Kaess, G. Majer, *J. Magn. Reson.* **2001**, *151*, 260.
28 T. Springer, *Quasielastic Neutron Scattering for the Investigation of Diffusive Motions in Solids and Liquids*, Springer Tracts in Modern Physics, Vol. 64, Springer, Berlin, **1972**.
29 M. Bée, *Quasielastic Neutron Scattering. Principles and Applications in Solid State Chemistry, Biology and Materials Science*, Adam Hilger, Bristol, **1988**.
30 N. G. van Kampen, *Stochastic Processes in Physics and Chemistry*, Elsevier, Amsterdam, **1992**.
31 K. S. Singwi, A. Sjölander, *Phys. Rev.* **1960**, *119*, 863–871.
32 D. Richter, T. Springer, *Phys. Rev. B* **1978**, *18*, 126–140.
33 R. Hempelmann, *J. Less-Common Met.* **1984**, *101*, 69.
34 D. Richter, G. Driesen, R. Hempelmann, I. S. Anderson, *Phys. Rev. Lett.* **1986**, *57*, 731–734.
35 K. Watanabe, Y. Fukai, *J. Phys. F: Met. Phys.* **1980**, *10*, 1795–1801.
36 H. Kronmüller, in *Hydrogen in Metals I*, G. Alefeld, J. Völkl (Eds.), Springer, Berlin, **1978**, p. 291.
37 J. Crank, *The Mathematics of Diffusion*, Oxford University Press, Oxford, **1975**.
38 E. Wicke, H. Brodowsky, in *Hydrogen in Metals II*, G. Alefeld, J. Völkl (Eds.), Springer, Berlin, **1978**, pp. 73–155.
39 R. Kirchheim, *Acta Metall.* **1982**, *30*, 1069.
40 H. Züchner, in *Hydrogen in Metals (Japan Institute of Metals, Sendai 1980): Trans. JIM (Suppl.)* **1980**, *21*, 101.

41 R. Kirchheim, R. B. McLellan,
    J. Electrochem. Soc. **1980**, *127*, 2419.
42 H. Wipf, in *Hydrogen in Metals III*,
    H. Wipf (Ed.), Springer, Berlin, **1997**,
    pp. 51–91.
43 H. Grabert, H. R. Schober, in *Hydrogen
    in Metals III*, H. Wipf (Ed.), Springer,
    Berlin, **1997**, pp. 5–49.
44 J. Völkl, H. C. Bauer, U. Freudenberg,
    K. Kokkinidis, G. Lang,
    K. A. Steinhauser, G. Alefeld, in *Internal
    Friction and Ultrasonic Attenuation in
    Solids*, R. R. Hasiguti, N. Mikoshiba
    (Eds.), University of Tokyo Press, Tokyo,
    1977, p. 485.
45 A.V. Skripov, *Defect Diffus. Forum* **2003**,
    224–225, 75–92.
46 Zh. Qi, J. Völkl, R. Lässer, H. Wenzl,
    J. Phys. F **1983**, *13*, 2053–2062.
47 W. Renz, G. Majer, A. V. Skripov,
    A. Seeger, *J. Phys.: Condens. Matter*
    1994, *6*, 6367–6374.
48 A. Klamt, H. Teichler, *Phys. Status Solidi
    (b)* **1986**, *134*, 503.
49 H. R. Schober, A. M. Stoneham, *Phys.
    Rev. Lett.* **1988**, *60*, 2307–2310.
50 J. Völkl, G. Wollenweber, K. H. Klatt,
    G. Alefeld, *Z. Naturforsch.A* **1971**, *26*,
    922.
51 G. Sicking, M. Glugla, B. Huber, *Ber.
    Bunsenges. Phys. Chem.* **1983**, *87*, 418.
52 L. Katz, M. Guinan, R. J. Borg, *Phys.
    Rev. B* **1971**, *4*, 330–341.
53 K. Kehr, in *Hydrogen in Metals I*,
    G. Alefeld, J. Völkl (Eds.), Springer,
    Berlin, **1978**, pp. 197–226.
54 A. V. Skripov, S. V. Rychkova, M.Yu.
    Belyaev, A. P. Stepanov, *J. Phys.:
    Condens. Matter* **1990**, *2*, 7195–7208.
55 A. V. Skripov, M.Yu. Belyaev,
    S. V. Rychkova, A. P. Stepanov, *J. Phys.:
    Condens. Matter* **1991**, *3*, 6277–6291.
56 A. V. Skripov, Yu. G. Cherepanov,
    B. A. Aleksashin, S. V. Rychkova,
    A. P. Stepanov, *J. Alloys Compd.* **1995**,
    227, 28–31.
57 G. Majer, W. Renz, A. Seeger,
    R. G. Barnes, J. Shinar, A. V. Skripov,
    J. Alloys Compd. **1995**, *231*, 220–225.
58 K. Foster, R. G. Leisure, A. V. Skripov,
    Phys. Rev. B **2001**, *64*, 214302.
59 C. T. Chudley, R. J. Elliott, *Proc. Phys.
    Soc.* **1961**, *77*, 353.

60 K. Sköld, G. Nelin, *J. Phys. Chem. Solids*
    1967, *28*, 2369.
61 J. M. Rowe, J. J. Rush, L. A. de Graaf,
    G. A. Ferguson, *Phys. Rev. Lett.* **1972**, *29*,
    1250–1253.
62 J. M. Rowe, J. J. Rush, H. E. Flotow,
    Phys. Rev. B **1974**, *9*, 5039–5045.
63 J. M. Rowe, K. Sköld, H. E. Flotow,
    J. J. Rush, *J. Phys. Chem. Solids* **1971**, *32*,
    41–54.
64 V. Lottner, J. W. Haus, A. Heim,
    K. W. Kehr, *J. Phys. Chem. Solids* **1979**,
    40, 557.
65 V. Lottner, H. R. Schober,
    W. J. Fitzgerald, *Phys. Rev. Lett.* **1979**, *42*,
    1162–1165.
66 V. A. Somenkov, I. R. Entin,
    A. Y. Chervyakov, S. Sh. Shil'stein,
    A. A. Chertkov, *Sov. Phys. Solid State*
    1972, *13*, 2178.
67 D. Richter, S. Mahling-Ennaoui,
    R. Hempelmann, *Z. Phys. Chem. N. F.*
    1989, *164*, 907–920.
68 V. A. Somenkov, A. V. Irodova,
    J. Less-Common Met. **1984**, *101*, 481–492.
69 K. Yvon, P. Fischer, in *Hydrogen in
    Intermetallic Compounds I*, L. Schlapbach
    (Ed.), Springer, Berlin, **1988**,
    pp. 87–138.
70 R. C. Bowman, B. D. Craft, A. Attalla,
    J. R. Johnson, *Int. J. Hydrogen Energy*
    1983, *8*, 801–808.
71 J. Shinar, D. Davidov, D. Shaltiel, *Phys.
    Rev. B* **1984**, *30*, 6331–6341.
72 A. V. Skripov, A. V. Soloninin,
    V. N. Kozhanov, *Solid State Commun.*
    2002, *122*, 497–501.
73 A. L. Buzlukov, A. V. Skripov, *J. Alloys
    Compd.* **2004**, *366*, 61–66.
74 P. Fischer, F. Fauth, A. V. Skripov,
    A. A. Podlesnyak, L. N. Padurets,
    A. L. Shilov, B. Ouladdiaf, *J. Alloys
    Compd.* **1997**, *253–254*, 282–285.
75 U. Eberle, G. Majer, A. V. Skripov,
    V. N. Kozhanov, *J. Phys.: Condens. Matter*
    2002, *14*, 153–164.
76 A. V. Skripov, J. C. Cook, D. S. Sibirtsev,
    C. Karmonik, R. Hempelmann, *J. Phys.:
    Condens. Matter* **1998**, *10*, 1787–1801.
77 A. V. Skripov, M. Pionke, O. Randl,
    R. Hempelmann, *J. Phys.: Condens.
    Matter* **1999**, *11*, 1489–1502.
78 A. V. Skripov, J. C. Cook, T. J. Udovic,
    M. A. Gonzalez, R. Hempelmann,

V. N. Kozhanov, *J. Phys.: Condens. Matter* **2003**, *15*, 3555–3566.

79 A. V. Skripov, J. C. Cook, C. Karmonik, V. N. Kozhanov, *Phys. Rev. B* **1999**, *60*, 7238–7244.

80 A. V. Skripov, J. C. Cook, T. J. Udovic, V. N. Kozhanov, *Phys. Rev. B* **2000**, *62*, 14099–14104.

81 D. J. Bull, D. P. Broom, D. K. Ross, *Chem. Phys.* **2003**, *292*, 153–160.

82 G. Majer, U. Kaess, M. Stoll, R. G. Barnes, J. Shinar, *Defect Diffus. Forum* **1997**, *143–147*, 957–962.

83 R. Hempelmann, D. Richter, A. Heidemann, *J. Less-Common Met.* **1982**, *88*, 343–351.

84 A. V. Skripov, A. V. Soloninin, A. L. Buzlukov, L. S. Voyevodina, J. C. Cook, T. J. Udovic, R. Hempelmann, *J. Phys.: Condens. Matter* **2005**, *17*, 5011–5025.

85 J. J. Didisheim, K.Yvon, D. Shaltiel, P. Fischer, *Solid State Commun.* **1979**, *31*, 47–50.

86 C. Schönfeld, R. Hempelmann, D. Richter, T, Springer, A. J. Dianoux, J. J. Rush, T. J. Udovic, S. M. Bennington, *Phys. Rev. B* **1994**, *50*, 853–865.

87 A. V. Skripov, A. V. Soloninin, A. P. Stepanov, V. N. Kozhanov, *J. Phys.: Condens. Matter* **2000**, *12*, 9607–9616.

88 D. Noréus, L. G. Olsson, *J. Chem Phys.* **1983**, *78*, 2419–2427.

89 A. L. Buzlukov, A. V. Soloninin, A. V. Skripov, *Solid State Commun.* **2004**, *129*, 315–318.

90 E. Mamontov, T. J. Udovic, O. Isnard, J. J. Rush, *Phys. Rev. B* **2004**, *70*, 214305.

91 L. Lichty, J. Shinar, R. G. Barnes, D. R. Torgeson, D. T. Peterson, *Phys. Rev. Lett.* **1985**, *55*, 2895–2898.

92 D. S. Sibirtsev, Yu. G. Cherepanov, A. V. Skripov, *J. Alloys Compd.* **1998**, *278*, 21–28.

93 R. Kirchheim, *Defect Diffus. Forum* **1997**, *143–147*, 911–925.

94 V. V. Kondratyev, A. V.Gapontsev, A. N.Voloshinskii, A. G.Obukhov, N. I.Timofeyev, *Int. J. Hydrogen Energy* **1999**, *24*, 819–824.

95 W. Schirmacher, M. Prem, J.-B. Suck, A. Heidemann, *Europhys. Lett.* **1990**, *13*, 523.

96 R. Kutner, A. Pekalski, K. Sznajd-Weron (Eds.), *Anomalous Diffusion: From Basics to Applications, Lecture Notes in Physics Vol. 519*, Springer, Berlin, **1999**.

97 Y. Gefen, A. Aharony, S. Alexander, *Phys. Rev. Lett.* **1983**, *50*, 77–80.

98 J. J. Rush, T. J. Udovic, R. Hempelmann, D. Richter, G. Driesen, *J. Phys.:Condensed Matter* **1989**, *1*, 1061–1070.

99 W. Schirmacher, *Ber. Bunsenges. Phys. Chem.* **1991**, *95*, 368.

100 M. Wagner, W. Schirmacher, *Ber. Bunsenges. Phys. Chem.* **1991**, *95*, 983.

101 T.Apih, M. Bobnar, J. Dolinsek, L. Jastrow, D. Zander, U. Koester, *Solid State Commun.* **2005**, *134*, 337–341.

102 T. P. Das, *J. Chem. Phys.*, **1957**, *27*, 763.

103 W. A. Phillips, *Amorphous Solids: Low Temperature Properties*, Springer, Berlin, **1981**.

104 H. Wipf, A. Magerl, S.M. Shapiro, S.K. Satija, W. Thomlinson, *Phys. Rev. Lett.* **1981**, *46*, 947–950.

105 A. Magerl, A. J. Dianoux, H. Wipf, K. Neumaier, I. S. Anderson, *Phys. Rev. Lett.* **1986**, *56*, 159–162.

106 G. J. Sellers, M. Paalanen, A. C. Anderson, *Phys. Rev. B* **1974**, *10*, 1912–1915.

107 G. J. Sellers, A. C. Anderson, H. K. Birnbaum, *Phys. Rev. B* **1974**, *10*, 2771–2776.

108 C. Morkel, H. Wipf, K. Neumaier, *Phys. Rev. Lett.* **1978**, *40*, 947–950.

109 H. Wipf, K. Neumaier, *Phys. Rev. Lett.* **1984**, *52*, 1308–1311.

110 P. Gutsmiedl, M. Schikhofer, K. Neumaier, H. Wipf, in *Quantum Aspects of Molecular Motions in Solids*, A. Heidemann, A. Magerl, M. Prager, D. Richter, T. Springer (Eds.), Springer, Berlin, **1987**, p. 158.

111 W. Morr, A. Müller, G. Weiss, H. Wipf, B. Golding, *Phys. Rev. Lett.* **1989**, *63*, 2084–2087.

112 H. Wipf, D. Steinbinder, K. Neumaier, P. Gutsmiedl, A. Magerl, A. J. Dianoux, *Europhys. Lett.* **1987**, *4*, 1379–1384.

113 K. Neumaier, D. Steinbinder, H. Wipf, H. Blank, G. Kearley, *Z. Phys. B* **1989**, *76*, 359–363.

114 D. Steinbinder, H. Wipf, A. Magerl, D. Richter, A. J. Dianoux, K. Neumaier, *Europhys. Lett.* **1988**, *6*, 535–540.

115 D. Steinbinder, H. Wipf, A. J. Dianoux, A. Magerl, K. Neumaier, D. Richter, R. Hempelmann, *Europhys. Lett.*, **1991**, *16*, 211–216.

116 J. Kondo, *Physica B* **1984**, *125*, 279.

117 K. Yamada, *Progr. Theor. Phys.* **1984**, *72*, 195–201.

118 K. Cornell, H. Wipf, J. C. Cook, G. J. Kearley, K. Neumaier, *J. Alloys Compd.* **1999**, *293–295*, 275–278.

119 G. Cannelli, R. Cantelli, M. Koiwa, *Philos. Mag. A* **1982**, *46*, 483.

120 G. Cannelli, R. Cantelli, F. Cordero, K. Neumaier, H. Wipf, *J. Phys. F* **1984**, *14*, 2507–2515.

121 K. Neumaier, H. Wipf, G. Cannelli, R. Cantelli, *Phys. Rev. Lett.* **1982**, *49*, 1423–1426.

122 S. G. O'Hara, G. J. Sellers, A. C. Anderson, *Phys. Rev. B* **1974**, *10*, 2777–2781.

123 V. K. Fedotov, V. E. Antonov, K. Cornell, G. Grosse, A. I. Kolesnikov, V. V. Sikolenko, V. V. Sumin, F. E. Wagner, H. Wipf, *J. Phys.: Condens. Matter* **1998**, *10*, 5255–5266.

124 A. I. Kolesnikov, V. E. Antonov, S. M. Bennington, B. Dorner, V. K. Fedotov, G. Grosse, J. C. Li, S. F. Parker, F. E. Wagner, *Physica B* **1999**, *263–264*, 421–423.

125 V. E. Antonov, T. E. Antonova, N. A. Chirin, E. G. Ponyatovsky, M. Baier, F. E. Wagner, *Scr. Mater.* **1996**, *34*, 1331–1336.

126 A. I. Kolesnikov, V. E. Antonov, V. K. Fedotov, A. J. Dianoux, B. Dorner, A. Hewat, A. S. Ivanov, G. Grosse, F. E. Wagner, *ILL Annual Report* **1999**.

127 S. I. Drechsler, G. M. Vujicic, N. M. Plakida, *J. Phys. F* **1984**, *14*, L243–L246.

128 V. E. Antonov, V. P. Glazkov, D. P. Kozlenko, B. N. Savenko, V. A. Somenkov, V. K. Fedotov, *JETP Lett.* **2002**, *76*, 318–320.

129 M. W. McKergow, D. K. Ross, J. E. Bonnet, I. S. Anderson, O. Schärpf, *J. Phys. C* **1987**, *20*, 1909–1923.

130 O. Blaschko, J. Pleschiutschnig, P. Vajda, J. P. Burger, J. N. Daou, *Phys. Rev. B* **1989**, *40*, 5344–5349.

131 C. K. Saw, B. J. Beaudry, C. Stassis, *Phys. Rev. B* **1983**, *27*, 7013–7017.

132 L. R. Lichty, J. W. Han, R. Ibanez-Meier, D. R. Torgeson, R. G. Barnes, E. F. W. Seymour, C. A. Sholl, *Phys. Rev. B* **1989**, *39*, 2012–2021.

133 I. S. Anderson, N. F. Berk, J. J. Rush, T. J. Udovic, R. G. Barnes, A. Magerl, D. Richter, *Phys. Rev. Lett.* **1990**, *65*, 1439–1442.

134 N. F. Berk, J. J. Rush, T. J. Udovic, I. S. Anderson, *J. Less-Common Met.* **1991**, *172–174*, 496–508.

135 J. J. Balbach, M. S. Conradi, R. G. Barnes, D. S. Sibirtsev, A. V. Skripov, *Phys. Rev. B* **1999**, *60*, 966–971.

136 A. V. Skripov, A. V. Soloninin, D. S. Sibirtsev, A. L. Buzlukov, A. P. Stepanov, J. J. Balbach, M. S. Conradi, R. G. Barnes, R. Hempelmann, *Phys. Rev. B* **2002**, *66*, 054306.

137 R. G. Leisure, R. B. Schwarz, A. Migliori, D. R. Torgeson, I. Svare, *Phys. Rev. B* **1993**, *48*, 893–900.

138 J. E. Atteberry, R. G. Leisure, A. V. Skripov, J. B. Betts, A. Migliori, *Phys. Rev. B* **2004**, *69*, 144110.

139 K. Foster, J. E. Hightower, R. G. Leisure, A. V. Skripov, *J. Phys.: Condens. Matter* **2001**, *13*, 7327–7341.

140 Yu. Kagan, *J. Low Temp. Phys.* **1992**, *87*, 525.

141 P. G. Sundell. G. Wahnström, *Phys. Rev. Lett.* **2004**, *92*, 155901.

Part VII
Special Features of Hydrogen-Transfer Reactions

Volume 2 of this work is closed by Part VII, three articles that address special features of hydrogen-transfer reactions. Truhlar and Garrett review the application of variational transition-state theory to hydrogen transfer in Ch. 27. This modern extension and improvement of more traditional forms of the transition-state theory has proven powerful in the development of deep understanding of how the reactions occur, whether in simple gas-phase reactions of small molecules or in large enzyme active sites. The subject thus spans the entire range of levels of organization and complexity that this work encompasses. In Ch. 28, K. U. Ingold leads the reader through the experimental and theoretical developments, largely from his own laboratory, that put a firm underpinning beneath the phenomenology of hydrogen-atom tunneling under highly characterized conditions. The precision and reliability of the observations were such as to provide a good foundation for the theoretical views that continue to play an important role. Smedarchina, Siebrand, and Fernández-Ramos in Ch. 29 then introduce a novel theoretical approach to the problem of multiple proton transfers and the degree to which proton motions are coupled, making use of experimental results described in other chapters of these volumes.

Hydrogen-Transfer Reactions. Edited by J. T. Hynes, J. P. Klinman, H. H. Limbach, and R. L. Schowen
Copyright © 2007 WILEY-VCH Verlag GmbH & Co. KGaA, Weinheim
ISBN: 978-3-527-30777-7

27
Variational Transition State Theory in the Treatment of Hydrogen Transfer Reactions

Donald G. Truhlar and Bruce C. Garrett

27.1
Introduction

Transition state theory (TST) [1–4] is a widely used method for calculating rate constants for chemical reactions. TST has a long history, which dates back 70 years, including both theoretical development and applications to a variety of reactions in the gas phase, in liquids, at interfaces, and in biological systems. Its popularity and wide use can be attributed to the fact that it provides a theoretical framework for understanding fundamental factors controlling chemical reaction rates and an efficient computational tool for accurate predictions of rate constants.

TST provides an approximation to the rate constant for a system where reactants are at equilibrium constituted by either a canonical ensemble (thermal equilibrium) or a microcanonical ensemble (corresponding to a fixed total energy). Two advances in TST have contributed significantly to its accuracy: (i) the variational form of TST in which the optimum dividing surface is determined to minimize the rate constant and (ii) the development of consistent methods for treating quantum mechanical effects, particularly tunneling. TST in a classical mechanical world can be derived by making one approximation – Wigner's fundamental assumption [3]. With this assumption, the net reactive flux through a dividing surface separating reactants and products is approximated by the equilibrium one-way flux in the product direction. In a classical world this approximation leads to an overestimation of the rate constant, since all reactive trajectories are counted as reactive, but some nonreactive ones also contribute to the one-way flux. In variational TST (VTST), the dividing surface is optimized to give the lowest upper bound to the true rate constant [5–7].

The need to include quantum mechanical effects in reaction rate constants was realized early in the development of rate theories. Wigner [8] considered the lowest order terms in an $\hbar$-expansion of the phase-space probability distribution function around the saddle point, resulting in a separable approximation, in which bound modes are quantized and a correction is included for quantum motion along the reaction coordinate – the so-called Wigner tunneling correction. This separable approximation was adopted in the standard *ad hoc* procedure for quan-

Hydrogen-Transfer Reactions. Edited by J. T. Hynes, J. P. Klinman, H. H. Limbach, and R. L. Schowen
Copyright © 2007 WILEY-VCH Verlag GmbH & Co. KGaA, Weinheim
ISBN: 978-3-527-30777-7

tizing TST [1]. In this approach, partition functions for bound modes are treated quantum mechanically, usually within a harmonic approximation, and a correction for tunneling through the potential barrier along the reaction coordinate is included. Even though more accurate treatments of tunneling through parabolic barriers have been presented [9], beyond the expansion through $\hbar^2$ of Wigner, tests of TST using accurate quantum mechanical benchmarks have shown that this nonseparable approximation is inadequate for quantitative predictions of rate constants when quantum mechanical effects are important [10, 11].

Breakthroughs in the development of quantum corrections for VTST, particularly tunneling, came from comparing the adiabatic theory of reactions [12–14] with VTST for a microcanonical ensemble (microcanonical variational theory). Even though the two theories are based upon very different approximations, they predict the same reaction rates when the reaction coordinate is defined in the same way in both theories and motion perpendicular to it is treated classically [7] or by an identical quantum mechanical approximation [15]. Consequently, quantum mechanical treatments of reaction-coordinate motion for the adiabatic theory provide a starting point for developing quantal corrections for VTST that include the nonseparable, multidimensional nature of tunneling [16–24]. In these approaches, the multidimensional character of the tunneling is included by specifying an optimal tunneling path through the multidimensional space. This approach was pioneered by Marcus and Coltrin [25], who developed the first successful nonseparable tunneling correction for the collinear $H + H_2$ reaction, for which only one mode is coupled to the reaction coordinate.

H-transfer reactions are of great interest because they play important roles in a variety of systems, from gas-phase combustion and atmospheric reactions of small molecules to complex catalytic and biomolecular processes. Many enzyme reactions involve proton or hydride transfer in the chemical step [26], and we know from experience with simpler reactions that multidimensional treatments of the tunneling process are essential for quantitative accuracy and sometimes even for qualitative understanding.

From a theoretical point of view H transfer reactions are of great interest because they provide opportunities to study the importance of quantum mechanical effects in chemical reactions and are a good testing ground to evaluate approximate theories, such as TST approaches. There is a long history of applications of TST to H-transfer reactions starting with the $H + H_2$ reaction and its isotopic variants. A comprehensive review of the early literature for the $H + H_2$ reaction appeared over 30 years ago [27]. In the 1960s it was being debated whether quantum mechanical tunneling was important in the $H + H_2$ reaction (see Ref. [28], pp. 204–206]). Since then studies on gas-phase H-atom transfer reactions, particularly using TST methods, have shown definitively that treatment of quantum mechanical effects on both bound modes and reaction coordinate motion is important in the treatment of light-atom transfer reactions, such as hydrogen atom, proton, and hydride transfers. This point is strongly supported by two comprehensive reviews of TST and applications of TST approaches to chemical reactions, including H transfer, that have been published over the past 20 years [4].

Kinetic isotope effects (KIEs) have played an important role in using experiments to unravel mechanisms of chemical reactions from experimental data [29]. The primary theoretical tool for interpreting KIEs is TST [29, 30]. The largest KIEs occur for hydrogen isotopes, for which it is critical to consider quantum mechanical effects. It has also been shown that effects of variationally optimizing the dividing surface can have significant effects on primary hydrogen KIEs [31]. VTST with multidimensional tunneling (MT) provides a more complete theory of kinetic isotope effects, which has recently been demonstrated by its ability to predict kinetic isotope effects in complex systems, such as hydride transfer in an enzymatic reaction [32].

Calculations of reaction rates with variationally determined dynamical bottlenecks and realistic treatments of tunneling require knowledge of an appreciable, but still manageably localized, region of the potential energy surface [33]. In this chapter we assume that such potentials are available or can be modeled or calculated by direct dynamics, and we focus attention on the dynamical methods.

In this chapter we provide a review of variational transition state theory with a focus on how quantum mechanical effects are incorporated. We use illustrative examples of H-transfer reactions to assist in the presentation of the concepts and to highlight special considerations or procedures required in different cases. The examples span the range from simple gas-phase hydrogen atom transfer reactions (triatomic to polyatomic systems), to solid-state and liquid-phase reactions, including complex reactions in biomolecular enzyme systems.

27.2
Incorporation of Quantum Mechanical Effects in VTST

An important consideration in developing variational transition state theory is the definition of the dividing surface separating reactants from products. A convenient choice is to consider a one-parameter sequence of dividing surfaces that are defined to be orthogonal to a reaction path [7, 34], rather than to allow more arbitrary definitions. This procedure has a few advantages. First, the variational optimization is performed for one parameter defining the dividing surfaces, even for complex, multidimensional reactions. Second, the reaction path can be uniquely defined as the path of steepest descent in a mass-weighted or mass-scaled coordinate system [35, 36], e.g., the minimum energy path (MEP), and this choice of reaction path has further advantages as discussed below. Third, use of a reaction path allows connection to the adiabatic theory of reactions, which provides the basis for including consistent, multidimensional tunneling corrections into VTST.

With this choice of dividing surfaces, a generalized expression for the transition state theory rate constant for a bimolecular reaction is given by:

$$k^{GT}(T, s) = \sigma \frac{k_B T}{h} \frac{Q^{GT}(T, s)}{\Phi^R(T)} \exp\left[-\frac{V_{MEP}(s)}{k_B T}\right] \tag{27.1}$$

where T is temperature, s is the distance along the reaction path with the convention that $s = 0$ at the saddle point and $s < 0$ (> 0) on the reactant (product) side of the reaction path, σ is a symmetry factor, k_B is Boltzmann's constant, h is Planck's constant, $Q^{GT}(T, s)$ is the generalized transition state partition function for the bound modes orthogonal to the reaction path at s, $\Phi^R(T)$ is the reactant partition function per unit volume and includes the translational partition function per unit volume for the relative motion of the two reacting species, and $V_{MEP}(s)$ is the potential evaluated on the MEP at s. The symmetry factor σ accounts for the fact that the generalized transition state partition function is computed for one reaction path, and for reactions with equivalent reaction paths this partition function needs to be multiplied by the number of equivalent ways the reaction can proceed. In computing the vibrational frequencies that are required to evaluate $Q^{GT}(T,s)$ at a value s of the reaction coordinate, we use a projection operator to project seven degrees of freedom out of the system's Hessian so that the frequencies correspond to a space orthogonal to the reaction coordinate and to three overall translations and three overall vibrations [37–40].

Canonical variational theory (CVT) is obtained by minimizing the generalized transition state rate expression $k^{GT}(T, s)$ with respect to the location s of the dividing surface along the reaction coordinate:

$$k^{CVT}(T) = \min_{s} k^{GT}(T, s) = k^{GT}\left[T, s^{CVT}(T)\right] \tag{27.2}$$

where $s^{CVT}(T)$ is the location of the dividing surface that minimizes Eq. (27.1) at temperature T.

Sometimes it is convenient to write Eq. (27.1) as [16]

$$k^{GT}(T, s) = \frac{k_B T}{h} K^{\ddagger o} \exp\left[-\Delta G_T^{GT,o}(s)\Big/RT\right] \tag{27.3}$$

where $K^{\ddagger o}$ is the reciprocal of the standard-state concentration for bimolecular reactions (it is unity for unimolecular reactions), R is the gas constant, and $\Delta G_T^{GT,o}$ is the standard-state generalized-transition-state theory molar free energy of activation. Then Eq. (27.2) becomes

$$k^{CVT}(T) = \frac{k_B T}{h} K^{\ddagger o} \exp\left(-\Delta G_T^{CVT,o}\Big/RT\right) \tag{27.4}$$

where

$$\Delta G_T^{CVT,o} = \Delta G_T^{GT,o}\left(s^{CVT}\right) \tag{27.5}$$

is the standard-state quasithermodynamic molar free energy of activation at temperature T.

Although we have described the theory in terms of taking the reaction path as the MEP in isoinertial coordinates, this can be generalized to arbitrary paths by methods discussed elsewhere [41]. The treatment of the reaction coordinate at geometries off the reaction path also has a significant effect on the results; one can use either rectilinear [34, 37–39] or curvilinear [40, 42] coordinates for this purpose, where the latter are more physical and more accurate. In particular, the vibrational modes are less coupled in curvilinear coordinates, and therefore anharmonic mode–mode coupling, which is hard to include, is less important. Recently a method has been presented for including anharmonicity in rate constant calculations of general polyatomics using curvilinear coordinates [43].

An expression similar to Eq. (27.1), but for a microcanonical ensemble (fixed energy instead of temperature) can be obtained for the generalized transition-state microcanonical rate constant. Optimizing the location of the dividing surface for this microcanonical expression at each energy and then performing a Boltzmann average of the microcanonical rate constants yields a microcanonical variational theory expression for the temperature-dependent rate constant. Canonical variational theory optimizes the dividing surface for each temperature, whereas microcanonical variational theory optimizes the dividing surface for each energy, and gives a rate constant that is lower than or equal to the CVT one [7, 15]. The improved canonical variational theory (ICVT) [16] is a compromise between CVT and the microcanonical theory, which only locates one optimum dividing surface for each temperature, but removes contributions from energies below the maximum in the ground-state adiabatic potential. The proper treatment of the reaction threshold in the ICVT method recovers most of the differences between the CVT and microcanonical theory. In most cases, CVT and ICVT give essentially the same predicted rate constants.

Equations (27.1) and (27.2) are "hybrid" quantized expressions in which the bound modes orthogonal to the reaction coordinate are treated quantum mechanically, that is, the partition functions $Q^{GT}(T, s)$ and $\Phi^R(T)$ are computed quantum mechanically for the bound degrees of freedom, although the reaction coordinate is still classical. In recent work we often use the word "quasiclassical" to refer to this hybrid. Others, mainly organic chemists and enzyme kineticists, often call this "semiclassical," but chemical physicists eschew this usage because "semiclassical" is often a good description for the WKB-like methods that are used to *include* tunneling.

The "hybrid" or "quasiclassical" approach is very old [1]. As the next step we go beyond the standard treatment, and we discuss using the adiabatic theory to develop a procedure for including quantum effects on reaction coordinate motion. A critical feature of this approach is that it is only necessary to make a partial adiabatic approximation, in two respects. First, one needs to assume adiabaticity only locally, not globally. Second, even locally, although one uses an adiabatic effective potential, one does not use the adiabatic approximation for all aspects of the dynamics.

27.2.1
Adiabatic Theory of Reactions

The adiabatic approximation for reaction dynamics assumes that motion along the reaction coordinate is slow compared to the other modes of the system and the latter adjust rapidly to changes in the potential from motion along the reaction coordinate. This approximation is the same as the Born–Oppenheimer electronically adiabatic separation of electronic and nuclear motion, except that here we use the vibrationally adiabatic approximation for an adiabatic separation of one coordinate, the reaction coordinate, from all other nuclear degrees of freedom [1, 14, 44]. The Born–Oppenheimer approximation is justified based on the large mass difference between electrons and atoms. It is less clear that the adiabatic approximation should be valid for separating different nuclear degrees of freedom, although a general principle for applying this to chemically reactive systems near the dynamical bottleneck would be that near the reaction threshold energy (which is the important energy range for thermally averaged rate constants), the reaction coordinate motion is slow because of the threshold condition. As we discuss below, the vibrationally adiabatic approximation can provide a useful framework for treating quantum mechanical tunneling.

For chemical reactions, the adiabatic approximation is made in a curvilinear coordinate system where the reaction coordinate s measures progress along a curved path through Cartesian coordinates, and the remainder of the coordinates $\mathbf{u}$ are locally orthogonal to this path. Note that the effective mass for motion along the reaction coordinate is unambiguously defined by the transformation from Cartesian to curvilinear coordinates. The effective mass may be further changed by scaling of the coordinates and momenta. When one scales a coordinate u_m by a constant $c^{1/2}$, one must change the reduced mass μ_m for that coordinate by a factor of c^{-1} so that the kinetic energy, $\frac{1}{2}\mu_m\dot{u}_m^2$, where an overdot denotes a time derivative, stays the same [45]. We are free to choose the value of the reduced mass for each coordinate, and we choose it consistently to be the same value μ for all coordinates because this makes it easier to write the kinetic energy for curved paths and for paths at arbitrary orientations with respect to the axes, and it makes it easier to make physical dynamical approximations. Coordinate systems in which the reduced masses for all motions are the same are called isoinertial. In the present article we call the constant reduced mass μ and set it equal to the mass of the hydrogen atom to allow easier comparison of intermediate quantities that depend on the reaction coordinate.

A convenient choice of the reaction path is the MEP in isoinertial coordinates, because by construction the gradient of the potential $V(s,\mathbf{u})$ is tangent to s and there is no coupling between s and $\mathbf{u}$ through second order. Therefore the potential can be conveniently approximated by

$$V(s, \mathbf{u}) = V_{\mathrm{MEP}}(s) + V_u(\mathbf{u}; s) \approx V_{\mathrm{MEP}}(s) + \sum_{i,j} u_i H_{ij}(s) u_j \qquad (27.6)$$

where the Hessian matrix for a location s along the reaction coordinate is given by

$$H_{ij}(s) = \left. \frac{\partial^2 V}{\partial u_i \partial u_j} \right|_{s, u_i = u_j = 0} \qquad (27.7)$$

and we choose the origin for the $\mathbf{u}$ coordinates to be on the MEP. Although the potential energy term in these coordinates is simple, the kinetic energy term is complicated by factors dependent upon the curvature of the reaction path [13, 14,

37, 46]. As a first approximation we will assume that the reaction-path curvature can be neglected, but we will eliminate this approximation after Eq. (27.22) because the curvature of the reaction path is very important for tunneling.

Treating bound modes quantum mechanically, the adiabatic separation between s and $\mathbf{u}$ is equivalent to assuming that quantum states in bound modes orthogonal to s do not change throughout the reaction (as s progresses from reactants to products). The reaction dynamics is then described by motion on a one-mathematical-dimensional vibrationally and rotationally adiabatic potential

$$V_a(\mathbf{n}, \Lambda, s) = V_{\mathrm{MEP}}(s) + \varepsilon_{\mathrm{int}}^{\mathrm{GT}}(\mathbf{n}, \Lambda, s) \qquad (27.8)$$

where $\mathbf{n}$ and Λ are quantum numbers for vibrations and rotations, respectively, and $\varepsilon_{\mathrm{int}}^{\mathrm{GT}}(n, \Lambda, s)$ is the vibrational-rotational energy of quantum state $(\mathbf{n}, \Lambda)$ of the generalized transition state at s. In a rigid-body, harmonic approximation, the generalized transitive-state energy level is given by

$$\varepsilon_{\mathrm{int}}^{\mathrm{GT}}(\mathbf{n}, \Lambda, s) = \sum_m \hbar\omega_m(s)\left(n_m + \frac{1}{2}\right) + \varepsilon_{\mathrm{rot}}^{\mathrm{GT}}(\Lambda, s) \qquad (27.9)$$

where the harmonic vibrational frequencies $\omega_m(s)$ are obtained from the non-zero eigenvalues of the Hessian matrix in Eq. (27.7), and the rotational energy level $\varepsilon_{\mathrm{rot}}^{\mathrm{GT}}(\Lambda, s)$ is determined for the rigid-body geometry of the MEP at location s. Six of the eigenvalues of the Hessian will be zero (for a nonlinear system), corresponding to three rotations and three translations of the total system.

If the reaction coordinate is treated classically, the probability for reaction on a state $(\mathbf{n}, \Lambda)$ at a total energy E is zero if the energy is below the maximum in the adiabatic potential for that state, and 1 otherwise:

$$P_C^A(\mathbf{n}, \Lambda, E) = \theta\left[E - V_a^A(\mathbf{n}, \Lambda)\right] \qquad (27.10)$$

where $V_a^A(\mathbf{n}, \Lambda)$ is the absolute maximum of the adiabatic potential $V_a(\mathbf{n}, \Lambda, s)$ and $\theta(x)$ is a Heaviside step function such that $\theta(x) = 0$ (1) for $x < 0$ (> 0). Since the classical reaction probability is determined entirely by whether the energy is above the adiabatic barrier or not, the neglect of reaction-path curvature in the kinetic energy term does not matter. We shall see below that this is not true when the reaction path is treated quantum mechanically, in which case the curvature of the reaction path must be included.

An expression for the rate constant can be obtained by the proper Boltzmann average over total energy E and sum over vibrational and rotational states

$$k^A(T) = \left[h\Phi^R(T)\right]^{-1}\int dE \, \exp(-E/k_B T)\sum_{\mathbf{n},\Lambda} P_C^A(\mathbf{n}, \Lambda, E) \qquad (27.11)$$

which can be reduced for a bimolecular reaction to

$$k^A(T) = \frac{k_B T}{h\Phi^R(T)}\sum_{\mathbf{n},\Lambda}\exp\left[-V_a^A(\mathbf{n}, \Lambda)/k_B T\right] \qquad (27.12)$$

Like Eq. (27.2), Eqs. (27.11) and (27.12) are also hybrid quantized expressions in which the bound modes are treated quantum mechanically but the reaction coordinate motion is treated classically. Whereas it is difficult to see how quantum mechanical effects on reaction coordinate motion can be included in VTST, the path forward is straightforward in the adiabatic theory, since the one-dimensional scattering problem can be treated quantum mechanically. Since Eq. (27.12) is equivalent to the expression for the rate constant obtained from microcanonical variational theory [7, 15], the quantum correction factor obtained for the adiabatic theory of reactions can also be used in VTST.

27.2.2
Quantum Mechanical Effects on Reaction Coordinate Motion

A fully quantum mechanical expression for the rate constant within the adiabatic approximation is given by replacing the classical reaction probabilities in Eq. (27.11) with quantum mechanical ones $P_Q^A(\mathbf{n}, \Lambda, E)$ corresponding to one-dimensional transmission through the potential $V_a(\mathbf{n}, \Lambda, s)$. Note that, at the energies of interest, tunneling and nonclassical reflection by this potential are controlled mainly by its shape near the barrier top, that is, near the variational transition state. Thus $P_Q^A(\mathbf{n}, \Lambda, s)$ only requires the assumption of local vibrational adiabaticity along with the observation that reactive systems pass through the dynamical bottleneck region in quantized transition states [47]. The quantum mechanical vibrationally and rotationally adiabatic rate constant can be expressed in terms of the hybrid expression in Eq. (27.12) by

$$k^{\text{VA}}(T) = \kappa^{\text{VA}}(T)\, k^A(T) \tag{27.13}$$

where the transmission coefficient is defined by

$$\kappa^{\text{VA}}(T) = \frac{\int\limits_0^{\infty} dE\, \exp(-E/k_B T) \sum\limits_{\mathbf{n}, \Lambda} P_Q^A(\mathbf{n}, \Lambda, E)}{\int\limits_0^{\infty} dE\, \exp(-E/k_B T) \sum\limits_{\mathbf{n}, \Lambda} P_C^A(\mathbf{n}, \Lambda, E)} \tag{27.14}$$

Rather than compute the reaction probabilities for all quantum states that contribute significantly to the sum in Eq. (27.14), we approximate the probabilities for all excited states by the probabilities for the ground state with the energy shifted by the difference in adiabatic barrier heights (relative to a single overall zero of energy) for the excited state, $V_a^A(\mathbf{n}, \Lambda)$, and ground state, V_a^{AG} [16]:

$$P_Q^A(\mathbf{n}, \Lambda, E) = P_Q^{\text{AG}}\left[E - V_a^A(\mathbf{n}, \Lambda) - V_a^{\text{AG}}\right] \tag{27.15}$$

where $P_Q^{\text{AG}}(E)$ is the reaction probability for the ground-state adiabatic potential. This approximation assumes that adiabatic potentials for excited states are similar in shape to the ground-state potential. Although this approximation is not valid in general, it works surprisingly well for calculating the transmission coefficient because at low temperatures the transmission coefficient is dominated by contributions from the ground state or states energetically similar to the ground state, and at high temperatures, where classical mechanics becomes valid, it correctly goes to a value of one. With this approximation the transmission coefficient takes the form

$$
\kappa^{\mathrm{VAG}}(T) = \frac{\int\limits_{0}^{\infty} dE \, \exp(-E/k_{\mathrm{B}}T) P_{\mathrm{Q}}^{\mathrm{AG}}(E)}{\int\limits_{0}^{\infty} dE \, \exp(-E/k_{\mathrm{B}}T) P_{\mathrm{C}}^{\mathrm{AG}}(E)}
\tag{27.16}
$$

where G in general denotes the ground state, $P_{\mathrm{Q}}^{\mathrm{AG}}(E)$ is $P_{\mathrm{Q}}^{\mathrm{A}}(\mathbf{n}, \varLambda, E)$ with $\mathbf{n}, \varLambda$ in the ground state, and $P_{\mathrm{C}}^{\mathrm{AG}}(E)$ is like $P_{\mathrm{Q}}^{\mathrm{AG}}(E)$ for all degrees of freedom except the reaction coordinate, but with the reaction coordinate motion classical. Then

$$
P_{\mathrm{C}}^{\mathrm{AG}}(E) = \theta\!\left(E - V_{\mathrm{a}}^{\mathrm{AG}}\right)
\tag{27.17}
$$

where

$$
V_{\mathrm{a}}^{\mathrm{AG}} = V_{\mathrm{a}}^{\mathrm{A}}[(\mathbf{n}, \varLambda) = G]
\tag{27.18}
$$

This yields

$$
\kappa^{\mathrm{VAG}}(T) = (k_{\mathrm{B}}T)^{-1} \exp\!\left(V^{\mathrm{AG}}/k_{\mathrm{B}}T\right) \int\limits_{0}^{\infty} dE \, \exp(-E/k_{\mathrm{B}}T) P_{\mathrm{Q}}^{\mathrm{AG}}(E)
\tag{27.19}
$$

We first consider the case where the reaction probabilities are computed for the adiabatic model with the reaction-path curvature neglected, the so-called vibrationally adiabatic zero-curvature approximation [36]. We approximate the quantum mechanical ground-state probabilities $P_{\mathrm{Q}}^{\mathrm{AG}}(E)$ for the one-dimensional scattering problem by a uniform semiclassical expression [48], which for $E < V^{\mathrm{AG}}$ is given by

$$
P^{\mathrm{SAG}}(E) = \{1 + \exp[2\theta(E)]\}^{-1}
\tag{27.20}
$$

where the imaginary action integral is

$$
\theta(E) = \hbar^{-1} \int\limits_{s_<}^{s_>} ds \{2\mu [V_{\mathrm{a}}^{\mathrm{G}}(s) - E]\}^{\frac{1}{2}}
\tag{27.21}
$$

μ is the mass for motion along the reaction coordinate, $V_{\mathrm{a}}^{\mathrm{G}}(s)$ is the ground-state adiabatic potential, that is $V_{\mathrm{a}}(\mathbf{n}, \varLambda, s)$ for $\mathbf{n} = \varLambda = 0$, and $s_<$ and $s_>$ are the classical turning points, that is the locations where $V_{\mathrm{a}}^{\mathrm{G}}(s) = E$. The uniform semiclassical approximation can be extended to energies above the ground-state barrier maximum for a parabolic barrier [48]

$$P^{SAG}\left(V^{AG} + \Delta E\right) = 1 - P^{SAG}\left(V^{AG} - \Delta E\right) \tag{27.22}$$

and we use this method to obtain reaction probabilities for energies above the barrier maximum up to $2V^{AG} - E_0$, where E_0 is the maximum of the reactant and product zero-point energies. For higher energies the probability is set to one.

When one uses CVT, one must replace $P_C^{AG}(E)$ by an approximation that is consistent with the threshold implicitly assumed by CVT. In particular we replace V_a^{AG} by $V_a^G[s^{CVT}(T)]$ in Eq. (27.17). This then yields for the rate constant with tunneling

$$k^{CVT/MT}(T) = \kappa(T)k^{CVT}(T) \tag{27.23}$$

where

$$\kappa(T) = \frac{\int\limits_0^\infty d(E/k_B T)\exp(-E/k_B T)P_Q^{AG}}{\exp\left\{-V_a^G[s^{CVT}(T)]\right\}} \tag{27.24}$$

and where MT can be SCT, LCT, or OMT.

The inability of the zero-curvature tunneling (ZCT) approximation to provide reliable rate constants has been known for over 30 years [10, 36], and over the last 25 years significant progress has been made in developing approaches to treat the multidimensional effect of reaction-path curvature in adiabatic calculations of reaction probabilities. The most successful methods for including the multidimensional effect of reaction-path curvature in adiabatic calculations of reaction probabilities specify a tunneling path that 'cuts the corner' and shortens the tunneling length [18]. Marcus and Coltrin [25] found the optimum tunneling path for the collinear H + H$_2$ reaction by finding the path that gave the least exponential damping. General multidimensional tunneling (MT) methods, applicable to polyatomic reactions, have been developed that are appropriate for systems with both small [17, 18, 22, 24] and large [20, 23, 34] reaction path curvature, as well as more general methods that optimize tunneling paths by a least-imaginary-action principle [20, 39]. In practice it is usually sufficient to optimize the imaginary action from among a small set of choices by choosing either the small-curvature tunneling (SCT) approximation or the large-curvature tunneling (LCT) approximation, whichever gives more tunneling at a given tunneling energy; this is called microcanonical optimized multidimensional tunneling (μOMT), or, for short, optimized multidimensional tunneling (OMT) [23, 49]. These methods are discussed below in more detail in the context of illustrative examples of H-transfer processes, but we anticipate that discussion and the later discussion of condensed-phase reactions by noting that all MT approximations generalize Eq. (27.24) to

$$
\kappa(T) = \frac{\int_0^\infty d(E/k_B T)\exp(-E/k_B T)P(V_1, V_2|E)}{\exp\{-V_2\left[s_*(T)\right]/k_B T\}}
\tag{27.25}
$$

where V_1 is the effective multidimensional potential energy surface, V_2 is the effective one-dimensional adiabatic potential energy curve, and $s_*(T)$ is the variational transition state location at temperature T. In the ZCT and SCT approximations, one need not specify V_1, that is the tunneling depends only on the effective one-dimensional adiabatic potential energy curve, but in the LCT and OMT approximations we need to know more about the potential energy surface than just the information contained in $V_2(s)$. For gas-phase reactions, $V_2(s)$ is just $V_a^G(s)$, and V_1 is the full potential energy surface.

27.3
H-atom Transfer in Bimolecular Gas-phase Reactions

Gas-phase reactions of two interacting reactants have provided a fertile ground for developing and testing methods for treating H-transfer reactions. In particular, triatomic reactions like $H + H_2$ have been instrumental in this development and in helping us understand the limits of validity of the approximations used in these methods, because accurate quantum mechanical results are available for comparison. We present three examples of reactions that help us present details of the methods as well as features displayed by H-atom transfers.

27.3.1
H + H$_2$ and Mu + H$_2$

The $H + H_2$ reaction and its isotopic variants have been extensively studied over the years. Muonium (Mu) is one of the most interesting isotopes studied for this reaction because Mu (consisting of a positive muon and an electron) has a mass that is about 8.8 times smaller than that of H and has the potential to exhibit very large kinetic isotope effects. Even though both reactions, $H + H_2$ and $Mu + H_2$, involve the transfer of a H atom, the presence of the much lighter Mu atom drastically changes the nature of the quantum mechanical effects on the H-transfer process. Calculations of H- and Mu-transfer rate constants are illustrated here using the Liu–Siegbahn–Truhlar–Horowitz (LSTH) potential energy surface [50].

The traditional treatment of KIEs is based upon conventional TST [29, 30], in which the dividing surface is placed at the saddle point of the reaction, with tunneling effects generally included by a separable approximation such as the Wigner correction or Bell parabolic tunneling. Using this approach with the harmonic approximation, the Mu/H KIE is determined by

$$\frac{k_{\text{Mu}}}{k_{\text{H}}} = \frac{\kappa_{\text{Mu}}^{\text{W}}}{\kappa_{\text{H}}^{\text{W}}} \frac{\Phi_{\text{H}}^{\text{R}}}{\Phi_{\text{Mu}}^{\text{R}}} \frac{Q_{\text{Mu}}}{Q_{\text{H}}} \tag{27.26}$$

The Wigner correction for tunneling depends only on the imaginary frequency $\omega_{s,X}$ for the unbound mode at the saddle point [8]

$$\kappa_{X} = 1 + \frac{1}{24} \left(\frac{\hbar |\omega_{s,X}|}{k_{\text{B}} T} \right)^{2} \tag{27.27}$$

Reactant vibrational and rotational partition functions are the same for both reactions (i.e., those for H_2) and the ratio of reactant partition functions reduces to the ratio of translational partition functions, which depends only on the reduced masses for the relative motion of the reactants

$$\frac{\Phi_{\text{H}}^{\text{R}}}{\Phi_{\text{Mu}}^{\text{R}}} = \left[\frac{m_{\text{H}}(m_{\text{Mu}} + 2m_{\text{H}})}{m_{\text{Mu}}(3m_{\text{H}})} \right]^{\frac{3}{2}} \approx 15.5 \tag{27.28}$$

The ratio of partition functions for bound modes at the saddle point is determined by the frequencies for those modes,

$$\frac{Q_{\text{Mu}}}{Q_{\text{H}}} = \prod_{i=1}^{F-1} \frac{\sinh \left(\hbar \omega_{\text{Mu},i}/2k_{\text{B}} T \right)}{\sinh \left(\hbar \omega_{\text{H},i}/2k_{\text{B}} T \right)} \tag{27.29}$$

where F is the number of vibrational modes. With the rate constant for the reaction with the light mass in the numerator, a KIE is termed 'normal' if it is greater than one. Because of the large difference in masses, the saddle point frequencies for the Mu reaction are larger than those of the H reaction, and the ratio of saddle point partition functions is less than one. The imaginary frequency for the reaction coordinate motion is also higher for the Mu reaction than for the H reaction, so that the ratio of tunneling factors is greater than one as well as the ratio of reactant partition functions. The saddle point frequencies for these two reactions using the LSTH potential energy surface are (2059, 909, and 1506 *i* cm⁻¹) and (4338, 1382, and 1784 *i* cm⁻¹) and for the (stretch, bend, and unbound) modes for H + H_2 and Mu + H_2, respectively. Using these frequencies in harmonic transition state theory with Wigner tunneling gives a KIE less than one, which is termed inverse, as shown in Fig. 27.1, where the TST/W results are compared with experiment [51]. Although TST with Wigner tunneling gives the right qualitative trend, both the magnitude and slope are inaccurate. We next discuss the other curves in Fig. 27.1, which present improvements in the treatment of quantum mechanical effects for the hydrogen transfer process.

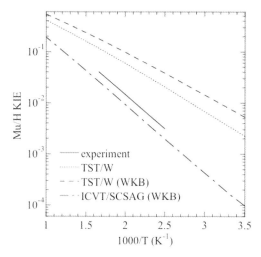

Figure 27.1 Kinetic isotope effects for the Mu/H + H$_2$ reaction as a function of temperature.

A first consideration is the treatment of the bound vibrational modes, which in the TST/W results shown in Fig. 27.1, use the harmonic approximation. The total harmonic zero-point energy at the saddle point (for stretch and bend vibrations) is much higher for the Mu reaction, 10.2 kcal mol^{-1}, than for the H reaction, 5.5 kcal mol^{-1}. As shown in Fig. 27.2, the stretching vibration extends to larger distances and higher energies for the Mu reaction than for the H reaction, and therefore accesses more anharmonic parts of the potential. In this situation methods for including anharmonicity must be considered [52, 53].

The straight lines through the saddle point end at the classical turning points for the harmonic approximation to the stretch potential at the saddle point. For the symmetric H + H$_2$ reaction the harmonic turning points extend just past the 12 kcal mol^{-1} contour, and on the concave side, it is very close to the accurate anharmonic turning point, calculated using a WKB approximation [53]. For the Mu + H$_2$ reaction the harmonic turning point on the concave side falls short of the anharmonic turning point, which is near the 16 kcal mol^{-1} contour, and it extends past the 20 kcal mol^{-1} contour on the convex side of the turning point, clearly indicating that the potential for this mode is quite anharmonic. Comparison of the curves label TST/W (harmonic treatment) and TST/W (WKB) in Fig. 27.1 shows the importance of anharmonicity in the quantum treatment of bound states. When the vibrational modes at the saddle point are treated more accurately using a WKB method [53], the Mu rate constants are increased by about a factor of two, while the H rate constants change only slightly, leading to a larger disagreement with experiment for the KIE.

A more accurate treatment of the reaction uses variational TST, in which the dividing surface is allowed to move off the saddle point, or equivalently, uses the adiabatic theory as described in Section 27.2. The vibrationally adiabatic potential

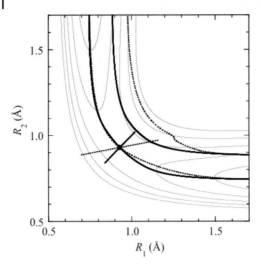

Figure 27.2 Potential energy contours (thin solid curves) at 4, 8, 12, 16, and 20 kcal mol^{-1} are shown for collinear A–HH geometries (A = H or Mu) with HH and AH distances represented by R_1 and R_2, respectively. The solid diamond denotes the saddle point. Thick solid and dashed curves are for the H and Mu reactions, respectively. Harmonic stretch vibrational modes are the straight lines through the saddle point. Minimum energy paths are the curved lines through the saddle point. The curved lines on the concave side of the MEP are paths of turning points for the anharmonic stretch vibration.

curves for the two reactions are shown in Fig. 27.3 and are compared with the potential along the MEP. The MEPs for the two reactions, as shown in Fig. 27.2, are very close to each other, so that the potentials along the MEP are also about the same. Note that the MEPs are paths of steepest descent in a mass-weighted or mass-scaled coordinated system, and therefore, the MEPs for the H and Mu reactions are slightly different. The largest differences are seen in the entrance channel (large R_1) where the reaction coordinate is dominated by either H or Mu motion relative to H$_2$, while in the exit channel (large R_2) the reaction coordinate for both reactions is an H atom moving relative to the diatomic product (either H$_2$ or MuH). The reaction coordinate in Fig. 27.3 is defined as the arc length along the MEP through mass-weighted coordinates. To facilitate comparisons of potential curves for the two reactions, we use the same effective mass for the mass weighting – the mass of the hydrogen atom. Because the mass of Mu is so much lighter than H, the scale of s on the reactant side is contracted when the same mass weighting is used for both reactions, leading to a steeper increase for Mu.

The ground-state adiabatic potential curves in Fig. 27.3 are constructed by adding accurate anharmonic zero-point energies for the stretch and bend modes to V_{MEP}. On the reactant side the shapes of V_{MEP} and V_a^G are very similar with the adiabatic potential being shifted up by approximately the zero-point energy for the H$_2$ stretch vibrations, 6.2 kcal mol^{-1}. Near the saddle point this contribution decreases markedly for the H + H$_2$ reaction, to 2.9 kcal mol^{-1}, causing the adia-

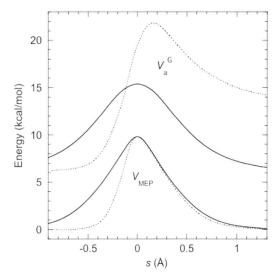

Figure 27.3 Potential along the MEP (V_{MEP}, lower pair of curves) and ground-state adiabatic potential (V_a^G, upper pair of curves) as a function of reaction coordinate s for the H + H$_2$ reaction (solid lines) and Mu + H$_2$ reaction (dashed line).

batic potential to be less peaked than V_{MEP} for this reaction. Contributions from the bending vibration near the saddle point are about 2.6 kcal mol^{-1}, otherwise the adiabatic potential curve would be even flatter near the saddle point. The zero-point energy for MuH is 13.4 kcal mol^{-1}, which accounts for the large difference in the H and Mu adiabatic curves in the product region and the shift of its maximum toward products. The difference in the maximum of the adiabatic curve and its value at the saddle point is about 2.3 kcal mol^{-1}, which leads to a decrease by about a factor of 10 in the Mu rate constant at 500 K. This is the main reason for the large shift in the curve labeled ICVT/SCSAG (WKB) relative to the TST/W (WKB) curve in Fig. 27.1.

We now turn our attention to the issue of quantum mechanical tunneling in these H-atom transfer reactions. The Wigner and Bell tunneling methods use the shape of V_{MEP} at the saddle point to estimate the tunneling correction. The effective mass for the reaction coordinate in Fig. 27.3 is the same for both reactions, therefore, tunneling is treated as the motion of a particle with the mass of a hydrogen atom through the potentials in the figure. The similarity in the V_{MEP} curves for H and Mu indicates why the tunneling correction using these methods gives similar results for the H and Mu reactions. For example, Wigner tunneling gives corrections for Mu that are less than 30% higher those for H for 300 K and higher temperatures. The shapes of the adiabatic curves exhibit greater differences with the curve for the Mu reaction having a narrow barrier near the maximum. When reaction-path curvature is neglected, the tunneling correction factors for Mu are factors of 2.5 and 1.6 higher than those for H at temperatures of 300 and 400 K.

As discussed above, the most accurate methods for treating tunneling include the effects of reaction-path curvature. The original small-curvature tunneling (SCT) method [17] provides an accurate description of the H-transfer process in these triatomic H-atom transfer reactions. In the Marcus–Coltrin method [25] the tunneling occurs along the path of concave-side turning points for the stretch vibration orthogonal to the reaction coordinate. Figure 27.2 shows paths of turning points $t_{str}(s)$ for the stretch vibration for the H/Mu + H$_2$ reactions, where the turning points are obtained for the anharmonic potential at the WKB zero-point energy. Tunneling along this path shortens the tunneling distance and the effect of the shortening of the path can be included in the calculation of the action integral by replacing the arc length along the MEP ds in Eq. (27.15) by the arc length along this new path $d\xi$, or equivalently by including the Jacobian $d\xi/ds$ in the integrand of Eq. (27.15). An approximate expression for $d\xi/ds$ can be written in terms of the curvature of the MEP and vibrational turning points [17, 18]. The MEP is collinear for the H + H$_2$ reaction and the curvature coupling the bend vibration to the reaction coordinate is zero for collinear symmetry. Therefore, the Jacobian can be written just in terms of the one mode

$$\left(\frac{d\xi}{ds}\right)^2 \approx [1 - \kappa(s)t_{str}(s)]^2 + \left(\frac{dt_{str}}{ds}\right)^2 \tag{27.30}$$

The SCT method extends the Marcus–Coltrin idea in a way that eliminates problems with the Jacobian becoming unphysical. Rather than including the Jacobian factor, the reduced mass for motion along the reaction coordinate μ is replaced by $\mu_{eff}(s)$ in Eq. (27.21), where $\mu_{eff}(s)$ is given by

$$\frac{\mu_{eff}(s)}{\mu} \approx \min\left\{1, \exp\left[-2a(s) - [a(s)]^2 + \left(\frac{dt_{str}}{ds}\right)^2\right]\right\} \tag{27.31}$$

$$a(s) = \kappa(s)t_{str}(s) \tag{27.32}$$

where $\kappa(s)$ is the curvature coupling between the reaction coordinate motion and the stretch vibrational motion [37]. Note that the signs of $\kappa(s)$ and $t_{str}(s)$ are chosen so that the path lies on the concave side of the path and their product $a(s)$ is positive.

The reaction-path curvature is given by the coupling of the stretch vibration to the reaction coordinate in the mass-weighted coordinate system, not the coordinate system used to display the paths in Fig. 27.2. The reaction Mu + H$_2$ has smaller reaction-path curvature than the H + H$_2$ reaction, by about a factor of two in the region near the peak of the adiabatic barriers, and the enhancement of the tunneling from corner cutting is much less for the Mu reaction. Neglect of reaction-path curvature gave tunneling factors for the Mu reaction that are much higher than those for the H reaction and including the effects of the curvature greatly reduces this large difference. In fact, at 300 K without curvature the Mu reaction has a tunneling factor that is 2.5 times larger than the H reaction and this is reduced to an enhancement of only 2% when curvature is included with the SCT method. For temperatures from 400 to 600 K, the SCT tunneling factors for the Mu reaction are lower than those for the H reaction by about 9%. When these effects are included the predicted KIEs are in good agreement with the experimental results, being only 30–40% low, as shown in Fig. 27.1.

The H/Mu + H_2 reactions are examples of H-atom transfers with relatively small reaction-path curvature and provide a good example of how the description of the hydrogen transfer process is effected by quantization of bound modes, variational optimization of the location of the dividing surface, and inclusion of quantum mechanical effects on reaction coordinate motion. The magnitude of the reaction-path curvature for an H-atom transfer reaction is often correlated with the skew angle, where the skew angle is defined as the angle between the gradient along the reaction path in the product channel with that in the reactant channel. For the H-atom transfer reaction AH + B → A + BH this angle is defined by

$$\cos \beta_{\text{skew}} = \left[\frac{m_A m_B}{(m_A + m_H)(m_B + m_H)} \right]^{1/2} \tag{27.33}$$

where A and B can be atomic or polyatomic moieties with masses m_A and m_B. Skew angles for the H and Mu reactions are 60° and 77°, respectively, and we saw above how the larger curvature in the system with the smaller skew-angle system resulted in greater tunneling. When the masses of A and B are much larger than the mass of H, the skew angle can become very small, resulting in large reaction-path curvature. These systems require tunneling methods that go beyond the small-curvature approach used here.

27.3.2
Cl + HBr

The collinear Cl + HBr reaction provides an example of a system with a very small skew angle. Figure 27.4 shows potential energy contours for this collinear reaction in mass-scaled coordinates x and y for the potential energy surface of Babamov et al. [54], where x is the distance from Cl to the center of mass of HBr and y is a scaled HBr distance. The kinetic energy is diagonal in this coordinate system and the scaling of y is chosen so that the effective masses for x and y motion are the same. Therefore, reaction dynamics in this coordinate system can be viewed as a single mass point moving on the potential energy contours in Fig. 27.4. The skew angle, which in this coordinate system is the angle between the minimum energy path in the asymptotic reactant channel and the asymptotic product channel, is only 12°. Regions of large reaction-path curvature, which can be seen near the saddle point, lead to a breakdown of the approximations used in the SCSAG method. The approximation of vibrational adiabaticity is valid in the entrance and exit channels where the stretch vibration is dominated by motion of the hydrogen

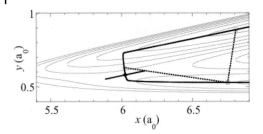

Figure 27.4 Potential energy contours (thin solid curves) from −10 to 20 kcal mol⁻¹ (spaced every 5 kcal mol⁻¹) are shown for the collinear Cl–H–Br reaction as a function of internal coordinates x and y (see text). A solid diamond denotes the saddle point and the thick straight line through the saddle point is the anharmonic vibrational mode. The thick solid curve is the minimum energy path. The classical turning point on the ground-state adiabatic potential energy curve at 9 kcal mol⁻¹ is indicated by the unfilled symbol in the entrance channel. Turning points for adiabatic potential curves with the stretch vibration in its ground state ($n = 0$) and excited state ($n = 2$) are shown as an unfilled circle ($n = 0$) and square ($n = 2$) in the exit channel. Dashed lines connect the turning point for the ground-state adiabatic potential curve in the entrance channel with the turning points for $n = 0$ and 2 in the exit channel.

atom. At the saddle point the vibrational motion more nearly resembles the relative motion of the two heavy atoms leading to a low vibrational energy (e.g., only 0.4 kcal mol⁻¹ at the saddle point compared to 3.8 kcal mol⁻¹ for reactants). The thick straight line through the saddle point shows the extent of the vibrational motion. Because of the large reaction-path curvature, the regions of vibrational motion in the reactant valley on the concave side of the MEP, just before the bend in the MEP, overlap with the vibrational motion at the saddle point. This complication, and the strong coupling of the reaction coordinate motion to the vibration orthogonal to it, argues against an adiabatic treatment of hydrogen atom tunneling in the saddle point region. For this type of system the LCT method is more appropriate [19, 20, 22, 39, 49, 55], and we describe it briefly here. A key aspect of LCT methods is that the tunneling depends on more aspects of the potential energy surface than just $V_a^G(s)$, and that is why we introduced the multidimensional potential energy surface in Eq. (27.25).

In the vibrationally adiabatic approximation, tunneling at a fixed total energy is promoted by motion along the reaction coordinate and initiates from the classical turning point on the adiabatic potential. The physical picture in the LCT method is that the rapid vibration of the hydrogen atom promotes transfer of the hydrogen atom between the reactant and product valleys and this hopping begins from turning points in the vibrational coordinate on the concave side of the MEP. For a given total energy, tunneling can take place all along the entrance channel, up to the adiabatic turning point, as the reactants approach and recede. Tunneling is assumed to occur along straight-line paths from the reactant to product valleys, subject to the constraint that adiabatic energies in the reactant and product channels are the same. Figure 27.4 illustrates the LCT tunneling paths, where the straight dashed line is the tunneling path connecting points, denoted by open circles, along the MEP for which the ground-state adiabatic potential curve has an energy of 9 kcal mol^{-1}. In the large-curvature ground state approximation, version 3 or version 4 [22, 39, 49], the straight-line path used to specify a given tunneling path initiates and terminates on the MEP rather than at the turning points for the vibrational motion. This assumption simplifies the extension of the method to polyatomic reactions, and it yields results that are similar to earlier versions of the method with more complicated specifications. In current work one should always use the latest version (version 4) of the LCT method because it incorporates our most complete experience on how to embed the physical approximations in a stable algorithm, although the differences between the versions are small in most cases.

Although the LCT method does not rely on the adiabatic approximation in the region where it breaks down (i.e., the nonadiabatic region), it does use the approximation in the reactant and product channels to determine the termini for the straight-line tunneling paths. Figure 27.5 shows adiabatic potential curves in the reactant and product regions and the potential along the MEP. This reaction is exoergic by about 16 kcal mol^{-1} with a barrier over 10 kcal mol^{-1} higher than the minimum of the reactant valley. Because of the large exoergicity and rapid decrease of the potential along the MEP on the product side, the value of the

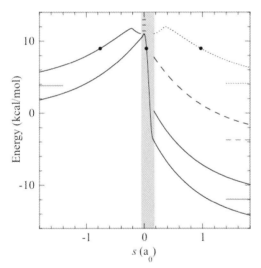

Figure 27.5 Potential along the minimum energy path (lowest continuous curve) and adiabatic potential segments in the reactant and product regions for $n = 0$ (solid curve), 1 (long dashed curve) and 2 (short dashed curve) as a function of reaction coordinate for the collinear Cl + HBr reactions. The values of the adiabatic potential curves in the asymptotic reactant and product regions are shown as short straight-line segments on the left and right of the plot. The gray shaded area around the saddle point is a region where the adiabatic approximation is not valid (see text). The 3 small tick marks at $s = 0$ are the values the 3 adiabatic potential curves would have at the saddle point. The bullets are turning points for a total energy of 9 kcal mol^{-1}.

ground-state adiabatic potential at the edge of the nonadiabatic region on the product side is already quite low (near 0).

The vibrationally adiabatic approximation requires that vibrational quantum numbers remain constant throughout a reaction. The strong coupling induced by the reaction-path curvature can lead to appreciable nonadiabaticity and population of excited states in the product channel, and this effect is included in the LCT method. (As mentioned above, the SCT approximation does not imply global vibrational adiabaticity either, but it does assume adiabaticity for the effective potential during the entire tunneling event itself; the LCT approximation includes vibrational nonadiabaticity even for the effective potential during the tunneling event.) Figure 27.5 shows the product-channel segments of the adiabatic potential curves for the ground and first two excited states. The product-side turning point for the first excited adiabatic curve also falls within the nonadiabatic region like the ground-state one, and on the scale of the plot is not discernible from the ground-state turning point. The energies of the $n = 2$ adiabatic curve are sufficiently high that the turning points occur well out into the product region (around $s = 1$ a_0 for a tunneling energy of 9 kcal mol^{-1}). Figure 27.4 shows the tunneling path corresponding to these turning points. This path is seen to cut the corner significantly. The barrier to tunneling along this path is comparable to the adiabatic barrier and the shorter tunneling path offered by this corner cutting greatly enhances the tunneling. The LCT method was extended to account for contributions from tunneling into excited states of products [55], and for this reaction, the contribution to the tunneling correction factor is dominated by tunneling into the $n = 2$ state.

The small-curvature (SC) and large-curvature (LC) methods were developed to treat tunneling in the cases of two extremes of reaction-path curvature. In the SC methods, the effective tunneling path (which is implicit but never constructed and not completely specified, since it need not be) is at or near the path of concave-side turning points for the bound vibrational motions that are coupled to the reaction coordinate motion. In the LC methods, the effective tunneling paths (which are explicit) are straight-line paths between the reactant and product valleys. The optimum tunneling paths for reactions with intermediate reaction-path curvature may be between these two extremes, and for these reactions the least-action tunneling (LAT) method [20, 39, 56] is most appropriate. In the LAT method, we consider a sequence of tunneling paths depending on a single parameter α such that for $\alpha = 0$ the tunneling path is the MEP and for $\alpha = 1$ it is the LCT tunneling path. The optimum value of α (yielding the optimum tunneling path) for each tunneling energy is determined to minimize the imaginary-action integral and thereby maximize the tunneling probability. Figure 27.6 compares rate constants computed by the ICVT method, including tunneling by SCT, LCT, and LAT methods, with accurate quantum mechanical ones [55] for the collinear Cl + HBr reaction. The adiabatic method (SCT) cannot account for the large probability of populating the $n = 2$ excited product state and underestimates the accurate rate constants by factors of 3 to 6 for temperatures from 200 to 300 K. The LCT and LAT methods agree to within plotting accuracy, and are therefore shown as one

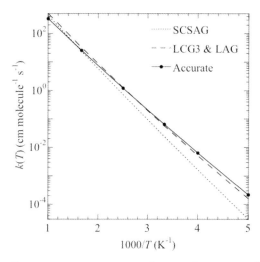

Figure 27.6 Rate constants as a function of temperature for the collinear Cl + HBr reaction. Accurate quantum mechanical rate constants (solid line with bullets) are compared with those computed using improved canonical variational theory (ICVT) with tunneling included by SCSAG (dotted line) and LCG3 and LAG (long dashed line).

curve, indicating that the optimum tunneling paths for this reaction are the straight-line paths connecting the reactant and product valleys. These methods underestimate the accurate rate constants by only 10–25% for T from 200 to 300 K and agree to within 50% over the entire temperature range from 200 to 1000 K. The excellent agreement with accurate rate constants for this model system indicates the good accuracy provided by the LCT and LAT methods for this type of small skew angle reaction.

The physical picture of tunneling in this system provided by the approximate, yet accurate, tunneling methods is very different than descriptions of tunneling in simpler conventional models of tunneling. In the Wigner and Bell tunneling approximations, properties of the potential near the saddle point determine the tunneling correction factors. As illustrated in Fig. 27.4, barriers along straight-line paths, which connect the reactant and product channels, control the actual tunneling in this small-skew angle system, and these paths are significantly displaced from the saddle point.

27.3.3
Cl + CH$_4$

The higher dimensionality of polyatomic reactions makes them more of a challenge to treat theoretically. Variational transition state theory with multidimensional tunneling has been developed to allow calculations for a wide variety of polyatomic systems. In this section we consider issues that arise when treating polyatomic systems. The Cl + CH$_4$ reaction provides a good system for this purpose because the accurate SPES potential energy surface of Corchado et al. [57] is available and a variety of experimental results [58] exist to validate the methods.

Hydrogen transfer between Cl and CH_3 corresponds to a heavy–light–heavy mass combination and this reaction has a small skew angle of about 17° and regions of large curvature along the reaction path. The reaction is endoergic by 6.1 kcal mol^{-1} on the analytical potential energy surface with its barrier in the product valley (the HCl bond length at the saddle point is only 0.08 Å longer than HCl in products while the CH bond length is 0.30 Å longer at the saddle point than in the reactants). Figure 27.7 shows the potential along the MEP and the ground-state adiabatic potential for this reaction, harmonic frequencies $\omega_i(s)$, and components $\kappa_i(s)$ of reaction-path curvature along the reaction coordinate. Although there are 11 vibrational modes orthogonal to the reaction path, only 3 have significant curvature components. Relative motion of the two heavy moieties CH_3 and Cl dominates the reaction coordinate in the reactant and product regions, while in the interaction region, where the curvature is largest, motion of the H atom between CH_3 and Cl characterizes reaction coordinate motion. The mode that couples most strongly to reaction coordinate motion mirrors this behavior and is denoted the reactive mode. It originates as a nondegenerate CH stretch in the reactants, transforms into motion that is dominated by C–Cl vibration in the region of strong coupling, and ends as an HCl stretch in the products. Note that regions of large reaction-path curvature also coincide with regions where the har-

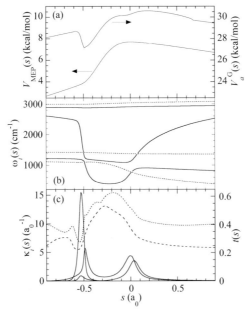

Figure 27.7 (a) Potential energy and ground-state adiabatic potential curves as a function of reaction coordinate for the Cl + CH_4 reaction using the SPES surface. (b) Nine highest harmonic frequencies for modes orthogonal to the reaction coordinate. Doubly degenerate modes are shown as dashed curves. The two lowest frequency transition modes are not shown. (c) Components of reaction-path curvature (solid lines) for 3 vibrational modes and two approximations for turning points along the curvature vector (dashed curves) as a function of reaction coordinate. Short dashed curve is the $\hat{\kappa}(s) \cdot \mathbf{t}(s)$ approximation and the long dashed curve is $\bar{t}(s)$ (see text).

monic frequencies change rapidly and we observe crossings of modes. Because of the late barrier, the largest curvature occurs well before the saddle point (between $s = -0.6$ and -0.4 a_0), where the potential along the MEP is only about half the value at the barrier maximum and the adiabatic potential exhibits a local minimum. The peak in the curvature and dip in the adiabatic potential are a result of the transformation of the reactive mode from a high-frequency CH stretching mode (2870 cm^{-1} in the reactants) to a lower frequency mode (~1300 cm^{-1} at $s = -0.5$ a_0) with contributions from CCl motion. A second region of large curvature occurs near the saddle point where this low-frequency mode transforms into a high frequency HCl vibration (2990 cm^{-1} at products). A second mode, corresponding to a methyl umbrella mode, also shows significant coupling to the reaction coordinate, and the value of curvature coupling for this mode is larger than the reactive mode near $s = 0$). A third mode, corresponding to a high-frequency CH stretching mode throughout the reaction, exhibits much smaller, but still significant, coupling near $s = -0.5$ a_0).

Accurate treatment of tunneling in this reaction requires consideration of how the curvature in multiple dimensions is taken into account. First we consider how the SCT method is defined to consistently treat reactions with curvature coupling in multiple modes. In SCT, we assume that the corner cutting occurs in the direction along the curvature vector $\boldsymbol{\kappa}(s)$ in the space of the local vibrational coordinates $\mathbf{Q}$. To emphasize this, the final version of the SCT method was originally called the centrifugal-dominant small-curvature approximation [22]. In this method, we make a local rotation of the vibrational axes so that $\boldsymbol{\kappa}(s)$ lies along one of the axes, u_1, and by construction the curvature coupling in all other vibrational coordinates, u_i, $i = 2$ to $F - 1$, are zero in this coordinate system. Defining $\bar{t}(s)$ as the turning point for zero-point motion in the potential for the u_i coordinate, the effective mass in the imaginary action integral is given by the SCT expression for one mode coupled to the reaction coordinate, as written in Eq. (27.31), with $a(s)$ replaced by

$$\bar{a}(s) = \left(\sum_{i=1}^{F-1} [\kappa_i(s)]^2 \right)^{\frac{1}{2}} \bar{t}(s) \tag{27.34}$$

where F is the number of vibrational modes, $t_i(s)$ is the turning point for mode i on the concave side of the MEP. The definition of $\bar{t}(s)$ is provided in previous work for a harmonic description of the vibrational modes [22]. We illustrate here how it works for the Cl + CH$_4$ reaction.

As discussed above, only three modes contribute significantly to the reaction-path curvature in the Cl + CH$_4$ reaction, and the coupling for two of the modes is much greater than for the third. Figure 27.8 shows a contour plot of the two harmonic vibrational modes with the largest coupling where the frequencies are those at $s = -0.49$ a_0. Turning points in these modes, $t_1(s)$ and $t_2(s)$, are indicated by the square and triangle. The direction $\hat{\boldsymbol{\kappa}}(s)$, a unit vector, of the curvature-coupling vector $\boldsymbol{\kappa}(s)$ is shown as a straight line and this line defines the u_1 axis. The line extends out to a value equal to $\hat{\boldsymbol{\kappa}}(s) \cdot \mathbf{t}(s) = \kappa_1(s)t_1(s) + \kappa_2(s)t_2(s)$. This approximation to the turning point in u_1 (which is what one would use if one allowed independent corner cutting in every generalized normal mode) gives a value that is too

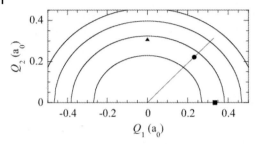

Figure 27.8 Potential energy contours for two harmonic vibrational modes, which are orthogonal to the reaction coordinate, for the Cl + CH$_4$ reaction at $s = -0.49$ a_0 on the reaction coordinate. The straight line is the direction u_1 of the reaction-path curvature vector and the symbols are turning points for zero-point harmonic motion along Q_1 (square), Q_2 (triangle), and u_1 (circle).

high when compared with value of $\bar{t}(s)$ for the SCT method, which is indicated by the circle. Figure 27.7 presents a comparison of $\bar{t}(s)$ and $\hat{\kappa}(s) \cdot \mathbf{t}(s)$ along the reaction path and shows that the value obtained using the $\hat{\kappa}(s) \cdot \mathbf{t}(s)$ approximation is consistently larger than $\bar{t}(s)$. For systems with many modes contributing significantly to the reaction-path curvature, the overestimate of the turning point that one would obtain by allowing independent corner cutting in every generalized normal mode is even larger and unphysical. Equation (27.34) gives a consistent procedure to extend the SCT approach to multidimensional systems.

As was the case with the Cl + HBr reaction, the small skew angle and concomitant large reaction-path curvature in the Cl + CH$_4$ reaction require consideration of methods beyond the small-curvature approximation. It might be argued that the SCT method is adequate because the region of largest curvature falls outside the region where tunneling contributes significantly to the thermal rate constant. However, the only true test is to perform calculations that treat corner cutting more accurately for large-curvature systems. Previous work on this system has shown that the optimum tunneling paths are the straight-line paths used in the LCT method [57]. Consistent procedures have been presented for extending large-curvature methods to multidimensional systems [22, 39, 49]. In the LCG3 and LCG4 versions of the LCT method the tunneling paths are uniquely defined as straight lines between points on the MEP in the reactant and product valleys, and the key to their success is the definition of the effective potentials along these tunneling paths. As mentioned previously, the SCT and LCT methods represent approaches that are most appropriate for two extremes and the most general and optimal way to interpolate between these extremes is the least-action method. A simpler optimized tunneling (OMT) approach [23, 59] is obtained by using the SCT and LCT reaction probabilities and choosing the one that gives the largest tunneling probability at each energy. In this case the OMT probability is given by

$$P^{\text{OMT}}(E) = \max \left\{ \begin{array}{l} P^{\text{SCT}}(E) \\ P^{\text{LCT}}(E) \end{array} \right. \tag{27.35}$$

and the microcanonical optimized multidimensional tunneling (μOMT) tunneling correction factor is obtained by substituting this expression for the probability into Eq. (27.13). Rate constants and kinetic isotope effects for the Cl + CH$_4$ reaction, its reserve, and its isotopic variants, computed using μOMT on the potential of Corchado et al. agree well with experiment [57].

27.4
Intramolecular Hydrogen Transfer in Unimolecular Gas-phase Reactions

Intramolecular hydrogen transfer is another important class of chemical reactions that has been widely studied using transition state theory. Unimolecular gas-phase reactions are most often treated using RRKM theory [60], which combines a microcanonical transition state theory treatment of the unimolecular reaction step with models for energy redistribution within the molecule. In this presentation we will focus on the unimolecular reaction step and assume that energy redistribution is rapid, which is equivalent to the high-pressure limit of RRKM theory.

Unimolecular hydrogen transfer reactions require additional considerations beyond those discussed for bimolecular reactions. The expression for the thermal rate constant takes the same form as Eq. (27.1), but the reactant partition function per unit volume in the bimolecular expression is replaced by a unitless partition function for the vibrations and rotations of the reactant molecule. More serious considerations are required in treating quantum mechanical effects, particularly tunneling. For bimolecular reactions, quantum mechanical tunneling can be initiated by relative translational motion along the reaction coordinate or by vibrational motion in small skew-angle systems. For unimolecular reactions, vibrational motion alone promotes tunneling. For bimolecular reactions, heavy–light–heavy mass combinations require the reaction coordinate to have regions of large reaction-path curvature to connect the reaction paths in the asymptotic entrance and exit channels. (If the barrier occurs in the region of high curvature, large-curvature tunneling may dominate small-curvature tunneling.) Such a general statement cannot be made for unimolecular reactions, and the type of reaction-path curvature in unimolecular H-transfer reactions can vary from small-curvature to large-curvature.

Initiation of tunneling by vibrational motion in the reaction coordinate motion requires modification to the expression used to obtain the thermally averaged tunneling correction factor, Eq. (27.19). For unimolecular processes tunneling does not occur for a continuum of translational energies, but from discrete energy levels in the bound wells of the adiabatic potential. In this case the integral in Eq. (27.19) should be replaced by a sum over discrete states plus contributions from continuum energies above the barrier [61, 62]

$$\kappa^{\mathrm{VA}}(T) = (k_{\mathrm{B}}T)^{-1}\exp\left(-V^{\mathrm{AG}}/k_{\mathrm{B}}T\right) \sum_{v} \frac{\mathrm{d}\varepsilon_v}{\mathrm{d}v}\, P_{\mathrm{Q}}^{\mathrm{AG}}(\varepsilon_v)\, \exp(-\varepsilon_v/k_{\mathrm{B}}T)$$

$$+ (k_{\mathrm{B}}T)^{-1}\exp\left(-V^{\mathrm{AG}}/k_{\mathrm{B}}T\right) \int_{V^{\mathrm{AG}}}^{\infty} \mathrm{d}E\; P_{\mathrm{Q}}^{\mathrm{AG}}(E)\exp(-E/k_{\mathrm{B}}T) \qquad (27.36)$$

where the sum is over all bound states along the reaction coordinate motion in the reactant well, ε_v is the energy of state v in the reactants, and $P_Q^{AG}(E)$ for energies above the barrier are given by Eq. (27.22) up to $2V^{AG} - E_0$ and are set to one above that energy. Equations (27.19) and (27.36) are equivalent for a sufficiently large density of states in the reactant well.

We provide two examples of intramolecular hydrogen transfer reactions in polyatomic systems to illustrate the convenience and value of VTST methods for treating these types of reactions.

27.4.1
Intramolecular H-transfer in 1,3-Pentadiene

The [1,5] sigmatropic rearrangement reaction of *cis*-1,3-pentadiene proceeds via hydrogen transfer from C-5 to C-1, and a primary kinetic isotope effect has been observed experimentally [63]. The large number of degrees of freedom (33 vibrational modes at the reactants) and types of motions involved in the rearrangement process, including torsional motions and vibrations of the carbon skeletal modes, as well as H atom motions, complicate theoretical treatment of this reaction. For this reason, approaches based on reduced-dimensional models [64] have difficulty capturing the correct dynamics of the rearrangement process. Variational transition state theory with multidimensional tunneling has been applied to this reaction in its full dimensionality to provide a complete understanding of the dynamics of the rearrangement process and the importance of tunneling in it [24, 65]. These studies used the direct dynamics approach [33, 66] in which electronic structure calculations of energies, gradients, and Hessians are performed as needed.

The reactant configuration of 1,3-pentadiene is the *s-trans* conformer. Denoting the dihedral angles for C1–C2–C3–C4 and C2–C3–C4–C5 as ϕ_1 and ϕ_2, respectively, motion along the MEP out of the reactant well corresponds to rotation of ϕ_1 around the C2–C3 single bond from 180° to a value of about 30°. The change in energy for this motion along the reaction coordinate is relatively small compared to the barrier height of 39.5 kcal mol^{-1}. Once the ethylene group (C1–C2) approaches the C5 methyl group, the second dihedral angle changes in a concerted manner with ϕ_1, that is, ϕ_2 increases from a value of zero as ϕ_1 continues to decrease. The potential along the MEP is shown in Fig. 27.9, and the left most extreme of the reaction coordinate ($s = -3\ a_0$) is approximately the value of the reaction coordinate where ϕ_2 starts to change. Much closer to the saddle point (within about 0.5 a_0) the reaction coordinate motion is characterized by H-atom motion (relative to C1 and C5) accompanied by rearrangement in the C–C distances, with the largest changes in the C1–C2 and C4–C5 distances. The saddle point is a cyclic structure with C_s symmetry; the transferring H atom is equidistant from the C_1 and C_5 carbon atoms with a bent C–H–C configuration.

Analysis of the frequencies along the minimum energy path allows identification of the modes that are most strongly coupled to the reaction coordinate and have the largest participation in the tunneling process. Figure 27.9 shows all 32

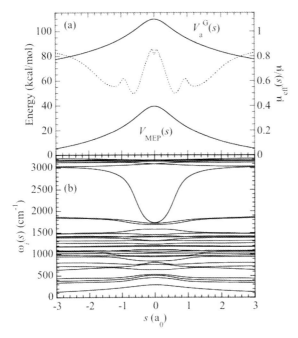

Figure 27.9 (a) Potential energy and ground-state adiabatic potential curves (solid curves) and SCT effective mass (dashed curve) as a function of reaction coordinate for the intramolecular H-tranfer in 1,3-pentadiene. (b) Harmonic frequencies for modes orthogonal to the reaction coordinate.

frequencies and the one mode that shows the most rapid change near the saddle point is that with the largest curvature coupling. This reactive mode starts as a CH vibration in reactants and transforms into the C2–C3–C4 asymmetric stretch near the saddle point. This mode accounts for the largest single component to the reaction-path curvature at the saddle point, varying from about 1/3 to 2/3 of the total contribution in the region between $s = -0.3$ and 0.3 a_0. The highest frequency modes (those above 3000 cm^{-1} at the saddle point) contribute less than a couple of percent and the lowest frequency modes (those below about 700 cm^{-1} at the saddle point) only contribute between 15 and 20% to the reaction-path curvature. This analysis shows the shortcomings of simple reduced-dimensional models of this complicated rearrangement and tunneling process. First, it is difficult to guess, *a priori*, the mode or modes that are critical for an accurate description of the multidimensional tunneling process [64]. Second, even when the dominant mode is discovered, there are 20 other modes in the range 800 to 2000 cm^{-1} that do contribute significantly to the curvature, and an accurate treatment of tunneling needs to account for motion in those degrees of freedom.

For this reaction, both LCG3 and SCT methods were applied to calculate tunneling corrction factors [24]. The SCT method gave larger tunneling probabilities, indicating that reaction-path curvature is small to intermediate for the region of the reaction coordinate over which tunneling is important. The adiabatic potential used in the SCT ground-state tunneling calculations is shown in Fig. 27.9. For temperatures around 460 K (the bottom of the range for experimental measurements) the maximum contribution to the tunneling integral occurs about 2.5 kcal mol^{-1} below the adiabatic barrier maximum. At this energy the turning points in the adiabatic potential occur for $s = \pm 0.26\ a_0$. These values are inside the two places where the maximum curvature occurs (around $s = \pm 0.6\ a_0$) and this is reflected in the values of the effective mass (also shown in Fig. 27.9), which are in the range 0.7–0.9 of the reduced mass in this region. Even with these moderate values for the effective mass, the SCT tunneling factor at 460 K is over 70% larger than the one neglecting reaction-path curvature.

27.4.2
1,2-Hydrogen Migration in Methylchlorocarbene

The 1,2-hydrogen migration in methylchlorocarbene converts it to chloroethene: $H_3CCCl \rightarrow H_2CC(H)Cl$. Calculations were carried out using direct dynamics [67]. At 365 K, tunneling lowers the gas-phase Arrhenius activation energy from 10.3 kcal mol^{-1} to 8.5 kcal mol^{-1}, and at 175 K the drop is even more dramatic, from 10.2 kcal mol^{-1} to 2.0 kcal mol^{-1}.

27.5
Liquid-phase and Enzyme-catalyzed Reactions

Placing the reagents in a liquid or an enzyme active site involves new complications. Since it is not presently practical to treat an entire condensed-phase system quantum mechanically one begins by dividing the system into two subsystems, which may be called solute and solvent, reactive system and bath, or primary subsystem and secondary subsystem. The "primary/secondary" language is often preferred because it is most general. For example, in a simple liquid-phase reaction the primary subsystem might consist of the reactive solute(s) plus one or more strongly coupled solvent molecules, and the secondary subsystem would be the rest of the solvent. In an enzyme-catalyzed reaction, the primary subsystem might be all or part of the substrate plus all or part of a cofactor and possibly a part of the enzyme and even one or a few solvent molecules, whereas the secondary subsystem would be all the rest. The solvent, bath, or secondary subsystem is sometimes called the environment.

The secondary subsystem might be treated differently from the primary one both in terms of the potential energy surface and the dynamics. For example, with regard to the former aspect, the primary subsystem might be treated by a quantum mechanical electronic structure calculation, and the secondary subsystem might be treated by molecular mechanics [68] or even approximated by an electrostatic field or a continuum model, as in implicit solvation modeling [69]. The partition into primary and secondary subsystems need not be the same for the potential energy surface step and the dynamics step. Since the present chapter is mainly concerned with the dynamics, we shall assume that a potential energy function is somehow available, and when we use the "primary/secondary" language, we refer to the dynamics step. Nevertheless the strategy chosen for the dynamics may be influenced by the methods used to obtain the potential function. This is of course true even for gas-phase reactions, but the interface between the two steps often needs to be tighter when one treats condensed-phase systems, because of their greater complexity.

We will distinguish six levels of theory for treating environmental aspects of condensed-phase reactions. These levels may be arranged as follows in a hierarchy of increasingly more complete coupling of primary and secondary subsystems:

- separable equilibrium solvation VTST (SES-VTST)
- potential-of-mean-force VTST (PMF-VTST) based on a distinguished reaction-coordinate, which is also called single-reaction-coordinate PMF-VTST (SRC-PMF-VTST)
- equilibrium solvation path VTST (ESP-VTST)
- nonequilibrium solvation path VTST (NES-VTST)
- ensemble-averaged VTST with static secondary zone (EA-VTST-SSZ)
- ensemble-averaged VTST with equilibrium secondary zone (EA-VTST-ESZ)

In practical terms, though, it is easier to consider these methods in terms of two parallel hierarchies. The first contains SES, ESP, and NES; the second contains PMF-VTST, EA-VTST-SSZ, and EA-VTST-ESZ. There is, however, a complication. While the first five rungs on the ladder correspond to successively more complete theories, the final rung (ESZ) may be considered an alternative to the fifth rung (SSZ), which may be better or may be worse, depending on the physical nature of the dynamics.

An example of a system in which both solute coordinates and solvent coordinates must be treated in a balanced way is the autoionization of water. One way to describe this process is to consider a cluster of at least a half dozen water molecules as the solute, and the rest of the water molecules as the solvent. One requires solute coordinates to describe the nature of the hydrogen bond network in the solute plus at least one solvent fluctuation coordinate; the latter may describe the direction and strength of the electric field on a critical proton or protons of the solute [70] as quantified, for example, by the energy gap between arranging the solvent to solvate the reactant and arranging it to solvate the product. Molecular dynamics simulations, though, indicate that a conventional energy gap coordinate is not necessarily the best way to describe the collective solvent re-organization. A detailed comparison of different kinds of collective solvent coordinates is given elsewhere [71]. The NES-VTST method is well suited to using collective solvent coordinates whereas EA-VTST is more convenient when explicit solvent is used. The SES, ESP, and PMF methods can easily be used with either kind of treatment of the solvent.

The PMF-VTST approach may be understood in terms of a general molecular dynamics calculation of the equilibrium one-way flux from a reactant region of phase space through a dividing surface [5–7, 72, 73]. When the no-recrossing approximation is valid at the dividing surface and when one neglects quantum effects, it may be viewed as the most efficient way to calculate the rate constant from an ensemble of trajectories. However, for reactions involving hydrogenic motion in the reaction coordinate, classical mechanics is not quantitatively accurate, and the transition state formulation provides a much more convenient way to include quantum effects than does a trajectory calculation. (Note that many workers use the term "molecular dynamics" to refer to classical trajectory calculations.)

In the rest of this section we briefly review the six rungs of the condensed-phase VTST ladder. In Section 27.6 we provide two examples that illustrate the application of the general theory.

27.5.1
Separable Equilibrium Solvation

The simplest way to include solvation effects is to calculate the reaction path and tunneling paths of the solute in the gas phase and then add the free energy of solvation at every point along the reaction path and tunneling paths. This is equivalent to treating the Hamiltonian as separable in solute coordinates and solvent coordinates, and we call it separable equilibrium solvation (SES) [74]. Adding tunneling in this method requires a new approximation, namely the canonical mean shape (CMS) approximation [75].

The gas-phase rate constant of Eqs. (27.4) and (27.23) is replaced in the SES approximation by

$$k^{\text{SES/MT}}(T) = \kappa(T) k^{\text{SES}}(T) \tag{27.37}$$

and

$$k^{\text{SES}}(T) = \frac{k_{\text{B}} T}{h} K^{\ddagger \text{o}} \min_{s} \exp\left\{-\left[\Delta G_T^{\text{GT,o}}(s) + \Delta\Delta G_{\text{S}}^{\text{o}}(s|T)\right]\right\} \tag{27.38}$$

where $\Delta\Delta G_{\text{S}}^{\text{o}}(s|T)$ is the difference between the standard-state free energy of solvation of the generalized transition state at s and that of the reactants. The transmission coefficient is given by Eq. (27.25), and all that is done to extend the SCT, LCT, and OMT approximations from the gas phase to liquid reactions is to generalize V_1 and V_2.

In the SES approximation, V_1 is taken as

$$V_1(\mathbf{R}|T) = U(\mathbf{R}|T) \tag{27.39}$$

where $\mathbf{R}$ denotes the complete set of solute coordinates, and $U(\mathbf{R}|T)$ is the CMS potential given by

$$U(\mathbf{R}|T) = W(\mathbf{R}|T) + (1/T)\frac{\partial W(\mathbf{R}|T)}{\partial(1/T)} \tag{27.40}$$

and $W(\mathbf{R}|T)$ is the potential of mean force (PMF) on the primary subsystem, which will be called the solute in the rest of this Subsection and in Subsection 27.5.3. The PMF is defined by

$$e^{-W(\mathbf{R}'|T)/k_B T} = \left\langle e^{-H/k_B T}\delta(\mathbf{R} - \mathbf{R}')\right\rangle_T \tag{27.41}$$

where H is the total system Hamiltonian, $\delta(\mathbf{R} - \mathbf{R}')$ is a multidimensional delta function that holds the solute coordinates fixed at $\mathbf{R}'$, and $\langle L\rangle_T$ denotes a normalized average over the phase space of the entire system. Colloquially, $W(\mathbf{R}|T)$ is the free energy surface of the solute. The function $U(\mathbf{R}|T)$ is the enthalpy-like component of $W(\mathbf{R}|T)$. In practice the second term of Eq. (27.40) is harder to approximate than the first term, and we can use the zero-order CMS approximation (CMS-0), which is

$$U(\mathbf{R}|T) \cong W(\mathbf{R}|T) \tag{27.42}$$

In the SES approximation,

$$W(\mathbf{R}|T) = V(\mathbf{R}) + \Delta G_S^o(\mathbf{R}|T) \tag{27.43}$$

where $\Delta G_S^o(\mathbf{R}|T)$ is the standard-state free energy of solvation. Since we will only need differences of W, e.g., its $\mathbf{R}$ dependence, it is not a matter of concern that different standard state choices correspond to changing the zero of $W(\mathbf{R}|T)$ by $\mathbf{R}$-independent amounts.

Finally, the SES approximation for the effective adiabatic potential is

$$V_2(s|T) = U_{RP}(s|T) + \varepsilon_{int}^{GT}(G, s) \tag{27.44}$$

where $U_{RP}(s|T)$ is $U(\mathbf{R}|T)$ evaluated along the reaction path, and $\varepsilon_{int}^{GT}(G, s)$ is the ground-state value of the second term of Eq. (27.8) for the solute modes. As for V_1, one can use any convenient zero of energy for $V_2(s|T)$ since the results are independent of adding a quantity independent of s.

The final protocol for an SES calculation with the CMS-0 approximation reduces to the following: Calculate a gas-phase MEP and carry out generalized normal mode analyses along the MEP to obtain $\varepsilon_{int}^{GT}(G, s)$ for the solute. (In an LCT calculation one also requires $\varepsilon_{int}^{GT}(\mathbf{n} \neq G, \Lambda = G, s)$.) Now add the free energy of solvation along the MEP to find the variational transition state rate constant and tunneling paths, and add the free energy of solvation along the tunneling paths to obtain an effective potential that is used to calculate the tunneling probabilities.

27.5.2
Equilibrium Solvation Path

In the equilibrium solvation path (ESP) approximation [74, 76], we first find a potential of mean force surface for the primary subsystem in the presence of the secondary subsystem, and then we finish the calculation using this free energy surface. Notice a critical difference from the SES in that now we find the MEP on U rather than V, and we now find solute vibrational frequencies using U rather than V.

27.5.3
Nonequilibrium Solvation Path

The SES and ESP approximations include the dynamics of solute degrees of freedom as fully as they would be treated in a gas-phase reaction, but these approximations do not address the full complexity of condensed-phase reactions because they do not allow the solvent to participate in the reaction coordinate. Methods that allow this are said to include nonequilibrium solvation. A variety of ways to include nonequilibrium solvation within the context of an implicit or reduced-degree-of-freedom bath are reviewed elsewhere [69]. Here we simply discuss one very general such NES method [76–78] based on collective solvent coordinates [71, 79]. In this method one replaces the solvent with one or more collective solvent coordinates, whose parameters are fit to bulk solvent properties or molecular dynamics simulations. Then one carries out calculations just as in the gas phase but with these extra one or more degrees of freedom. The advantage of this approach is its simplicity (although there are a few subtle technical details).

A difficulty with the nonequilibrium approach is that one must estimate the time constant or time constants for solvent equilibration with the solvent. This may be estimated from solvent viscosities, from diffusion constants, or from classical trajectory calculations with explicit solvent. Estimating the time constant for solvation dynamics presents new issues because there is more than one relevant time scale [69, 80]. Fortunately, though, the solvation relaxation time seems to depend mostly on the solvent, not the solute. Thus it is very reasonable to assume it is a constant along the reaction path.

Another difficulty with the NES model is not knowing how reliable the solvent model is and having no systematic way to improve it to convergence. Furthermore this model, like the SES and ESP approximations, assumes that the reaction can be described in terms of a reaction path residing in a single free energy valley or at most a small number of such valleys. The methods discussed next are designed to avoid that assumption.

The ESP method was applied to the reaction mentioned in Subsection 27.4.2, namely 1,2-hydrogen migration in chloromethylcarbene. Tunneling contributions are found to be smaller in solution than in the gas phase, but solvation by 1,2-dichloroethane lowers the Arrhenius activation energy at 298 K from 7.7 kcal mol^{-1} to 6.0 kcal mol^{-1} [67].

27.5.4
Potential-of-mean-force Method

In the PMF method one identifies a reaction coordinate on physical grounds rather than by calculating an MEP. For example, the reaction coordinate might be

$$z = r_{DH} - r_{AH} \qquad (27.45)$$

where r_{DH} is the distance from the transferred hydrogen to the donor atom, and r_{AH} is the distance from the transferred hydrogen to the acceptor atom. Then one calculates a one-dimensional potential of mean force ($W(z|T)$), and the classical mechanical rate constant for a unimolecular reaction in solution is given by Eq. (27.4) with [81]

$$\Delta G_T^{CVT,o} = \max_z \; [W(z|T) + W_{curv}(z|T)] - G_\zeta^R \qquad (27.46)$$

where $W_{curv}(z|T)$ is a kinematic contribution [81], usually small, at least when the reaction coordinate is a simple function of valence coordinates as in Eq. (27.45), and G_ζ^R [82] is the free energy of the reaction-coordinate motion of the reactant. Like the SES and ESP approximations, PMF-VTST involves a single reaction coordinate.

Even within the equilibrium-solvation approximations and neglecting recrossing effects, the classical mechanical result of Eq. (27.46) needs to be improved in two ways. First one needs to quantize the vibrations transverse to the reaction coordinate. A method for doing this has been presented [83], and including this step converts Eq. (27.46) to a quasiclassical result. Second, one must include tunneling. The inclusion of tunneling is explained in the next subsection, and it involves partitioning the system into primary and secondary subsystems. Note that any reasonable definition of the primary subsystem would include the three atoms involved in the definition of the reaction coordinate given in Eq. (27.45). Thus, in the present section, if one uses Eq. (27.45), the secondary subsystem does not participate in the reaction coordinate.

27.5.5
Ensemble-averaged Variational Transition State Theory

Ensemble-averaged VTST [82, 84] provides a much more complete treatment of condensed-phase reactions. Originally developed in the context of enzyme kinetics, it is applicable to any reaction in the liquid or solid state. First one carries out a quasiclassical PMF-VTST calculation as explained in Subsection 27.5.4. This is called stage 1, and it involves a single, distinguished reaction coordinate. Then, in what is called stage 2, one improves this result with respect to the quality of the reaction coordinate (allowing the secondary subsystem to participate), with respect to averaging over more than one reaction coordinate, and by including tunneling.

Stage 2 consists of a series of calculations, each one of which corresponds to a randomly chosen member of the transition state ensemble. For this purpose the transition state ensemble consists of phase points from the quasiclassical PMF calculation with the value of z in a narrow bin centered on the variational transition state, which is the value of z that maximizes the quantized version of the right-hand side of Eq. (27.46). In practice one uses the version of Eq. (27.46) in which quantization effects of modes orthogonal to z [83] are added to W. For each member of the transition state ensemble, one now optimizes the primary subsystem to the nearest saddle point in the field of the frozen secondary subsystem, and then one computes a minimum energy path through the isoinertial coordinates of the primary subsystem, with the secondary subsystem frozen. Based on this MEP one carries out a VTST/MT rate constant calculation, just as in the gas phase except for three differences. First, one does not need the reactant partition function. Second, one freezes the secondary subsystem throughout the entire calculation. Third, the projection operator discussed below Eq. (27.1) is replaced by one that just projects out the reaction coordinate because the frozen secondary subsystem removes translational invariance, converting the overall translations and rotations to librations. (The same simplified projection operator is also used for treating solid-state reactions [61].)

The calculations described in the previous paragraph yield, for each ensemble member ℓ, a free energy of activation profile $\Delta G_\ell^{GT}(T, s)$ and a transmission coefficient $\kappa_\ell(T)$, where $\ell = 1,2,...,L$, and L is the number of MEPs computed. The standard EA-VTST/MT result, called the static-secondary-zone result, is then given by

$$k^{\text{EA-VTST/MT}} = \gamma(T)k^{\text{QPMF}}(T) \tag{27.47}$$

where k^{QPMF} is the result from stage 1, based on the quantized PMF and identical to the result of Subsection 27.5.4, and γ is a transmission coefficient given by

$$\gamma = \frac{1}{L}\sum_{\ell=1}^{L}\Gamma_\ell(T)\kappa_\ell(T) \tag{27.48}$$

where

$$\Gamma_\ell = \exp\left\{-\left[\Delta G_\ell^{GT}(T, s_{*,\ell}) - \Delta G_\ell^{GT}(T, s_{0,\ell})\right]/RT\right\} \tag{27.49}$$

where $s_{*,\ell}$ is the value of s that maximizes $\Delta G_\ell^{GT}(T, s)$, and $s_{0,\ell}$ is the value of s corresponding to the value of z that maximizes the PMF of stage 1. The physical interpretation of Γ_ℓ is that, by using a more appropriate reaction coordinate for each secondary-zone configuration, one is correcting for recrossing of the original, less appropriate dividing surface defined by $z = $ constant. An alternative, more expensive way to do this is by starting trajectories at the dividing surface and counting their recrossings, if any [6, 15, 72]. More expensive is not necessarily more accurate though because the trajectories may lose their quantization before they recross.

In the equilibrium-secondary-zone approximation [82, 85] we refine the effective potential along each reaction path by adding the charge in secondary-zone free energy. Thus, in this treatment, we include additional aspects of the secondary subsystem. This need not be more accurate because in many reactions the solvation is not able to adjust on the time scale of primary subsystem barrier crossing [86].

27.6
Examples of Condensed-phase Reactions

27.6.1
H + Methanol

References for a large number of SES calculations are given in a previous review [69], but there have been far fewer calculations using the ESP and NES approximations. The ESP and NES approximations based on collective solvent coordinates have, however, been applied [78] to (R1) H + $CH_3OH \rightarrow H_2$ + CH_4OH, (R2) D + $CH_3OD \rightarrow DH + CH_2OD$, and (R3) H + $CD_3OH \rightarrow HD + CD_2OH$.

The resulting rate constants for reaction (R1) are shown in Table 27.1. In this particular case the NES results are accidentally similar to the SES ones, but that is not of major importance. What is more significant is that the true equilibrium solvation results differ from the SES ones by about a factor of two, and nonequilibrium solvation decreases the rate constants in solution by more than a factor of two as compared to the equilibrium solvation effect. If the solute–solvent coupling is decreased, the NES result becomes closer to the equilibrium solvation result, and it is difficult to ascertain how realistic the best estimates of the coupling strength actually are. Perhaps more interesting though is that if the coupling is made four times stronger, the calculated rate constant drops by another factor of three. Since ionic reactions might have much stronger solute–solvent coupling than this free radical reaction, we conclude that nonequilibrium effects might be larger for many reactions in aqueous solution.

Table 27.2 shows the kinetic isotope effects [78, 87]. Although the solvation effects are smaller than for the rate constants themselves, they are not negligible.

Tab. 27.1 Rate constants (10^{-15} cm^3 molecule^{-1} s^{-1}) at various levels of dynamical theory for H + $CH_3OH \rightarrow H_2$ + CH_2OH in aqueous solution at 298 K [78].

	Gas	SES	ESP	NES
CVT	0.7	0.9	1.9	0.81
CVT/SCT	8.3	8.7	16.6	6.5
CVT/OMT	12.9	12.7	25.9	12.4

Tab. 27.2 Kinetic isotope effects for $H + CH_3OH \rightarrow H_2 + CH_2OH$ at various levels of dynamical theory in aqueous solution at 298 K (CVT/OMT [78, 87]).

	Gas	SES	ESP	NES
R1/R2	0.68	0.48	0.51	0.37
R1/R3	21.1	21.3	20.2	19.5

Although a more recent calculation [88] indicates a barrier height about 2 kcal mol^{-1} higher than that on the potential energy surface used for these studies, the qualitative conclusions still hold if they are regarded as based on a realistic model reaction.

27.6.2
Xylose Isomerase

Xylose isomerase catalyzes a hydride transfer reaction as part of the conversion of xylose to xylulose. This reaction has been calculated [32] by the EA-VTST/MT method using Eq. (27.45) as the reaction coordinate and using $L = 5$ in Eq. (27.48). The primary zone had 32 atoms, and the secondary zone had 25 285 atoms. The average value of Γ_ℓ was 0.95. The fact that this is so close to unity indicates that the reaction coordinate of Eq. (27.45) is very reasonable for this reaction, even though the reaction coordinate is strongly coupled to a Mg–Mg breathing mode. The transmission coefficient γ was calculated to be 6.57, with about 90% of the reactive events calculated to occur by tunneling.

Calculations were also carried out for deuteride transfer. The kinetic isotope effect was calculated to be 1.80 without tunneling and 3.75 with tunneling. The latter is within the range expected from various experimental [89] determinations.

27.6.3
Dihydrofolate Reductase

The ensemble-averaged theory has also been applied to several other enzyme reactions involving transfer of a proton, hydride ion, or hydrogen atom, and the results are reviewed elsewhere [84, 90]. More recently than these reviews, the method has been applied to calculate [91] the temperature dependence of the rate constant and kinetic isotope effect for the hydride transfer catalyzed by *E. coli* dihydrofolate reductase (*ec*DHFR). In earlier work [92] we had calculated a primary KIE in good agreement with experiment [93] and also predicted a secondary KIE that turned out to be in good agreement with a later [94] experiment. In both studies [91, 92], we treated the dynamics of 31 atoms quantum mechanically. The primary KIE had also been calculated by Agarwal et al. [95], also in good agreement with experiment, but they could not calculate the secondary KIE because they treated the dynamics of only one atom quantum mechanically. In the new work [91] we predicted the temperature dependence of the KIE and found that it is small. In previous work by other groups, new mechanisms had been invoked when temperature-independent or nearly temperature-independent KIEs had been observed. The importance of the new work [91] is not so much the actual predicted small temperature dependence of the KIE (because the quantitative results may be sensitive to improving the calculation) but rather the demonstration that even nearly temperature-independent KIEs can be accommodated by VTST/MT theory, and one need not invoke new theoretical concepts.

27.7
Another Perspective

For another perspective we mention a second approach of which the reader should be aware. In this approach the dividing surface of transition state theory is defined not in terms of a classical mechanical reaction coordinate but rather in terms of the centroid coordinate of a path integral (path integral quantum TST, or PI-QTST) [96–99] or the average coordinate of a quantal wave packet. In model studies of a symmetric reaction, it was shown that the PI-QTST approach agrees well with the multidimensional transmission coefficient approach used here when the frequency of the bath is high, but both approaches are less accurate when the frequency is low, probably due to anharmonicity [98] and the path centroid constraint [97]. However, further analysis is needed to develop practical PI-QTST-type methods for asymmetric reactions [99].

Methods like PI-QTST provide an alternative perspective on the quasithermodynamic activation parameters. In methods like this the transition state has quantum effects on reaction coordinate motion built in because the flux through the dividing surface is treated quantum mechanically throughout the whole calculation. Since tunneling is not treated separately, it shows up as part of the free energy of activation, and one does not obtain a breakdown into overbarrier and tunneling contributions, which is an informative interpretative feature that one gets in VTST/MT.

Other alternative approaches for approximating the quantum effects in VTST calculations of liquid-phase [4] and enzyme reactions [90] are reviewed elsewhere.

27.8
Concluding Remarks

In the present chapter, we have described a formalism in which overbarrier contributions to chemical reaction rates are calculated by variational transition state theory, and quantum effects on the reaction coordinate, especially multidimensional tunneling, have been included by a multidimensional transmission coefficient. The advantage of this procedure is that it is general, practical, and well validated.

It is sometimes asked if a transmission coefficient is a "correction" and therefore less fundamental than other ways of including tunneling in the activation free energy. In fact, this is not the case. The transmission coefficient is a general way to include tunneling in the flux through the dividing surface. We can see this by writing the exact rate constant as a Boltzmann average over the exact rate constants for each of the possible initial states (levels) of the system, where these initial levels are labeled as $n(\text{initial})$:

$$k \equiv \left\langle k_{n(\text{initial})} \right\rangle \tag{27.50}$$

We can then replace this average by an average over systems that cross the transition state in various levels of the transition state, each labeled by $n(\text{VTS})$:

$$k \equiv \left\langle k_{n(\text{VTS})} \right\rangle \tag{27.51}$$

We can write this as

$$k \equiv \frac{\left\langle k_{n(\text{VTS})} \right\rangle}{\left\langle k_n^{\text{TST}} \right\rangle} \left\langle k_n^{\text{TST}} \right\rangle \tag{27.52}$$

where we have multiplied and divided by an average over transition-state-theory rates for each $n(\text{VTS})$. The VTST rate constant can easily be written [15] in the form of the average that we have inserted into Eq. (27.52), so $\left\langle k_n^{\text{TST}} \right\rangle$ is just k^{VTST}. The fraction in Eq. (27.52) is easily recognized as the transmission coefficient κ, and therefore we have the following expression, which is exact:

$$k \equiv \kappa \, k^{\text{VTST}} \tag{27.53}$$

In practice, we approximate the exact transmission coefficient by a mean-field-type of approximation; that is we replace the ratio of averages by the ratio for an "average" or effective potential. For gas-phase reactions with small reaction-path curvature, this effective potential would just be the vibrationally adiabatic ground-state potential. In the liquid phase and enzymes we generalize this with the canonical mean-shape approximation. In any event, though, the transmission coefficient should not be thought of as a perturbation. The method used here may be thought of as an approximate full-dimensional quantum treatment of the reaction rate.

At the present stage of development, we have well validated methods available for calculating reactive rates of hydrogen atom, proton, and hydride transfer reactions in both gaseous and condensed phases, including reliable methods for multidimensional tunneling contributions. The accuracy of calculated rate constants is often limited more by the remaining uncertainties in potential energy surfaces and practical difficulties in including anharmonicity than by the dynamical formalism *per se*.

Acknowledgments

This work was supported in part at both Pacific Northwest National Laboratory (PNNL) and the University of Minnesota (UM) by the Division of Chemical Sciences, Office of Basic Energy Sciences, U. S. Department of Energy (DOE), and it was supported in part (condensed-phase dynamics) at the University of Minnesota by the National Science Foundation. Battelle operates PNNL for DOE.

References

1 H. Eyring, *J. Chem. Phys.* 3 (1935) 107; S. Glasstone, K. J. Laidler, H. Eyring, *The Theory of Rate Processes*, McGraw-Hill, New York, 1941.

2 H. Eyring, Trans. Faraday Soc. 34 (1938) 41; B. C. Garrett, D. G. Truhlar, in *Encyclopedia of Computational Chemistry*, P. v. R. Schleyer, N. L. Allinger, T. Clark, J. Gasteiger, P. A. Kollman, H. F. Schaefer, III (Eds.), John Wiley & Sons, Chichester, 1998.

3 E. Wigner, *Trans. Faraday Soc.* 34 (1938) 29.

4 D. G. Truhlar, W. L. Hase, J. T. Hynes, *J. Phys. Chem.* 87 (1983) 2664; D. G. Truhlar, B. C. Garrett, S. J. Klippenstein, *J. Phys. Chem.* 100 (1996) 12771.

5 E. Wigner, *J. Chem. Phys.* 5 (1937) 720; J. Horiuti, *Bull. Chem. Soc. Jpn.* 13 (1938) 210; J. C. Keck, *J. Chem. Phys.* 32 (1960) 1035; D. G. Truhlar, B. C. Garrett, *Annu. Rev. Phys. Chem.* 35 (1984) 159.

6 J. C. Keck, *Adv. Chem. Phys.* 13 (1967) 85.

7 B. C. Garrett, D. G. Truhlar, *J. Phys. Chem.* 83 (1979) 1052.

8 E. Wigner, *Z. Phys. Chem. B* 19 (1932) 203.

9 R. P. Bell, *The Proton in Chemistry*, Chapman and Hall, London, 1973.

10 D. G. Truhlar, A. Kuppermann, *Chem. Phys. Lett.* 9 (1971) 269.

11 W. H. Miller, *Acc. Chem. Res.* 9 (1976) 306.

12 J. O. Hirschfelder, E. Wigner, *J. Chem. Phys.* 7 (1939) 616; H. Eyring, J. Walter, G. E. Kimball, *Quantum Chemistry*, Wiley, New York, 1944; R. A. Marcus, *J. Chem. Phys.* 43 (1965) 1598; R. A. Marcus, *J. Chem. Phys.* 45 (1966) 4450; D. G. Truhlar, *J. Chem. Phys.* 53 (1970) 2041.

13 G. L. Hofacker, *Z. Naturforsch. A* 18 (1963) 607.

14 R. A. Marcus, *J. Chem. Phys.* 45 (1966) 4493.

15 B. C. Garrett, D. G. Truhlar, *J. Phys. Chem.* 83 (1979) 1079.

16 B. C. Garrett, D. G. Truhlar, R. S. Grev, A. W. Magnuson, *J. Phys. Chem.* 84 (1980) 1730.

17 R. T. Skodje, D. G. Truhlar, B. C. Garrett, *J. Phys. Chem.* 85 (1981) 3019.

18 R. T. Skodje, D. G. Truhlar, B. C. Garrett, *J. Chem. Phys.* 77 (1982) 5955.

19 B. C. Garrett, D. G. Truhlar, A. F. Wagner, T. H. Dunning, Jr., *J. Chem. Phys.* 78 (1983) 4400; D. K. Bondi, J. N. L. Connor, B. C. Garrett, D. G. Truhlar, *J. Chem. Phys.* 78 (1983) 5981.

20 B. C. Garrett, D. G. Truhlar, *J. Chem. Phys.* 79 (1983) 4931.

21 B. C. Garrett, T. Joseph, T. N. Truong, D. G. Truhlar, *Chem. Phys.* 136 (1989) 271.

22 D.-h. Lu, T. N. Truong, V. S. Melissas, G. C. Lynch, Y.-P. Liu, B. C. Garrett, R. Steckler, A. D. Isaacson, S. N. Rai, G. C. Hancock, J. G. Lauderdale, T. Joseph, D. G. Truhlar, *Comput. Phys. Commun.* 71 (1992) 235.

23 Y.-P. Liu, D.-h. Lu, A. Gonzalez-Lafont, D. G. Truhlar, B. C. Garrett, *J. Am. Chem. Soc.* 115 (1993) 7806.

24 Y.-P. Liu, G. C. Lynch, T. N. Truong, D.-h. Lu, D. G. Truhlar, B. C. Garrett, *J. Am. Chem. Soc.* 115 (1993) 2408.

25 R. A. Marcus, M. E. Coltrin, *J. Chem. Phys.* 67 (1977) 2609.

26 C. J. Walsh, *Enzyme Reaction Mechanisms*, Freeman, New York, 1979.

27 D. G. Truhlar, R. E. Wyatt, *Annu. Rev. Phys. Chem.* 27 (1976) 1.

28 H. S. Johnston, *Gas Phase Reaction Rate Theory*, Ronald, New York, 1966.

29 L. Melander, *Isotope Effects on Reaction Rates*, Ronald Press, New York, 1960; L. Melander, W. H. Saunders, *Reaction Rates of Isotopic Molecules*, Krieger Publishing, Malabar, FL, 1987.

30 J. Bigeleisen, M. Wolfsberg, *Adv. Chem. Phys.* 1 (1958) 15.

31 B. C. Garrett, D. G. Truhlar, *J. Am. Chem. Soc.* 102 (1980) 2559.

32 M. Garcia-Viloca, C. Alhambra, D. G. Truhlar, J. Gao, *J. Comput. Chem.* 24 (2003) 177.

33 D. G. Truhlar, M. S. Gordon, *Science* 249 (1990) 491.

34 B. C. Garrett, D. G. Truhlar, *J. Chem. Phys.* 70 (1979) 1593.

35 R. E. J. Weston, *J. Chem. Phys.* 31 (1959) 892; I. Shavitt, *J. Chem. Phys.* 49 (1968) 4048; R. A. Marcus, *J. Chem. Phys.* 49 (1968) 2617; K. Fukui, in *The World of Quantum Chemistry*, R. Daudel, B. Pullman (Eds.), Reidel, Dordrecht, 1974; H. F. Schaefer, III, *Chem. Brit.* 11 (1975) 227.

36 D. G. Truhlar, A. Kuppermann, *J. Am. Chem. Soc.* 93 (1971) 1840.

37 W. H. Miller, N. C. Handy, J. E. Adams, *J. Chem. Phys.* 72 (1980) 99.

38 A. D. Isaacson, D. G. Truhlar, *J. Chem. Phys.* 76 (1982) 1380.

39 D. G. Truhlar, A. D. Isaacson, B. C. Garrett, in *Theory of Chemical Reaction Dynamics*, M. Baer (Ed.), CRC Press, Boca Raton, FL, 1985.

40 C. F. Jackels, Z. Gu, D. G. Truhlar, *J. Chem. Phys.* 102 (1995) 3188.

41 S. J. Klippenstein, R. A. Marcus, *J. Chem. Phys.* 87 (1987) 3410; J. Villa, D. G. Truhlar, *Theor. Chem. Acc.* 97 (1997) 317; P. L. Fast, D. G. Truhlar, *J. Chem. Phys.* 109 (1998) 3721; J. R. Pliego, W. B. De Almeida, *Phys. Chem. Chem. Phys.* 1 (1999) 1031.

42 B. C. Garrett, D. G. Truhlar, *J. Chem. Phys.* 72 (1980) 3460; G. A. Natanson, B. C. Garrett, T. N. Truong, T. Joseph, D. G. Truhlar, *J. Chem. Phys.* 94 (1991) 7875; G. A. Natanson, *Theor. Chem. Acc.* 112 (2004) 68.

43 A. D. Isaacson, *J. Phys. Chem. A* 110 (2006) 379.

44 R. A. Marcus, *J. Chem. Phys.* 46 (1967) 959; R. A. Marcus, *J. Chem. Phys.* 49 (1968) 2610.

45 D. G. Truhlar, B. C. Garrett, *J. Phys. Chem. A* 107 (2003) 4006.

46 R. A. Marcus, *J. Chem. Phys.* 41 (1964) 2614; G. L. Hofacker, N. Rösch, *Ber. Bunsen- Ges. Phys. Chem.* 77 (1973) 661.

47 D. C. Chatfield, R. S. Friedman, D. G. Truhlar, B. C. Garrett, D. W. Schwenke, *J. Am. Chem. Soc.* 113 (1991) 486; D. C. Chatfield, R. S. Friedman, D. W. Schwenke, D. G. Truhlar, *J. Phys. Chem.* 96 (1992) 2414; D. C. Chatfield, R. S. Friedman, S. L. Mielke, G. C. Lynch, T. C. Allison, D. G. Truhlar, D. W. Schwenke, in *Dynamics of Molecules and Chemical Reactions*, R. E. Wyatt, J. Z. H. Zhang (Eds.), Marcel Dekker, New York, 1996, p. 323.

48 J. N. L. Connor, *Mol. Phys.* 15 (1968) 37; B. C. Garrett, D. G. Truhlar, *J. Phys. Chem.* 83 (1979) 2921.

49 A. Fernandez-Ramos, D. G. Truhlar, *J. Chem. Phys.* 114 (2001) 1491.

50 D. G. Truhlar, C. J. Horowitz, *J. Chem. Phys.* 68 (1978) 2466.

51 A. A. Westenberg, N. De Haas, *J. Chem. Phys.* 47 (1967) 1393; K. A. Quickert, D. J. Leroy, *J. Chem. Phys.* 53 (1970) 1325; I. D. Reid, L. Y. Lee, D. M. Garner, D. J. Arseneau, M. Senba, D. G. Fleming, *Hyperfine Interact.* 32 (1986) 801.

52 B. C. Garrett, D. G. Truhlar, *J. Phys. Chem.* 83 (1979) 1915; D. G. Truhlar, A. D. Isaacson, R. T. Skodje, B. C. Garrett, *J. Phys. Chem.* 86 (1982) 2252; D. G. Truhlar, *J. Comput. Chem.* 12 (1991) 266.

53 B. C. Garrett, D. G. Truhlar, *J. Chem. Phys.* 81 (1984) 309.

54 V. K. Babamov, V. Lopez, R. A. Marcus, *Chem. Phys. Lett.* 101 (1983) 507; V. K. Babamov, V. Lopez, R. A. Marcus, *J. Chem. Phys.* 78 (1983) 5621;

V. K. Babamov, V. Lopez, R. A. Marcus, *J. Chem. Phys.* 80 (1984) 1812.

55 B. C. Garrett, N. Abusalbi, D. J. Kouri, D. G. Truhlar, *J. Chem. Phys.* 83 (1985) 2252.

56 T. C. Allison, D. G. Truhlar, in *Modern Methods for Multidimensional Dynamics Computations in Chemistry*, D. L. Thompson (Ed.), World Scientific, Singapore, 1998.

57 J. C. Corchado, D. G. Truhlar, J. Espinosa-Garcia, *J. Chem. Phys.* 112 (2000) 9375.

58 M. S. Zahniser, B. M. Berquist, F. Kaufman, *Int. J. Chem. Kinet.* 10 (1978) 15; M. L. Pohjonen, J. Koskikallio, *Acta Chem. Scand. Ser. A* 33 (1979) 449; J. J. Russell, J. A. Seetula, S. M. Senkan, D. Gutman, *Int. J. Chem. Kinet.* 20 (1988) 759; T. J. Wallington, M. D. Hurley, *Chem. Phys. Lett.* 189 (1992) 437; W. R. Simpson, A. J. Orr-Ewing, R. N. Zare, *Chem. Phys. Lett.* 212 (1993) 163; Y. Matsumi, K. Izumi, V. Skorokhodov, M. Kawasaki, N. Tanaka, *J. Phys. Chem. A* 101 (1997) 1216; A. J. Orr-Ewing, W. R. Simpson, T. P. Rakitzis, S. A. Kandel, R. N. Zare, *J. Chem. Phys.* 106 (1997) 5961; S. A. Kandel, R. N. Zare, *J. Chem. Phys.* 109 (1998) 9719; G. D. Boone, F. Agyin, D. J. Robichaud, F. M. Tao, S. A. Hewitt, *J. Phys. Chem. A* 105 (2001) 1456.

59 D. G. Truhlar, D.-h. Lu, S. C. Tucker, X. G. Zhao, A. Gonzalez-Lafont, T. N. Truong, D. Maurice, Y.-P. Liu, G. C. Lynch, in *Isotope Effects in Gas-Phase Chemistry*, J. A. Kay (Ed.), American Chemical Society, Washington, D. C., 1992.

60 R. G. Gilbert, S. C. Smith, *Theory of Unimolecular and Recombination Reactions*, Blackwell, Oxford, 1990; T. Baer, W. L. Hase, *Unimolecular Reaction Dynamics*, Oxford University Press, New York, 1996.

61 J. G. Lauderdale, D. G. Truhlar, *Surf. Sci.* 164 (1985) 558.

62 G. C. Hancock, C. A. Mead, D. G. Truhlar, A. J. C. Varandas, *J. Chem. Phys.* 91 (1989) 3492; B. M. Rice, B. C. Garrett, M. L. Koszykowski, S. M. Foiles, M. S. Daw, *J. Chem. Phys.* 92 (1990) 775; S. E. Wonchoba, W. P. Hu,

D. G. Truhlar, *Phys. Rev. B* 51 (1995) 9985; P. S. Zuev, R. S. Sheridan, T. V. Albu, D. G. Truhlar, D. A. Hrovat, W. T. Borden, *Science* 299 (2003) 867.

63 W. R. Roth and J. König, *Liebigs Ann. Chem.* 699 (1966) 24.

64 L. Chantranupong, T. A. Wildman, *J. Am. Chem. Soc.* 112 (1990) 4151.

65 Y. Kim, J. C. Corchado, J. Villa, J. Xing, D. G. Truhlar, *J. Chem. Phys.* 112 (2000) 2718.

66 K. K. Baldridge, M. S. Gordon, R. Steckler, D. G. Truhlar, *J. Phys. Chem.* 93 (1989) 5107; B. C. Garrett, M. L. Koszykowski, C. F. Melius, M. Page, *J. Phys. Chem.* 94 (1990) 7096; D. G. Truhlar, in *The Reaction Path in Chemistry: Current Approaches and Perspectives*, D. Heidrich (Ed.), Kluwer, Dordrecht, 1995.

67 T. V. Albu, B. J. Lynch, D. G. Truhlar, A. C. Goren, D. A. Hrovat, W. T. Borden, R. A. Moss, *J. Phys. Chem. A* 106 (2002) 5323.

68 J. Gao, M. A. Thompson, *Combined Quantum Mechanical and Molecular Mechanical Methods*, American Chemical Society, Washington, DC, 1998; H. Lim, D. G. Truhlar, *Theor. Chem. Acc.* in press, online DOI: 10.1007/s00214-006-0143-z.

69 C. J. Cramer, D. G. Truhlar, *Chem. Rev.* 99 (1999) 2161.

70 R. A. Marcus, *J. Chem. Phys.* 24 (1956) 966; P. L. Geissler, C. Dellago, D. Chandler, J. Hutter, M. Parrinello, *Science* 291 (2001) 2121.

71 G. K. Schenter, B. C. Garrett, D. G. Truhlar, *J. Phys. Chem. B* 105 (2001) 9672.

72 C. H. Bennett, in *Algorithms for Chemical Computation*, R. E. Christofferson (Ed.) American Chemical Society, Washington, DC, 1977.

73 K. Hinsen, B. Roux, *J. Chem. Phys.* 106 (1997) 3567.

74 Y.-Y. Chuang, C. J. Cramer, D. G. Truhlar, *Int. J. Quantum Chem.* 70 (1998) 887.

75 D. G. Truhlar, Y.-P. Liu, G. K. Schenter, B. C. Garrett, *J. Phys. Chem.* 98 (1994) 8396.

76 B. C. Garrett, G. K. Schenter, *Int. Rev. Phys. Chem.* 13 (1994) 263.

77 B. C. Garrett, G. K. Schenter, in *Structure and Reactivity in Aqueous Solution*, C. J. Cramer, D. G. Truhlar (Eds.), American Chemical Society, Washington, DC, 1994.

78 Y.-Y. Chuang, D. G. Truhlar, *J. Am. Chem. Soc.* 121 (1999) 10157.

79 S. Lee, J. T. Hynes, *J. Chem. Phys.* 88 (1988) 6863; D. G. Truhlar, G. K. Schenter, B. C. Garrett, *J. Chem. Phys.* 98 (1993) 5756; M. V. Basilevsky, G. E. Chudinov, D. V. Napolov, *J. Phys. Chem.* 97 (1993) 3270.

80 I. Ohmine, H. Tanaka, *J. Chem. Phys.* 93 (1990) 8138; G. K. Schenter, R. P. McRae, B. C. Garrett, *J. Chem. Phys.* 97 (1992) 9116; M. Maroncelli, *J. Mol. Liq.* 57 (1993) 1; M. Maroncelli, V. P. Kumar, A. Papazyan, *J. Phys. Chem.* 97 (1993) 13; M. Cho, G. R. Fleming, S. Saito, I. Ohmine, R. M. Stratt, *J. Chem. Phys.* 100 (1994) 6672; M. F. Ruiz-Lopez, A. Oliva, I. Tunon, J. Bertran, *J. Phys. Chem. A* 102 (1998) 10728; B. Bagchi, R. Biswas, *Adv. Chem. Phys.* 109 (1999) 207; J. Li, C. J. Cramer, D. G. Truhlar, *Int. J. Quantum Chem.* 77 (2000) 264.

81 G. K. Schenter, B. C. Garrett, D. G. Truhlar, *J. Chem. Phys.* 119 (2003) 5828.

82 C. Alhambra, J. Corchado, M. L. Sanchez, M. Garcia-Viloca, J. Gao, D. G. Truhlar, *J. Phys. Chem. B* 105 (2001) 11326.

83 M. Garcia-Viloca, C. Alhambra, D. G. Truhlar, J. Gao, *J. Chem. Phys.* 114 (2001) 9953.

84 D. G. Truhlar, J. Gao, C. Alhambra, M. Garcia-Viloca, J. Corchado, M. L. Sanchez, J. Villa, *Acc. Chem. Res.* 35 (2002) 341; D. G. Truhlar, J. L. Gao, M. Garcia-Viloca, C. Alhambra, J. Corchado, M. L. Sanchez, T. D. Poulsen, *Int. J. Quantum Chem.* 100 (2004) 1136.

85 T. D. Poulsen, M. Garcia-Viloca, J. Gao, D. G. Truhlar, *J. Phys. Chem. B* 107 (2003) 9567.

86 G. van der Zwan, J. T. Hynes, *J. Chem. Phys.* 78 (1983) 4174; J. T. Hynes, in *Solvent Effects and Chemical Reactivity*, O. Tapia, J. Bertrán (Eds.), Kluwer, Dordrecht, 1996.

87 Y.-Y. Chuang, M. L. Radhakrishnan, P. L. Fast, C. J. Cramer, D. G. Truhlar, *J. Phys. Chem. A* 103 (1999) 4893.

88 J. Z. Pu, D. G. Truhlar, *J. Phys. Chem. A* 109 (2005) 773.

89 C. Lee, M. Bagdasarian, M. Meng, J. G. Zeikus, *J. Biol. Chem.* 265 (1990) 19082; H. van Tilbeurgh, J. Jenkins, M. Chiadmi, J. Janin, S. J. Wodak, N. T. Mrabet, A.-M. Lambeir, *Biochemistry* 31 (1992) 5467; P. B. M. van Bastelaere, H. L. M. Kerstershilderson, A.-M. Lambeir, *Biochem. J.* 307 (1995) 135.

90 J. Gao, D. G. Truhlar, *Annu. Rev. Phys. Chem.* 53 (2002) 467.

91 J. Pu, S. Ma, J. Gao, D. G. Truhlar, *J. Phys. Chem. B* 109 (2005) 8851.

92 M. Garcia-Viloca, D. G. Truhlar, J. L. Gao, *Biochemistry* 42 (2003) 13558.

93 C. A. Fierke, K. A. Johnson, S. J. Benkovic, *Biochemistry* 26 (1987) 4085.

94 R. S. Sikorski, L. Wang, K. A. Markham, P. T. R. Rajagopalan, S. J. Benkovic, A. Kohen, *J. Am. Chem. Soc.* 126 (2004) 4778.

95 P. K. Agarwal, S. R. Billeter, S. Hammes-Schiffer, *J. Phys. Chem. B* 106 (2002) 3283.

96 M. J. Gillan, *J. Phys. C* 20 (1987) 3621; G. A. Voth, D. Chandler, W. H. Miller, *J. Chem. Phys.* 91 (1989) 7749; G. R. Haynes, G. A. Voth, *Phys. Rev. A* 46 (1992) 2143; G. A. Voth, *J. Phys. Chem.* 97 (1993) 8365; M. Messina, G. K. Schenter, B. C. Garrett, *J. Chem. Phys.* 98 (1993) 8525; G. K. Schenter, G. Mills, H. Jónsson, *J. Chem. Phys.* 101 (1994) 8964; G. Mills, G. K. Schenter, D. E. Makarov, H. Jonsson, *Chem. Phys. Lett.* 278 (1997) 91.

97 R. P. McRae, G. K. Schenter, B. C. Garrett, G. R. Haynes, G. A. Voth, G. C. Schatz, *J. Chem. Phys.* 97 (1992) 7392.

98 R. P. McRae, B. C. Garrett, *J. Chem. Phys.* 98 (1993) 6929.

99 S. Jang, C. D. Schwieters, G. A. Voth, *J. Phys. Chem. A* 103 (1999) 9527.

28
Quantum Mechanical Tunneling of Hydrogen Atoms in Some Simple Chemical Systems

K. U. Ingold

> *Not only is it difficult to prove that a given reaction proceeds via tunneling, it is even difficult to define the term tunneling unambiguously. The hydrogen atoms themselves, one presumes, are unaware that they are tunneling: from their vantage point the barrier is, like beauty, only present in the eye of the observer. Moreover, not all observers will see the barrier but only those that have not yet overcome their classical prejudices.*
>
> Willem Siebrand 1984

28.1
Introduction

In 1933 Bell [1] predicted that, due to quantum mechanical effects, the rate of transfer of a hydrogen atom (H-atom) or proton would become temperature independent at low temperatures. Since that time, kineticists have embraced the concept of quantum mechanical tunneling (QMT) so enthusiastically that it is frequently invoked on the flimsiest of experimental evidence, often using data obtained at, or above, room temperature. At such elevated temperatures, conclusive evidence that the rate of an H-atom or proton transfer is enhanced above that due to "over the top of the barrier" thermal activation, and can only be explained by there being a *significant* contribution from QMT, is rare. Significant has been italicized in the foregoing sentence because QMT will *always* make some contribution to the rate of such transfers. The QMT contribution to the transfer rate becomes more obvious at low temperatures. For this reason, the *unequivocal* identification of QMT in simple chemical systems requires that their rates of reaction be measured at low temperatures.

In this chapter, a few simple unimolecular and bimolecular reactions will be described in which the rates of H-atom motion were measured down to very low temperatures. These kinetic measurements provide unequivocal evidence that QMT dominates the reaction rates over a wide range of temperatures. There are two common themes. First, all the experimental data were generated in my own

Hydrogen-Transfer Reactions. Edited by J. T. Hynes, J. P. Klinman, H. H. Limbach, and R. L. Schowen
Copyright © 2007 WILEY-VCH Verlag GmbH & Co. KGaA, Weinheim
ISBN: 978-3-527-30777-7

laboratory where QMT became one research focus between 1974 and 1990. The second theme is that all the kinetic data were generated using electron paramagnetic resonance (EPR) spectroscopy.

28.2
Unimolecular Reactions

28.2.1
Isomerization of Sterically Hindered Phenyl Radicals

28.2.1.1 2,4,6-Tri–*tert*–butylphenyl

The lifetimes of many classes of free radicals can be dramatically increased by attaching two (or more) *tert*-butyl groups close to the radical center [2]. This sterically induced increase in free radical persistence was subject to a severe challenge with phenyl radicals because both their H-atom abstraction and addition reactions with organic substrates are exothermic and very rapid. Nevertheless, 2,4,6-tri–*tert*–butylphenyl, $1_H^\bullet$, produced by reaction of $1_H Br$ with photochemically generated $Me_3Sn^\bullet$ (or $Me_3Si^\bullet$) radicals in liquid cyclopropane in the cavity of an EPR spectrometer was found to be fairly long-lived at ambient and lower temperatures [3]

$$Me_3SnSnMe_3 \xrightarrow{h\nu} 2\ Me_3Sn^\bullet \qquad (28.1)$$

$$Me_3Sn^\bullet + \text{[structure]}-Br \longrightarrow Me_3SnBr + \text{[structure]}^\bullet \qquad (28.2)$$

$$1_H Br\ (1_D Br) \qquad\qquad 1_H^\bullet\ (1_D^\bullet)$$

On interrupting the photolysis, $1_H^\bullet$ decayed with first-order kinetics and with rate constants from −30 to −90 °C that were independent of the concentrations of $1Br$, ditin and cyclopropane [3]. The product, also identified by EPR, was the 3,5-di-*tert*-butylneophyl radical, $2^\bullet$, that itself decayed with second-order kinetics.

$$\text{[structure]}^\bullet \xrightarrow{k_3} \text{[structure]}^\bullet \qquad (28.3)$$

$$1_H^\bullet\ (1_D^\bullet) \qquad\qquad 2_H^\bullet\ (2_D^\bullet)$$

$$2\,2_H^\bullet \longrightarrow 2_H{-}2_H \qquad (28.4)$$

Reaction (28.3) was found to have a surprisingly low Arrhenius pre-exponential factor (log (A_H/s^{-1}) = 5.3). 1_DBr was synthesized in which the three *tert*-butyl groups had been essentially fully deuterated (^{2}H content > 99%) [4]. Under similar conditions in the EPR an even more persistent $1_D^\bullet$ radical was obtained. This also decayed with first-order kinetics and yielded $k_3^H/k_3^D \approx 50$ at −30 °C. It was thought probable that reaction (28.3) would provide one of the first clear and unequivocal examples of QMT in an H-atom transfer. This reaction and related reactions were therefore examined in considerable detail [4, 5].

The decay of $1_H^\bullet$ generated from 1_HBr with Me$_3$Sn$^\bullet$ (reaction (28.2)) and by the photolysis of **3** (reaction (28.5)) occurred with "clean" first-order kinetics and at

$$\text{3: } \quad \overset{}{\underset{\textbf{3}}{}} \quad -\text{C(O)OOBu}^t \quad \xrightarrow{h\nu} \quad \left(\quad -\text{CO}_2^\bullet \quad \right) \quad \longrightarrow \quad 1_H^\bullet \qquad (28.5)$$

identical rates in liquid cyclopropane, propane, isopentane and toluene and at temperatures from 247 to 113 K [4]. $1_D^\bullet$ also decayed by reaction (28.3) with "clean" first-order kinetics from 293 to 124 K [4]. Arrhenius plots of both sets of kinetic data are curved (see Fig. 28.1) implying QMT of both H- and D-atoms. When $1_H^\bullet$ or $1_D^\bullet$ were generated from **1**Br using Me$_3$Si$^\bullet$ radicals formed via reaction (28.6)

$$\text{Bu}^t\text{OOBu}^t \xrightarrow{h\nu} \text{Bu}^t\text{O}^\bullet \xrightarrow{\text{Me}_3\text{SiH}} \text{Bu}^t\text{OH} + \text{Me}_3\text{Si}^\bullet \qquad (28.6)$$

the decay rates were the same as for the Me$_3$Sn$^\bullet$ method at 245 K but were significantly greater at lower temperatures. These faster reactions were attributed to the intermolecular reaction (28.7) becoming competitive with the intramolecular isomerization, reaction (28.3) [4].

$$\textbf{1}^\bullet + \text{Me}_3\text{SiH} \longrightarrow \textbf{1}\text{-H} + \text{Me}_3\text{Si}^\bullet \qquad (28.7)$$

The unimolecular isomerization of $\textbf{1}^\bullet$ to $\textbf{2}^\bullet$ involves a relatively inflexible and nonpolar species. This suggested that the isomerization rate might not be affected by changing from the liquid to the solid phase. If this were the case, the reaction could be studied at temperatures below 113–K where the effects of QMT should become even more pronounced. Benzene, perdeuteriobenzene and neopentane were used as the solid matrices and kinetic measurements on the isomerization of $1_H^\bullet$ were made over a range of temperatures down to the boiling points of liquid nitrogen (77 K) and liquid neon (28 K) [5]. In the overlapping temperature range, the rate constants for isomerization of $1_H^\bullet$ were essentially the same for reactions in the solid state and in solution, see Fig. 28.1. Measurements of the rates of isomerization of $1_D^\bullet$ were made only after sufficient time had elapsed for any incompletely deuterated radical to have decayed completely (Fig. 28.1). Unfortunately, the $1_D^\bullet$ isomerization became too slow to measure (in any reasonable time) at temperatures below 123 K.

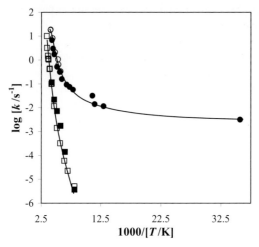

Figure 28.1 Arrhenius plots of the rate constants for the isomerization of 2,4,6-tri-*tert*-butylphenyl. Key: $1^{\bullet}_H$ isomerization in solution (○) and in matrices, ●; $1^{\bullet}_D$ isomerization in solution (□) and in matrices. (■)

The isomerization reaction: $\mathbf{1}^{\bullet} \rightarrow \mathbf{2}^{\bullet}$, exhibits all the phenomena characteristic of a hydrogen atom motion in which QMT dominates the rate. These phenomena are:

1. Large deuterium kinetic isotope effects (DKIEs). When QMT makes only an insignificant contribution to the rate, the DKIE arises only from differences in the zero-point energies (ZPEs) of the H- and D-containing reactants and their transition states. The DKIE will be maximized when all the ZPE in the reactants is lost in the transition state. In such a case, the difference in the ZPEs of the H- and D-containing reactants will equal the difference in the activation enthalpies, $E_D - E_H$. For the breaking of a C–H/C–D bond, $E_D - E_H$ ≤ 1354 cal mol^{-1}, provided that the ZPEs of both stretching and bending vibrations are lost in the transition state [6]. Thus, for the "classical" rupture of a C–H/C–D bond, i.e., in the absence of significant QMT, the maximum possible DKIEs (k_H/k_D) are, for example, 17, 53 and 260 at 243, 173 and 123 K, respectively. The experimentally measured DKIEs for the $\mathbf{1}^{\bullet} \longrightarrow \mathbf{2}^{\bullet}$ isomerization were always much larger than these calculated values, viz., 80, 1400 and 13000 at 243, 173 and 123 K. Admittedly, in the case of the $1^{\bullet}_D$ isomerization there will be a small additional contribution to the DKIE from secondary DKIEs but these are unlikely to be greater than 2 at 243 K and 6 at 123 K [4].

2. Nonlinear Arrhenius plots. QMT will become relatively more important as the temperature is decreased and this will lead to curved Arrhenius plots with the curvature being more pronounced for H transfer than for D transfer. This is clearly seen in Fig. 28.1. Moreover, if the reaction rate can be monitored at sufficiently low temperatures, there will be little or no thermal activation and the reaction will only occur because of QMT. This means that at very low temperatures the rate should become essentially independent of the temperature. This is clearly the case for $1_H^\bullet$ isomerization (see Fig. 28.1) and presumably would also have been true for $1_D^\bullet$ had it been possible to make measurements at lower temperatures.

3. Large differences in Arrhenius activation enthalpies and pre-exponential factor for H and for D transfer. Both of these criteria of QMT are commonly employed when the kinetic measurements are confined to such a narrow range that curvature of the Arrhenius plots is not as obvious as it is for $1_H^\bullet$ and $1_D^\bullet$ (see Fig. 28.1). Although neither of these criteria is required to conclude that QMT plays a dominant role in the isomerization of **1•**, it is worth noting that the "least-squares" Arrhenius plots using only the kinetics measured in solution yielded E_D= 6.4, E_H= 2.5 kcal mol^{-1} (difference: 3.9 kcal mol^{-1}), and A_D= 10$^{5.1}$, A_H= 10$^{3.1}$ s^{-1} [4].

The experimental rate constants for the isomerization of **1•** were analyzed [4, 5] in the manner customary in the 1970s [7]. This assumed that passage through the transition state could be described by the motion of a particle of constant mass along a single, separable, coordinate. According to this one-dimensional model [7], the temperature dependence of the rate constant, $k(T)$, could be represented by:

$$k(T) = A\Gamma(T)e^{-V_o/RT} \tag{28.I}$$

where V_o is the height of the potential barrier, R is the gas constant, A is the approximate temperature-independent frequency of mass point collisions with the barrier, and $\Gamma(T)$ is the ratio of the quantum mechanical to the classical barrier transmission rates of a Boltzmann distribution of incident mass-point kinetic energies. The barrier heights and widths were determined by finding the best fit of both the $1_H^\bullet$ and $1_D^\bullet$ kinetic data to a common barrier using three differently shaped potential barriers known as the Eckart, Gaussian and truncated parabolic barriers.

The results of these computations [4, 5] will not be reported here because they are irrelevant from today's perspective. The problem is that these types of potential energy barriers are single "bumps" on an otherwise flat, constant energy, reaction coordinate extending from minus to plus infinity. These are certainly not realistic barriers for any H-atom transfer reaction.

Fortunately, Siebrand and coworkers [8] developed a new and much more satisfying approach to processes involving QMT that avoided the usual tunneling formalism. Instead of formulating the H-atom transfer in terms of barrier penetration, it was described as a radiationless transition between potential energy surfaces. An explicit barrier shape is not employed, though one can be obtained from the vibrational potentials determining the initial (reactant) and final (product) states and the interaction operator which allows the H-atom transfer to occur. These are also one-dimensional barriers like those described above and, not too surprisingly, they can fail to account for the experimental transfer kinetics. These inconsistencies were removed by including low-frequency, nonhydrogenic modes which cause barrier oscillations and periodically favorable transfer conditions.

This generalized model yielded satisfactory descriptions of the temperature and isotope dependence of some reported transfer rate constants [9]. The calculated rate constants, plotted as an exponential function of T rather than $1/T$, show a constant part at low temperatures followed by a quasi-linear part at higher temperatures. Siebrand et al. [9] applied their procedure to a number of intramolecular H- and D-atom transfers for which some experimental data were available. This included the isomerizations of the tri-*tert*-butylphenyl radicals, $1^\bullet_H$ and $1^\bullet_D$. The curves obtained [9] using an anharmonic low frequency motion (which was superior to the harmonic version) are shown in Fig. 28.2. These curves give very satisfactory fits to the experimental rate constants. The calculated, limiting, low temperature QMT-only, DKIE is ~ 50000! It appears to be worthy of the Guinness Book of Records [28].

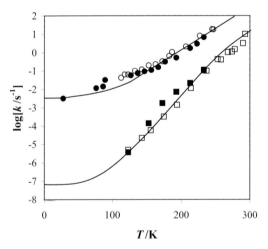

Figure 28.2 Plots of the rate constants for the isomerization of 2,4,6-tri-*tert*-butylphenoxyl, $1^\bullet_H$ and $1^\bullet_D$, against T (K). The symbols are the same as those in Fig. 28.1. The solid lines depict the calculated rate constants for H-atom and D-atom transfer in these $1^\bullet \longrightarrow 2^\bullet$ reactions.

28.2.1.2 Other Sterically Hindered Phenyl Radicals

2,4,6-Tri(1'-adamantyl)phenyl, [4] $4_H^\bullet$, and octamethyloctahydroanthracen-9-yl, [5] $5_H^\bullet$, have also been generated from their parent bromides and observed by EPR spectroscopy. Both of these radicals decayed with "clean" first-order kinetics.

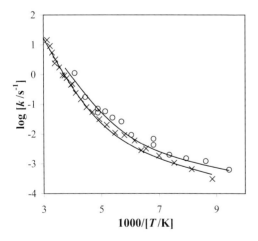

$$(28.8)$$

$4_H^\bullet$ (Ad = 1-adamantyl) $5_H^\bullet$

Arrhenius plots of their decay rate constants show pronounced curvature, see Fig. 28.3. By analogy with the $1_H^\bullet \rightarrow 2_H^\bullet$ isomerization it can be concluded that $4_H^\bullet$ and $5_H^\bullet$ decay by intramolecular H-atom transfers involving 5-center cyclic transition states with QMT playing a dominant role. Unfortunately, product radicals could only be observed in the case of $2_H^\bullet$, those from $4_H^\bullet$ and $5_H^\bullet$ could not be detected at any temperature. This is because these two product radicals have a very large number of individual EPR lines which would have made their detection extremely difficult.

At the same temperature, $4_H^\bullet$ and $5_H^\bullet$ are more persistent than $1_H^\bullet$ [4, 5] (cf. Fig. 28.1 and 28.3). This was attributed to the fact that the minimum distance that the hydrogen atom must "jump" is considerably less for $1_H^\bullet$ (1.34 Å, assuming normal bond lengths and angles) than for the other two hindered phenyl radicals (e. g., 1.84 Å for $4_H^\bullet$) [4, 5]. Attempts to further confirm the importance of QMT in the isomerization of $4_H^\bullet$ and $5_H^\bullet$ by studying these reactions in frozen matrices at really low temperatures were frustrated by poor resolution of the phenyl radicals' EPR spectra in the

Figure 28.3 Arrhenius plots of the rate constants for the isomerization of 2,4,6-tri(1'-adamantyl)phenyl, $4_H^\bullet$ (○) and for the isomerization of octamethyloctahydroanthracen-9-yl, $5_H^\bullet$ (×).

solids. The compound $5_D Br$ in which all eight methyl groups were perdeuterated was therefore synthesized and converted to $5_D^\bullet$ using tri-*n*-butyltin radicals (generated by photolysis of hexa-*n*-butyl ditin) [5]. As expected, $5_D^\bullet$ decayed with first-order kinetics and more slowly than $5_H^\bullet$. However, the DKIE increased only from ca. 20 to ca. 50 over a temperature range from 323 to 173 K. Moreover, in contrast to the isomerization of $5_H^\bullet$ where the rates were independent of the ditin concentration and solvent (cylcopropane, propane, isopentane, isooctane), the rates of "isomerization" of $5_D^\bullet$ were *dependent* on the ditin concentration and solvent (faster in isopentane than in cyclopropane) [5]. Obviously, $5_D^\bullet$ decays not by an intramolecular D-atom QMT but by an intermolecular H-atom abstraction from the surrounding medium. This is unfortunate but does have a positive side: Because $5_H^\bullet$ is so much shorter lived than $5_D^\bullet$ it cannot be reacting with the surrounding medium to any significant extent. That is, $5_H^\bullet$ must decay by intramolecular H-atom QMT.

The much less sterically hindered phenyl radicals $6_H^\bullet$, $7_H^\bullet$ and $8_H^\bullet$ were generated

$$\text{(28.9)}$$

$6_H^\bullet \qquad 7_H^\bullet \qquad 8_H^\bullet \qquad 9_H^\bullet \qquad 10_H^\bullet$

from their parent bromides in the usual ways but did not give EPR detectable signals [4, 5]. That $6_H^\bullet$ and $7_H^\bullet$ were being produced was demonstrated by the addition of tetramethylgermane. The (very sharp) EPR signals due to $(CH_3)_3 GeCH_2^\bullet$ could then be readily detected, e.g.,

$$7_H^\bullet + (CH_3)_4 Ge \longrightarrow 7_H H + (CH_3)_3 GeCH_2^\bullet \qquad \text{(28.10)}$$

That $8_H^\bullet$ was formed was obvious from the appearance, even at 113 K, of the EPR signal due to a neophyl-type radical. This must arise from a fast intramolecular H-atom abstraction via a 6-center cyclic transition state.

$8_H^\bullet \longrightarrow$

$$\text{(28.11)}$$

The phenyl radicals $9_H^\bullet$ and $10_H^\bullet$ could be detected by EPR spectroscopy but were much less persistent than $1_H^\bullet$ [5]. They decayed with first-order kinetics in cyclopropane but their rates of decay were proportional to the concentration of hexabutyl ditin. Generation of $9_H^\bullet$ by direct photolysis of $9_H Br$ decreased the rate of its decay significantly and the addition of $(CH_3)_4 Ge$ gave rise to the $(CH_3)_3 GeCH_2^\bullet$ EPR signal.

Radicals $9_H^\bullet$ and $10_H^\bullet$ were prepared with the hope of increasing phenyl radical persistence by increasing the distance the H-atom must "jump" in a 5-center intramolecular H-atom abstraction in comparison with the distances involved for $1_H^\bullet$ and $4_H^\bullet$ e.g., 2.25 Å for $9_H^\bullet$. This approach was unsuccessful (in solution) because of intermolecular H-atom abstraction by the more exposed radical centers. However, these experiments did serve to define the requirements necessary to most readily observe unequivocal examples of QMT in intermolecular H-atom abstractions, viz., relatively "fixed" and relatively close spatial coordinates for the H-atom donor and the H-atom receiver, see final section in this chapter.

28.2.2
Inversion of Nonplanar, Cyclic, Carbon-Centered Radicals

The vast majority of trivalent carbon-centered radicals are planar (single energy minimum) or, if not strictly planar, they generally have negligible barriers to inversion, e.g., $Me_3C^\bullet$ [10]. However, nonplanar trivalent carbon-centered radicals with significant barriers to inversion exist when the radical center has strongly electron-withdrawing atoms directly attached, e.g., $F_3C^\bullet$ [11], or forms a part of a three-membered ring, e.g. cyclopropyl [12]. Although an inversion may not always be considered to be a chemical "reaction" they can provide very nice examples of systems in which QMT plays a dominant role in the inversion kinetics.

28.2.2.1 Cyclopropyl and 1-Methylcyclopropyl Radicals
Both of these cyclopropyls are nonplanar at their radical centers and have a similar degree of deviation from planarity [12]. They are therefore likely to have similar "classical" barriers to inversion. For cyclopropyl, $11_H^\bullet$, the four ring hydrogen atoms are magnetically equivalent on

$$11^\bullet \; (R_\alpha = H, D, Me) \tag{28.12}$$

the EPR time scale at the lowest temperatures explored (89 K in ethane as solvent) [13]. This was also true for $11_D^\bullet$ [13] In contrast, for 1-methylcyclopropyl the four ring hydrogen atoms are only equivalent at temperatures down to 183 K. At still lower temperatures, the syn and anti hydrogen become magnetically unequivalent [13]. The experimental spectra were simulated through the coalescence temperature range. The calculated values of the inversion rate constant, k_{12}, gave an excellent (linear) Arrhenius plot:

$$\log (k_{12}^{Me}/s^{-1}) = 13.1 - 3.1/2.3RT \text{ (kcal mol}^{-1}) \tag{28.II}$$

The magnitude of the pre-exponential factor provides further confirmation that methyl inversion is essentially a "classical" process, as expected for such a massive group. However, because $11_H^\bullet$ and $11_{Me}^\bullet$ are expected to have roughly the same inversion barriers, the failure to resolve the syn and anti hydrogen atoms in $11_H^\bullet$ and even in $11_D^\bullet$ is a strong indication that QMT dominates both H and D inversions.

28.2.2.2 The Oxiranyl Radical

As was the case with $11_{Me}^\bullet$, the rates of inversion of the oxiranyl radicals, $12_H^\bullet$ and $12_D^\bullet$ could be measured by EPR line broadening over a wide range of temperatures

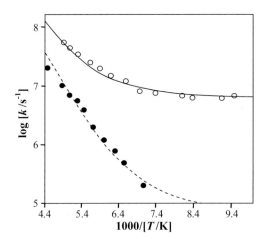

$$\tag{28.13}$$

$12^\bullet$ (R$_\alpha$ = H, D)

[14]. The results are shown as Arrhenius plots in Fig. 28.4. In view of the large DKIE, there can be little doubt that QMT dominates the H-atom inversion. Furthermore, the Arrhenius plot for H-atom inversion is curved and the rate con-

Figure 28.4 Arrhenius plots of the rate constants for the inversion of oxiranyl, $12_H^\bullet$ (○) and α-deuteriooxiranyl, $12_D^\bullet$ (●). The solid and dashed lines depict the calculated rate constants for H-atom and D-atom inversion, respectively.

stant for inversion, k_{11}^H, reaches a limiting value of ca. 7×10^6 s^{-1} at temperatures below ca. 140 K. Unfortunately, the rate of D-atom inversion became too slow to measure at temperatures lower than ca. 140 K. The available data for D-atom inversion can be fitted reasonably well to the Arrhenius equation:

$$\log (k_{13}^D/s^{-1}) = 10.9 - 3.6/2.3RT \text{ (kcal mol}^{-1}) \tag{28.III}$$

However, the pre-exponential factor in this equation is roughly 2 orders of magnitude smaller than would be expected for a "classical" (over the barrier) inversion (compare with Eq. (28.II)). It is therefore highly probable that QMT is also important in the D-atom inversion over the temperature range covered.

To interpret the kinetic results for oxiranyl inversion quantitatively, the structure and vibrational force field were, in the absence of experimental data, determined by *ab initio* molecular orbital calculations [14]. The minimum-energy path for inversion was found to involve not only components perpendicular to the COC plane but also components parallel to the plane. Moreover, atoms other than the inverting hydrogen were found to undergo substantial displacements so that the calculation of accurate inversion rate constants would have required a multidimensional potential energy surface. For practical reasons only a one-dimensional effective potential was employed. The frequency associated with the effective potential was taken to be that of the inverting H-atom, H$_a$, in $12_H^\bullet$ because this has the lowest frequency and will contribute the most to the temperature dependence of k_{13}^H at low temperatures. This yielded only a partial potential, defined only at stationary points and could not be used to calculate $k_{13}^H(T)$ directly. The observed inversion rate constants were therefore employed to derive a one-dimensional empirical, double-minimum potential energy surface that was reasonably close to the theoretical potential at its stationary points. The empirical barrier height was 6.8 kcal mol^{-1} in excellent agreement with the experimental barrier height for inversion of 2,3-dimethyloxiranyl [15], which amounts to 7.0 kcal mol^{-1} after correction for the zero point energy [14].

It is noteworthy that only a one-dimensional potential is required for the inversion of **12$^\bullet$** rather than the two-dimensional potential required for the isomerization of **1$^\bullet$**. Note also that the rate constants for D-atom inversion smoothly fit a curved Arrhenius plot that has its strongest curvature at lower temperatures than for the H-atom inversion (see Fig. 28.4). This is because the frequency of the out-of-plane bending mode is lower for $12_D^\bullet$ than for $12_H^\bullet$, so that $12_D^\bullet$'s excited states start contributing to the tunneling at lower temperatures. The limiting rate constant, k_{13}^D, at 0 K was calculated to be 6.9×10^4 s^{-1} [14]. Thus, the limiting, QMT-only, DKIE for oxiranyl inversion is $7 \times 10^6/6.9 \times 10^4 \approx 100$. This limiting DKIE is comparable to that calculated for the inversion of the dioxolanyl radical [16] (vide infra), but is much smaller than that obtained for 2,4,6-tri-*tert*-butyl-phenyl isomerization (~50000, vide supra) and that which could be estimated for reaction (28.14) in matrices (vide infra).

$$CH_3^\bullet + CH_3OH \text{ (CD}_3\text{OD)} \rightarrow CH_4 \text{ (CH}_3\text{D)} + {}^\bullet CH_2OH \text{ ($^\bullet$CD}_2\text{OD)} \tag{28.14}$$

28.2.2.3 The Dioxolanyl Radical

The EPR spectrum of 1,3-dioxolan-2-yl, $13_H^\bullet$, showed no detectable line broadening

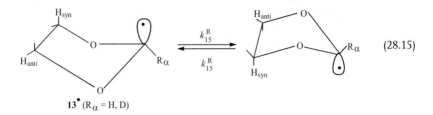

$$(28.15)$$

down to 93 K [16]. However, for $13_D^\bullet$, the syn and anti hydrogen atoms at C-4 and C-5 are not magnetically equivalent at low temperatures and values of k_{15}^D have been calculated from 191 to 99 K [16]. An Arrhenius plot of these data provides a very nice example of D-atom tunneling in an inversion. The methods of calculation used for the $12_H^\bullet$ and $12_D^\bullet$ oxiranyl inversions were applied to the dioxolanyl radicals and provided the lines shown in Fig. 28.5. These calculations imply that the QMT-only inversion rate constant for $13_H^\bullet$, is only just too great for measurement by EPR line broadening. During inversion the C_a and H_a atoms move in opposite directions while the remaining atoms move very little. This inversion is therefore "double-hinged" so that the C_a motion reduces the H_a (or D_a) tunneling path length compared, for example, to the oxiranyl radical, leading to faster tunneling for $13^\bullet$, than for $12^\bullet$ although the barrier heights for these two radicals are rather similar (vide infra).

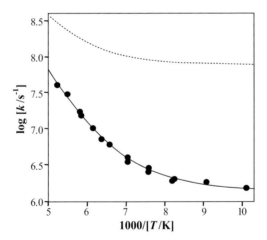

Figure 28.5 Arrhenius plot of the rate constants for the inversion of [2D]-1,3-dioxolan-2-yl, $13_D^\bullet$. The solid line depicts the calculated rate constant for this process. The broken line depicts the calculated rate constants for inversion of the nondeuterated radical, $13_H^\bullet$.

28.2.2.4 Summary

The inversion of cyclic carbon-centered radicals provides some very nice examples of H- and D-atom QMT. For purposes of comparison [16], approximate values of: (i) the pyramidyl angles at the radicals' centers are 39, 45, and 42°; (ii) the barrier heights are 3.0, 6.8, and 7.7 kcal mol^{-1}; and (iii) the limiting, QMT-only, DKIEs are 8, 100, and 50; for cyclopropyl, oxiranyl and dioxolan-2-yl, respectively. The EPR line broadening method allowed the rate constants for H- and D-atom inversion in oxiranyl and D-atom inversion in dioxolan-2-yl to be measured over a wide range of temperatures. However, as a consequence of QMT, both H- and D-atom inversion in cyclopropyl and H-atom inversion in dioxolanyl occurred too rapidly for their rate constants to be determined by this technique even at the lowest temperatures.

28.3
Bimolecular Reactions

28.3.1
H-Atom Abstraction by Methyl Radicals in Organic Glasses

Methyl radical decay in simple organic glasses made from acetonitrile [7, 17], methyl isocyanide [18] and methanol [19], at low temperatures, e.g., 77 K (liquid N$_2$) commonly occur by H-atom abstraction from a molecule in the glass:

$$CH_3^{\bullet} + CH_3G \rightarrow CH_4 + {}^{\bullet}CH_2G \tag{28.16}$$

These processes were generally regarded as outstanding examples of reactions in which QMT dominates the thermally activated process. This is undoubtedly true (vide infra) but the interpretation of the experimental results from all these systems suffered from two difficulties. First, in these glasses reaction (28.16) did not follow (pseudo-)first-order kinetics. Second, the rates of deuterium transfer from CD$_3$G were generally too slow to measure accurately. Indeed $^{\bullet}$CD$_3$ radicals in CD$_3$CN [17] and in CD$_3$NC [18] disappear by reactions other than D-atom abstraction from the glass, leading to the term "all-or-nothing isotope effects" [18].

If methyl radical decay in glasses followed first-order kinetics its concentration would decrease according to $\exp(-k_{16}t)$. However, it has been convincingly demonstrated [20] that in several glasses the decay actually follows a rate law of the form $\exp(-ct^{1/2})$. The meaning of the constant c was not made clear and no plausible kinetic scheme supporting such a rate law had been put forward.

This unsatisfactory state of affairs changed in 1984 with a proposal [21] that these abnormal kinetics were a consequence of the inhomogeneity of the matrix. In the experiments that were carried out [21], the methyl radicals were produced in a methanol glass from methyl halides by photo-induced electron capture using traces of diphenylamine, reaction (28.17). It was suggested [21] that the radicals

$$\text{Ph}_2\text{NH} \xrightarrow{\text{hv}} e^- \xrightarrow{\text{CH}_3\text{X}} {}^{\bullet}\text{CH}_3 + \text{X}^- \qquad (28.17)$$

were formed in a distribution of sites in the glass. This meant that decay would occur with a distribution of first-order rate constants arising from a distribution of H-atom transfer distances. It was found that when the irradiation time used to generate the ${}^{\bullet}\text{CH}_3$ was short (10 s) relative to the decay times, the plot of $\log[{}^{\bullet}\text{CH}_3]$ vs. $t^{1/2}$ at 77 K was indeed linear for a 200-fold decrease in the radical's concentration, a decrease which occurred over 1600 s [21]. However, when a reduced light intensity and a long irradiation time (~10000 s) were employed the plot of $\log[{}^{\bullet}\text{CH}_3]$ vs. $t^{1/2}$ was nonlinear, being strongly curved downwards towards the $t^{1/2}$ axis. These preliminary results [21] were satisfactorily accommodated within a simple model. The range of first-order rate constants in the different sites in the glass was found to extend over roughly two orders of magnitude and was attributed to a distribution of H-atom transfer distances (and thus was indirectly related to the structure of the glass). The different decay kinetics found for long and short irradiation times arise because the methyls in the more reactive sites had mostly decayed before the measurements were started.

In subsequent publications [22] this model was refined and a great deal more kinetic data on reaction (28.14) and partially deuterated versions of this reaction were reported. Measurements were made over as wide a range of temperatures as was experimentally possible. This was 5–89 K for CH_3OH and 77–97 K for CD_3OD glasses, the upper temperature being set by the softening of the glass (phase transition at 103 K) and the 77 K lower limit for CD_3OD glass by the extreme slowness of the D-atom abstraction at lower temperatures [22b].

For each site where the methyl radicals are trapped, the H- or D-atom transfer rate is, of course, governed only by the local properties of the glass, irrespective of other sites. The first-order rate constant for each site is determined by the distance from the center of the radical to the nearest methanolic methyl hydrogen atom. Since short-range order is conserved in the glass, this distance, governed by the van der Waals radii of the methyl group and methyl radical, will vary only slightly from site to site. Since, on the time scale of the experiments, methyl rotation and tumbling are rapid at these temperatures, the distribution of transfer distances will be narrow and random. The rate constants of this distribution were analyzed theoretically to obtain a quantitative relation between rate and equilibrium distance for H- and D-atom transfer. The model used was based on QMT and a two-dimensional barrier. One dimension, associated with C–H stretching in the CH_3OH, accounts for the observed large DKIE, the other associated with the lattice mode, is mainly responsible for the temperature dependence. Some of the parameters employed in these computations were independently known, e.g., the C–H stretching frequency. Other parameters were computed by determining the best fit to the experimental data. The numerical values of these "adjustable" parameters were found to be physically reasonable, e.g., 3.9 Å for the most probable equilibrium distance between the methyl radical and the methyl group, a distance that can be compared with 4.0 Å for the sum of their van der Waals radii (therefore, the most probable equilibrium tunneling distance is $\approx 3.9 - 2 \times 1.09 \approx 1.7$ Å), and a lattice frequency of 140 cm^{-1} which is close to the Debye frequency for ice of 133 cm^{-1}.

The experimental nonexponential decay of the EPR signal due to the methyl radical yield k_0, the maximum in the distribution of first-order rate constants, i.e., the "most probable" rate constant [22]. Values of k_0 plotted against T and the computed rate constants are shown in Fig. 28.6 for the CH$_3$OH and CD$_3$OH glasses [22b,c] The good fit of theory to experiment using realistic parameters lends credence to the validity of the interpretation and to the tunneling distances deduced from it. The strongest deviations occur at very low temperatures where methyl rotations may reduce to librations with the favored orientations relatively unfavorable to QMT. There was also evidence that methyl radical generation was not instantaneous near 5 K [22b].

The EPR spectra of methyl radicals trapped in methanol glasses show "forbidden" lines as satellites of the main 1:3:3:1 quartet lines [23]. These are due to dipolar coupling of the unpaired electron with protons of neighboring methanol molecules. Comparison of the relative intensities of these satellites in CH$_3$OH, CH$_3$OD, CHD$_2$OD, CD$_3$OH and CD$_3$OD indicate that around the trapped methyl radical the structure is similar to the (disordered) β-phase crystal structure of methanol, with the radical replacing a methanol molecule and occupying a position close to its methyl position [23]. The calculated methyl–methyl distances from these experiments [23] are compatible with the distance previously calculated from the methyl radical decay kinetics [22].

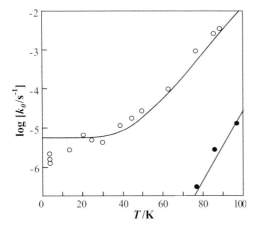

Figure 28.6 Plots of the "most probable" rate constants for H-atom (○) and D-atom (●) abstraction by methyl radicals in CH$_3$OH and CD$_3$OD glasses, respectively, against T (K). The solid lines depict the calculated rate constants for the two reactions.

The measured rates of decay of methyl radicals embedded in glasses made from CH_2DOD over a range of temperatures [22c], and from mixtures of CH_3OH and CD_3OH at 77 K [24] were also consistent with the distribution of trapping sites/distribution of first-order kinetics model. However, decay rates in the isotopomerically mixed glasses showed that the static picture was inadequate [24]. At 77 K, the radical had to be able to diffuse through the glass on the time scale of the kinetic measurements. Such diffusion allows the radical to encounter more CH_3OH molecules than would be expected for the static structure on a statistical basis. That is, the effective (reactive) mole fraction of CH_3OH in the mixtures was higher than the analytical concentration. For example, with 5% CH_3OH in CD_3OH, the radical encounters, on average, ca. 26 methanol molecules before abstraction occurs which corresponds to diffusion over roughly 11 Å.

28.3.2
H-Atom Abstraction by Bis(trifluoromethyl) Nitroxide in the Liquid Phase

The convincing evidence given above for the dominant role of QMT in the rates of H-atom abstractions, both intramolecular, e.g., $1^{\bullet}_H \rightarrow 2^{\bullet}_H$, and intermolecular, e.g., methyl radical decay in glassy methanol, were obtained in systems involving only a limited number of transferable hydrogen atoms around the radical center. Furthermore, those transferable H-atoms were fairly rigidly held (at tunneling distances) away from the radical center, and the transfers were strongly exothermic.

Convincing experimental evidence for QMT's involvement in any intermolecular H-atom abstraction in the liquid phase had not been presented and represented an interesting challenge in the 1980s (and to this day, so far as the author is aware). If H-atom tunneling is to be identified in the generalized reaction (28.18)

$$X^{\bullet} + RH \longrightarrow XH + R^{\bullet} \tag{28.18}$$

in solution, considerable difficulties would have to be overcome. A kinetic EPR spectroscopic study would have to meet the following criteria:

1. $X^{\bullet}$ must be observable by EPR.
2. Reaction (28.18) must be (relatively) rapid, i.e. $X^{\bullet}$ must be highly reactive and/or RH must be a "good" H-atom donor.
3. Reaction (28.18) must be irreversible
4. $X^{\bullet}$ must be destroyed only by reaction (28.18), i.e. there must be no bimolecular self-reaction of $X^{\bullet}$ ($2\,X^{\bullet} \xrightarrow{\quad//\quad}$ products),
   no unimolecular rearrangement or scission of
   $X^{\bullet}(X^{\bullet} \xrightarrow{\quad//\quad}$ products), and no reaction of $X^{\bullet}$ with
   the solvent ($X^{\bullet} +$ solvent $\xrightarrow{\quad//\quad}$ products).
5. The solvent must have an extremely low freezing point, as close to 77 K as is compatible with it being liquid over a wide range of temperatures.

The radical that appeared to meet the criteria for $X^\bullet$ most closely was $(CF_3)_2NO^\bullet$ and the desired solvent properties were most closely met by some of the Freons (chlorofluorocarbons). The radical was generated photochemically via reactions (28.19) and (28.20) in solvent containing the RH substrate. Initial experiments

$$CF_3OOCF_3 \xrightarrow{h\nu} 2\,CF_3O^\bullet \tag{28.19}$$

$$CF_3O^\bullet + (CF_3)_2NOH \longrightarrow CF_3OH + (CF_3)_2NO^\bullet \tag{28.20}$$

[25] were rather encouraging despite the difficulties experienced in keeping the samples at a constant low temperature for times between 2 weeks and 4 months! In these initial experiments, rate constants for reaction (28.21) were measured from 327 down to 123 K for toluene and from 345 to 183 K for toluene-d_8.

$$(CF_3)_2NO^\bullet + C_6H_5CH_3(d_8) \xrightarrow{k_{21}^H(k_{21}^D)} (CF_3)_2NOH(D) + C_6H_5CH_2^\bullet(d_7^\bullet) \tag{28.21}$$

$$(CF_3)_2NO^\bullet + C_6H_5CH_2^\bullet(d_7^\bullet) \xrightarrow{\text{v fast}} (CF_3)_2NOCH_2C_6H_5(d_7) \tag{28.22}$$

Arrhenius plots exhibited slight curvature for the H-atom transfer at low temperatures. If these low temperature points were ignored, the plots yielded $A_{21}^H \approx A_{21}^D = 10^4$ M^{-1} s^{-1}, a pre-exponential factor that is well below the $10^{8.5\pm0.5}$ M^{-1} s^{-1} found for the vast majority of intermolecular H-atom abstractions [26], and $E_{21}^D - E_{21}^H = 1.6$ kcal mol^{-1} [25]. The (possibly) curved Arrhenius plot, the low A factors and the differences in activation enthalpies between D-and H-atom abstraction all suggested a significant role for QMT in reaction (28.21).

Follow-up work [27] revealed that the $CF_3OOCF_3/(CF_3)_2NOH$ method for generating $(CF_3)_2NO^\bullet$ radicals that had been employed [25] had problems. This method necessarily involves low $[(CF_3)_2NO^\bullet]/[(CF_3)_2NOH]$ ratios and not all $R^\bullet$ were trapped by $(CF_3)_2NO^\bullet$ (e.g., reaction (28.22)). Some $R^\bullet$ radicals were lost by reaction (28.23).

$$R^\bullet + (CF_3)_2NOH \longrightarrow RH + (CF_3)_2NO^\bullet \tag{28.23}$$

This leads to a decrease in the measured rate constant as the reaction progresses (and the $[(CF_3)_2NO^\bullet]/[(CF_3)_2NOH]$ ratio decreases. The loss of $(CF_3)_2NO^\bullet$ could be seen not to follow (pseudo-)first-order kinetics if the decay was monitored for 3 or more half-lives.

To overcome this problem, in the new work the $(CF_3)_2NO^\bullet$ radical itself was employed [27]. Rate constants for H-atom abstraction from 11 new substrates yielded Arrhenius pre-exponential factors ranging from a low of $10^{4.2}$ M^{-1} s^{-1} for diethyl ether (temperature range 297–178 K) to a high of $10^{6.5}$ M^{-1} s^{-1} for 1,4-cyclohexadiene (296– 192 K). In addition, the rate constants were measured for a reaction in which QMT could not be involved. This was the addition of $(CF_3)_2NO^\bullet$ to $CH_2=CCl_2$.

$$(CF_3)_2NO^\bullet + CH_2=CCl_2 \longrightarrow (CF_3)_2NOCH_2CCl_2{}^\bullet \qquad\qquad (28.24)$$

The Arrhenius plot yielded a pre-exponential factor, A_{24}, that was only $10^{5.3}$ M^{-1} s^{-1} that is also much lower than the expected [26] $10^{8.5}$ M^{-1} s^{-1}.

The measured rate constants [27] should be reliable (except for the earlier toluene data [25]) and they yielded Arrhenius plots that were linear over a wide range of temperatures. Moreover, for H- and D-atom abstraction from C_6H_5CHO and C_6H_5CDO the pre-exponential factors were equal within experimental error ($A^H = 10^{5.3\pm0.3}$, $A^D = 10^{5.1\pm0.5}$ M^{-1} s^{-1}) and, although the DKIE was large, viz. 15 at 298 K, it was not so large that it could only be accounted for by there being a significant role for QMT in the H-atom abstraction. It was, therefore, (reluctantly) concluded that in none of these, quite extensive, H-atom abstraction experiments with $(CF_3)_2NO^\bullet$ in the liquid phase was there *unequivocal* evidence that QMT played a *significant* role. *Another conclusion that could be drawn is that it is going to be extremely difficult, if not impossible, to prove that QMT is truly important in any bimolecular H-atom abstraction in the liquid phase by the one certain test, a curved Arrhenius plot and a rate constant that is independent of the temperature.*

References

1 . Bell, R. P. *Proc. R. Soc. London, Ser. A* **1933**, *139*, 466–474; see also: Bell, R. P. *Proc. R. Soc. London, Ser.A* **1935**, *148*, 241–250.

2 Griller, D.; Ingold, K. U. *Acc. Chem. Res.* **1976**, *9*, 13–19.

3 Barclay, L. R. C.; Griller, D.; Ingold, K. U. *J. Am. Chem. Soc.* **1974**, *96*, 3011–3012.

4 Brunton, G.; Griller, D.; Barclay, L. R. C.; Ingold, K. U. *J. Am. Chem. Soc.* **1976**, *98*, 6803–6811.

5 Brunton, G.; Gray, J. A.; Griller, D.; Barclay, L. R. C. ; Ingold, K. U. *J. Am. Chem. Soc.* **1978**, *100*, 4197–4200.

6 Bell, R. P. *Chem. Soc. Rev.* **1974**, *3*, 513–544.

7 LeRoy, R. J.; Sprague, E. D.; Williams, F. *J. Phys. Chem.* **1972**, *76*, 546–551.

8 Laplante, J.-P.; Siebrand, W. *Chem. Phys. Lett.* **1978**, *59*, 433–436; Siebrand, W.; Wildman, T. A.; Zgierski, M. Z. *J. Am. Chem. Soc.* **1984**, *106*, 4083–4089.

9 Siebrand, W.; Wildman, T. A.; Zgierski, M. Z. *J. Am. Chem. Soc.* **1984**, *106*, 4089–4096.

10 Griller, D; Ingold, K. U.; Krusic, P. J.; Fischer, H. *J. Am. Chem. Soc.* **1978**, *100*, 6750–6752.

11 Griller, D; Ingold, K. U.; Krusic, P. J.; Smart, B. E.; Wonchoba, E. R. *J. Phys. Chem.* **1982**, *86*, 1376–1377.

12 Johnston, L. J.; Ingold, K. U. *J. Am. Chem. Soc.* **1986**, *108*, 2343–2348.

13 Deycard, S.; Hughes, L.; Lusztyk, J.; Ingold, K. U. *J. Am. Chem. Soc.* **1987**, *109*, 4954–4960.

14 Deycard, S.; Lusztyk, J.; Ingold, K. U.; Zerbetto, F.; Zgierski, M. Z.; Sieband, W. *J. Am. Chem. Soc.* **1988**, *110*, 6721–6726.

15 Itzel, H.; Fischer, H. *Helv. Chim. Acta.* **1976**, *59*, 880–901.

16 Deycard, S.; Lusztyk, J.; Ingold, K. U.; Zerbetto, F.; Zgierski, M. Z.; Siebrand, W. *J. Am. Chem. Soc.* **1990**, *112*, 4284–4290.

17 Sprague, E. D.; Williams, F. *J. Am. Chem. Soc.* **1971**, *93*, 787–788.

18 Wang, J-T.; Williams, F. *J. Am. Chem. Soc.* **1972**, *94*, 2930–2934.

19 Campion, A.; Williams, F. *J. Am. Chem. Soc.* **1972**, *94*, 7633–7637.

20 Bol'shakov, B. V.; Tolkatchev, V. A. *Chem. Phys. Lett.* **1976**, *40*, 468–470; Stepanov, A. A.; Tkatchenko, V. A.; Bol'shakov, B. V; Tolkatchev, V. A. *Int. J. Chem. Kinet.* **1978**, *10*, 637–648; Bol'shakov, B. V; Doktorov, A. B.; Tolkatchev, V. A.; Burshtein, A. I. *Chem. Phys. Lett.* **1979**, *64*, 113–115; Bol'shakov, B. V; Stepanov, A. A.; Tolkatchev, V. A. *Int. J. Chem. Kinet.* **1980**, *12*, 271–281.

21 Doba, T.; Ingold, K. U.; Siebrand, W. *Chem. Phys. Lett.* **1984**, *103*, 339–342.

22 (a) Doba,T.; Ingold, K. U.; Siebrand, W.; Wildman, T. A. *Chem. Phys. Lett.* **1985**, *115*, 51–54; (b) Doba, T.; Ingold, K. U.; Siebrand, W.; Wildman, T.A. *Chem. Phys. Lett.* **1984**, *88*, 3165–3167; (c) Doba,T.; Ingold, K. U.; Siebrand, W.; Wildman, T. A. *Faraday Discuss. Chem. Soc.* **1984**, *78*, 175–191.

23 Doba,T.; Ingold, K. U.; Reddoch, A. H.; Siebrand, W.; Wildman, T. A. *J. Chem. Phys.* **1987**, *86*, 6622–6630.

24 Doba, T.; Ingold, K. U.; Lusztyk, J.; Siebrand, W.; Wildman, T. A. *J. Chem. Phys.* **1993**, *98*, 2962–2970.

25 Malatesta, V.; Ingold, K. U. *J. Am. Chem. Soc.* **1981**, *103*, 3094–3098.

26 Benson, S. W. *Thermochemical Kinetics*, 2nd edn., Wiley, New York, 1976

27 Doba, T.; Ingold K. U. *J. Am. Chem. Soc.* **1984**, *106*, 3958–3963.

28 Computational methods now exist that include contributions from all vibrational modes to the H/D-transfer process, thus eliminating the need to introduce any empirical parameters, e.g., variational transition state theory with semiclassical tunneling corrections (Truhlar, D. G.; Garett, B. C.; Klippenstein, S. J. *J. Phys. Chem.* **1996**, *100*, 12771) and the approximate instanton method (Siebrand, W.; Smedarchina, Z.; Zgierski, M. Z.; Fernández-Ramos, A. *Int. Rev. Chem. Phys.* **1999**, *18*, 5).

29
Multiple Proton Transfer: From Stepwise to Concerted

Zorka Smedarchina, Willem Siebrand, and Antonio Fernández-Ramos

29.1
Introduction

It is well recognized that reactions involving the transfer of a proton or hydrogen atom are special in that these particles can tunnel through classically forbidden regions [1]. The wave-like properties also add a new element to reactions in which two or more protons transfer, since under appropriate conditions, they may allow the protons to move as a single particle. In this contribution we review the dynamics of such reactions [2], focusing on double proton transfer for simplicity. In particular, we probe how the motion of one proton influences that of the other and which conditions lead to weak or strong correlation between their motions. Generally speaking, no correlation results in independent transfer and weak correlation in stepwise transfer. Strengthening the correlation will ultimately lead to concerted transfer and may give rise to synchronous transfer if the transferred particles are equivalent.

While proton–proton correlation is a unifying concept that allows us to classify and understand the various multiproton transfer mechanisms, it is not a quantity that is easily measured or calculated. In dealing with a specific reaction, one tends to use a simpler approach based on the search for a transition state, i.e. a configuration along the transfer path characterized by a first-order saddle point representing a vibrational force field with one imaginary frequency. More generally, the presence of two mobile particles implies that the potential energy surface contains stationary states with zero, one, or two imaginary frequencies, representing, respectively, a stable intermediate, a transition state, and a state with a second-order saddle point.

A stable intermediate roughly halfway along the trajectory implies barriers separating it from the equilibrium configurations. Such a potential favors stepwise transfer under conditions where the intermediate is thermally accessible. This basically reduces the dynamics to that appropriate for single proton transfer, but leaves open the question of how to deal with transfer at low temperature. A barrier corresponding to a single transition state, similar to that observed for single proton transfer, implies concerted transfer of the two protons. This again can

Hydrogen-Transfer Reactions. Edited by J. T. Hynes, J. P. Klinman, H. H. Limbach, and R. L. Schowen
Copyright © 2007 WILEY-VCH Verlag GmbH & Co. KGaA, Weinheim
ISBN: 978-3-527-30777-7

be treated by the methods developed for single proton dynamics [1, 3]. However, if, instead, this barrier corresponds to a second-order saddle point, it represents concerted motion along *two* proton coordinates. This situation does not immediately reveal the nature of the corresponding transfer process, but it drives home the point that the presence of two mobile protons allows at least two transfer mechanisms. On a potential energy surface with more than one saddle point, there will in general be multiple pathways along which the potential has a double-minimum profile. To analyze the transfer dynamics governed by such a potential, we need an approach that goes beyond the question whether the transfer is concerted or stepwise.

To elucidate the effect of proton–proton correlation on the transfer mechanism, we approach the problem from two sides. On the one hand, we develop in Section 29.2 a theoretical model based on two identical single proton transfer potentials subject to proton–proton coupling represented by a simple bilinear function. On the other hand, we consider a representative range of two- and multi-proton transfer processes for which the proton dynamics has been studied experimentally and/or theoretically, typical examples being dimeric carboxylic acids [4–7], porphyrins [8–11] and porphycenes [12–14], naphthazarins [15], and transfer catalyzed by proton conduits such as chains of water molecules [16–18]. Our own calculations on several of these systems are based on the approximate instanton method (AIM) [3, 19], which we review in Section 29.3 and extend to transfer along two reaction coordinates in Section 29.4.

Experimental information on multiple proton transfer is available in the form of (state-specific) level splittings and/or thermal rate constants. The observation of a level splitting implies concerted transfer between levels of the same energy. It does not necessarily imply that the potential has a single saddle point or that the transfer can be described in terms of a single trajectory. This question is analyzed in Section 29.4 for a model two-dimensional potential relevant to double proton transfer, with explicit evaluation of tunneling probabilities along various paths. The observation of deuterium (and tritium) isotope effects on level splittings and thermal rate constants provides valuable information on the transfer mechanism [2]. However, whereas in single proton transfer a large isotope effect indicates transfer dominated by tunneling, such an observation may be ambiguous for multiple proton transfer because of the proportionality of zero-point energy shifts with the number of protons. In this connection, generalization of the Swain–Schaad relation between H, D, and T transfer [20] so as to include tunneling [21], offers an alternative method. In the remaining sections, we apply these theoretical approaches to systems for which transfer data are available. We use these results to probe to what extent it is possible to predict the mechanisms contributing to multiproton transfer reactions in a given system, on the basis of known physical properties such as symmetry, geometry, transfer distance and hydrogen bond strength, in an attempt to arrive at a coherent picture of the present state of our understanding of these reactions.

29.2
Basic Model

Our basic model for double proton transfer is a molecule or complex in which two equivalent protons (hydrogen atoms) can transfer between equivalent positions. We arrange the four carrying atoms X (oxygen, nitrogen, carbon, etc.) in a rectangle with sides a and b, such that any hydrogen bonding takes place along a, the hydrogen bonds being separated by b, as illustrated in Fig. 29.1. This allows us to vary the strength of the hydrogen bonds by varying a, and the strength of their correlation by varying b. The separation a may vary *during* the transfer in the case of hydrogen bonding and this in turn may affect the correlation. In Fig. 29.1(a) the protons are arranged as in the formic acid dimer, i.e. according to an equilibrium structure belonging to point group C_{2h}. The alternative structure of Fig. 29.1(b) belonging to point group C_{2v} would be realized in the case of double proton transfer between ethane and ethylene moieties in a parallel arrangement. The two arrangements transform into each other through single proton transfer. To keep the notation simple, we use in this section dimensionless units by expressing the coordinates in units $r/2$, where r is the transfer distance of each proton, i.e. the distance between its two equilibrium positions, and expressing the energies in units $2U_0$, where U_0 is the barrier height for single proton transfer along the XH$\cdots$X bonds. For any symmetric double-minimum potential we have $U_0 \propto r^2$ and for the linear X–H$\cdots$X bond of Fig. 29.1 we have $r = a - 2R_{XH}$, where R_{XH} is the X–H bond length.

First we consider the case where b is so large that the correlation between the two protons is negligible. Then we can write the transfer potential as the sum of two double minimum potentials, which we represent by quartic potentials

$$U(x_1, x_2) = \frac{1}{2}[(1 - x_1^2)^2 + (1 - x_2^2)^2]$$ (29.1)

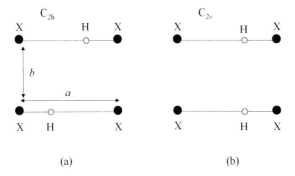

Figure 29.1 Schematic representation of the basic model of double-proton transfer along parallel equivalent hydrogen bonds in a symmetric system with C_{2h} (a) or C_{2v} (b) symmetry.

The corresponding potential-energy surface has a maximum $U = 1$ at $x_1 = x_2 = 0$, where the two protons are in the center of their paths, and four equivalent minima $U = 0$ at $|x_1| = |x_2| = 1$ where the protons are in their equilibrium positions $x_{1,2} = \pm 1$; two of these correspond to the structure of Fig. 29.1(a) and two to that of Fig. 29.1(b). The minima along each coordinate are separated by a barrier $U = 1/2$ at $x_1 = 0, |x_2| = 1$ and $|x_1| = 1, x_2 = 0$.

We now introduce a weak coupling between the two protons. To derive a functional form for this coupling, we note that it should be symmetric in the two coordinates and sensitive to their sign. The simplest coupling term that meets these requirements will be proportional to $x_1 x_2$. This bilinear term is likely to be the leading term in any expansion of the coupling between symmetric local potentials, irrespective of the coupling mechanism. It will create a difference in energy between the structures of Figs. 29.1(a) and 29.1(b). We choose the structures $x_1 = x_2 = \pm 1$ as those of the equilibrium configurations. Adopting this coupling and omitting constant terms, we arrive at the potential

$$U(x_1, x_2) = \frac{1}{2}[(1 - x_1^2)^2 + (1 - x_2^2)^2] - 2Gx_1 x_2 \tag{29.2}$$

where $G \geq 0$ is the coupling parameter (in units $2U_0$), which formally represents the interaction between the two $XH \cdots X$ hydrogen bonds.

To illustrate the nature of this coupling, we first consider the simple case that it is dominated by electrostatic interactions between the $XH \cdots X$ hydrogen bonds with dipole moments $|\mu_{1,2}| = f|x_{1,2}|$, where f is the fractional charge of the proton. For the model of Fig. 29.1(a), the interaction is attractive when both protons are in the equilibrium position ($x_1 = x_2 = \pm 1$), repulsive when one of the protons has transferred ($x_1 = -x_2 = \pm 1$), and zero when the protons are in the center of their path ($x_1 = x_2 = 0$). For two dipoles separated by a distance R this behavior can be simulated by the function

$$K \simeq \frac{\vec{\mu_1}\vec{\mu_2}}{R^3} = \frac{|\mu_1||\mu_2|\cos\phi}{R^3} = -\frac{f^2 r^2}{R^3}x_1 x_2 \tag{29.3}$$

where we used the fact that the angle ϕ between the dipoles is 0 or π when they are parallel or antiparallel, respectively. The separation R is expressed in the distances b and a according to

$$\frac{1}{R^3} \simeq \frac{1}{b^3}[1 - a(x_1 - x_2)^2]; \quad a = 3(a - 2R_{XH})^2/2b^2. \tag{29.4}$$

Addition of this coupling to Eq. (29.1) yields

$$U(x_1, x_2) = \frac{1}{2}[(1 - x_1^2)^2 + (1 - x_2^2)^2] - 2Gx_1 x_2[1 - a(x_1 - x_2)^2] \tag{29.5}$$

where $G = f^2 r^2 / 16 b^3 U_0$ may be a weak function of a through the implicit r-dependence of U_0 noted above. Equation (29.5) reduces to Eq. (29.2) when b exceeds a such that $b >> a - 2R_{XH}$.

While such a simple electrostatic picture yields qualitatively correct results for, e.g., dimeric formic acid, it is clearly inadequate for, e.g., naphthazarin, illustrated in Fig. 29.2, where the interaction is governed by the skeletal π-electrons. Nevertheless, the work of de la Vega et al. [15] indicates that a potential of the form of Eq. (29.2) remains a suitable first approximation for this molecule as well. Although, in general, higher-order interaction terms will be necessary to obtain

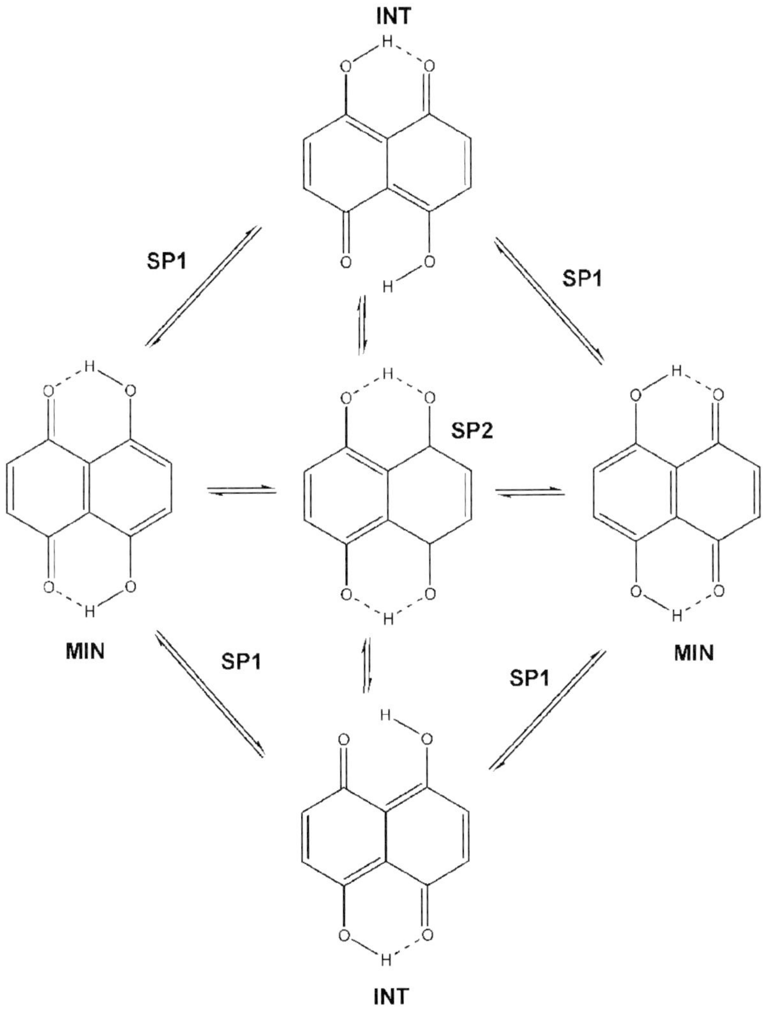

Figure 29.2 Illustration of stepwise and concerted double proton transfer in naphthazarin, showing the equilibrium configurations (MIN), the (unstable) intermediates (INT), the second-order saddle point (SP2), and the location of the first-order saddle points (SP1).

quantitative results for specific molecules, especially if the coupling is strong, the leading bilinear coupling term used in Eq. (29.2) should be adequate to map out the various distinguishable transfer mechanisms.

To obtain the stationary points and their curvature for the potential (29.2), we calculate the first and second derivatives with respect to the two coordinates. From the sums and differences of the first derivatives

$$\frac{\partial U}{\partial x_1} = 2[x_1(x_1^2 - 1) - Gx_2] = 0; \qquad \frac{\partial U}{\partial x_2} = 2[x_2(x_2^2 - 1) - Gx_1] = 0 \qquad (29.6)$$

we obtain the expressions

$$(x_1 + x_2)[(x_1 - x_2)^2 + x_1 x_2 - (1 + G)] = 0;$$
$$(x_1 - x_2)[(x_1 + x_2)^2 - x_1 x_2 - (1 - G)] = 0 \qquad (29.7)$$

which define the following stationary points:

1. two global minima $x_1 = x_2 = \pm\sqrt{1 + G}$, with zero energy
2. the global maximum $x_1 = x_2 = 0$, which is a saddle point of second order, with energy $(1 + G)^2$
3. two minima $x_1 = -x_2 = \pm\sqrt{1 - G}$ with energy $4G$, which represent stable intermediates with one proton transferred
4. two transition states $x_{1,2}^2 = (1\pm\sqrt{1 - 4G^2})/2$ with energy $(1 + 2G)^2/2$, which represent the barriers for single proton transfer

The minima (3) disappear for $G \geq 1$, at which point they coincide with the absolute maximum (1). From the second derivatives

$$\frac{\partial^2 U}{\partial x_1^2} = 6x_1^2 - 2; \qquad \frac{\partial^2 U}{\partial x_2^2} = 6x_2^2 - 2 \qquad (29.8)$$

it follows that the extrema (3) are minima only for $G < 2/3$. The energy of the transition states (4) exceeds that of the extrema (3) if $G \geq 1/2$, a limit that contradicts the limit obtained for the existence of a stable minimum (3). Clearly, the local representation $\{x_1, x_2\}$ becomes inadequate for couplings $G \geq 1/2$.

For coupling in this range we therefore switch to collective coordinates. Defining normal coordinates as plus and minus combination of the local coordinates:

$$x_p = (x_1 + x_2)/2, \qquad x_m = (x_1 - x_2)/2 \qquad (29.9)$$

we obtain the potential (29.1) for uncoupled protons in the form

$$U(x_m, x_p) = (1 - x_p^2)^2 + (1 - x_m^2)^2 + 6x_m^2 x_p^2 - 1$$
$$= (1 - x_p^2)^2 + 6x_m^2(x_p^2 - 1/3) + x_m^4 \qquad (29.10)$$

Introduction of the coupling of Eq. (29.2) leads to

$$U(x_m, x_p) = (x_p^2 - \xi_p^2)^2 + 6x_m^2(x_p^2 - \xi_m^2) + x_m^4 \qquad (29.11)$$

where $\xi_p^2 = 1 + G$ and $\xi_m^2 = (1 - G)/3$. This yields first derivatives

$$\frac{\partial U}{\partial x_p} = 12x_p[(x_p^2 - \xi_p^2)/3 + x_m^2] = 0;$$

$$\frac{\partial U}{\partial x_m} = 12x_m[(x_p^2 - \xi_m^2) + x_m^2/3] = 0 \qquad (29.12)$$

and second derivatives

$$\frac{\partial^2 U}{\partial x_p^2} = 12[(x_p^2 + x_m^2) - \xi_p^2/3]; \quad \frac{\partial^2 U}{\partial x_m^2} = 12[(x_p^2 + x_m^2) - \xi_m^2] \qquad (29.13)$$

In this representation the global minima retain the same form as in the local coordinate representation, (1) turning into $x_m = 0, x_p = \pm\xi_p = \pm\sqrt{1 + G}$. The global maximum (2), i.e. $x_p = x_m = 0$, retains its character as a second-order saddle point up to $G = 1$. For larger coupling, it turns into a simple first-order saddle point when the derivative with respect to x_m goes to zero. This eliminates the second reaction coordinate x_m and results in one-dimensional synchronous motion of the two protons. The intermediate configuration (3) corresponding to $x_p = 0$, $x_m = \pm\xi_m\sqrt{3} = \pm\sqrt{1 - G}$ is a minimum for $G < 1$; it is separated from the global minima by transition states (4) at $x_p = \pm\sqrt{1 - 2G}/2, x_m = \pm\sqrt{1 + 2G}/2$. It ceases to be a stable minimum and turns into a saddle point with $\partial^2/\partial x_p^2 < 0$ for $1/2 < G < 1$, under which conditions the two transition states (4) disappear. The coordinates and energies of these extrema are collected in Table 29.1.

Table 29.1 Parameters of the most characteristic configurations of model surface (29.11).

Parameter	Global minima (1)	Maximum (2)	Intermediate (3)	Transition state (4)	Bifurcation point
$2\|x_p\|/r$	$\sqrt{1 + G}$	0	0	$\sqrt{1 - 2G}/2$	$\sqrt{(1 - G)/3}$
$2\|x_m\|/r$	0	0	$\sqrt{1 - G}$	$\sqrt{1 + 2G}/2$	0
$U/2U_0$	0	$(1 + G)^2$	$4G$	$(1/2)(1 + 2G)^2$	$(4/9)(1 + 2G)^2$

It follows that the model potential (29.11) can reproduce three types of potential energy surfaces relevant to double proton transfer, as illustrated in Fig. 29.3(a)–(c). For $G < 1/2$ ("weak coupling") it leads to surface (a) that supports a stable intermediate. For $1/2 < G < 1$ ("intermediate coupling") it leads to surface (b) without

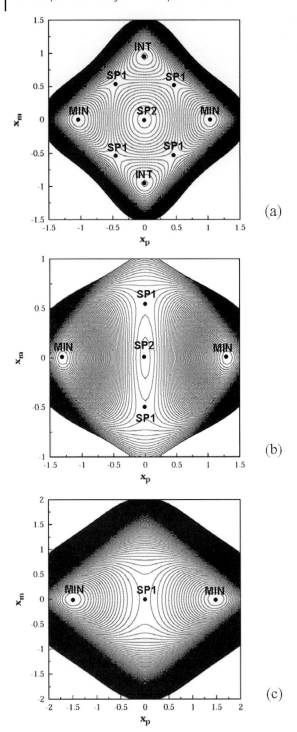

◀ Figure 29.3 Schematic two-dimensional potential energy surfaces appropriate to double proton transfer for weak (a), intermediate (b) and strong (c) proton–proton correlation, respectively, where the labeling is that used in Fig. 29.2.

a stable intermediate, but with two types of saddle points, one of first and one of second order. For $G > 1$ ("strong coupling") it leads to surface (c) with a single transition state along the coordinate for concerted motion.

The model surface (29.11) is defined by three parameters: r, U_0 and G. In practice, the potential energy surface and the vibrational force field will be calculated quantum-chemically, which should lead to a more accurate description than that given by the present model potential. We note, however, that the calculated surfaces known to date fall indeed into the three types described above. In the following sections we address in detail specific examples for each case.

If the calculations lead to a stable intermediate of type (3) that can be reached via two transition states of type (4), i.e. if the weak coupling regime ($G < 1/2$) applies, stepwise transfer of the protons is possible, provided the intermediate state is thermally accessible. In that case the dynamics calculations can be carried out consecutively along the one-dimensional reaction coordinates x_1 and x_2. If the temperature approaches zero, this mechanism will fade out. Transfer is still possible at $T = 0$ but will require a degree of coordination between the motions of the two protons. Two pathways stand out: synchronous tunneling through the barrier of type (2) with energy $(1 + G)^2$ and concerted but asynchronous tunneling through the minima (3) with energy $4G$ and transition states (4) with energy $(1 + 2G)^2/2$ (all in units $2U_0$). The former pathway involves a tunneling distance of $2\sqrt{1 + G}$ along x_p and the latter is longer by an amount of roughly $2\sqrt{1 - G}$ along x_m (both in units $r/2$). The same competition between these pathways, the one with the higher barrier and the one with the longer transfer distance, will govern transfer in the intermediate coupling region ($1/2 < G < 1$), characterized by very shallow minima of type (3) along x_m or flat barrier tops of type (2). If the quantum-chemical calculations lead to a single strong maximum of type (2), the strong coupling regime ($G > 1$) applies and the dynamics are then governed by the synchronous motion of the two protons along a one-dimensional reaction coordinate x_p.

Chemically, the coupling regime can be altered by modification of the geometry of the transfer site. It is obvious that an increase in the separation between the protons by means of an increase in the parameter b in Fig. 29.1 will reduce the coupling. Alternatively, an increase in the parameter a will correspond to a weakening of the hydrogen bonding and a loss of XH$\cdots$X polarity. This may require the introduction of additional coupling terms such as those in the potential (29.5). For this potential, the expressions for the extrema in Table 29.1 will get additional terms proportional to aG. In particular, for the intermediate (3) we obtain

$$x_m^2 \simeq 1 - G + 2aG; \quad U \simeq 4G(1 - a/2) \tag{29.14}$$

which implies that the intermediate exists if

$$a \equiv \frac{3}{2}[(a - 2R_{XH})/b]^2 > (G - 1)/2G \tag{29.15}$$

Thus even for strong coupling there may be an intermediate if the hydrogen bond is long enough.

To verify these results for rate constants and tunneling splittings in real systems, we need to investigate the dynamics explicitly. As a preliminary to the proposed treatment of tunneling along two reaction coordinates, we review in the next section an approach that has been applied successfully to tunneling along a single reaction coordinate.

29.3
Approaches to Proton Tunneling Dynamics

In general, proton transfer occurs via a combination of over-barrier and through-barrier pathways. The rate constant of over-barrier transfer is usually calculated by standard transition state theory (TST) [22] by separating the reaction coordinate from the remaining degrees of freedom. If tunneling effects and the curvature of the reaction path are neglected, this leads to the expression

$$k_{TST}(T) = (k_B T/h) \frac{Q^\ddagger}{Q^R} e^{-U_A/k_B T} \tag{29.16}$$

where U_A is the energy of the transition state (adiabatic barrier height), and Q^R and $Q^\ddagger$ are the partition functions for the reactant and the transition state, respectively. Various methods are available to calculate the rate constant of through-barrier transfer or tunneling. Most of our evaluations are based on the approximate instanton method (AIM) [3, 19], an adaptation of instanton theory to proton transfer in large molecular systems. Instanton theory [1, 23–26] is a semiclassical approach in which, below a certain temperature, the manifold of tunneling trajectories for a system with a Hamiltonian $\mathcal{H}$ is replaced by a single, least-action trajectory, the instanton or bounce path. This instanton represents a trajectory that is periodic in imaginary time $\tau = it$ in the inverted potential, namely the trajectory for which the Euclidian action, defined by

$$S_I = \int_{-\beta/2}^{\beta/2} \mathcal{H} d\tau, \quad \beta = 1/k_B T \tag{29.17}$$

has an extremum. The instanton equations are defined by

$$\frac{\delta S_I}{\delta\{x\}} = 0 \tag{29.18}$$

where $\{x\}$ represents the system of coordinates. Their solutions are subject to the periodic boundary conditions $\{x(\beta)\} = \{x(0)\}$.

The instanton method defines the thermal rate constant for tunneling transfer in terms of the action $S_I(T)$ (expressed hereafter in units $\hbar$) along this extremal path:

$$k_{\text{tun}}(T) = \mathcal{A}(T)e^{-S_I(T)} \tag{29.19}$$

where the preexponential factor $\mathcal{A}(T)$ accounts for fluctuations about the path. AIM is designed to allow direct application of this methodology to proton transfer in multidimensional systems for which the structure and vibrational force field of the stationary configurations along the reaction path can be evaluated quantum-chemically. For simplicity, we focus on symmetric systems. For background on the application of instanton theory to chemical processes we refer to Benderskii et al. [1].

To outline the method used in our calculations, we first consider transfer along a single reaction coordinate involving a potential energy barrier. For double proton transfer, this may refer to the case where the two protons transfer synchronously. To calculate $S_I(T)$, AIM [27–30] generates a full-dimensional potential energy surface in terms of the normal (mass-weighted) coordinates $\{x, y_j\}$ of the transition state configuration. The AIM Hamiltonian separates the tunneling mode x, taken to be the mode with imaginary frequency $i\omega^*$, from the other (transverse) modes $\{y_j\}$ with frequencies $\{\omega_j\}$. If there is a single tunneling coordinate, i.e. if x is one-dimensional, it has the form

$$H = T + U, \quad T = \frac{1}{2}\dot{x}^2 + \frac{1}{2}\sum_j \dot{y}_j^2,$$

$$U = U_C(x) + \frac{1}{2}\sum_j \omega_j^2(y_j^2 - \Delta y_j^2) - x^2 \sum_s C_s(y_s - \Delta y_s) \tag{29.20}$$

$$-x\sum_a C_a(y_a \pm \Delta y_a) + x^2 \sum_j D_j(y_j^2 - \Delta y_j^2)$$

where the transition state corresponds to $\{x, y_j\} = 0$ and the $+$ and $-$ signs represent the minima of the reactant and the product state, respectively. The kinetic energy, which is diagonal in the stationary configurations, is taken to be diagonal throughout. The Hamiltonian (29.20) accounts for the mixing of the normal modes between the stationary points and can be shown [27] to reduce to the Hamiltonian of the transition state and the minima for $x = 0$ and $x = \pm \Delta x$, respectively. The subscripts s and a refer to transverse modes that are symmetric and antisymmetric, respectively, with respect to the dividing surface in the transition state, which is perpendicular to x. Their coupling terms with the tunneling mode x are taken to be linear in the transverse mode coordinates, except for a biquadratic term that accounts for the frequency differences between the reactant and the transition state. They are calculated from the displacements $\Delta y_{a,s}$ between these states and, in the case of the biquadratic term, from the corresponding frequency shifts $\Omega_j - \omega_j$:

$$C_a = \omega_a^2 \Delta y_a / \Delta x, \quad C_s = \omega_s^2 \Delta y_s / \Delta x^2, \quad D_j = (\Omega_j^2 - \omega_j^2)/2\Delta x^2 \tag{29.21}$$

where the Ω_j denote the frequencies in the equilibrium configuration. The anti-symmetric modes have the same symmetry as the tunneling mode and undergo reorganization between reactant and product during the proton transfer while the symmetric modes do not undergo such reorganization but can be displaced between the equilibrium configurations and the transition state. Modes that are not displaced between the reactant and the transition state are not coupled linearly in the harmonic approximation and can contribute only via the biquadratic term of Eq. (29.20).

The one-dimensional potential along the tunneling coordinate, represented by $U_C(x)$ in Eq. (29.20), is a "crude-adiabatic" potential evaluated with the heavy atoms fixed in the equilibrium configuration, i.e. with $y_a = \pm \Delta y_a, y_s = \Delta y_s$; it is equivalent to the potential along the linear reaction path. This symmetric double-minimum potential has a maximum $U_C(0) \equiv U_0$ at $x = 0$, minima $U_C(\pm \Delta x) = 0$ at $x = \pm \Delta x$, and a curvature in the minima given by the effective frequency Ω_0 which accounts for the contribution of the normal modes of the minima to the reaction coordinate [27]. For the shape of the potential in the intermediate points we use an interpolation formula based on the calculated energies and curvatures near the stationary points. We have found that in many cases the simple quartic potential of the form

$$U_C(x) = U_0[1 - (x/\Delta x)^2]^2 \tag{29.22}$$

used in the model potentials (29.1) and (29.2) is satisfactory.

To calculate the parameters governing the Hamiltonian, we use an approximation that amounts to separating the transverse modes into "high-frequency" (HF) modes, treated adiabatically, and "low-frequency" (LF) modes, treated in the sudden approximation. This separation is based on the value of the "zeta factor" [27]

$$\zeta_{a,s} = \frac{\omega_{a,s}}{\sqrt{\Omega^2 - C_{a,s}^2 / 2\omega_{a,s}^2}} \tag{29.23}$$

where Ω, the "scaling" frequency, is defined by $\Omega^2 \Delta x^2 = U_0$. Modes are treated as HF or LF depending on whether $\zeta_{a,s} \gg 1$ or $\ll 1$. Modes for which $\zeta_{a,s} \sim 1$ require special treatment as discussed elsewhere [7].

Coupling to HF modes leads to an effective one-dimensional motion with renormalized potential $U_C^{eff}(x)$ and coordinate-dependent mass $m^{eff}(x)$. Since each HF mode y_i is assumed to follow the reaction coordinate x adiabatically, we have $\partial U / \partial y_i = 0$, so that $y_s^{HF} = C_s x^2 / \omega_s^2$ and $y_a^{HF} = C_a x / \omega_a^2$. Substitution in the Hamiltonian (29.20) provides a correction to the mass of the tunneling particle and thus modifies the kinetic energy operator:

$$T^{\text{eff}} \equiv \frac{1}{2}\dot{x}^2 + \frac{1}{2}\sum_{a,s}^{(\text{HF})} \dot{y}^2_{a,s} = \frac{1}{2}m^{\text{eff}}(x)\,\dot{x}^2 \tag{29.24}$$

Using Eq. (29.21), we obtain for the dimensionless renormalized mass

$$m^{\text{eff}}(x) = 1 + (x/\Delta x)^2 \sum_s^{(\text{HF})} \Delta m_s + \sum_a^{(\text{HF})} \Delta m_a \tag{29.25}$$

$$\Delta m_s = (2\Delta y_s^{\text{HF}}/\Delta x)^2, \quad \Delta m_a = (\Delta y_a^{\text{HF}}/\Delta x)^2$$

Renormalization of the one-dimensional potential (29.22) yields a potential of the same shape but with a barrier height corrected by the standard vibrational-adiabatic correction for the HF modes only.

The preexponential factor $\mathcal{A}(T)$ in the rate expression (29.19) consists of a longitudinal component $\mathcal{A}_{\|}$ and a perpendicular component $\mathcal{A}_{\perp}$, representing the contributions of fluctuations that are, respectively, parallel and perpendicular to the instanton path. If in the one-dimensional potential $U_C(x)$ the "long" tunneling action evaluated at $E = 0$ for $T = 0$ is replaced by the "short" action evaluated at the zero-point energy $\hbar\Omega_0/2$, we can use the approximation $\mathcal{A}_{\|} \simeq \Omega_0/2\pi$ for the longitudinal component. The perpendicular component is treated in the adiabatic approximation, which yields $\mathcal{A}_{\perp} \simeq 1$ if $U_C(x)$ is replaced by its vibrationally adiabatic counterpart. Using these approximations, we obtain

$$\mathcal{A}_{\text{AIM}}(T) \simeq \Omega_0/2\pi; \quad U_C^{\text{eff}}(x) = U_{0,\text{VA}}[1 - (x/\Delta x)^2]^2 \tag{29.26}$$

where the effective one-dimensional potential retains the form (29.22) but includes vibrationally-adiabatic correction over *all* modes.

To obtain an expression for the multidimensional instanton action, we generalize [29] the analytical solutions obtained [21, 31] for two- and three-dimensional Hamiltonians in the form

$$S_I(T) = \frac{S_I^0(T)}{1 + \sum_s^{(\text{LF})} \delta_s(T)} + a \sum_a^{(\text{LF})} \delta_a(T) \tag{29.27}$$

where $S_I^0(T)$ is the corresponding instanton action for the one-dimensional Hamiltonian $H^0 = \frac{1}{2}m^{\text{eff}}(x)\dot{x}^2 + U_C^{\text{eff}}(x)$. The δ_a-terms of the form [3, 19]

$$\delta_a(T) = \frac{(2C_a\Delta x)^2}{\omega_a^3}\tanh\frac{\hbar\omega_a}{4k_{\text{B}}T} \tag{29.28}$$

lead to a Franck–Condon factor arising from the reorganization of antisymmetric LF modes, which act similarly to a thermal heat bath. Symmetric LF modes, represented by δ_s terms of the form [3, 19]

$$\delta_s(T) = \frac{\Omega_0}{\omega_s}\left(\frac{C_s \Delta x^2}{2U_0}\right)^2 \coth\frac{\hbar\omega_s}{2k_B T} \tag{29.29}$$

effectively reduce the tunneling distance and thus enhance tunneling; the factor a < 1 in Eq. (29.27) is the square of this reduced distance (in dimensionless units) [29].

All parameters needed to generate the Hamiltonian (29.20) are obtained from the standard electronic structure and vibrational force field output of conventional quantum chemistry programs. If $U_C(x)$ is approximated by a quartic potential of the form (29.22), these calculations need to be carried out only for the stationary configurations along the reaction path. The resulting output is fed directly as input into the DOIT program [30] to calculate the instanton action $S_I(T)$ from Eqs. (29.27–29.29). This is a very efficient procedure since it does not require explicit knowledge of the instanton path. Once the instanton action is calculated, evaluation of rate constants and tunneling splittings is straightforward.

For $T \leq T^*$, where $T^* \sim \omega^*/2\pi$, the total transfer rate constant is the sum of the tunneling and over-barrier rate constants

$$k(T) = k_{\text{tun}}(T) + k_{\text{TST}}(T); \quad k_{\text{tun}}(T) \simeq k_{\text{AIM}}(T) = (\Omega_0/2\pi)e^{-S_I(T)} \tag{29.30}$$

where the over-barrier rate constant $k_{\text{TST}}(T)$, given by Eq. (29.16), and Ω_0 is the effective frequency of the tunneling mode in the reactant state. For $T \geq 2T^*$, we ignore the tunneling contribution and in the intermediate temperature range we interpolate. The zero-point tunneling splitting is related to the low-temperature limit of the rate constant by

$$\Delta_0 = (\hbar\Omega_0/\pi)e^{-S_I(0)/2} \tag{29.31}$$

AIM can also be used to calculate tunneling splittings of vibrationally excited levels [3, 6]. This approach, as implemented in the computer program DOIT [19, 30], has been applied to a variety of tunneling potentials with a single imaginary frequency. It needs generalization for tunneling through barriers with more than one imaginary frequency or more than one maximum. Such a generalization is presented in the next section.

29.4
Tunneling Dynamics for Two Reaction Coordinates

Because we are dealing with double proton transfer, we need a more general Hamiltonian than Eq. (29.20). Following the approach introduced in the preceding section, we can construct such a Hamiltonian as the sum of two local Hamiltonians describing the transfer of each of the protons together with a proton–proton coupling:

$$\mathcal{H}(x_1, x_2) = \mathcal{H}_1(x_1) + \mathcal{H}_2(x_2) + \mathcal{K}(x_1, x_2) \tag{29.32}$$

where $\mathcal{H}_1(x_1)$ and $\mathcal{H}_2(x_2)$, represent Hamiltonians of the form (29.20) with indices 1 and 2, respectively, added to the couplings C_a, C_s, and D_j, and $\mathcal{K}(x_1, x_2)$ represents the the proton–proton correlation, i.e. the coupling between the proton transfer modes x_1 and x_2. The form of the coupling is discussed in Section 29.2. In practice, we use normal coordinates $x_{p,m}$ obtained by diagonalizing the total Hamiltonian $\mathcal{H}(x_p, x_m)$. In the limiting cases where the resulting potential is governed by a single transition state, we can obtain the transfer rate constant in the same way as for single proton transfer. Thus for concerted double proton transfer Eqs. (29.30, 29.31) can be used directly to obtain the overall rate constant $k(T)$ and/or the tunneling splitting. Similarly, it can be used to obtain the rate constant for a single step in a stepwise process if the temperature is high enough to make this process dominant; the total rate constant is obtained by combining the steps according to standard kinetic procedures. However, we need a more general approach when there is more than one reaction coordinate. This happens for weak and intermediate proton–proton coupling, when apart form the saddle point of second order along x_p, the transfer potential has additional saddle point(s).

For weak coupling, these are the two transition states that separate the stable intermediate from the global minima. At temperatures high enough to make the intermediate accessible, such a potential will give rise to thermally activated stepwise transfer via this intermediate. However, there will be additional contributions from tunneling transfer, and at low temperatures these contributions will become dominant. This may be transfer along the coordinate for synchronous transfer x_p through a barrier of energy $U(x_p, x_m) = U(0,0) = 2U_0(1 + G)^2$, but we should also consider the possibility of transfer along a longer two-dimensional pathway through the transition state whose barrier is lower by $U_0/2 \le U_0(1 - 2G^2) \le U_0$. Such a two-dimensional pathway will also be possible in the intermediate coupling region; then the stable intermediate will be replaced by a saddle point, lower by $0 \le 2U_0(1 - G)^2 \le 2U_0$ than the maximum, thus eliminating the stepwise mechanism, but retaining tunneling through different paths. Only when the coupling becomes strong, so that these saddle points disappear and the second-order saddle point in the center turns into a first-order saddle point, does this ambiguity disappear, leaving synchronous tunneling as the only viable transfer mechanism, at least for the model of Section 29.2. In principle, the asynchronous mechanism may still play a role through coupling with skeletal modes that increase the tunneling distance. For practical reasons such modes are not included in the model; their effect on the relation between the two transfer mechanism deserves further study.

For the time being we ignore these modes and use a Hamiltonian based on the potential (29.11)

$$H(x_m, x_p) = T + U = \frac{1}{2}(\dot{x}_p^2 + \dot{x}_m^2) + (x_p^2 - \xi_p^2)^2 + 6x_m^2(x_p^2 - \xi_m^2) + x_m^4. \tag{29.33}$$

The instanton equations (29.18), which take the form

$$\frac{\delta S_I}{\delta x_p} = 0; \quad \frac{\delta S_I}{\delta x_m} = 0 \tag{29.34}$$

are normally written as "equations of motion" in imaginary time in the inverted potential $U \rightarrow -U$:

$$\ddot{x}_p = \frac{\partial U}{\partial x_p}; \quad \ddot{x}_m = \frac{\partial U}{\partial x_m} \tag{29.35}$$

to be solved with periodic boundary conditions $x_{p,m}(\beta) = x_{p,m}(0)$. This leads to equations

$$\ddot{x}_p = \frac{\partial U_{1D}}{\partial x_p} + 12 x_p x_m^2; \quad \ddot{x}_m = 12 x_m (x_p^2 - \xi_m^2 + x_m^2/3) \tag{29.36}$$

where U_{1D} is the one-dimensional (1D) double-minimum potential along the coordinate of synchronous motion, here represented by a quartic potential $(x_p^2 - \xi_p^2)^2$. The potential in Eq. (29.33) always allows such a solution in the form of a 1D instanton, since for $x_m \equiv 0$, Eqs. (29.36) reduce to $\ddot{x}_p = \partial U_{1D}/\partial x_p$. Under certain conditions this 1D instanton becomes unstable, so that two-dimensional (2D) solutions of Eqs. (29.36) are required. Such solutions were obtained by Bend-erskii and coworkers [32–37] for a model surface of type (b) in Fig. 29.3. Here we are interested in the broader issue of establishing the dynamics features common to all three types of surfaces in Fig. 29.3, in order to relate the labels "stepwise" and "concerted" to the potential-energy surfaces obtained by quantum-chemical calculations for realistic systems.

Solutions involving both coordinates exist in the weak and intermediate coupling regions as discussed in Section 29.2; specifically they exist if $\xi_m^2 > 0$, i.e. $G < 1$. These solutions correspond to paths of lower energy but greater length than the 1D instanton path. First we consider the weak-coupling limit $G < 1/2$, where there is a stable intermediate. The high-temperature case where this intermediate is thermally accessible and supports stepwise transfer has been discussed in Section 29.2. Here we consider the low-temperature case where tunneling prevails and focus on the zero temperature limit for simplicity. To illustrate the interplay between barrier height and path length, we choose among the many possible 2D trajectories the one with lowest energy, i.e. the minimum barrier path (MBP). From the expression for $U(x_p, x_m)$ in Eq. (29.33) and the extremum condition $\partial U(x_p, x_m)/\partial x_m = 0$ in Eq. (29.12), it follows that the MBP is a parabola of the form

$$x_m^2 = 3(\xi_m^2 - x_p^2) \tag{29.37}$$

depicted in Fig. 29.4. It starts on the x_p axis at the "bifurcation point" $|x_p| = \xi_m$ and passes through the transition state (4) and the intermediate state (3), located,

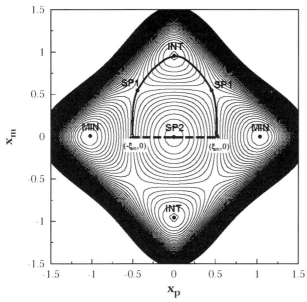

Figure 29.4 Competing tunneling paths on the two-dimensional surface of Fig. 29.3(a): the one-dimensional instanton $(-\xi_m, \xi_m)$ along x_p, illustrated by a dashed line and one of a family of two-dimensional paths, here arbitrarily represented by the minimum barrier path (MBP), illustrated by a solid line. The two pathways have parts from MIN to $\pm\xi_m$ (not illustrated) in common.

respectively, at $x_p = -\sqrt{1-2G}/2$, $x_m = \sqrt{1+2G}/2$ and $x_p = 0$, $x_m = \sqrt{1-G}$, all characteristic points being indentified in Table 29.1.

The tunneling probability at $T = 0$ is proportional to $\exp[-S_I(0)]$; to compare the contributions of the 1D and 2D paths to this probability, we calculate the respective actions after their bifurcation. Since at $T = 0$ the energy $E = 0$ is an integral of motion, the corresponding actions are given by [39, 40]

$$S_{1D}(E = 0) = C \int_0^{\xi_m} dx_p \sqrt{U_{1D}(x_p)},$$

$$S_{2D}(E = 0) = C \int_0^{\xi_m} dx_p \sqrt{U_{MBP}(x_p)m^{eff}(x_p)} \tag{29.38}$$

where the coordinates are in units $r/2$, the energies in units $2U_0$, and C is a constant in units $\hbar$, dimensioned as $C = 4r\sqrt{2U_0 m^H}$, m^H being the hydrogen mass. For $U_{1D} = (x_p^2 - \xi_p^2)^2$ the action along the 1D instanton is easily evaluated:

$$S_{1D}(E = 0) = C\xi_m(\xi_p^2 - \xi_m^2/3) \tag{29.39}$$

$U_{MBP}(x_p)$ is obtained by substituting Eq. (29.37) into the potential of Eq. (29.33):

$$U_{MBP}(x_p) = (x_p^2 - \xi_p^2)^2 - 9(x_p^2 - \xi_m^2)(2x_p^2 + 2\xi_m^2 - 1) \tag{29.40}$$

The contribution of the x_m coordinate is formally included as an effective mass $m^{eff} = 1 + \Delta m(x_p)$, where the extra term

$$\Delta m(x_p) = \frac{3x_p^2}{\xi_m^2 - x_p^2} \tag{29.41}$$

is due to the fact that x_m from Eq. (29.37) modifies $\dot{x}_p^2$ so as to yield a new term proportional to $\dot{x}_m^2$. The resulting 2D action requires numerical evaluation. Here we limit ourselves to an analytical estimate based on the near-constancy of the potential along the MBP between the bifurcation point where $U = (4/9)(1 + 2G)^2$ and the transition state where $U = (1/2)(1 + 2G)^2$ (in units $2U_0$). We therefore divide the relevant section of the parabola (29.37) into two parts, each of which we approximate by a straight line. To the section from the bifurcation point $|\xi_m|$ to the transition state (4), we assign a constant energy $(1/2)(1 + 2G)^2$, and to that from (4) to the intermediate state (3), we assign an energy equal to the average of the energies of (3) and (4). Instead of integrating S_{2D} along x_p with an effective mass, we integrate along the MBP coordinate s; the result is

$$S_{2D}(0) = C \int_{MBP} ds \sqrt{U_{MBP}(s)} > \sqrt{\{[x_p^{(4)} - \xi_m]^2 + (x_m^{(4)})^2\}U^{(4)}}$$
$$+ \sqrt{\{[x_m^{(3)} - x_m^{(4)}]^2 + (x_p^{(4)})^2\}[U^{(3)} + U^{(4)}]/2} \tag{29.42}$$

where the coordinates and energies of the extrema (3) and (4), expressed in the coupling coupling constant G, are listed in Table 29.1.

This allows a direct comparion of the mechanisms represented by Eqs. (29.39) and (29.42). As illustrated in Fig. 29.5, for the whole range of G values from 0 to 1, $S_{2D}(0)$ is found to be substantially larger than $S_{1D}(0)$. In view of the fact that for relevant parameter values the constant C in Eq. (29.38) is at least 10, this implies that the contribution of the two-dimensional concerted but asynchronous mechanism to the transfer is a negligible fraction of that of the one-dimensional synchronous mechanism. The fact, implied by Eq. (29.42), that the calculated value of $S_{2D}(0)$ is smaller than the actual value because of the linearization of the MBP, further strengthens this argument. In addition, it receives support from the hitherto neglected effect of promoting vibrations on the two mechanisms. In the presence of hydrogen bonding, the transition state for synchronous transfer will give rise to formation of two symmetric hydrogen bonds, leading to strong contraction, an effect that will be weaker for asynchronous transfer.

These results single out the one-dimensional synchronous mechanism as dominant at low temperatures for model potential (29.11), regardless of the strength of the proton–proton correlation. It follows then that for systems with a stable intermediate the critical temperature where the mechanism changes from concerted to stepwise can be estimated by comparing the rate constant for synchronous dou-

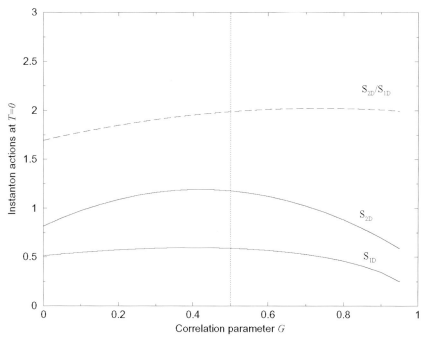

Figure 29.5 Instanton actions at $T = 0$ along the one-dimensional instanton $(-\xi_m, \xi_m)$ (S_{1D}) and the MBP (S_{2D}) (solid lines, in units C), and their relation (dashed line) as function of the correlation parameter G. The vertical line divides the regions of weak ($G < 1/2$) and intermediate ($1/2 < G < 1$) coupling corresponding to surfaces (a) and (b) in Fig. 29.3, respectively.

ble proton tunneling, $k_2(T)$, with twice that for the single proton step, $k_1(T)\exp(-E_i/k_B T)$, where $E_i(= 8G U_0)$ is the energy of the intermediate. A more rigorous analysis of the temperature dependence in the region of the critical temperature should involve 2D instantons. We note, however, that no such solutions have been obtained to date for surfaces of the type of Fig. 29.3(a).

The model potential (29.11) can reproduce all the types of surfaces found to date and yields correct relations between the frequencies along the collective coordinates. Therefore the basic conclusions for the dynamics obtained above should remain valid. Since the potential is defined by three parameters and is, admittedly, simplified, the quantitative relations need further testing. As noted earlier, in practice the potential energy surface and the vibrational force field for any system of interest will be calculated quantum-chemically, which should lead to more accurate dynamics. In later sections, we discuss specific examples where this issue has been raised such as porphine, naphthazarin and dimeric formic acid, which together cover the range from weak via intermediate to strong coupling.

29.5
Isotope Effects

Rate constants and tunneling splittings associated with proton transfer are sensitive to deuterium (and tritium) isotope effects resulting from the large difference in mass between the isotopes. Deuterium (and tritium) substitution thus provides a tool for studying these processes experimentally. The corresponding kinetic isotope effect (KIE) for a single proton transfer reaction is defined by

$$\eta^{H/D} = k^{H}(T)/k^{D}(T) \tag{29.43}$$

An analogous expression applies to tritium substitution. If the reaction is obstructed by a potential-energy barrier, two modes of transfer may be distinguished, "classical" transfer over the barrier and quantum-mechanical tunneling through the barrier, the latter mode resulting from the small mass of the proton and its isotopes. For over-barrier transfer the rate constants can be obtained from standard TST leading to Eq. (29.16). To calculate the corresponding KIE, we can usually neglect rotational and translational partition functions and consider only vibrational partition functions of the form [22]

$$Q = \prod_{j} \frac{1}{2\sinh\left(\hbar\omega_j/2k_B T\right)} \tag{29.44}$$

where the product runs over all normal modes except the reaction coordinate for the transition state. If $\hbar\omega_j \gg k_B T$ for all modes affected by the isotopic substitution, the KIE can be obtained from the familiar simplified formula

$$k_{TST}(T) = (k_B T/h)e^{-(U_A - \Delta U)/k_B T} \tag{29.45}$$

where U_A is the adiabatic barrier height and ΔU the correction for zero-point energy changes between the reactant and the transition states. Thus the corresponding KIE will be governed by the effect of isotopic substitution on the difference in zero-point energy between the initial state and the transition state. If the effect of this substitution on modes other than the reaction coordinate is neglected, the KIE for deuterium substitution of the transfering hydrogen, expressed in logarithmic form, reduces to

$$\ln \eta_{TST}^{H/D} \simeq \frac{\hbar(\omega^H - \omega^D)}{2k_B T} \simeq \frac{\hbar\omega^H(1 - \sqrt{1/2})}{2k_B T} \tag{29.46}$$

where ω is the frequency of the transfer mode in the initial well and $\omega^H : \omega^D \simeq \sqrt{1/m^H} : \sqrt{1/m^D}$ if we neglect the small difference between the atomic and the reduced mass ratio. Together with the same calculation for the tritium isotope effect, we obtain the well-known Swain–Schaad relation for through-barrier transfer [20, 38]

$$\frac{\ln \eta_{TST}^{H/T}}{\ln \eta_{TST}^{H/D}} \simeq \frac{1 - \sqrt{1/3}}{1 - \sqrt{1/2}} = 1.44 \qquad (29.47)$$

or its alternative $\ln \eta_{H/T}/\ln \eta_{D/T} = 3.26$. In practice, these Swain–Schaad exponents are usually calculated from the partition functions (29.44) rather than from approximate relations of the type (29.45, 29.46) but the result was found to be remarkably close to the approximation (29.47) [21].

For tunneling we use the AIM formalism developed in the preceding section. Combining Eqs. (29.19) and (29.27), we have for single proton transfer

$$k_{AIM}(T) = (\Omega_0/2\pi)\exp\left[\frac{S_I^0(T)}{1 + \sum_s^{(LF)} \delta_s(T)} + a\sum_a^{(LF)} \delta_a(T)\right] \qquad (29.48)$$

To calculate the KIE, we neglect the weak isotope dependence of the δ_a term and write the δ_s term in the simplified forms $\delta_s^{H,D}$. This yields

$$\eta^{H/D} \simeq \sqrt{2}\exp[S_I^{0,D}/(1 + \delta_s^D) - S_I^{0,H}/(1 + \delta_s^H)] \qquad (29.49)$$

where normally $\delta_s^D \geq \delta_s^H$. If the coupling is weak or moderate and $\delta_s^D \approx \delta_s^H = \delta_s$, we can write $1/(1 + \delta_s) \simeq 1 - \delta_s$, so that

$$\eta^{H/D} \simeq \eta_0^{H/D} \cdot \exp[-\delta_s S_I^{0,H}(\sqrt{m^D/m^H} - 1)] \qquad (29.50)$$

where $\eta_0^{H/D} = \sqrt{2}\exp[S_I^{0,D} - S_I^{0,H}] \simeq \exp[S_I^{0,H}(\sqrt{m^D/m^H} - 1)]$ represents the KIE in the absence of coupling. This confirms that coupling to promoting modes, which increases the rate constants, reduces the KIE, an effect that is always present and depends only on the frequency of the promoting mode and the strength of its coupling to the tunneling mode.

In the absence of coupling, i.e. for one-dimensional tunneling, we obtain [21] instead of Eq. (29.47)

$$\frac{\ln \eta_{tun}^{H/T}}{\ln \eta_{tun}^{H/D}} \simeq \frac{\sqrt{3} - 1}{\sqrt{2} - 1} = 1.77 \qquad (29.51)$$

It follows that tunneling increases the Swain–Schaad exponent, contrary to what is usually assumed [38]. This increase is mitigated by coupling to promoting modes. It has been shown [21], that larger and larger contributions of these modes to the point where they are effectively taking over proton transfer leads to a limiting value of 1.44 for the exponent, i.e. the value originally derived for over-barrier transfer as given by Eq. (29.47). This means that we can combine Eqs. (29.47) and (29.51) by introducing a parameter ρ^H that varies from 0, when over-barrier transfer dominates, to 1 for unassisted tunneling:

$$\frac{\ln \eta^{H/T}}{\ln \eta^{H/D}} \simeq \frac{\sqrt{2}/(\sqrt{2}-1) - \rho^H}{\sqrt{3}/(\sqrt{3}-1) - \rho^H} = \frac{3.414 - \rho^H}{2.366 - \rho^H} \tag{29.52}$$

The resulting generalized Swain–Schaad exponent varies from 1.44 for $\rho^H = 0$ to 1.77 for $\rho^H = 1$. A more realistic upper limit that takes into account coupling to promoting modes will be 1.65 corresponding to $\rho^H \sim 0.75$. Such generalized Swain–Schaad exponents can be used to estimate tunneling contributions to proton transfer [21].

The conclusion that tunneling *increases* the Swain–Schaad exponent runs opposite to that reported by Saunders [38], which was based on the (valid) argument that the tunneling contribution to the probability of transfer through a static one-dimensional barrier decreases with the mass of the transferring particle. However, the model used to apply tunneling corrections to the rate constant for proton transfer in complex systems has serious deficiencies. It ignores the fact that in the one-dimensional model in which the reaction coordinate represents a hydrogenic mode, the energy spectrum is not a continuum starting at zero, but consists of discrete levels of which only one, namely the lowest at an energy $\hbar\omega_0/2$ is significantly populated at most temperatures of interest. If treated quantum-mechanically such a model leads to the Swain–Schaad exponent 1.77 shown in Eq. (29.51) rather than a value < 1.44. The continuum of energy levels implicit in the model presupposes the presence of other degrees of freedom. However, these will interact with the reaction coordinate, implying that the reaction path is not the same for different hydrogen isotopes; therefore applying the tunneling corrections to an isotope-independent barrier, as done in the model, is not appropriate. Explicit introduction of additional modes, as in Eq. (29.52) or, more elaborately, in our earlier treatment [21] and that of Tautermann et al. [42], both applying multidimensional tunneling dynamics to high-level potential energy surfaces, yields exponents in the range $1.44 \le e_1 \le 1.77$. We therefore conclude that Saunders' conclusion is an artifact resulting from the model used.

To interpret ρ^H in physical terms, we note that, for a barrier formed by the crossing of two equivalent harmonic potentials, the instanton action can be approximated by [21]

$$S_I(T) = \frac{r_0^2}{2a_0^2 + A_s^2(T)} \tag{29.53}$$

where r_0 is the equilibrium transfer distance, $a_0 = (\hbar/\mu^H \omega^H)^{1/4}$ is the zero-point amplitude of the proton, and $A_s(T)$ is the thermal amplitude of the promoting vibration(s). In this approximation ρ^H can be defined as $\rho^H = 2a_0^2/[2a_0^2 + A_s^2(T)]$; a more accurate definition, based on the use of a quartic potential of the form (29.22) for the proton, is

$$\rho^H = \frac{4a_0^2}{4a_0^2 + A_s^2(T)} \tag{29.54}$$

Thus ρ^H measures the fraction of the transfer trajectory traveled by the protons rather than the atoms of the promoting modes. This fraction decreases with increasing temperature because excitation of the promoting mode increases the part traveled by them. At high temperature, we must include the temperature dependence of a_0, which will increase the over-barrier contribution to the transfer. If tunneling is strongly dominant in the temperature region of interest, the first of these effects should dominate, implying that the temperature dependence of ρ^H should be proportional to that of the amplitude of the promoting mode(s):

$$A_0^2(T) = A_0^2 \coth \frac{\hbar \omega_s}{2k_B T} \tag{29.55}$$

where ω_s is the frequency of the promoting mode with effective mass μ_s and $A_0^2 = \sqrt{\hbar/\mu_s \omega_s}$ is the zero-point amplitude.

The generalized Swain–Schaad exponent (29.52) is directly applicable to most two-proton transfers; obviously, this holds true for each step of a stepwise process, but it also applies to concerted processes in which the two proton isotopes are the same, since the effective mass and harmonic frequency of the relevant symmetric or antisymmetric XH-stretch modes are essentially the same as those of their one-dimensional components. However, if the two proton isotopes are different, this argument no longer suffices because the normal mode that represents the frequency and effective mass of the transfer coordinate in the transition state correlates with two distinct normal modes in the equilibrium configuration. Hence there is no unambiguous Swain–Schaad type exponent relating HD to HH and DD transfer.

However, we can relate the rate for stepwise HD transfer to the rates of HH and DD transfer through standard kinetics. In the case of a symmetric potential, we have $k^{HD} = k^{DH}$ and thus [2]

$$\eta^{HH/DD} = \eta^{H/D}; \quad \eta^{HH/HD} = \frac{1}{2}(1 + \eta^{H/D}); \quad \eta^{HD/DD} = \frac{2\eta^{H/D}}{1 + \eta^{H/D}} \tag{29.56}$$

It follows that for $\eta^{H/D} \gg 1$ the Arrhenius curve for HD transfer will be much closer to the DD than to the HH curve. No simple rules can be given if the protons are not equivalent and face different barriers. In that case the relation between the KIEs depends on the relation between the rate of the proton step across the higher barrier and the deuteron step across the lower barrier [2].

To relate the rate of concerted HD transfer to the rates of HH and DD transfer, we use [2] a generalization of Eq. (29.49)

$$\ln \eta^{AB/CD} \simeq \frac{S_I^{0,CD} - S_I^{0,AB}}{1 + \delta_s} \tag{29.57}$$

where the capital superscripts are either H or D. Using the approximate but quite robust relation $S_I^{0,AB} \simeq S_I^0 \sqrt{m^A + m^B}$, where $S_I^0 = S_I^{0,H}$ is the action corresponding to a single-proton transfer and $m_{H/D} = 1/2$, one obtains

$$\ln \eta^{AB/CD} \approx (\sqrt{m^C + m^D} - \sqrt{m^A + m^B}) \frac{S_I^0}{1 + \delta_s} \qquad (29.58)$$

where the expression in parenthesis equals 0.32 for HH/HD, 0.27 for HD/DD and thus 0.59 for HH/DD. It follows that in the Arrhenius plot the HD curve is roughly halfway between the HH and DD curves but slightly closer to the latter. It follows also that for large KIEs and symmetric potentials, the spacing of the Arrhenius curves for HH, HD, and DD transfer can be an indication whether the transfer is concerted or stepwise, but that for small KIEs and asymmetric potentials the results are likely to be ambiguous.

29.6
Dimeric Formic Acid and Related Dimers

We begin our study of actual systems with an analysis of observed and calculated tunneling splittings in dimeric formic acid and related dimers. Dimeric formic acid, the simplest carboxylic acid dimer, is depicted in Fig. 29.6, top. It has been studied extensively, but only recently did Madeja and Havenith [4] succeed in observing level splittings characteristic of double proton tunneling. The observation was made by high-resolution infrared spectroscopy in a cold beam, the transition being between the zero-point level and a vibrationally excited level of the electronic ground state. The observed splitting of about 400 MHz (0.013 cm^{-1}) could be divided into two components. The authors tentatively assigned the larger component to the vibrationally excited level corresponding to a CO-stretch fundamental, but our calculations [6] revise this assignment. Based on symmetry considerations and backed by detailed calculations, we adopt the alternative assignment of Madeja and Havenith, leading to a zero-point splitting of 375 MHz and a splitting of 94 MHz for the CO-stretch fundamental. However, the precise magnitude of the splitting is immaterial in the following analysis.

The observation of a tunneling splitting implies concerted although not necessarily synchronous transfer; the alternative possibility of transfer to an intermediate state with the same energy as the equilibrium state, corresponding to the complete absence of proton–proton correlation, can be excluded on elementary grounds. High-level calculations lead to a single transition state with an imaginary frequency of about half that of the symmetric CH-stretch vibration in the equilibrium configuration, consistent with synchronous transfer. The structure of the equilibrium configuration and transition state is illustrated in Fig. 29.6 (top). The two hydrogen bonds are almost linear; although relatively weak in the equilibrium configuration, they are strong and short in the transition state, as expected for symmetric hydrogen bonds. As a result, the transition state has a lower energy than the state corresponding to single proton transfer, which would correspond to a complex between the anion and the cation of formic acid.

Earlier, Shida et al. [41] concluded, on the basis of MCSCF calculations, that in addition to the synchronous path, an asynchronous pathway appeared if the inter-

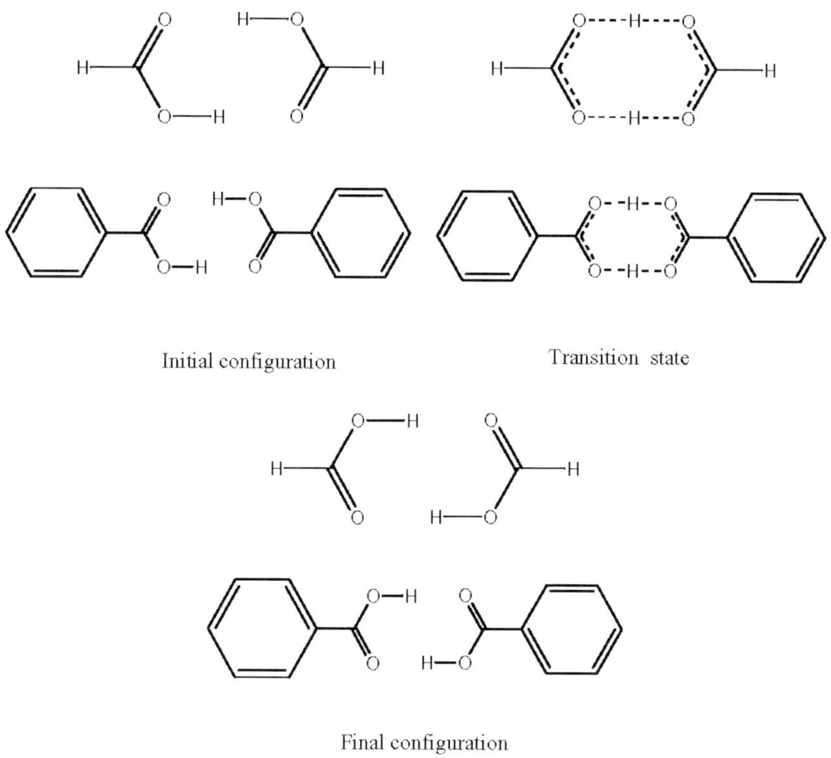

Initial configuration · · · · · · · · · · · · · · · · Transition state

Final configuration

Figure 29.6 Schematic representation of the synchronous double proton transfer process in the formic acid dimer (top) and benzoic acid dimer (bottom).

monomer distance was kept fixed at appropriate large values. In their calculation this bifurcation occurred at a distance slightly below the equilibrium separation. The opening of the asynchronous path reduced the calculated tunneling splitting by about 20%. A rough indication of the proton positions at the two equivalent transition states for asynchronous transfer is that if one proton had not moved significantly, the other would have overshot the midpoint by about 0.12 Å. Equation (29.15) indicates that such an asynchronous pathway will always appear for sufficiently large transfer distances in systems dominated by synchronous transfer. Shida et al. [41] have pointed out that such large distances may be reached during the $O \cdots O$ vibration that modulates the tunneling distance. To probe this effect, we have repeated their calculations at the B3LYP/6-31+G(d) level, which yields a more accurate value for the tunneling splitting and compares well with high-level methods [7, 42]. It leads to a $C \cdots C$ separation of 3.85 Å in the equilibrium configuration, which shrinks to 3.57 Å in the transition state, whose imaginary symmetric OH-stretch frequency amounts to $1322i$ cm^{-1} and whose real antisymmetric OH-stretch frequency amounts to 1197 cm^{-1}. If the transition state calculation is repeated with the C···C separation constrained, bifurcation characterized by a zero antisymmetric OH-stretch frequency appears at a separation of 3.79 Å, marginally smaller than the equilibrium separation, a result comparable to that of Shida et al. This effect, viz. weakening the coupling by restraining the optimization of the transition state, is represented by Eqs. (29.4, 29.5) in Section 29.2. In the present case it is brought about by coupling with promoting modes. However, at our level of calculation, the energy of the bifurcation point is found to be about 5.6 kcal mol^{-1} higher than that of the optimized transition state, so that the asynchronous path will have a negligible effect on the transfer dynamics, including the tunneling splitting. It seems likely that this conclusion, which is based on a method that includes electron correlation, and contradicts the results of Shida et al., based on a method without such correlation, can be generalized to all dimers held together by hydrogen bonding, but it may not apply to systems where the motion of the groups carrying the mobile protons is constrained.

The methods used to calculate tunneling splittings have been applied also to the calculation of transfer rate constants as a function of temperature in the formic acid dimer and its DD and TT isotopomers in order to test the validity of the Swain–Schaad relation between HH, DD, and TT transfer [7, 42]. As expected, the conventional relation (29.47), derived for over-barrier transfer is inadequate for this system, but the generalized relation (29.52) yields good agreement over the entire temperature range for which calculations are available. The tunneling parameter ρ^H is found to decrease with increasing temperature, in response to the increasing amplitude of the promoting mode and the increasing contribution of over-barrier transfer.

It is instructive to compare the results for dimeric formic acid with calculations on structural analogs in which the O atoms are gradually replaced by NH groups. This changes the strength of the hydrogen bonds in the order OH–N > OH–O > NH–O > NH–N. The calculated equilibrium structures and transition states for four analogs of the formic acid dimer, calculated at the DFT-B3LYP/6-31G(d,p) level [43–46], are illustrated in Fig. 29.7. Comparison with Fig. 29.6 shows that the symmetric dimer, namely dimeric formamidine with a D_{2h} transition state [43], shows the same pattern as the formic acid dimer, i.e. a transition state corresponding to synchronous transfer with tightened hydrogen bonding, despite the weakened hydrogen bonds. The higher barrier and longer transfer distance indicate that the rate of transfer will be much lower and the level splitting much smaller than in the formic acid dimer. Simultaneously, the proton–proton coupling parameter G is expected to be smaller than that of dimeric formic acid since the dipolar couplings will be weaker. However, the calculated potential energy surface indicates that it will still be strong enough to be in the strong coupling regime.

The other dimers in Fig. 29.7 show asymmetry. If the proton transfer potential is asymmetric, there will be no synchronous transfer and at most accidental level splittings. In the formamide dimer [44] the two hydrogen bonds are the same but the donor and acceptor groups are different. The available calculations indicate that the structure of the equilibrium configuration is intermediate between the equilibrium configurations of dimeric formic acid and dimeric formamidine, as

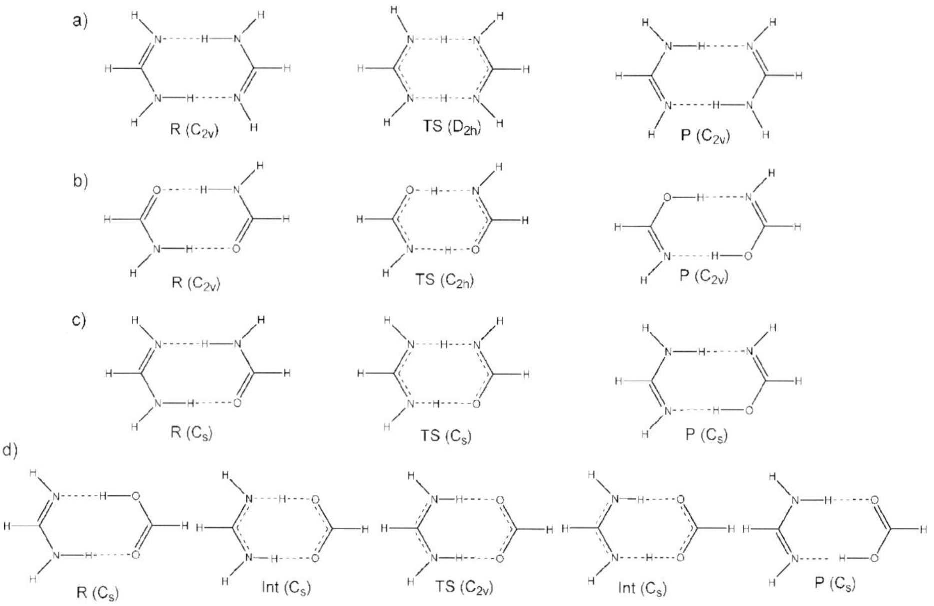

Figure 29.7 Stationary points representing the reactant (R), transition state (TS), product (P), and (where applicable) intermediate (Int) related to double proton transfer in analogs of the formic acid dimer: (a) formamidine dimer [43]; (b) formamide dimer [44]; (c) formamidine–formamide complex [45] and (d) formic acid–formamidine complex [46].

expected. However, the transition state belongs to the C_{2h} rather than the D_{2h} point group and the transfer is endothermic by about 10 kcal mol^{-1}. No stable intermediate has been found; estimates of the barrier height vary widely. Concerted but asynchronous transfer is expected to be the dominant mechanism.

Stepwise transfer is more likely to be favored when the two hydrogen bonds are different. This is the case in the mixed formamide–formamidine dimer [45], which has only C_s symmetry. In one of the hydrogen bonds the donor and acceptor groups are the same and in the other they are different. Calculations at the HF level yield a marginally stable zwitterionic intermediate; it disappears, however, at higher levels, where a single transition state appears with a barrier height of about 13 kcal mol^{-1} and an endothermicity that is smaller by only 1–2 kcal mol^{-1}. In particular, it is found that the proton in the NH$\cdots$N hydrogen bond moves first, leading to a shortening of the NH$\cdots$O hydrogen bond, which allows the second proton to move. The reverse order is not observed, in keeping with the smaller proton affinity of formamide relative to formamidine. The probable mechanism for double proton transfer in this mixed dimer is thus concerted but highly asynchronous transfer along a single pathway. This mechanism may readily turn into stepwise transfer in a dielectric medium that can stabilize the intermediate zwitterion.

An intermediate zwitterionic structure that corresponds to a minimum in the potential energy in the gas phase is found for the formic acid–formamidine complex [46], as illustrated in Fig. 29.7. Extrapolating back from this dimer via the formamide–formamidine dimer to the formamidine dimer, we observe that the replacement N→O in one monomer reduces the synchronous transfer to highly asynchronous transfer, so that it may be expected that a second replacement in the same molecule will lead to stepwise transfer. High-level calculations have partly confirmed this expectation. Although the equilibrium structure of this dimer has C_s symmetry, the double proton transfer potential is symmetric since the two hydrogen bonds are complementary, i.e. they turn into each other by proton transfer. Putting the two protons in the center of the hydrogen bonds, in an effort to construct a transition state for synchronous transfer of C_{2v} symmetry, gives rise to a second-order saddle point. Moving along the symmetric ON-stretch coordinate leads to the equilibrium configuration and moving along the antisymmetric coordinate to the "stable" zwitterionic intermediate of C_{2v} symmetry. The intermediate is very shallow, however, probably too shallow to support the dimer during a vibrational period. Hence this system, which deserves further investigation, seems to be on the border between weak and intermediate coupling.

With one exception, these results are based solely on quantum-chemical calculations of the potential energy surface. Theoretical evaluation of the transfer dynamics has been attempted only for the formic acid dimer, for which two general level splittings have been observed and assigned to synchronous double proton tunneling in the ground state and a vibrational excited state, respectively.

29.7
Other Dimeric Systems

Experimental data relevant to double proton transfer are available for several other dimeric systems. Remmers et al. [5], who investigated the high-resolution ultraviolet spectrum of dimeric benzoic acid, shown in Fig. 29.6 (bottom), observed a splitting of 1107 MHz, which they assigned to double proton transfer in the ground state and/or the excited state. They left open the question of the relative contributions of the two states to the splitting. According to our analysis [7], it should be assigned to splitting of the zero-point level of the ground state rather than of the electronically excited state, the main reason being that the excitation is localized on one of the two monomers. The corresponding deformation of the symmetry of the dimer will tend to slow down proton transfer. This is supported by the observation that hydrogen bonding is weakened in the excited state. The larger splitting with respect to the formic acid dimer does not contradict this; on the contrary, it is in good agreement with the calculated hydrogen bond strengths in the two dimers. The ground-state assignment is supported by detailed calculations [7], which show that the coupling pattern closely resembles that of the formic acid dimer and leads to the conclusion that the transfer occurs by synchronous tunneling.

Additional kinetic data are available for benzoic crystals [47, 48], in which the benzoic acid molecules are stacked as dimers. Since crystal forces reduce the inter-molecular separation, the coupling causing the transfer is stronger in the crystal than in the gas-phase dimer. However, in the crystal no splitting can be observed since the protons move in an asymmetric potential. The asymmetry, which is much larger than this coupling, can be reduced at particular sites by doping the crystal with thioindigo; for dimers adjacent to the dopant the asymmetry assumes a value small enough to allow indirect observation of low-temperature tunneling splitting; the measured value extrapolates to about 8 GHz in the limit of zero asymmetry. The implied large value of the coupling has made it possible to measure the rate of proton transfer in neat crystals down to very low temperatures by NMR relaxometry. The observed low-temperature rate constant is in rough agreement with the extrapolated tunneling splitting, if the effect of doping on the adjacent dimer structure is taken into account [7]. These observations indicate that double proton transfer follows the same basic mechanism in the crystal as in the isolated dimer, although the asymmetry of the crystal potential will introduce a measure of asynchronicity in the tunneling. Recently these measurements were extended to crystals in which all or part of the benzoic acid molecules carry a mobile deuteron instead of a mobile proton. In this way it was possible to measure rate constants as a function of temperature not only for HH but also for HD and DD transfer [49–51]. The resulting Arrhenius plots, illustrated in Fig. 29.8,

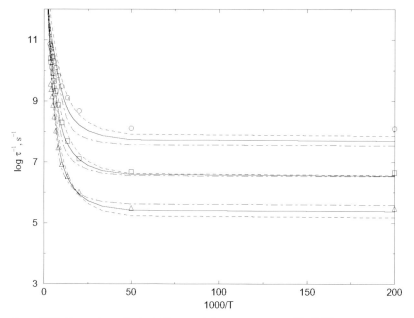

Figure 29.8 Comparison of spin–lattice correlation rates measured by NMR relaxometry [51] with those calculated by AIM/DOIT [7] for solid benzoic acid isotopomers with mobile HH (top), HD (center), and DD (bottom) pairs. Measurements are depicted by symbols; the broken and dot-dash lines represent theoretical results for two limiting cases, the solid curve being their geometric mean [7].

show strong curvature, the rate constants for all three isotopomers being constant below 20 K and showing classical activated behavior above 200 K, as expected for tunneling assisted by thermal excitation of promoting vibrations and, at higher temperatures, by over-barrier transfer. Correspondingly the ratio of DD/HH rate constants decreased from about 500 below 20 K to about 25 at room temperature. The HH, HD, and DD Arrhenius curves were evenly spaced, which, according to the analysis of Section 29.5, is consistent with concerted transfer.

Observations analogous to those for carboxylic acid dimers have been reported for 2-pyridone·2-hydroxypyridine (2PY·2HP) [52, 53], illustrated in Fig. 29.9. This dimer, formed from two isomers, is asymmetric, but it has a symmetric double proton transfer potential since the transfer interchanges the isomers. In a collaborative effort, Pratt, Zwier, Leutwyler and their coworkers [52] measured and analyzed the high-resolution fluorescence-excitation spectrum of the $S_1 \leftarrow S_0$ origin and observed a level splitting of 527 MHz, which they assigned to double proton

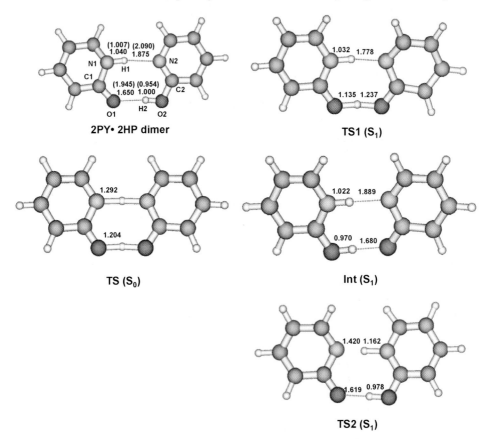

Figure 29.9 Calculated structures of the equilibrium configuration and the transition state of the 2-pyridone · 2-hydroxypyridine dimer in the ground state and the excited state [29]. Pertinent bond lengths are given in Å; the numbers in parentheses refer to the equilibrium configuration of the excited state.

transfer. Although the two protons are not equivalent, one being part of an OH$\cdots$O and the other of an N$\cdots$HN hydrogen bond, the observation of a splitting implies that the transfer is concerted. Evidently, it cannot be fully synchronous; however, the magnitude of the observed splitting, which is intermediate between the splittings observed for dimeric formic acid and dimeric benzoic acid, shows that this does not seriously interfere with the rate of transfer. The assignment leaves open the question whether the level splitting is mainly due to the ground state or the excited state. The conclusion that it is mainly due to the ground state [29] is based on arguments similar to those made for dimeric benzoic acid. The excitation is essentially localized on the 2PY component. That this tends to distort the dimer so as to trap the protons follows from the observation that hydrogen bonds are longer and thus weaker in the excited state than in the ground state.

The lower symmetry of 2PY·2HP compared to carboxylic acid dimers is reflected in the deuterium isotope effect on the tunneling splitting, since it implies that the splitting in 2PY-d_1·2HP, where the isotopic substitutions is on N, will differ from that in 2PY·2HP-d_1, where it is on O. This is confirmed by direct measurements, which give rise to splittings of ≤ 10 and 62 MHz, respectively, for these two isotopomers. This order of splittings is confirmed by high-level dynamics calculations, which yield splittings of 12 and 43 MHz, respectively [53]. It means that deuteration of the NH group in 2PY has a larger effect on the transfer dynamics than deuteration of the OH group in 2HP. This agrees with the notion that the NH$\cdots$N hydrogen bond is longer than the OH$\cdots$O hydrogen bond and thus makes a larger contribution to the transfer path and consequently to the KIE. On the basis of these arguments and calculations we propose that, as a general rule, the weaker link in a double proton transfer process can be recognized by its greater sensitivity to isotopic substitution.

It is instructive to apply this rule to the dimer of 7-azaindole, illustrated in Fig. 29.10. In contrast to 2PY·2HP, where the two monomers are different but the tunneling potential is symmetric, this dimer is formed from two identical monomers but exhibits an asymmetric tunneling potential, due to the different electronegativity of the two nitrogens in each monomer. The monomer exists in two tautomeric forms, depending on which nitrogen carries the hydrogen. The lower energy form in the ground state, i.e. the form with the NH group in the five-membered ring, becomes the higher-energy form in the excited state. By exciting the dimer in a cold beam and monitoring the fluorescence, one can follow the transformation of the higher-energy to the lower-energy tautomer in the excited state. In a careful study of HH, HD, and DD transfer, Sakota and Sekiya [54] showed convincingly that the transfer is concerted, contradicting earlier conclusions [55], and that HD transfer contains two components due to the fact that the coupling between the monomers is small (about 3.5 cm^{-1}) compared to the energy shift for NH→ND substitution (about 40 cm^{-1}), so that the excitation is effectively localized on one of the monomers in the mixed (i.e. d_1) isotopomer, a situation similar to that encountered in dimeric benzoic acid and in 2PY·2HP. Specifically, they found that H*D transfer is faster than HD* transfer, where the asterisk indicates the monomer on which the excitation is localized. By analogy to the 2PY·2HP

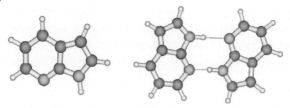

Figure 29.10 7-Azaindole and its dimer.

result, this implies that the NH*···N hydrogen bond is weaker than the N···HN bond in the higher-energy tautomer of the excited dimer. No theoretical study of the dynamics of these processes has been reported to date.

The apparent prevalence of concerted transfer whenever there is hydrogen bonding, even if the two protons are inequivalent and/or replaced by a proton–deuteron pair, is striking. However, it has to be borne in mind that these are gas-phase experiments at low temperatures. The observed splittings do not exclude the possibility of a stable intermediate with an energy that makes it thermally inaccessible under these conditions. However, the conclusions of Section 29.4 as well as the available calculations for specific systems argue strongly against such a possibility.

29.8
Intramolecular Double Proton Transfer

Intramolecular proton transfer between keto and enol functions is well known. An example of intramolecular double proton transfer of this kind is observed in naphthazarin, illustrated in Fig. 29.2. Using an SCF approach, de la Vega et al. [15] calculated the corresponding two-dimensional potential energy surface. No stable intermediate was found, but moving from the second-order saddle point along the antisymmetric OH-stretch coordinate led to two first-order saddle points, from where motion along a path that involves both coordinates led back to the equilibrium configuration, as sketched in Fig. 29.3b. Although the authors considered this a stepwise process, it remains concerted although not synchronous. It involves a lower barrier than the synchronous process along the symmetric OH-stretch coordinate, but a longer pathway.

To calculate the tunneling splitting, they approximated their potential analytically by a function with terms containing powers of $x_{p,m}$ of order four, including a biquadratic cross term, i.e. a term similar to the coupling term in our model potential (29.11). Both their analytical and our model potential depend on three parameters; in their case the distance between the minima, the energy of the maximum and the energy of the saddle point, found to be, respectively, 0.78 Å, 28.0 kcal mol^{-1} and 25.6 kcal mol^{-1}. Using numerical diagonalization, the authors found the lowest vibrational states of this two-dimensional potential energy surface and evaluated the zero-point tunneling splitting. They found that the two-dimensional pathway, representing asynchronous transfer along both x_p and x_m

contributed more to the splitting than the one-dimensional synchronous pathway along x_p. This result contradicts our calculations based on model potential (29.11), which leads to a dominant contribution from the synchronous path.

This difference can be traced back to the nature of the two potentials. Our potential (29.11) is derived from a model of two equivalent quartic potentials plus a coupling that is symmetric in the two local coordinates. If the potential of de la Vega et al. is transformed into local coordinates, it does not reduce to two equivalent potentials plus such a coupling. Hence it is not consistent with the symmetry of the system. As a result, it yields an inadequate force field in the equilibrium configuration, namely a frequency for the antisymmetric mode that is less than half that of the symmetric mode, instead of two frequencies of the same general magnitude, as appropriate in the present case where the localized coordinates $x_{1,2}$ are weakly coupled. This renders their analytical potential inadequate for dynamics, and especially for evaluation of tunneling splittings, which depend crucially on the quality of the force field in the minimum. We note also that the use of *adiabatic* energies for the saddle points in the two-dimensional potential in Ref. [15] is incorrect; the *crude-adiabatic* energies corresponding to the "frozen" skeleton are the required parameters.

As an example of double proton transfer in a molecule with little or no hydrogen bonding, we consider porphine [8–11], depicted in the insert of Fig. 29.11. In this molecule, belonging to the D_{2h} point group, the two equivalent hydrogens in the inner ring are bound to two of the four equivalent nitrogens. In the reactant they are in one of the two *trans* positions, in the product they are in the other *trans* position. B3LYP/6-31G(p)-level calculations [11] indicate that the intermediate *cis* position has an energy that is higher by about 8.3 kcal mol^{-1}, which can be reached via a transition state with an (adiabatic) energy of 16.7 kcal mol^{-1}. The *trans* to *trans* (adiabatic) barrier for synchronous double proton transfer with an energy of 25.3 kcal mol^{-1} corresponds to a second-order saddle point. The angle between the NH bonds in the initial and final position is 90°, which prevents effective hydrogen bonding. The structure of the molecule rules out deformations that reduce this angle significantly in the transition state. Hence the tightening of the hydrogen bonding in the transition states of the dimers discussed in Sections 29.6 and 29.7, which supports concerted transfer, is weak in porphine.

Two sets of experimental data are available, one set measured by NMR spectroscopy in the range 200–300 K [8, 9] and another set measured by optical spectroscopy in the range 95–130 K [10]. Arrhenius plots of the data shown in Fig. 29.11 are slightly curved; at low temperature the slope approaches a constant value close to the calculated *cis–trans* energy difference of 8.3 kcal mol^{-1}. The observation that double proton transfer remains thermally activated by roughly this amount down to low temperatures, immediately suggests that the process proceeds stepwise. The same conclusion follows from the observation that the HD Arrhenius curve is closer to the DD than the HH curve, in agreement with Eq. (29.56).

Using the approach of Section 29.2, we can estimate the proton–proton coupling parameter G by associating the extrema of the calculated potential with those of the model potential. However, for this we require crude-adiabatic poten-

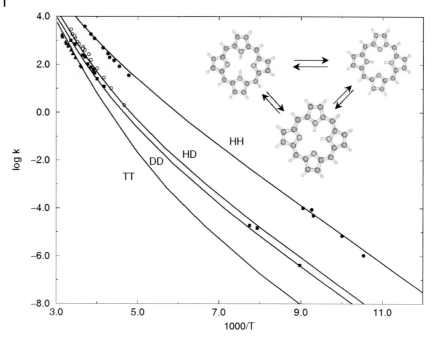

Figure 29.11 Temperature dependence of the rate constant (in s^{-1}) of double proton transfer in porphine-d_0, -d_1, -d_2 and -t_2 evaluated for the stepwise mechanism. The symbols represent observed rate constants from Refs. [8–10] and the curves represent the results of a multidimensional AIM/DOIT calculation [11]. The insert shows the stepwise and concerted transfer mechanisms.

tials rather than the adiabatic potentials calculated quantum-chemically [11]. To obtain a rough order-of-magnitude estimate, we assume that the ratio of the crude-adiabatic barrier heights between the extrema (2) and (4) of Table 29.1 is smaller but of the order of that between the corresponding adiabatic barrier heights, and that the ratio of the crude-adiabatic to the adiabatic barrier heights of extremum (2) is about 3, as previously found for carboxylic acid dimers. Assuming further that the minimum (3) will not be much affected, we obtain from the energy ratio (3)/(2) the estimate $G \approx 0.05$, which roughly satisfies the ratio (4)/(2). This estimate clearly indicates weak coupling.

Kinetic data are also available for porphine in which one or both inner protons are isotopically labeled [8–10]. The observed kinetic isotope effects are in good agreement with high-level dynamics calculations [11], which indicate stepwise transfer in the temperature regime where measurements are available. Although the kinetic data for HH, DD, and TT transfer do not cover a common range of temperatures, the availability of dynamics calculations that provide a good fit to the data allows extrapolations yielding sets of *cis* → *trans* KIEs for a range of temperatures. As shown in Fig. 29.12, they lead to Swain–Schaad exponents in excellent agreement with Eq. (29.52) and show the gradual decrease with increasing temperature of the part of the transfer path located under the barrier.

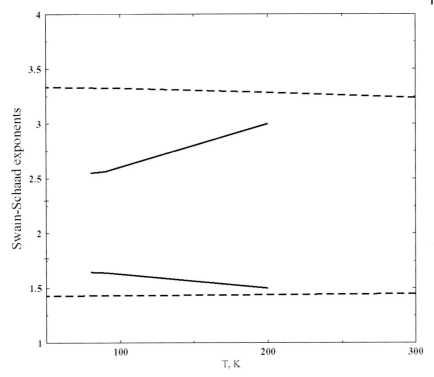

Figure 29.12 Swain–Schaad-type exponents observed for porphine. The solid lines are derived from observed rate constants [8–10] extrapolated by high-level multidimensional dynamics calculations [11]. The broken lines are derived from multidimensional TST calculations without tunneling.

It is interesting to compare the results for porphine with those for the related molecule porphycene [12–14], depicted in Fig. 29.13. In porphycene the four nitrogen atoms form a rectangle with sides of about 2.65 and 2.84 Å calculated at the B3LYP/6-31G(d,p) level. The preferred proton exchange will be between the more closely spaced nitrogens. The longer NH bond (1.043 vs. 1.027 Å) and the smaller N–H–N angle along the transfer path compared to porphine indicate substantial hydrogen bonding in porphycene. Hence one expects the double proton transfer to be considerably more rapid than in porphine. This is indeed observed; according to NMR measurements the transfer rate constant at 298 K in porphycene is larger by a factor of 3000. Since no low-temperature measurements are available and the observed activation energy in the range 228–355 K is about 6 kcal mol^{-1}, it is not immediately clear whether the transfer is stepwise or concerted. The calculations [12, 14] indicate an energy difference between the *cis* and *trans* configurations of about 2.3 kcal mol^{-1}, implying a weak proton–proton correlation. Attempts to fit the kinetic data to the calculated transfer potential have met with problems. The *trans–cis* barrier height of 4.7 kcal mol^{-1} is too low to fit the data. To obtain an acceptable fit, it is to be raised to about 11 kcal mol^{-1} with a *cis*

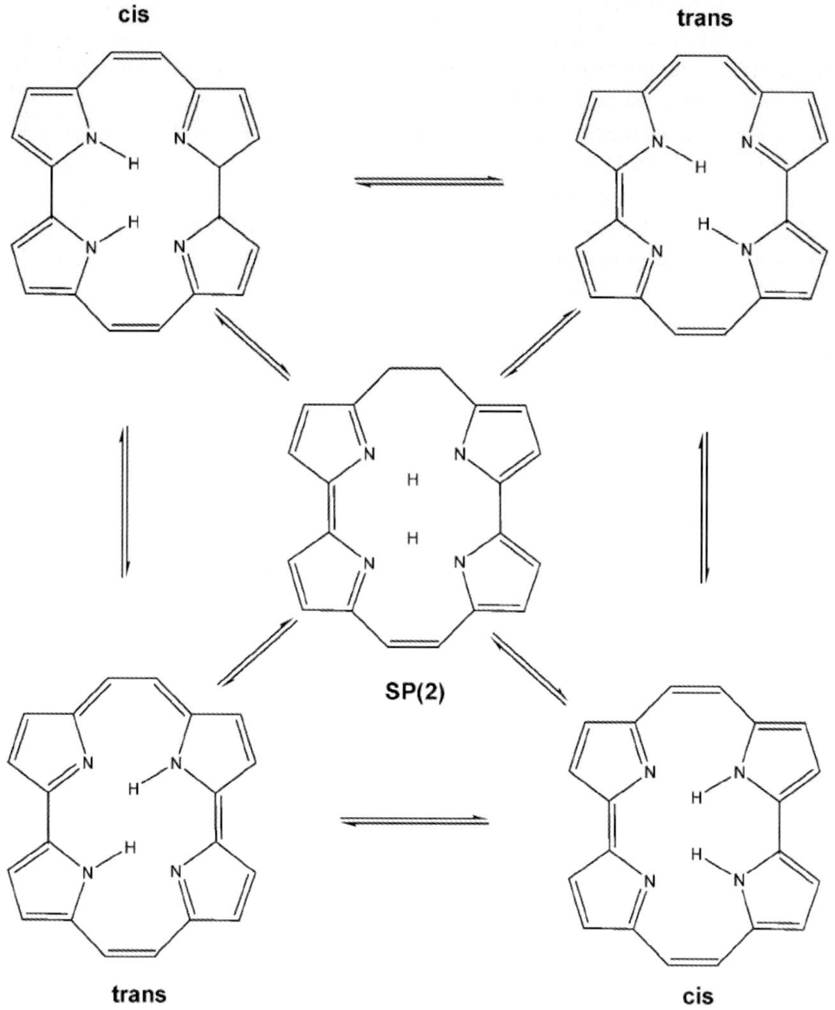

Figure 29.13 Illustration of stepwise and concerted double-proton transfer in porphycene. *Trans* indicates the stable configuration, *cis* the intermediate corresponding to single proton transfer and SP(2) the saddle point of second order corresponding to concerted transfer.

energy of 3 kcal mol^{-1}. If we assume concerted transfer, we can fit the data by a barrier corresponding to a second-order saddle point of about 14 kcal mol^{-1}, whereas the calculated value is 6.6 kcal mol^{-1}. These results suggest that porphycene is an intermediate case, where the mechanism combines aspects of stepwise transfer involving a very shallow intermediate with highly asynchronous concerted transfer. They also cast doubt on the ability of presently available DFT functionals to represent NH$\cdots$N hydrogen bonds.

Returning to systems without hydrogen bonding, we consider a group of molecules in which the hydrogen transfer occurs between carbon atoms, namely bridged ring compounds where the groups between which the two hydrogens are exchanged are of the general form illustrated in Fig. 29.14, the simplest example being exchange between (substituted) ethane and ethylene. In these compounds the ethane and ethylene analogs are oriented in a parallel fashion, their movements being restrained by stiff three-dimensional structures. Typical examples are the *syn*-sesquinorbornene disulfones of Fig. 29.14 studied by Paquette et al. [56, 57] and the compounds studied by Mackenzie et al. [58, 59]. The data available for these compounds are limited to temperatures near room temperature, the effective activation energies obtained from Arrhenius plots being in the range 24–30 kcal mol^{-1}. Such high activation energies suggests either classical over-barrier transfer or stepwise tunneling involving a high-energy stable intermediate structure. The relatively large kinetic isotope effects (of order 10 at 373 K for HH compared to DD transfer) suggest the latter alternative. Since the barriers faced by the protons are not symmetric, the relation between the rates of HH, HD, and DD transfer cannot provide unequivocal answers. Also, these complex structures cannot be studied theoretically at a level high enough to settle this problem. However, the available calculations [57, 60] indicate that the observed rate constants and activation energies are incompatible with concerted transfer.

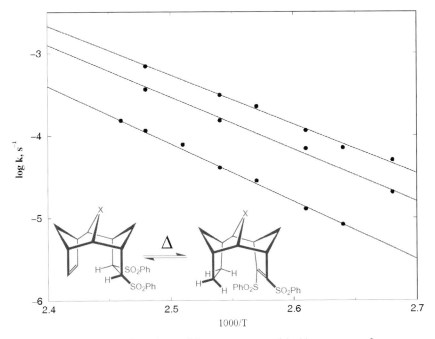

Figure 29.14 Temperature dependence of the rate constants of double proton transfer in *syn*-sesquinorbornene disulfone–d_0, -d_1, and -d_2, illustrated in the insert, evaluated for $R_{1,2}$=CH$_2$. The symbols represent observed rate constants [56, 57] and the solid lines the results of two-dimensional semi-empirical Golden Rule calculations [57].

Qualitative arguments in favor of stepwise transfer are the absence of signifi-cant hydrogen bonding and the stiffness of the bridged structures. The transfer is between CH bonds whose polarity does not favor hydrogen bonding and thus assumes the character of neutral hydrogen transfer. The intermediate state pro-duced by single hydrogen transfer will not be the ionic state typical of single pro-ton transfer and hence will not create a significant dipole moment. Under these conditions the coupling parameter G will be very small. Moreover, the stiffness of the structure does not favor a transition state in which the two $C \cdots H \cdots C$ bonds are significantly closer than in the equilibrium configuration while remaining aligned. Twisting appears the most likely deformation, which causes an asym-metric distortion of the active site that would favor single hydrogen transfer and thus stepwise transfer.

29.9
Proton Conduits

An OH or NH group carries a proton as well as a lone pair of electrons and can therefore form two connected hydrogen bonds through which protons can be con-ducted from a donor to an acceptor group by means of a relay mechanism. The resulting double proton transfer differs from most of the processes discussed in the previous sections in that it amounts to a net transport of charge. In biological systems, chains of OH groups, belonging to water molecules and amino acid resi-dues, are known to transport protons over considerable distances. To investigate the mechanism of these processes, we first consider a double proton transfer reac-tion in the acetic acid–methanol complex [61, 62], illustrated in Fig. 29.15, for which the potential is almost symmetric. Although reactant and product are iden-tical (except for HD transfer), the barrier is asymmetric because the two moving protons belong to different partners. To obtain transfer rate constants, Gerritzen and Limbach [61] carried out NMR measurements in tetrahydrofuran solution at 270–330 K. In Fig. 29.16 the measured rate constants are compared with the results of quantum dynamics calculations based on a high-level potential [62], which yield a single transition state and thus predict concerted transfer despite the relatively high temperature and the (weakly) polar medium. Because of the asymmetry of the barrier, the two protons move asynchronously; in the transition state they are closer to the methanol than to the acetic acid moiety. The asymmetry is clearly shown by the difference between the $k^{HD}(T)$ and $k^{DH}(T)$ values. Note that the $k^{HD}(T)$ curve in Fig. 29.16 is closer to the $k^{DD}(T)$ than the $k^{HH}(T)$ curve, rather than being in the center.

Proton conduits can also catalyze isomerization. A biologically interesting example is the isomerization of DNA bases such as guanine, as illustrated in Fig. 29.17. These molecules occur is several isomeric forms that differ in the position of one or more of the protons. Water or alcohol molecules can form an OH bridge connecting an occupied with an unoccupied position, which may lead to forma-tion of an isomer. For the isomerization depicted in Fig. 29.17, a single water mol-

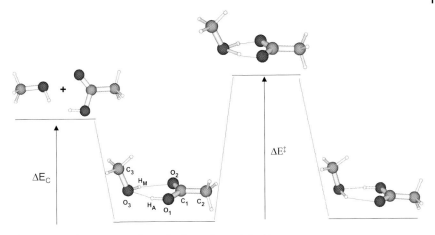

Figure 29.15 The two steps of complex formation and double proton exchange in the methanol–acetic acid complex [62].

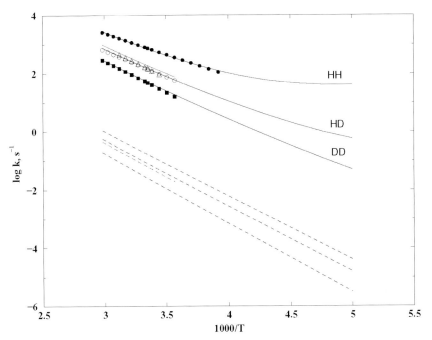

Figure 29.16 Comparison of experimental [61] and theoretical [62] bimolecular rate constants of proton exchange in the methanol–acetic acid complex in tetrahydrofuran-d_8, represented by symbols and solid lines, respectively, for HH, HD and DD exchange. The dashed lines represent the rate constants obtained by transition state theory. The two (close) sets of data for the mixed isotope combinations HD and DH reflect the asymmetry of the barrier.

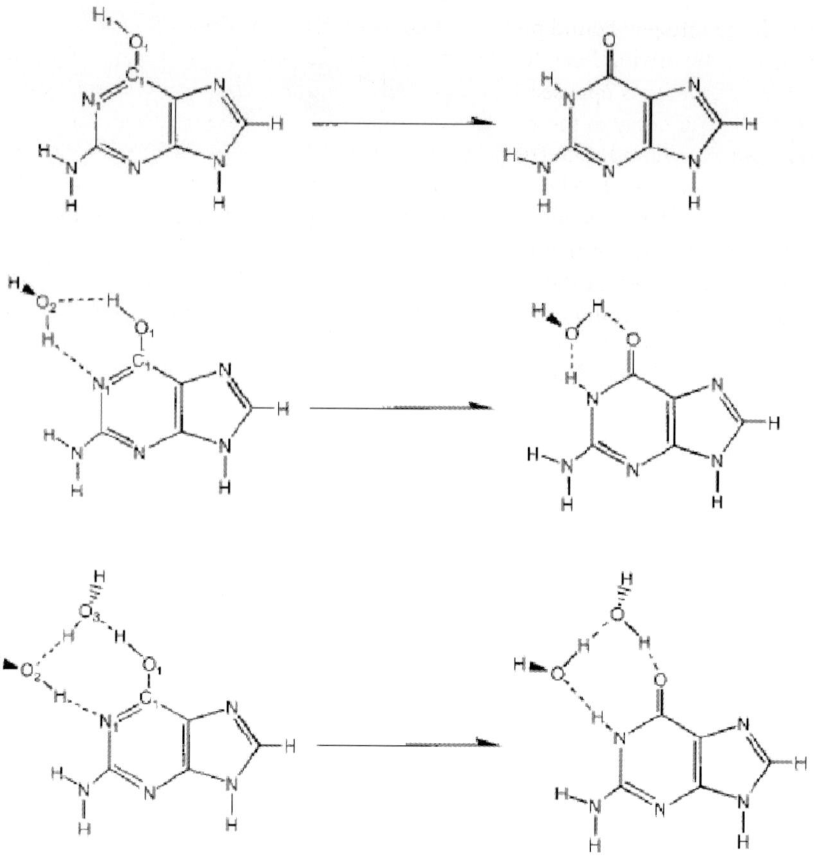

Figure 29.17 Schematic representation of the tautomerization of guanine and its 1:1 and 1:2 complexes with water [16].

ecule can increase the rate of isomerization by up to 12 orders of magnitude [16].
A chain of two water molecules is also active but turns out to be a less effective
catalyst. A noncanonical isomer can cause a GC→AT point mutation in the DNA
chain; in the case of guanine, the isomer may replace adenine and pair with thymine rather than cytosine, thus forming, after a subsequent cell division, an AT
base pair where a GC pair is required. An active hydroxyl group can catalyze such
processes in two directions: it may produce noncanonical isomers, but it also may
regenerate the canonical form once the unwanted isomer is produced. Hence it
may play a role in point mutation as well as in enzymatic DNA repair [63].

To illustrate the relative efficiency of long and short proton conduits, we return
to the 7-azaindole, molecule, whose dimer was briefly discussed in Section 29.7.
As pointed out in Section 29.7, excitation of this molecule to the fluorescent state
redistributes the charge between the two nitrogens. This redistribution renders
the tautomeric form produced by the excitation unstable relative to the form in

which the nitrogen-bound proton has changed place; in other words, the redistribution is the driving force for the observed proton transfer. In the excited monomer this transfer is opposed by a high barrier and as a result is much too slow to compete with decay of the excited state, which has a lifetime of about 8 ns. (It is faster in the ground state [16] where the barrier is lower). However, the example treated in Section 29.7 shows that transfer can be catalyzed by dimerization [54]. In the dimer the excitation is essentially localized on one partner and the unexcited partner serves as an effective proton conduit.

The same occurs in the monomer if water is present. Water molecules form hydrogen-bonded complexes with 7-azaindole, two of which are illustrated in Fig. 29.18 (top) [17, 64, 65]. The water molecules connect the NH group with the non-hydrogenated N atom, thereby facilitating proton exchange. It is well-known that tautomerization of 7-azaindole in aqueous solution is extremely rapid ($\geq 10^9$ s^{-1}) [66, 67]. To account for this effect of water, we calculated the rate of proton transfer through water chains connecting the two nitrogens [17]. The results show that a single water molecule does not form an effective proton conduit for neutralizing the effect of the excitation in a cold beam because the transfer is too slow to compete effectively with fluorescence decay. A chain of two water molecules is even less effective; in that case the transfer is still concerted but far from synchronous and the transition state shows a large charge separation, which makes the corresponding transfer subject to strong environmental effects. Although ineffective in

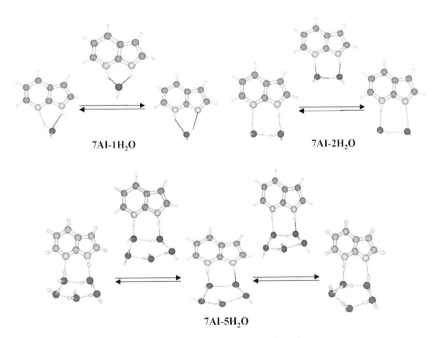

7AI-1H$_2$O   7AI-2H$_2$O

7AI-5H$_2$O

Figure 29.18 Calculated structures of the stationary points along the reaction coordinate for the excited state tautomerization in 7-azaindole complexes with one, two and five water molecules [17].

a cold beam, the transfer rates are calculated to increase rapidly with increasing temperature and to increase further if the complexes are embedded in a dielectric continuum; nevertheless they still fall short of the values observed in aqueous solution. However, appropriately high values have been obtained [17] for complexes with five water molecules of the form illustrated in Fig. 29.18 (bottom). In this system an ionic intermediate state is formed through with the transfer occurs stepwise by a classical over-barrier process.

Hence in the system excited 7-azaindole + water, we see a gradual change from concerted to stepwise transfer if more and more water molecules are added and if the temperature is increased. Short linear chains of water molecules tend to give give rise to concerted transfer, whereas branching or clustering of water chains seems to favor stepwise transfer. The apparent reason for this change of mechanism is the ability of the water cluster to stabilize an ionic structure. This flattens the barrier and ultimately generates a local minimum, which splits the barrier without reducing its width at the base. The ionic structure imposes a drastic rearrangement on the conducting water chain, which results in a large increase in the effective mass of the moving protons. This reduces the transfer rate at low temperatures but opens the possibility of proceeding stepwise via the intermediate minimum when the temperature increases. Such a transfer mechanism is just emerging in the 1:2 complex. The extreme form of this mechanism is proton transport in ice, which proceeds stepwise and classically, as indicated by the minimal isotope effect.

These calculations illustrate two aspects of the role of hydrogen-bonding solvents in proton transfer processes. On the one hand they catalyze these processes by forming proton conduits between donor and acceptor atoms, and on the other they provide a dielectric medium that stabilizes ionic or highly polar intermediates. Representation of the solvent by a dielectric continuum cannot account for this dual role. To treat the transfer process adequately, it will be necessary to introduce a primary shell of discrete solvent molecules into the calculations.

In the ground state of 7-azaindole the same proton transfer processes will occur but move in the opposite direction. Our calculations [16] indicate that the dependence of the transfer rate on the length of the water chain and on temperature is strikingly different from that in the excited state. The reduced polarity of the ground state prevents formation of an ionic structure in the 1:2 complex and keeps the protons moving more or less synchronously. As a result the transfer rates of the 1:1 and 1:2 complexes are calculated to be very similar. The corresponding deuterium isotope effects are calculated to be small, namely 6.6 and 2.6, respectively. Small isotope effects are typical for loose water chains, where the motions of the oxygen atoms contribute significantly to the transfer.

A small isotope effect for proton transport through a water chain is also observed for catalytic conversion of CO_2 to HCO_3^- by carbonic anhydrase II [68, 69]. The rate-determining step in this process is the transfer of a proton from the H_2O ligand of a four-coordinated zinc ion to a histidine residue located at a distance of about 8 Å. The proton conduit is known to consists of water molecules located in a pocket that can contain several such molecules, which are freely exchanged with embedding fluids. A minimum of two water molecules is sufficient to form a connecting chain, as indicated by a theoretical study of an active-site model of the enzyme with two water molecules forming the proton conduit in a pocket whose size is constrained to conform to available X-ray data [18]. The model is illustrated in Fig. 29.19; the structure and vibrational force field were calculated by a density functional method after extensive testing. A single transition state was found, corresponding to concerted triple proton transfer through an adiabatic barrier of about 6 kcal mol^{-1}; the normal mode with imaginary frequency is illustrated in

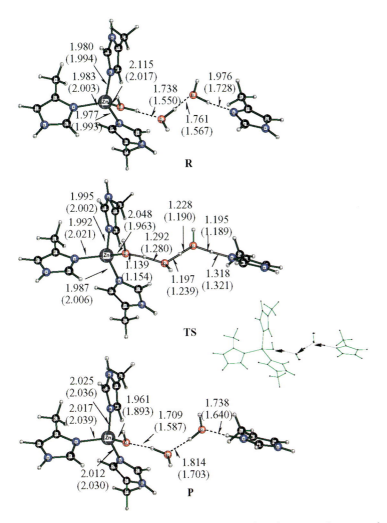

Figure 29.19 The reactant (R), transition state (TS) and product (P) configurations for the rate-determining triple proton transfer step of the 58-atom model used to represent the active site of carbonic anhydrase II [18]. The numbers denote bond distances (in Å) calculated at two different levels of theory. The arrows in the insert figure represent the tunneling mode and illustrate the degree of synchronicity of the transfer.

the insert to Fig. 29.19. Tunneling rate constants were evaluated with AIM/DOIT and, on a more limited scale, with TST with semiclassical tunneling corrections. The results indicate that concerted transfer through this short water chain is fast and can serve as a feasible if unconventional model for proton transport in this enzymatic reaction. Tunneling rates are found to be faster than classical over-barrier rates by almost two orders of magnitude. Tunneling rate constants calculated for partial and total deuteration along the water chain closely match the observed KIEs in the range 1–4 at room temperature [18, 70], as shown in Fig. 29.20. KIEs calculated by TST for over-barrier transfer are larger because of the the cumulative effect of zero-point energy differences. By contrast, tunneling KIEs are small because of strong participation of the motion of heavier atoms, especially the oxygen atoms of the water chain and the zinc ligand. This behavior is common for chains of two or more water molecules. Such a proton conduit is a loose structure and can be easily deformed without substantial expenditure of energy. Both symmetric and antisymmetric modes participate in the deformation, the former helping and the latter hindering tunneling. The transfer in the chain is not only concerted but actually highly synchronous without any indication that an ionic intermediate is being formed. The KIE of partially deuterated chains does not depend strongly on the place of deuteration and the rate constant varies smoothly with the number of deuterium substituents.

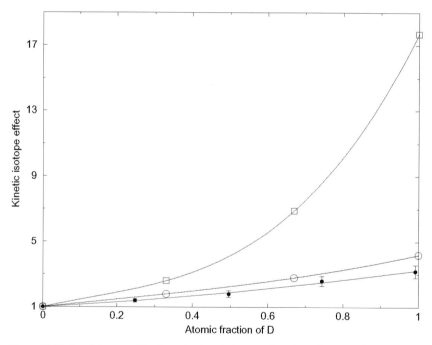

Figure 29.20 Kinetic isotope effect in carbonic anhydrase II plotted as a function of the atomic fraction of deuterium in water. The dots represent experimental results taken from Ref. [70], the circles and squares are the results of quantum and classical calculations, respectively [18].

In this example the two water molecules together with the zinc-coordinated water molecule make up a chain of three waters. Given the size of the pocket, it is clear that many other connecting water bridges are possible. Cui and Karplus [71] found that the single transition state is maintained if a fourth water is added to the linear chain, but that an intermediate minimum appears when the chain is branched. No reliable calculation of the rate of proton transfer through these structures is available to date.

29.10
Transfer of More Than Two Protons

In the preceding section we encountered several examples of transfer of three or more protons along chains of water molecules. There is as yet no experimental evidence that any of these transfer reactions is concerted or stepwise. Calculations support the intuitive notion that the longer the chain the smaller will be the probability that the transfer is concerted. Strong coupling of the proton motions to oxygen atom displacements lead to the prediction of small KIEs, which limits the usefulnes of deuterium substitution as a diagnostic tool. To get a clear picture of the concertedness of the transfer, one needs systems in which the transfer can give rise to level splittings. Two systems of this kind are trimeric water and calix[4]arene.

According to measurements of Pugliano and Saykally [72], the chiral water trimer, illustrated in Fig. 29.21 (top), occurs in two interconverting enantiomeric forms. Although this interpretation was not supported by Monte Carlo simulations [73], two recent multidimensional tunneling calculations [19, 74] confirm the interconversion by triple proton tunneling across a high barrier (about 26 kcal mol^{-1}). The transfer is predicted to be concerted and to produce a zero-point splitting in the range 1–10 MHz. This is a very small splitting that would be difficult to observe by presently available methods. The transfer rate is greatly enhanced by coupling to a symmetric breathing mode of about 730 cm^{-1}; the splitting of the fundamental of this mode is predicted to be 40 times as large as that of the zero-point level, which may put it within measureable range. Unfortunately, the breathing mode is not infrared active.

Calix[4]arene is a bowl-shaped chiral molecule with four moving protons, as shown in Fig. 29.21 (bottom). Inside the bowl there are four hydroxy groups, which form a chiral ring of symmetry C_4. Horsewill et al. [75, 76] used NMR relaxometry on crystalline powders to study proton exchange in calix[4]arene and p-tert-butyl calix[4]arene at low temperatures (30–80 and 15–21 K, respectively). They reported clear evidence for concerted quadruple proton transfer. Our AIM calculations for an isolated calix[4]arene molecule, based on a potential calculated with density functional theory at the B3LYP/cc-pVDZ level, are in agreement with this conclusion [77]. The calculated barrier height is about 17 kcal mol^{-1}, a high value, as expected for a process involving rupture of four OH bonds, as illustrated in Fig. 29.21. Nevertheless, the predicted splitting amounts to almost 40 MHz due

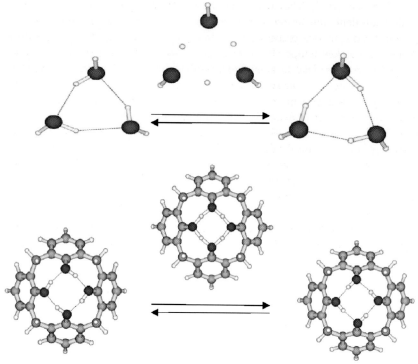

Figure 29.21 The two stereoisomers and the transition state corresponding to synchronous transfer of three and four protons in the water trimer (top) [19] and in calix[4]arene (bottom) [77], respectively.

to strong assistance of modes that modulate the O···O distances. Unfortunately the measured splitting of the same general magnitude turned out to be an experimental artifact, so that there is no definite proof that the four protons move concertedly.

It follows that thus far concerted multiple proton transfer has been demonstrated only for two protons.

29.11
Conclusion

The nonclassical dynamics of protons, which allows them to tunnel through barriers, also influences their collective behavior. Specifically, it gives them a propensity to transfer concertedly and even synchronously. The clearest evidence of concerted transfer is the observation of level splittings in high-resolution spectra. For the time being, these observations are limited to double proton transfer and single levels, mostly zero-point levels. The observed splittings are small, typically

≤ 1 GHz (0.04 cm^{-1}). They require symmetric transfer potentials but not necessarily equivalent transfering particles. The experimental evidence for concerted proton transfer in rate processes is based mostly on the vanishing temperature dependence at low temperatures. Deuterium isotope effects observed at higher temperatures often lead to ambiguous results, because for over-barrier transfer they have a tendency to *increase* with increasing numbers of protons while for tunneling they have a tendency to *decrease* because of increasing coupling to transvers modes. Although the evidence for concerted transfer in rate processes is less firm than that obtained from level splittings, it is strongly supported by theory, in particular quantum-chemical calculations leading to single transition states. On the other hand, there is also convincing experimental evidence for rate processes in which protons move separately rather than collectively. Such stepwise transfers imply the existence of stable intermediates, which indeed have been found in theoretical calculations.

In this contribution we have treated these two transfer mechanisms as limiting cases in a general picture of multiple proton transfer, represented by a model in which the degree of concertedness of the transfer is governed by specific parameters. The basic model concerns double proton transfer in a symmetric potential. The parameter controlling the concertedness is the proton–proton correlation represented by a term that is bilinear in the local proton coordinates. The model is combined with a previously developed approach to single proton transfer based on an approximate instanton method. This leads to the recognition of three coupling regimes governing the mechanism of nonclassical double proton transfer.

Strong coupling between the hydrogen bonds along which the transfer takes place leads to a potential characterized by a single transition state, whose imaginary frequency is along the symmetric component of the two-dimensional reaction coordinate, the frequency along the antisymmetric component being real. A single mechanism is operative in this limit, namely concerted transfer and, if the potential is symmetric, synchronous transfer of the two protons, leading to level splittings. The transfer dynamics is identical to that of a single particle. The rate of transfer will be independent of temperature at low temperatures and its temperature dependence at moderate temperatures will be governed by thermal excitation of the skeletal modes coupled to tunneling. The rate will be subject to a deuterium isotope effect that tends to be weaker than that of single proton transfer reactions due to larger contributions of promoting modes. For symmetric potentials, the HD rate constant will be close to the geometric mean of the HH and DD rate constants. This limiting situation, represented by the potential of Fig. 29.3(c), is favored if the hydrogen bonds are strong, (anti)parallel, and closely spaced. The formic acid dimer is a typical example.

Weakening of the coupling will lead to a point where the transition state that represents the barrier for synchronous double proton transfer turns into a second-order saddle point; an intermediate also appears, which is a minimum along the antisymmetric collective coordinate but a saddle point along the symmetric one, as illustrated in Fig. 29.3(b). We then enter the intermediate coupling region where a single trajectory can no longer account for the transfer dynamics. In addition to the one-dimensional synchronous path, there will be two-dimensional paths of concerted but asynchronous transfer that are longer but of lower energy. Hence, in addition to the synchronous mechanism that governs the strong-coupling regime, a two-dimensional concerted mechanism involving both reaction coordinates is possible. To compare the two concerted mechanisms, we have carried out simple model calculations, which suggest that the one-dimensional synchronous process will generally dominate at low temperature. This result requires further theoretical and experimental testing. Naphthazarin and porphycene may be suitable subjects for such an investigation.

Further weakening of the coupling will turn the intermediate into a stable minimum, separated by transition states from the global minima, as illustrated in Fig. 29.3(a). This opens a third transfer mechanism consisting of two independent one-dimensional steps to and from the intermediate. Stepwise double proton transfer along this path requires enough thermal energy to reach the intermediate; its rate will tend to show a much stronger temperature dependence than that of the two concerted mechanisms. It will therefore be favored by high temperatures. At temperatures low enough to render the intermediate inaccessible, the two concerted mechanism operative for intermediate coupling will take over. It follows that under weak-coupling conditions three independent mechanisms contribute to the transfer. This limiting situation is favored if the hydrogen bonds are weak, not parallel, and far apart; porphine is a typical example.

This classification scheme for multiple proton transfer remains incomplete. It needs to be extended to asymmetric transfer potentials, an aspect that has been briefly discussed only in the sections dealing with specific examples. Another aspect of symmetry breaking will arise when the effect of symmetric and anti-sysmmetric transverse vibrations is included. These vibrations should affect the symmetric and antisymmetric reaction coordinates differently, which will change the balance between the two concerted transfer mechanisms operative in the weak (Fig. 29.3(a)) and intermediate (Fig. 29.3(b)) coupling regions. Symmetric transverse modes tend to act as promoting modes for transfer along the symmetric reaction coordinate by shortening the effective tunneling distance, an effect illustrated by the corresponding hydrogen-bond contraction in the central maximum (2) of Table 29.1. This would favor synchronous transfer. On the other hand, antisymmetric transverse modes may favor the asynchronous transfer associated with the alternative lower-energy, longer-path mechanism since it breaks the instantaneous equivalence of the two protons. The degree of excitation of these transverse modes and thus the temperature will also play a part in these considerations. Obviously, this problem requires further study. The problem of how to distinguish these two concerted mechanism experimentally also remains unsolved.

For the moment these unsolved problems do not seem to seriously hamper our ability to interpret the available data on multiple proton transfer. Most of these data concern concerted transfer under strong-coupling conditions where alternative mechanisms do not contribute. The systems identified as undergoing stepwise transfer under weak-coupling conditions have only been studied at temperatures high enough to render the stepwise mechanism dominant. No system for which experimental data are available has been clearly identified as belonging to the intermediate-coupling regime. Finding such systems remains a challenge to both experimentalists and theorists.

Acknowledgment

A.F.-R. thanks the Ministerio de Educatión y Ciencia for a Ramon y Cajal Research Contract and for Project No. BQU2003-01639.

References

1 See, e.g., the review V. A. Benderskii, D. E. Makarov, C. H. Wight, *Adv. Chem. Phys.* **88**, 1 (1994), and the original literature cited therein.

2 Z. Smedarchina, W. Siebrand, A. Fernández-Ramos, Kinetics isotope effects in multiple proton transfer in *Isotope Effects in Chemistry and Biology*, Marcel Dekker, New York, 2005.

3 W. Siebrand, Z. Smedarchina, M. Z. Zgierski, A. Fernández-Ramos, *Int. Rev. Phys. Chem.* **18**, 5 (1999).

4 F. Madeja, M. Havenith, *J. Chem. Phys.* **117**, 7162 (2002).

5 K. Remmers, W. L. Meerts, I. Ozier, *J. Chem. Phys.* **112**, 10890 (2000).

6 Z. Smedarchina, A. Fernández-Ramos, W. Siebrand, *Chem. Phys. Lett.* **395**, 339 (2004).

7 Z. Smedarchina, A. Fernández-Ramos, W. Siebrand, *J. Chem. Phys.* **122**, 134309 (2005).

8 J. Braun, M. Schlabach, B. Wehrle, M. Kocher, E. Vogel, H.-H. Limbach, *J. Am. Chem. Soc.* **116**, 6593 (1994).

9 J. Braun, H.-H. Limbach, P.G. Williams, H. Morimoto, D. E. Wemmer, *J. Am. Chem. Soc.* **118**, 7231 (1996).

10 T. J. Butenhof, C. B. Moore, *J. Am. Chem. Soc.* **110**, 8336 (1988).

11 Z. Smedarchina, M. Z. Zgierski, W. Siebrand, P. M. Kozlowski, *J. Chem. Phys.* **109**, 1014 (1998).

12 M. F. Shibl, M. Tashikawa, O. Kühn, *Phys. Chem. Chem. Phys.* **7**, 1368 (2005).

13 U. Langer, Ch. Hoelger, B. Wehrle, L. Latanowicz, E. Vogel, H.-H. Limbach, *J. Phys. Org. Chem.* **13**, 23 (2000).

14 P. M. Kozlowski, M. Z. Zgierski, *J. Chem. Phys.* **109**, 5905 (1998).

15 J. R. de la Vega, J. H. Bush, J. H. Schauble, K. L. Kunze, B. E. Haggart, *J. Am. Chem. Soc.* **104**, 3295 (1982).

16 Z. Smedarchina, W. Siebrand, A. Fernández-Ramos, L. Gorb, J. Leszczynski, *J. Chem. Phys.* **112**, 566 (2000).

17 A. Fernández-Ramos, Z. Smedarchina, W. Siebrand, M. Z. Zgierski, *J. Chem. Phys.* **114**, 7518 (2001).

18 Z. Smedarchina, W. Siebrand, A. Fernández-Ramos, Q. Cui, *J. Am. Chem. Soc.* **125**, 243 (2003).

19 Z. Smedarchina, A. Fernández-Ramos, W. Siebrand, *J. Comp. Chem.* **22**, 787 (2001).

20 C. G. Swain, E. C. Stivers, J. F. Reuwer, Jr., L. J. Schaad, *J. Am. Chem. Soc.* **80**, 5885 (1958).

21 Z. Smedarchina, W. Siebrand, *Chem. Phys. Lett.* **410**, 370 (2005).

22 See, e.g., the review D. G. Truhlar, B. C. Garrett, S. J. Klippenstein, *J. Phys. Chem.* **100** 12771 (1996), and references therein.

23 W. H. Miller, *J. Chem. Phys.* **62**, 1899 (1975).

24 C. G. Callan, S. Coleman, *Phys. Rev. D* **16**, 1762 (1977).

25 G. S. Langer, *Ann. Phys.* **54**, 258 (1969).

26 A. M. Polyakov, *Nucl. Phys.* **121**, 429 (1977).

27 Z. Smedarchina, W. Siebrand, M. Z. Zgierski, *J. Chem. Phys.* **103**, 5326 (1995).

28 Z. Smedarchina, A. Fernández-Ramos, W. Siebrand, *Chem. Phys. Lett.* **395**, 339 (2004).

29 Z. Smedarchina, W. Siebrand, A. Fernández-Ramos, E. Martínez-Núñez, *Chem. Phys. Lett.* **386**, 396 (2004).

30 Z. Smedarchina, A. Fernández-Ramos, W. Siebrand, M. Z. Zgierski, DOIT 2.0, a computer program to calculate hydrogen tunneling rate constants and splittings.

31 V. A. Benderskii, V. I. Goldanskii, D. E. Makarov, *Chem. Phys.* **154**, 407 (1991).

32 V. A. Benderskii, V. I. Goldanskii, D. E. Makarov, *Chem. Phys. Lett.* **186**, 517 (1991).

33 V. A. Benderskii, V. I. Goldanskii, D. E. Makarov, *Chem. Phys.* **159**, 29 (1992).

34 V. A. Benderskii, S. Yu. Grebenshchikov, E. V. Vetoshkin, G. V. Mil'nikov, D. E. Makarov, *J. Phys. Chem.* **98**, 3300 (1994).

35 V.A. Benderskii, S.Yu. Grebenshchikov, D.E. Makarov, E.V. Vetoshkin, Chem. Phys. **185**, 101 (1994).

36 V. A. Benderskii, S. Yu. Grebenshchikov, G. V. Mil'nikov, E. V. Vetoshkin, *Chem. Phys.* **188**, 19 (1995); *Chem. Phys.* **194**, 1 (1995); *Chem. Phys.* **198**, 281 (1995).

37 V. A. Benderskii, E. V. Vetoshkin, E. I. Kats, H. P. Trommsdorff, *Phys. Rev. E* **67**, 26102 (2003).

38 W. H. Saunders, Jr., *J. Am. Chem. Soc.* **107**, 184 (1985).

39 J. P. Sethna, *Phys. Rev. B* **24**, 698 (1981).

40 J. P. Sethna, *Phys. Rev. B* **25**, 5050 (1982).

41 N. Shida, P. F. Barbara, J. Almlöf, *J. Chem. Phys.* **94**, 3633 (1991).

42 C. S. Tautermann, M. J. Loferer, A. F. Voegele, K. R. Liedl, *J. Chem. Phys.* **120**, 11650 (2004).

43 J.-H. Lim, E. K. Lee, Y. Kim, *J. Phys. Chem. A* **101**, 2233 (1997).

44 Y. Kim, S. Lim, H.-J. Kim, Y. Kim, *J. Phys. Chem. A* **103**, 617 (1999).

45 Y. Podolyan, L. Gorb, J. Leszczynski, *J. Phys. Chem. A* **106**, 12103 (2002).

46 Y. Kim, S. Lim, Y. Kim, *J. Phys. Chem. A* **103**, 6632 (1999).

47 A. Oppenländer, C. Rambaud, H. P. Trommsdorff, J.-C. Vial, *Phys. Rev. Lett.* **63**, 1432 (1989).

48 C. Rambaud, H. P. Trommsdorff, *Chem. Phys. Lett.* **306**, 124 (1999).

49 A. Heuer, U. Haeberlen, *J. Chem. Phys.* **95**, 4201 (1991).

50 D. F. Brougham, A. J. Horsewill, R. I. Jenkinson, *Chem. Phys. Lett.* **272**, 69 (1997).

51 Q. Xue, A. J. Horsewill, M. R. Johnson, H. P. Trommsdorff, *J. Chem. Phys.* **120**, 11107 (2004).

52 D. R. Borst, J. R. Roscoli, D. W. Pratt, G. M. Florio, T. S. Zwier, A. Müller, S. Leutwyler, *Chem. Phys.* **283**, 341 (2002).

53 J. R. Roscoli, D. W. Pratt, Z. Smedarchina, W. Siebrand, A. Fernández-Ramos, *J. Chem. Phys.* **120**, 11351 (2004).

54 K. Sakota, H. Sekiya, *J. Chem. Phys. A* **109**, 2722 (2005).

55 A. Douhal, S. K. Kim, A. H. Zewail, *Nature* **378**, 260 (1995).

56 G. A. O'Doherty, R. D. Rogers, L. A. Paquette, *J. Am. Chem. Soc.* **116**, 10883 (1994).

57 K. N. Houk, Y. Li, M. A. McAllister, G. A. O'Doherty, L. A. Paquette, W. Siebrand, Z. Smedarchina, *J. Am. Chem. Soc.* **116**, 10895 (1994).

58 K. Mackenzie, E. C. Gravett, R. J. Gregory, J. A. K. Howard, J. Maher, *Tetrahedron Lett.* **30**, 5005 (1989).

59 K. Mackenzie, *Recent Res. Devel. Org. Chem.* **4**, 295 (2000).

60 Z. Smedarchina, W. Siebrand, *J. Mol. Struct.* **297**, 207 (1993).

61 D. Gerritzen, H.-H. Limbach, *J. Am. Chem. Soc.* **106**, 869 (1984).

62 A. Fernández-Ramos, Z. Smedarchina, J. Rodríguez-Otero, *J. Chem. Phys.* **114**, 1567 (2001).

63 L. Gorb, Y. Podolyan, J. Leszczinski, W. Siebrand, A. Fernández-Ramos, Z. Smedarchina, *Biopolymers* **61**, 77 (2002).

64 M. S. Gordon, *J. Phys. Chem.* **100**, 3974 (1996).

65 G. M. Chaban, M. S. Gordon, *J. Phys. Chem. A* **103**, 185 (1999).

66 D. McMorrow, T. J. Aartsma, *Chem. Phys. Letters* **125**, 581 (1986).

67 R. S. Moog, M. Maroncelli, *J. Phys. Chem.* **95**, 10359 (1991).

68 D. W. Christianson, C. A. Fierke, *Acc. Chem. Res.* **29**, 331 (1996).

69 D. N. Silverman, S. Lindskog, *Acc. Chem. Res.* **21**, 30 (1988).

70 K. S. Venkatassuban, D. N. Silverman, *Biochemistry* **19**, 4984 (1980).

71 Q. Cui, M. Karplus, *J. Phys. Chem. B* **107**, 1071 (2003).

72 N. Pugliano, R. J. Saykally, *Science* **257**, 1938 (1992).

73 J. K. Gregory, D. C. Clary, *J. Chem. Phys.* **102**, 7817 (1995).

74 K. R. Liedl, S. Sekusak, R. T. Kroemer, B. M. Rode, *J. Phys. Chem. A* **101**, 4707 (1997).

75 D. F. Brougham, R. Caciuffo, A. J. Horsewill, *Nature* **397**, 241 (1999).

76 A. J. Horsewill, N. H. Jones, R. Caciuffo, *Science* **291**, 100 (2001).

77 A. Fernández-Ramos, Z. Smedarchina, F. Pichierri, *Chem. Phys. Lett.* **343**, 627 (2001).